Advanced Numerical Methods in Applied Sciences

Advanced Numerical Methods in Applied Sciences

Special Issue Editors

Luigi Brugnano
Felice Iavernaro

MDPI • Basel • Beijing • Wuhan • Barcelona • Belgrade

MDPI

Special Issue Editors
Luigi Brugnano
Università defli Studi di Firenze
Italy

Felice Iavernaro
Università degli Studi di Bari
Italy

Editorial Office
MDPI
St. Alban-Anlage 66
4052 Basel, Switzerland

This is a reprint of articles from the Special Issue published online in the open access journal *Axioms* (ISSN 2075-1680) from 2018 to 2019 (available at: https://www.mdpi.com/journal/axioms/special_issues/advanced_numerical_methods)

For citation purposes, cite each article independently as indicated on the article page online and as indicated below:

LastName, A.A.; LastName, B.B.; LastName, C.C. Article Title. *Journal Name* **Year**, *Article Number*, Page Range.

ISBN 978-3-03897-666-0 (Pbk)
ISBN 978-3-03897-667-7 (PDF)

Contents

About the Special Issue Editors

Luigi Brugnano is full Professor of Numerical Analysis, based at the Mathematics and Informatics Department of the University of Firenze, Italy. He is the author of 125 scientific publications, including 2 research and 4 didactical monographs. His research interests cover a wide range of subjects in Numerical Analysis and Scientific Computing, with a more recent focus on Geometric Integration.

Felice Iavernaro is associate Professor of Numerical Analysis at the Department of Mathematics of the University of Bari, Italy. His primary interests include the design and implementation of efficient methods for the numerical solution of differential equations, with emphasis on the simulation of dynamical systems with geometric properties.

axioms

Editorial

Advanced Numerical Methods in Applied Sciences

Luigi Brugnano [1,*] and Felice Iavernaro [2,*]

[1] Dipartimento di Matematica e Informatica "U. Dini", Università di Firenze, Viale Morgagni 67/A, 50134 Firenze, Italy
[2] Dipartimento di Matematica, Università di Bari, Via Orabona 4, 70125 Bari, Italy
* Correspondence: luigi.brugnano@unifi.it (L.B.); felice.iavernaro@uniba.it (F.I.)

Received: 25 January 2019; Accepted: 29 January 2019; Published: 31 January 2019

Abstract: The use of scientific computing tools is, nowadays, customary for solving problems in Applied Sciences at several levels of complexity. The great need for reliable software in the scientific community conveys a continuous stimulus to develop new and more performing numerical methods which are able to grasp the particular features of the problem at hand. This has been the case for many different settings of numerical analysis, and this Special Issue aims at covering some important developments in various areas of application.

Keywords: numerical analysis; numerical methods; scientific computing

---

1. Special Issue Overview

The special issue contains 15 contributions covering a number of areas of application in Numerical Analysis and Scientific Computing, which we can summarize as follows:

1. Numerical Linear Algebra [1–3];
2. Numerical solution of differential equations [4–10];
3. Geometric integration [11,12];
4. Computer graphics [13];
5. Optimization [14,15].

Below, we highlight the main results of the papers.

1.1. Numerical Linear Algebra

In [1], the authors study the generalized Schur algorithm (GSA), which allows to compute well-known matrix decompositions, such as the QR and LU factorizations. In particular, they use the GSA to obtain new theoretical insights on the bounds of the entries of the matrix R in the QR factorization of some structured matrices, with related applications.

In [2], the author deals with the definition of limited memory preconditioners for symmetric and positive definite matrices. The existing connections with similar preconditioners are also discussed, along with its efficient implementaion. Extensive numerical tests are reported.

The authors of [3] discuss block-generalized, locally Toeplitz sequences which arise, e.g., from the discretization of many kinds of differential equations. The theoretical framework is then recalled, also completing previous results from the same authors, and a number of examples derived from the numerical solution of differential equations are worked out.

1.2. Numerical Solution of Differential Equations

The author of [4], who pioneered the order analysis of Runge-Kutta methods based on the theory of trees, introduces here the more general concept of stump. Stumps are then applied to the analysis of B-series, and used to study the order of Runge-Kutta methods when applied to non-autonomous scalar problems.

In [5], the authors review recent findings on the use of collocation methods for numerically solving Volterra integral and integro-differential equations. Both one-step and multi-step methods are considered, studying their convergence and providing comparisons in terms of efficiency and accuracy.

The authors in [6] study systems of fractional differential equations, in which different equations may have a different fractional time derivative at the left-hand side term of the equation. The linear case is completely worked out, providing a theory which collapses to the well-known Mittag-Leffler solution in the case where the indices are the same.

Fractional differential equations are also studied in [7], where a numerical method based on B-splines is proposed for their solution. In particular, the fractional diffusion problem is considered, and its numerical solution worked out.

Stochastic differential equations are considered in [8], where the authors review stability issues related to stochastic ordinary and Volterra integral equations. Two-step methods are then considered for the numerical solution in the ordinary case, and the θ method in the case of Volterra equations.

The numerical solution of Black-Scholes-type partial differential equations is studied in [9], where the authors provide a numerical method, and a related Matlab® code, for pricing some kinds of Asian options.

Arbitrarily high-order schemes using derivatives discontinuous Galerkin (ADER-DG) finite element methods are studied in [10]. The proposed methods are applicable to a wide class of nonlinear systems of partial differential equations, and are aimed at efficiently scaling on massively parallel supercomputers, as is testified by the numerical tests.

1.3. Geometric Integration

In [11] the authors study a class of A-stable, symmetric, one-step Hermite-Obreshkov methods previously introduced by other authors, which are here proved to be conjugate-symplectic. Moreover, a new and efficient implementation of the corresponding continuous spline extension is introduced. Numerical tests on some Hamiltonian problems are reported.

The authors in [12] study the use of the so-called *Line Integral Methods* for numerically solving conservative problems. In particular, energy-conserving methods for Hamiltonian problems are reviewed, with a number of extensions to related problems, such as constrained Hamiltonian problems, highly-oscillatory problems, and Hamiltonian partial differential equations.

1.4. Computer Graphics

In [13], the authors study the efficient construction of (truncated) hierarchical B-splines. In particular, hierarchical refinement strategies are considered, to be used within the framework of the so-called *isogeometric analysis* for numerically solving partial differential equations. The theoretical properties of the refinement algorithms and the resulting meshes are thoroughly analyzed and presented together with extensive numerical testing.

1.5. Optimization

In [14], the authors describe a two-step procedure for solving the so-called *low-rank matrix completion problem*. In the first step, a one-dimensional optimization problem, which depends on a scalar parameter, is solved. In the second step, the same functional, now depending on a matrix, is minimized. This latter minimization is achieved by solving a related matrix ODE.

The authors of [15] study the so-called problem of *histogram specification*, one of the main important tools in image processing. In particular, they propose a convex model that can include additional constraints based on different applications in edge-preserving smoothing. The convexity of the model allows to compute the output efficiently by the Fast Iterative Shrinkage-Thresholding Algorithm or the Alternating Direction Method of Multipliers.

Conflicts of Interest: The authors declare no conflict of interest.

References

1. Laudadio, T.; Matronardi, N.; van Dooren, P. The Generalized Schur Algorithm and Some Applications. *Axioms* **2018**, *7*, 81. [CrossRef]
2. Morini, B. On Partial Cholesky Factorization and a Variant of Quasi-Newton Preconditioners for Symmetric Positive Definite Matrices. *Axioms* **2018**, *7*, 44. [CrossRef]
3. Garoni, C.; Mazza, M.; Serra-Capizzano, S. Block Generalized Locally Toeplitz Sequences: From the Theory to the Applications. *Axioms* **2018**, *7*, 49. [CrossRef]
4. Butcher, J.C. Trees, Stumps, and Applications. *Axioms* **2018**, *7*, 52. [CrossRef]
5. Cardone, A.; Conte, D.; D'Ambrosio, R.; Paternoster, B. Collocation Methods for Volterra Integral and Integro-Differential Equations: A Review. *Axioms* **2018**, *7*, 45. [CrossRef]
6. Burrage, K.; Burrage, P.; Turner, I.; Zheng, F. On the Analysis of Mixed-Index Time Fractional Differential Equation Systems. *Axioms* **2018**, *7*, 25. [CrossRef]
7. Pitolli, F. Optimal B-Spline Bases for the Numerical Solution of Fractional Differential Problems. *Axioms* **2018**, *7*, 46. [CrossRef]
8. Cardone, A.; Conte, D.; D'Ambrosio, R.; Paternoster, B. Stability Issues for Selected Stochastic Evolutionary Problems: A Review. *Axioms* **2018**, *7*, 91. [CrossRef]
9. Aimi, A.; Diazzi, L.; Guardasoni, C. Efficient BEM-Based Algorithm for Pricing Floating Strike Asian Barrier Options (with MATLAB® Code). *Axioms* **2018**, *7*, 40. [CrossRef]
10. Dumbser, M.; Fambri, F.; Bader, M.T.M.; Weinzierl, T. Efficient Implementation of ADER Discontinuous Galerkin Schemes for a Scalable Hyperbolic PDE Engine. *Axioms* **2018**, *7*, 63. [CrossRef]
11. Mazzia, F.; Sestini, A. On a Class of Conjugate Symplectic Hermite-Obreshkov One-Step Methods with Continuous Spline Extension. *Axioms* **2018**, *7*, 58. [CrossRef]
12. Brugnano, L.; Iavernaro, F. Line Integral Solution of Differential Problems. *Axioms* **2018**, *7*, 36. [CrossRef]
13. Bracco, C.; Giannelli, C.; Vásquez, R. Refinement Algorithms for Adaptive Isogeometric Methods with Hierarchical Splines. *Axioms* **2018**, *7*, 43. [CrossRef]
14. Scalone, C.; Guglielmi, N. A Gradient System for Low Rank Matrix Completion. *Axioms* **2018**, *7*, 51. [CrossRef]
15. Chan, K.C.K.; Chan, R.H.; Nikolova, M. A Convex Model for Edge-Histogram Specification with Applications to Edge-Preserving Smoothing. *Axioms* **2018**, *7*, 53. [CrossRef]

![axioms logo] **axioms**

MDPI

Article

The Generalized Schur Algorithm and Some Applications

Teresa Laudadio [1,*] [iD], **Nicola Mastronardi** [1] [iD] and **Paul Van Dooren** [2] [iD]

1   Istituto per le Applicazioni del Calcolo "M. Picone", CNR, Sede di Bari, via G. Amendola 122/D, 70126 Bari, Italy; n.mastronardi@ba.iac.cnr.it
2   Catholic University of Louvain, Department of Mathematical Engineering, Avenue Georges Lemaitre 4, B-1348 Louvain-la-Neuve, Belgium; paul.vandooren@uclouvain.be
*   Correspondence: t.laudadio@ba.iac.cnr.it; Tel.: +39-080-5929752

Received: 2 October 2018; Accepted: 7 November 2018; Published: 9 November 2018

✓ check for updates

Abstract: The generalized Schur algorithm is a powerful tool allowing to compute classical decompositions of matrices, such as the QR and LU factorizations. When applied to matrices with particular structures, the generalized Schur algorithm computes these factorizations with a complexity of one order of magnitude less than that of classical algorithms based on Householder or elementary transformations. In this manuscript, we describe the main features of the generalized Schur algorithm. We show that it helps to prove some theoretical properties of the R factor of the QR factorization of some structured matrices, such as symmetric positive definite Toeplitz and Sylvester matrices, that can hardly be proven using classical linear algebra tools. Moreover, we propose a fast implementation of the generalized Schur algorithm for computing the rank of Sylvester matrices, arising in a number of applications. Finally, we propose a generalized Schur based algorithm for computing the null-space of polynomial matrices.

Keywords: generalized Schur algorithm; null-space; displacement rank; structured matrices

1. Introduction

The generalized Schur algorithm (GSA) allows computing well-known matrix decompositions, such as QR and LU factorizations [1]. In particular, if the involved matrix is structured, i.e., Toeplitz, block-Toeplitz or Sylvester, the GSA computes the R factor of the QR factorization with complexity of one order of magnitude less than that of the classical QR algorithm [2], since it relies only on the knowledge of the so-called *generators* [2] associated to the given matrix, rather than on the knowledge of the matrix itself. The stability properties of the GSA are described in [3–5], where it is proven that the algorithm is weakly stable provided the involved hyperbolic rotations are performed in a stable way.

In this manuscript, we first show that, besides the efficiency properties, the GSA provides new theoretical insights on the bounds of the entries of the R factor of the QR factorization of some structured matrices. In particular, if the involved matrix is a symmetric positive definite (SPD) Toeplitz or a Sylvester matrix, we prove that all or some of the diagonal entries of R monotonically decrease in absolute value.

We then propose a faster implementation of the algorithm described in [6] for computing the rank of a Sylvester matrix $S \in \mathbb{R}^{(m+n) \times (m+n)}$, whose entries are the coefficients of two polynomials of degree m and n, respectively. This new algorithm is based on the GSA for computing the R factor of the QR factorization of S. The proposed modification of the GSA-based method has a computational cost of $O(rl)$ floating point operations, where $l = \min\{n, m\}$ and r is the computed numerical rank.

It is well known that the upper triangular factor R factor of the QR factorization of a matrix $A \in \mathbb{R}^{n \times n}$ is equal to the upper triangular Cholesky factor $R_c \in \mathbb{R}^{n \times n}$ of $A^T A$, up to a diagonal sign matrix D, i.e., $R = DR_c$, $D = \text{diag}(\pm1, \cdots, \pm1) \in \mathbb{R}^{n \times n}$. In this manuscript, we assume, without loss of generality, that the diagonal entries of R and R_c are positive and since the matrices are then equal, we denote both matrices by R.

Finally, we propose a GSA-based approach for computing a null-space basis of a polynomial matrix, which is an important problem in several systems and control applications [7,8]. For instance, the computation of the null-space of a polynomial matrix arises when solving the column reduction problem of a polynomial matrix [9,10].

The manuscript is structured as follows. The main features of the GSA are provided in Section 2. In Section 3, a GSA implementation for computing the Cholesky factor R of a SPD Toeplitz matrix is described, which allows proving that the diagonal entries of R monotonically decrease. In Section 4, a GSA-based algorithm for computing the rank of a Sylvester matrix S is introduced, based on the computation of the Cholesky factor R of $S^T S$. In addition, in this case, it is proven that the first diagonal entries of R monotonically decrease. The GSA-based method to compute the null-space of polynomial matrices is proposed in Section 5. The numerical examples are reported in Section 6 followed by the conclusions in Section 7.

2. The Generalized Schur Algorithm

Many of the classical factorizations of a symmetric matrix, e.g., QR and LDL^T, can be obtained by the GSA. If the matrix is Toeplitz-like, the GSA computes these factorizations in a fast way. For the sake of completeness, the basic concepts of the GSA for computing the R factor of the QR factorization of structured matrices, such as Toeplitz and block-Toeplitz matrices, are introduced in this Section. A comprehensive treatment of the topic can be found in [1,2].

Let $A \in \mathbb{R}^{n \times n}$ be a symmetric positive definite (SPD) matrix. The semidefinite case is considered in Sections 4 and 5. The *displacement* of A with respect to a matrix Z of order n, is defined as

$$\nabla_Z A = A - ZAZ^T, \tag{1}$$

while the *displacement rank* k of A with respect to Z is defined as the rank of $\nabla_Z A$. If $\text{rank}(\nabla_Z A) = k$, Equation (1) can be written as the sum of k rank-one matrices,

$$\nabla_Z A = \sum_{i=1}^{k_1} \mathbf{g}_i^{(p)} \mathbf{g}_i^{(p)T} - \sum_{i=1}^{k_2} \mathbf{g}_i^{(n)} \mathbf{g}_i^{(n)T},$$

where $(k_1, n - k_1 - k_2, k_2)$ is the inertia of $\nabla_Z A$, $k = k_1 + k_2$, and the vectors $\mathbf{g}_i^{(p)} \in \mathbb{R}^n$, $i = 1, \ldots, k_1$, $\mathbf{g}_i^{(n)} \in \mathbb{R}^n$, $i = 1, \ldots, k_2$, are called the *positive* and the *negative generators* of A with respect to Z, respectively, conversely, if there is no ambiguity, simply the positive and negative generators of A. The matrix $G \equiv [\mathbf{g}_1^{(p)}, \mathbf{g}_2^{(p)}, \ldots, \mathbf{g}_{k_1}^{(p)}, \mathbf{g}_1^{(n)}, \mathbf{g}_2^{(n)}, \ldots, \mathbf{g}_{k_2}^{(n)}]^T$ is called the *generator* matrix.

The matrix Z is a nilpotent matrix. In particular, for Toeplitz and block-Toeplitz matrices, the matrix Z can be chosen as the shift and the block shift matrix

$$Z_1 = \begin{bmatrix} 0 & 0 & \cdots & 0 \\ 1 & \ddots & \ddots & \vdots \\ \vdots & \ddots & \ddots & \vdots \\ 0 & \cdots & 1 & 0 \end{bmatrix}, \quad Z_2 = \begin{bmatrix} 0 & 0 & \cdots & 0 \\ Z_1 & \ddots & \ddots & \vdots \\ \vdots & \ddots & \ddots & \vdots \\ 0 & \cdots & Z_1 & 0 \end{bmatrix},$$

respectively.

The implementation of the GSA relies only on the knowledge of the generators of A rather than on the knowledge of the matrix itself [1].

Let

$$J = \text{diag}(\underbrace{1, 1, \ldots, 1}_{k_1}, \underbrace{-1, -1, \ldots, -1}_{k_2}).$$

5

Since

$$
\begin{aligned}
A - ZAZ^T &= G^T J G, \\
ZAZ^T - Z^2 A Z^{2^T} &= Z G^T J G Z^T, \\
\vdots\qquad &\qquad \vdots \\
Z^{n-2} A Z^{n-2^T} - Z^{n-1} A Z^{n-1^T} &= Z^{n-2} G^T J G Z^{n-2^T}, \\
Z^{n-1} A Z^{n-1^T} &= Z^{n-1} G^T J G Z^{n-1^T},
\end{aligned}
\tag{2}
$$

then, adding all members of the left and right-hand sides of Equation (2) yields

$$
A = \sum_{j=0}^{n-1} Z^j G^T J G Z^{j^T},
\tag{3}
$$

which expresses the matrix A in terms of its generators.

Exploiting Equation (2), we show how the GSA computes R by describing its first iteration. Observe that the matrix products involved in the right-hand side of Equation (2) have their first row equal to zero, with the exception of the first product, $G^T J G$.

A key role in GSA is played by J-orthogonal matrices [11,12], i.e., matrices Φ satisfying $\Phi^T J \Phi = J$.

Any such matrix Φ can be constructed in different ways [11–14]. For instance, it can be considered as the product of Givens and hyperbolic rotations. In particular, a Givens rotation acting on rows i and j of the generator matrix is chosen if $J(i,i)J(j,j) > 0, i,j \in \{1,\ldots,n\}, i \neq j$. Otherwise, a hyperbolic rotation is considered. Indeed, suitable choices of Φ allow efficient implementations of GSA, as shown in Section 4.

Let $G_0 \equiv G$ and Φ_1 be a J-orthogonal matrix such that

$$
\tilde{G}_1 = \Phi_1 G_0, \quad \tilde{G}_1 \mathbf{e}_1 = [\alpha_1, 0, \ldots, 0]^T, \quad \text{with} \quad \alpha_1 > 0,
\tag{4}
$$

and $\mathbf{e}_i, i = 1,\ldots,n$, be the ith column of the identity matrix. Furthermore, let $\tilde{\mathbf{g}}_1^T$ and $\tilde{\Gamma}_1$ be the first and last $k-1$ rows of $\tilde{G}_1$, respectively, i.e., $\tilde{G}_1 = \begin{bmatrix} \tilde{\mathbf{g}}_1^T \\ \tilde{\Gamma}_1 \end{bmatrix}$.

From Equation (4), it turns out that the first column of $\tilde{\Gamma}_1$ is zero. Let $\tilde{J}$ be the matrix obtained by deleting the first row and column from J. Then, Equation (2) can be written as follows,

$$
\begin{aligned}
A &= \sum_{j=0}^{n-1} Z^j G_0^T J G_0 Z^{j^T} \\[2mm]
&= \sum_{j=0}^{n-1} Z^j G_0^T \Phi_1^T J \Phi_1 G_0 Z^{j^T} \\[2mm]
&= \sum_{j=0}^{n-1} Z^j \begin{bmatrix} \tilde{\mathbf{g}}_1^T \\ \tilde{\Gamma}_1 \end{bmatrix}^T J \begin{bmatrix} \tilde{\mathbf{g}}_1^T \\ \tilde{\Gamma}_1 \end{bmatrix} Z^{j^T} \\[2mm]
&= \tilde{\mathbf{g}}_1 \tilde{\mathbf{g}}_1^T + \sum_{j=1}^{n-1} Z^j \tilde{\mathbf{g}}_1 \tilde{\mathbf{g}}_1^T Z^{j^T} + \sum_{j=0}^{n-2} Z^j \tilde{\Gamma}_1^T \tilde{J} \tilde{\Gamma}_1 Z^{j^T} + \underbrace{Z^{n-1} \tilde{\Gamma}_1^T \tilde{J} \tilde{\Gamma}_1 Z^{n-1^T}}_{=0} \\[2mm]
&= \tilde{\mathbf{g}}_1 \tilde{\mathbf{g}}_1^T + \sum_{j=0}^{n-2} Z^j \begin{bmatrix} \tilde{\mathbf{g}}_1^T Z^T \\ \tilde{\Gamma}_1 \end{bmatrix}^T J \begin{bmatrix} \tilde{\mathbf{g}}_1^T Z^T \\ \tilde{\Gamma}_1 \end{bmatrix} Z^{j^T} \\[2mm]
&= \tilde{\mathbf{g}}_1 \tilde{\mathbf{g}}_1^T + \sum_{j=0}^{n-2} Z^j G_1^T J G_1 Z^{j^T}, \\[2mm]
&= \tilde{\mathbf{g}}_1 \tilde{\mathbf{g}}_1^T + A_1,
\end{aligned}
$$

where $G_1 \equiv [Z\tilde{g}_1, \check{\Gamma}_1^T]^T$, that is, G_1 is obtained from $\tilde{G}_1$ by multiplying $\tilde{g}_1$ with Z, and $A_1 \equiv \sum_{j=0}^{n-2} Z^j G_1^T J G_1 Z^{j^T}$. If A is a Toeplitz matrix, this multiplication with Z corresponds to displacing the entries of $\tilde{g}_1$ one position downward, while it corresponds to a block-displacement downward in the first generator if A is a block-Toeplitz matrix.

Thus, the first column of G_1 is zero and, hence, $\tilde{g}_1^T$ is the first row of the R factor of the QR factorization of A. The above procedure is recursively applied to A_1 to compute the other rows of R.

The jth iteration of GSA, $j = 1, \ldots, n$, involves the products $\Phi_j G_{j-1}$ and $Z\tilde{g}_1$. The former multiplication can be computed in $O\left(k(n-j)\right)$ operations [11,12], and the latter is done for free if Z is either a shift or a block–shift matrix. Therefore, if the displacement rank k of A is small compared to n, the GSA computes the R factor in $O(kn^2)$ rather than in $O(n^3)$ operations, as required by standard algorithms [15].

For the sake of completeness, the described GSA implementation is reported in the following `matlab` style function. (The function `givens` is the `matlab` function having as input two scalars, x_1 and x_2, and as output an orthogonal 2×2 matrix Θ such that $\Theta \begin{bmatrix} x_1 \\ x_2 \end{bmatrix} = \begin{bmatrix} \sqrt{x_1^2 + x_2^2} \\ 0 \end{bmatrix}$. The function `Hrotate` computes the coefficients of the 2×2 hyperbolic rotation Φ such that, given two scalars x_1 and x_2, $|x_1| > |x_2|$, $\Phi \begin{bmatrix} x_1 \\ x_2 \end{bmatrix} = \begin{bmatrix} \sqrt{x_1^2 - x_2^2} \\ 0 \end{bmatrix}$. The function `Happly` applies Φ to two rows of the generator matrix. Both functions are defined in [12]).

```
function[R] = GSA(G, n);
for  i = 1 : n,
    for  j = 2 : k₁,
        Θ = givens(G(1, i), G(j, i));
        G([1, j], i : n) = Θ * G([1, j], i : n);
    end % for
    for  j = k₁ + 2 : k₁ + k₂,
        Θ = givens(G(k₁ + 1, i), G(j, i));
        G([k₁ + 1, j], i : n) = Θ * G[k₁ + 1, j], i : n);
    end % for
    [c₁, s₁] = Hrotate(G(1, i), G(k₁ + 1, i));
    G([1, k₁ + 1], i : n) = Happly(c₁, s₁, G([1, k₁ + 1], i : n), n - i + 1);
    R(i, i : n) = G(1, i : n);
    G(1, i + 1 : n) = G(1, i : n - 1);  G(1, i) = 0;
end % for
```

The GSA has been proven to be weakly stable [3,4], provided the hyperbolic transformations involved in the construction of the matrices Φ_j are performed in a stable way [3,11,12].

3. GSA for SPD Toeplitz Matrices

In this section, we describe the GSA for computing the R factor of the Cholesky factorization of a SPD Toeplitz matrix A, with R upper triangular, i.e., $A = R^T R$. Moreover, we show that the diagonal entries of R decrease monotonically.

Let $A \in \mathbb{R}^{n \times n}$ and $Z \in \mathbb{R}^{n \times n}$ be a SPD Toeplitz matrix and a shift matrix, respectively, i.e.,

$$A = \begin{bmatrix} t_1 & t_2 & \ddots & t_n \\ t_2 & \ddots & \ddots & \ddots \\ \ddots & \ddots & \ddots & t_2 \\ t_n & \ddots & t_2 & t_1 \end{bmatrix}, \quad Z_n = \begin{bmatrix} 0 & 0 & \cdots & 0 \\ 1 & \ddots & \ddots & \vdots \\ \vdots & \ddots & \ddots & \vdots \\ 0 & \cdots & 1 & 0 \end{bmatrix},$$

and let $\mathbf{t} = A(:, 1)$. Then,

$$\nabla_Z A = \begin{bmatrix} t_1 & t_2 & \cdots & t_n \\ \hline t_2 & 0 & \cdots & 0 \\ \vdots & \vdots & \vdots & \vdots \\ t_n & 0 & \cdots & 0 \end{bmatrix},$$

i.e., $\nabla_Z A$ is a symmetric rank-2 matrix. Moreover, the generator matrix G is given by

$$G = \begin{bmatrix} \mathbf{g}_1^T \\ \mathbf{g}_2^T \end{bmatrix}, \quad \text{with } \mathbf{g}_1 = \frac{\mathbf{t}}{\sqrt{t_1}}, \quad \mathbf{g}_2 = [0, \mathbf{g}_1(2:n)^T]^T.$$

In this case, the GSA can be implemented in `matlab`-like style as follows.

```
function[R] = GSA_chol(G_0)
for  i = 1 : n,
     [c_1, s_1] = Hrotate(G_{i-1}(1, i), G^{(i)}(2, i));   G_{i-1}(:, i : n) = Happly(c_1, s_1, G_{i-1}(:, i : n), n - i + 1);
     R(i, i : n) = G_{i-1}(1, i : n);
     G_i(1, i + 1 : n) = G_{i-1}(1, i : n - 1);   G_i(2, i + 1 : n) = G_{i-1}(2, i + 1 : n - 1);
end  % for
```

The following lemma holds.

Lemma 1. *Let A be a SPD Toeplitz matrix and let R be its Cholesky factor, with R upper triangular. Then,*

$$R(i - 1, i - 1) \geq R(i, i), \quad i = 2, \ldots, n.$$

Proof. At each step i of `GSA_chol`, $i = 1, \ldots, n$, first a hyperbolic rotation is applied to G_{i-1} in order to annihilate the element $G_i(2, i)$. Hence, the first row of G_{i-1} becomes the row i of R. Finally, $G_i(1, :)$ is obtained displacing the entries of the first row of G_{i-1} one position right, while $G_i(2, :)$ is equal to $G_{i-1}(2, :)$. Taking into account that $G_{i-1}(2, 1) = 0$, the diagonal entries of R are

$$R(1, 1) = G_0(1, 1)$$

$$R(2, 2) = \sqrt{G_1^2(1, 2) - G_1^2(2, 2)} = \sqrt{R^2(1, 1) - G_1^2(2, 2)} \leq R(1, 1);$$

$$\vdots$$

$$R(i, i) = \sqrt{G_{i-1}^2(1, i) - G_{i-1}^2(2, i)} = \sqrt{R^2(i - 1, i - 1) - G_{i-1}^2(2, i)} \leq R(i - 1, i - 1);$$

$$\vdots$$

$$R(n, n) = \sqrt{G_{n-1}^2(1, n) - G_{n-1}^2(2, n)} = \sqrt{R^2(n - 1, n - 1) - G_{n-1}^2(2, n)} \leq R(n - 1, n - 1).$$

□

4. Computing the Rank of Sylvester Matrices

In this section, we focus on the computation of the rank of Sylvester matrices. The numerical rank of a Sylvester matrix is a useful information for determining the degree of the greatest common divisor of the involved polynomials [6,16,17].

A GSA-based algorithm for computing the rank of S has been recently proposed in [6]. It is based on the computation of the Cholesky factor R of $S^T S$, with R upper triangular, i.e., $R^T R = S^T S$.

Here, we propose a more efficient variant of this algorithm that allows proving that the first entries of R monotonically decrease.

Let $w_i \in \mathbb{R}$, $i = 0, 1, \ldots, n$, and let $y_i \in \mathbb{R}$, $i = 0, 1, \ldots, m$. Denote by $w(x)$ and $y(x)$ two univariate polynomials,

$$
\begin{aligned}
w(x) &= w_n x^n + w_{n-1} x^{n-1} + \cdots + w_1 x + w_0, \quad w_n \neq 0, \\
y(x) &= y_m x^m + y_{m-1} x^{m-1} + \cdots + y_1 x + y_0, \quad y_m \neq 0.
\end{aligned}
\tag{5}
$$

Let $S \in \mathbb{R}^{(m+n) \times (m+n)}$ be the Sylvester matrix defined as follows,

$$
S = \left[\ W\ |\ Y\ \right], \quad
W = \begin{bmatrix}
w_n & & & \\
w_{n-1} & w_n & & \\
\vdots & w_{n-1} & \ddots & \\
w_1 & \vdots & \ddots & w_n \\
w_0 & w_1 & \ddots & w_{n-1} \\
& w_0 & \ddots & \vdots \\
& & \ddots & w_1 \\
& & & w_0
\end{bmatrix}, \quad
Y = \begin{bmatrix}
y_m & & & \\
y_{m-1} & y_m & & \\
\vdots & y_{m-1} & \ddots & \\
y_1 & \vdots & \ddots & y_m \\
y_0 & y_1 & \ddots & y_{m-1} \\
& y_0 & \ddots & \vdots \\
& & \ddots & y_1 \\
& & & y_0
\end{bmatrix},
\tag{6}
$$

with $W \in \mathbb{R}^{(m+n) \times m}$ and $Y \in \mathbb{R}^{(m+n) \times n}$ band Toeplitz matrices.

We now describe how the GSA-based algorithm proposed in [6] for computing the rank of S can be implemented in a faster way. This variant is based on the computation of the Cholesky factor $R \in \mathbb{R}^{(m+n) \times (m+n)}$ of $S^T S$, with R upper triangular, i.e., $R^T R = S^T S$.

Defining

$$
Z = \left[\begin{array}{c|c} Z_m & \\ \hline & Z_n \end{array}\right], \quad \text{with } Z_k = \begin{bmatrix}
0 & 0 & \cdots & 0 \\
1 & \ddots & \ddots & \vdots \\
\vdots & \ddots & \ddots & \vdots \\
0 & \cdots & 1 & 0
\end{bmatrix}_{k \times k}, \quad k \in \mathbb{N},
\tag{7}
$$

the generator matrix G of $S^T S$ with respect to Z is then given by [6]

$$
G = \left[\ \mathbf{g}_1 \quad \mathbf{g}_2 \quad \mathbf{g}_3 \quad \mathbf{g}_4\ \right]^T
$$

where

$$
\begin{aligned}
&\mathbf{g}_1 = \mathbf{x}_1 / \|S(:,1)\|_2, \\
&\mathbf{g}_2([2:n+m]) = \mathbf{x}_2([2:n+m]) / \|S(:,m+1)\|_2, \quad \mathbf{g}_2(1) = 0, \\
&\mathbf{g}_3(2:n+m) = \mathbf{g}_1(2:n+m), \quad \mathbf{g}_3(1) = 0, \\
&\mathbf{g}_4([1:m,m+2:n+m]) = \mathbf{g}_2([1:m,m+2:n+m]), \quad \mathbf{g}_4(m+1) = 0,
\end{aligned}
\tag{8}
$$

with $\mathbf{x}_1 = S^T S \mathbf{e}_1$, $\mathbf{x}_2 = S^T S \mathbf{e}_{m+1}$, $\mathbf{e}_j$ the jth vector of the canonical basis of $\mathbb{R}^{m+n}$, and $J = \mathrm{diag}(1, 1, -1, -1)$.

The algorithm proposed in [6] is based on the following GSA implementation for computing the R factor of the QR factorization of S.

```
function[R] = GSA_chol2(G)
for  i = 1 : n,
    Θ₁ =givens(G(1,i), G(2,i));          Θ₂ =givens(G(3,i), G(4,i));
    G(1:2,i:n) = Θ₁G(1:2,i:n);           G(3:4,i:n) = Θ₂G(3:4,i:n);
    [c₁,s₁] =Hrotate(G(1,i), G(3,i));
    G([1,3],i:n) = Happly(c₁,s₁,G([1,3],i:n),n-i+1);
    R(i,i:n) = Gᵢ(1,i:n);
    G(1,i+1:n) = G(1,i:n-1)Zᵀ;           G(2,i+1:n) = G(2:4,i+1:n-1);
end % for
```

At the ith iteration of the algorithm, $i = 1, \ldots, n$, the Givens rotations Θ_1 and Θ_2 are computed and applied, respectively, to the first and second generators, and to the third and fourth generators, to annihilate $G(2, i)$ and $G(4, i)$. Hence, the hyperbolic rotation $\begin{bmatrix} c_1 & -s_1 \\ -s_1 & c_1 \end{bmatrix}$ is applied to the first and the third row of G to annihilate $G(3, i)$. Finally, the first row of G becomes the ith row of R and the first row of G is multiplied by Z^T.

Summarizing, at the first step of the ith iteration of GSA, all entries of the ith column but the first one of G, are annihilated. If the number of rows of G is greater than 2, this can be accomplished in different ways (see [5,14]) .

Analyzing the pattern of the generators in Equation (8), we are able to derive a different implementation of GSA that costs $O(rl)$, with $l = \min\{n, m\}$. Moreover, this implementation allows proving that the first l diagonal entries of R are monotonically decreasing.

We observe that the matrix $W^T W$ in Equation (6) is the SPD Toeplitz matrix

$$
W^T W = \begin{bmatrix}
t_1 & t_2 & \cdots & t_n & t_{n+1} & & \\
t_2 & t_1 & t_2 & \ddots & t_n & \ddots & \\
\vdots & t_2 & \ddots & \ddots & \ddots & \ddots & t_{n+1} \\
t_n & \vdots & \ddots & \ddots & \ddots & \vdots & t_n \\
t_{n+1} & t_n & \ddots & \ddots & \ddots & t_2 & \vdots \\
& \ddots & \ddots & \vdots & t_2 & t_1 & t_2 \\
& & t_{n+1} & t_n & \cdots & t_2 & t_1
\end{bmatrix}_{m \times m},
\tag{9}
$$

with

$$
t_i = \sum_{j=i}^{n+1} w_{j-1} w_{j-i}, \quad i = 1, 2, \ldots, n+1.
$$

Since

$$
S^T S = \left[\begin{array}{c|c} W^T W & W^T Y \\ \hline Y^T W & Y^T Y \end{array} \right],
$$

if $n \ll m$, from Equation (9), it turns out that $G([1, 3], n+2 : m) = 0$. Moreover, the rows $G(2, :)$ and $G(4, :)$ have their first entry equal to zero and differ only in their entry in column $m + 1$. This particular pattern of G is close to the ones described in [13,14,18], allowing to design an alternative GSA implementation with respect to that considered in [6], and thereby reducing the complexity from $O(r(n + m))$ to $O(rl)$, where r is the computed rank of S and $l = \min\{n, m\}$.

Since the description of the above GSA implementation is quite cumbersome and similar to the algorithms reported in [13,14,18], we omit it here. The corresponding `matlab` pseudo–code can be obtained from the authors upon request.

If the matrix S has rank $r < (n + m)$, at the $k = (n + m - r + 1)$st iteration, it turns out that $G^2(1, k) - G^2(3, k) = 0$ in exact arithmetic [6]. Therefore, at each iteration of the algorithm we check whether

$$
G^2(1, k) - G^2(3, k) > tol,
\tag{10}
$$

where *tol* is a fixed tolerance. If Equation (10) is not satisfied, we stop the computation considering k as the computed numerical rank of S.

The R factor of the QR factorization of S is unique if the diagonal entries of R are positive. The considered GSA implementation, yielding the rank of S and based on computing the R factor of the QR factorization of S, allows us to prove that the first l entries of the diagonal of R are ordered in a decreasing order, with $l = \min\{m, n\}$. In fact, the following theorem holds.

Theorem 1. *Let $R^T R = S^T S$ be the Cholesky factorization of $S^T S$ with S the Sylvester matrix defined in Equation (6) with rank $r \geq l = \min\{m, n\}$. Then,*

$$R(i-1, i-1) \geq R(i, i) \geq 0, \quad i = 2, \ldots, l. \tag{11}$$

Proof. Each entry i of the diagonal of R is determined by the ith entry of the first row of G at the end of iteration i, for $i = 1, \ldots, m + n$. Let us define $\hat{G} \equiv G(:, 1 : l)$ and consider the following alternative implementation of the GSA for computing the first l columns of the Cholesky factor of $S^T S$.

```
for  i = 1 : l,
      Θ =givens(Ĝ(1, i), Ĝ(2, i));
      Ĝ(1 : 2, i : l) = Θ * Ĝ(1 : 2, i : l);
      [c₁, s₁] =Hrotate(Ĝ(1, i), Ĝ(4, i));  Ĝ([1, 4], :) = Happly(c₁, s₁, Ĝ([1, 4], :), l);
      [c₂, s₂] =Hrotate(Ĝ(1, i), Ĝ(3, i));  Ĝ([1, 3], :) = Happly(c₂, s₂, Ĝ([1, 3], :), l);
      R(i, i : l) = Ĝ(1, i : l);
      Ĝ(1, i + 1 : l) = Ĝ(1, i : l - 1);
      Ĝ(1, i) = 0;
end % for
```

We observe that, for $i = 1$, $\hat{G}(1, 1)$ is the only entry in the first column of $\hat{G}$ different from 0. Hence, $R(1, i) = \hat{G}(1, 1 : l)$ and the first iteration amounts only to shifting $\hat{G}(1, 1 : l)$ one position rightward, i.e., $\hat{G}(1, 2 : l) = \hat{G}(1, 1 : l - 1)$, $\hat{G}(1, 1) = 0$.

At the beginning of iteration $i = 2$, the second and the fourth row of $\hat{G}$ are equal Equation (8). Hence, when applying a Givens rotation to the first and the second row in order to annihilate the entry $\hat{G}(2, i)$ and when subsequently applying a hyperbolic rotation to the first and fourth row of $\hat{G}$ in order to annihilate $\hat{G}(4, i)$, it turns out that $\hat{G}(2, i : l)$ and $\hat{G}(4, i : l)$ are then modified but still equal to each other, while $\hat{G}(1, i : l)$ remains unchanged. The equality between $\hat{G}(2, :)$ and $\hat{G}(4, :)$ is maintained throughout the iterations $1, 2, \ldots, l$.

Therefore, the second and the fourth row of $\hat{G}$ do not play any role in computing $R(1 : l, 1 : l)$ and can be neglected. Hence, the GSA for computing $R(1 : l, 1 : l)$ reduces only to applying a hyperbolic rotation to the first and the third generators, as described in the following algorithm.

```
for  i = 1 : l,
      [c₂, s₂] =Hrotate(Ĝ(1, i), Ĝ(3, i));  Ĝ([1, 3], :) = Happly(c₂, s₂, Ĝ([1, 3], :), l);
      R(i, i : l) = Ĝ(1, i : l);
      Ĝ(1, i + 1 : l) = Ĝ(1, i : l - 1);
      Ĝ(1, i) = 0;
end % for
```

Since at the beginning of iteration i, $i = 2, \ldots, i$, $\hat{G}(1, i : l) = R(i - 1, i - 1 : l - 1)$, then the involved hyperbolic rotation $\Phi = \begin{bmatrix} c_2 & -s_2 \\ -s_2 & c_2 \end{bmatrix}$ is such that

$$\Phi \begin{bmatrix} \hat{G}(1, i) \\ \hat{G}(3, i) \end{bmatrix} = \Phi \begin{bmatrix} R(i - 1, i - 1) \\ \hat{G}(3, i) \end{bmatrix} = \begin{bmatrix} \hat{G}(1, i) \\ 0 \end{bmatrix} = \begin{bmatrix} R(i, i) \\ 0 \end{bmatrix},$$

where the updated $\hat{G}(1, i)$ is equal to $\sqrt{\hat{G}(1, i)^2 - \hat{G}(3, i)^2} \geq 0$. Therefore, $R(i, i) = \sqrt{R(i - 1, i - 1)^2 - \hat{G}(3, i)^2} \geq 0$, and thus $R(i, i) \leq R(i - 1, i - 1)$. □

Remark 1. *The above GSA implementation allows to prove the inequality Equation (11). This property is difficult to obtain if the QR factorization is performed via Householder transformations or if the classical Cholesky factorization of $S^T S$ is used.*

5. GSA for Computing the Null-Space of Polynomial Matrices

In this section, we consider the problem of computing a polynomial basis $X(s) \in \mathbb{R}^{n \times (n-\rho)}$ of the null-space of an $m \times n$ polynomial matrix of degree δ and rank $\rho \leq \min(m, n)$,

$$M(s) = \sum_{i=0}^{\delta} M_i s^i, \qquad M_i \in \mathbb{R}^{m \times n}, \ i = 0, \ldots, \delta. \tag{12}$$

As described in [8,19,20], the above problem is equivalent to that of computing the null-space of a related block-Toeplitz matrix. Algorithms to solve this problem are proposed in [8,19] but they do not explicitly exploit the structure of the involved matrix. Algorithms to solve related problems have also been described in the literature, e.g., in [8,19,21,22].

In this paper, we propose an algorithm for computing the null-space of polynomial matrices based on a variant of the GSA for computing the null-space of a related band block-Toeplitz matrix [8].

5.1. Null-Space of Polynomial Matrices

A polynomial vector $\mathbf{v}(s) = \sum_{i=0}^{\gamma} \mathbf{v}_i s^i$, $\mathbf{v}_i \in \mathbb{R}^n$, $i = 0, \ldots, \gamma$, $\gamma \in \mathbb{N}$, is said to belong to the null-space of (12) if

$$M(s)\mathbf{v}(s) = 0 \Leftrightarrow \sum_{j=0}^{\delta} M_j s^j \sum_{i=0}^{\gamma} \mathbf{v}_i s^i = 0.$$

The polynomial vector $\mathbf{v}(s)$ belongs to the null-space of $M(s)$ iff $\mathbf{v} = [\mathbf{v}_0^T, \mathbf{v}_1^T, \ldots, \mathbf{v}_\gamma^T]^T$, $\mathbf{v}_i \in \mathbb{R}^n$, $i = 0, \ldots, \gamma$, is a vector belonging to the null-space of the band block-Toeplitz matrix

$$T = \begin{bmatrix} M_0 & & & \\ M_1 & M_0 & & \\ \vdots & M_1 & \ddots & \\ M_\delta & \vdots & \ddots & M_0 \\ & M_\delta & \ddots & M_1 & M_0 \\ & & \ddots & \vdots & M_1 \\ & & & M_\delta & \vdots \\ & & & & M_\delta \end{bmatrix}_{\hat{m} \times \hat{n}}, \tag{13}$$

where $\hat{m} = m(\delta + n_b)$, $\hat{n} = n n_b$, with $n_b = \gamma + 1$ the number of block columns of T, that can be determined, e.g., by the algorithm described in [8]. Hence, the problem of computing the null-space of the polynomial matrix in Equation (12) is equivalent to the problem of computing the null-space of the matrix in Equation (13). To normalize the entries in this matrix, it is appropriate to first perform a QR factorization of each block column of T:

$$\begin{bmatrix} M_0 \\ M_1 \\ \vdots \\ M_\delta \end{bmatrix} = \begin{bmatrix} Q_0 \\ Q_1 \\ \vdots \\ Q_\delta \end{bmatrix} U, \quad \text{where} \quad \sum_i Q_i^T Q_i = I_n,$$

and to absorb the upper triangular factor U in the vector $\mathbf{u}(s) := U\mathbf{v}(s)$. The convolution equation $M(s)\mathbf{v}(s) = 0$ then becomes an equation of the type $Q(s)\mathbf{u}(s) = 0$, but where the coefficient matrices Q_i of $Q(s)$ form together an orthonormalized matrix.

Remark 2. *Above, we have assumed that there are no constant vectors $\mathbf{v}$ in the kernel of $M(s)$. If there are, then, the block column of M_i matrices has rank less than n and the above factorization will discover it in the sense that the matrix U is nonsquare and the matrices Q_i have less columns than M_i, $i = 0, 1, \ldots, \delta$. This trivial null-space can be eliminated and we therefore assume that the rank was full. For simplicity, from now on, we also assume that the coefficient matrices of the polynomial matrix $M(s)$ were already normalized in this way and the norm of the block columns of T are thus orthonormalized. This normalization proves to be very useful in the sequel.*

Denote by

$$
Z_{n_b} = \begin{bmatrix} 0_n & & & \\ I_n & 0_n & & \\ & \ddots & \ddots & \\ & & I_n & 0_n \end{bmatrix}_{\hat{n} \times \hat{n}} \quad \text{and} \quad Z = \begin{bmatrix} Z_{n_b} & \\ & Z_{n_b} \end{bmatrix}_{2\hat{n} \times 2\hat{n}},
$$

where 0_n is the null–matrix of order $n \in \mathbb{N}$.

If $\mathbf{v}_i \neq 0$, $0 < i < \gamma$, and $\mathbf{v}_j = 0$, $j = i+1, \ldots, \gamma$, i.e.,

$$
\mathbf{v} = [\mathbf{v}_0^T, \mathbf{v}_1^T, \ldots, \mathbf{v}_i^T, \underbrace{0, \ldots, 0}_{\gamma - i}]^T,
$$

and $\mathbf{v} \in \ker(T)$, then also $Z_{n_b}^k \mathbf{v} \in \ker(T)$, $k = 0, 1, \ldots, \gamma - i$. In this case, the vector $\mathbf{v}$ is said to be a *generator vector* of a chain of length $\gamma - i + 1$ of the null-space of T.

The proposed algorithm for the computation of the null-space of polynomial matrices is based on the GSA for computing the R factor of the QR-factorization of the matrix T in Equation (13) and, if R is full column rank, its inverse R^{-1}.

Let us first assume that the matrix T is full rank, i.e., $\text{rank}(T) = \rho = \min\{\hat{m}, \hat{n}\}$. Without loss of generality, we suppose $\hat{m} \geq \hat{n}$. If $\hat{m} < \hat{n}$, the algorithm still computes the R factor in trapezoidal form [23]. Moreover, in this case, we compute the first $\hat{m}$ rows of the inverse of the matrix obtained appending the last $\hat{n} - \hat{m}$ rows of the identity matrix of order $\hat{n}$ to R.

Let us consider the SPD block-Toeplitz matrix

$$
\hat{T} = T^T T = \begin{bmatrix} \hat{T}_0 & \hat{T}_1 & \cdots & \hat{T}_\delta & & \\ \hat{T}_1 & \hat{T}_0 & \hat{T}_1 & \ddots & \ddots & \\ \vdots & \hat{T}_1 & \ddots & \ddots & \ddots & \hat{T}_\delta \\ \hat{T}_\delta & \ddots & \ddots & \ddots & \ddots & \vdots \\ & \ddots & \ddots & \ddots & \hat{T}_0 & \hat{T}_1 \\ & & \hat{T}_\delta & \cdots & \hat{T}_1 & \hat{T}_0 \end{bmatrix} \in \mathbb{R}^{\hat{n} \times \hat{n}}, \tag{14}
$$

whose blocks are

$$
\hat{T}_{i-j} = \begin{cases} \sum_{k=0}^{\delta - |i-j|} M_{k+|i-j|}^T M_k, & \text{if } i \leq j \\ \sum_{k=0}^{\delta - |i-j|} M_k^T M_{k+|i-j|}, & \text{if } i > j. \end{cases} \tag{15}
$$

Notice that, because of the normalization introduced before, we have that $\hat{T}_0 = I_n$ and $\|\hat{T}_i\|_2 \leq 1$. This is used below. The matrix

$$W = \left[\begin{array}{c|c} \hat{T} & I_{\hat{n}} \\ \hline I_{\hat{n}} & 0_{\hat{n}} \end{array} \right] \tag{16}$$

can be factorized in the following way,

$$W = \hat{R}^T \hat{J} \hat{R} \equiv \left[\begin{array}{cc} R^T & \\ R^{-1} & R^{-1} \end{array} \right] \left[\begin{array}{cc} I_{\hat{n}} & \\ & -I_{\hat{n}} \end{array} \right] \left[\begin{array}{cc} R & R^{-T} \\ & R^{-T} \end{array} \right], \tag{17}$$

where $R \in \mathbb{R}^{\hat{n} \times \hat{n}}$ is the factor R of the QR-factorization of T, i.e., the Cholesky factor of $\hat{T}$. Hence, R and its inverse R^{-1} can be retrieved from the first $\hat{n}$ columns of the matrix $\hat{R}^T$.

The displacement matrix and the displacement rank of W with respect to Z, are given by

$$\nabla_Z(W) = W - ZWZ^T = \left[\begin{array}{ccccccc|cccc} I_n & \hat{T}_1 & \cdots & \hat{T}_\delta & 0_n & \cdots & 0_n & I_n & 0_n & \cdots & 0_n \\ \hat{T}_1 & & & & & & & & & & \\ \vdots & & & & & & & & & & \\ \hat{T}_\delta & & & & & & & & & & \\ 0_n & & & & & & & & & & \\ \vdots & & & & & & & & & & \\ 0_n & & & & & & & & & & \\ \hline I_n & & & & & & & & & & \\ 0_n & & & & & & & & & & \\ \vdots & & & & & & & & & & \\ 0_n & & & & & & & & & & \end{array} \right] \tag{18}$$

and $\rho(W, Z) = \mathrm{rank}(\nabla_Z(W))$, respectively, with $\nabla_Z(W) \in \mathbb{R}^{2\hat{n} \times 2\hat{n}}$.

Then, taking the order n of the matrices $\hat{T}_i, i = 0, 1, \ldots, \delta$, into account, it turns out that $\rho(W, Z) \leq 2n$. Hence, Equation (18) can be written as the difference of two matrices of rank at most n, i.e.,

$$\nabla_Z(W) = G^{(+)T} G^{(+)} - G^{(-)T} G^{(-)} = G^T J G, \quad \text{where} \quad G := \left[\begin{array}{c} G^{(+)} \\ G^{(-)} \end{array} \right] \quad \text{and} \quad J = \mathrm{diag}(I_n, -I_n).$$

Since $\hat{T}_0 = I_n$, the construction of G does not require any computation: it is easy to check that G is given by

$$G := \left[\begin{array}{c} G^{(+)} \\ G^{(-)} \end{array} \right] = \left[\begin{array}{ccccccc|cccc} I_n & \hat{T}_1 & \cdots & \hat{T}_\delta & 0_n & \cdots & 0_n & I_n & 0_n & \cdots & 0_n \\ 0_n & \hat{T}_1 & \cdots & \hat{T}_\delta & 0_n & \cdots & 0_n & 0_n & 0_n & \cdots & 0_n \end{array} \right]. \tag{19}$$

Remark 3. *Observe that increasing n_b, with $n_b \geq \delta + 1$, the structures of W and $\nabla_Z(W)$ do not change due to the block band structure of the matrix W. Consequently, the length of the corresponding generators changes but their structure remains the same since only $T_0, T_1, \ldots, T_\delta$ and I_n are different from zero in the first block row.*

The computation by the GSA of the R factor of T and of its inverse R^{-1} is made by only using the matrix G rather than the matrix T. Its implementation is a straightforward block matrix extension of the GSA described in Section 2.

Remark 4. *By construction, the initial generator matrix G_0 has the first $\delta + 1$ block rows and the block row $n_b + 1$ different from zero. Therefore, the multiplication of G_0 by the J-orthogonal matrix H_1 does not modify the structure of the generator matrix.*

Let $G_0 = G$. At each iteration i (for $i = 1, \ldots, n_b$,), we start from the generator matrix G_{i-1} having the blocks (of length n) $i, i+1, \ldots, i+\delta$ and $n_b + 1, \ldots, n_b + i$ different form zero. We then look for a J-orthogonal matrix H_i such that the product $H_i G_{i-1}$ has in position $(1 : n, (i-1)n + 1 : in)$ and $(n+1 : 2n, (i-1)n + 1 : in)$ a nonsingular upper triangular and zero matrix, respectively.

Then, G_i is obtained from $\begin{bmatrix} \tilde{G}_i^{(+)} \\ \tilde{G}_i^{(-)} \end{bmatrix} \equiv H_i G_{i-1}$ by multiplying the first n columns with Z, i.e.,

$$G_i = \begin{bmatrix} \tilde{G}_i^{(+)} Z^T \\ \tilde{G}_i^{(-)} \end{bmatrix}.$$

The computation of the J-orthogonal matrix H_i at the ith iteration of the GSA can be constructed as a product of n Householder matrices $\hat{H}_{i,j}$ and n hyperbolic rotations $\hat{Y}_{i,j}, j = 1, \ldots, n$.

The multiplication by the Householder matrices $\hat{H}_{i,j}$ modifies the last n columns of the generator matrix, annihilating the last n entries but the $(n+1)$st in the row $(i-1)n + j, j = 1, \ldots, n$, while the multiplication by the hyperbolic rotations $\hat{Y}_{i,j}$ acts on the columns i and $n+1$, annihilating the entry in position $((i-1)n + j, n+1)$.

Given $v_1, v_2 \in \mathbb{R}$, $|v_1| > |v_2|$, a hyperbolic matrix $Y \in \mathbb{R}^{2 \times 2}$ can be computed

$$Y = \begin{bmatrix} c & -s \\ -s & c \end{bmatrix}, \quad \text{with } c = \frac{v_1}{\sqrt{v_1^2 - v_2^2}}, \quad s = \frac{v_2}{\sqrt{v_1^2 - v_2^2}},$$

such that $[v_1, v_2]Y = [\sqrt{v_1^2 - v_2^2}, 0]$.

The modification of the sparsity pattern of the generator matrix after the first and ith iteration of the GSA are displayed in Figures 1 and 2, respectively.

The reliability of the GSA strongly depends on the way the hyperbolic rotation is computed. In [4,5,24], it is proven that the GSA is weakly stable if the hyperbolic rotations are implemented in an appropriate manner [3,11,12,24].

Let

$$H_{i,j} = \begin{bmatrix} I_n & \\ & \hat{H}_{i,j} \end{bmatrix} \qquad Y_{i,j} = \begin{bmatrix} I_{j-1} & & & & \\ & c_j & & -s_j & \\ & & I_{n-j} & & \\ & -s_j & & c_j & \\ & & & & I_{n-1} \end{bmatrix}.$$

Then,

$$H_i = H_{i,1} Y_{i,1} \cdots H_{i,n-1} Y_{i,n-1} H_{i,n} Y_{i,n}.$$

As previously mentioned, GSA relies only on the knowledge of the generators of W rather than on the matrix $\hat{T}$ itself. Its computation involves the product $\hat{T}^T \hat{T}$, which can be accomplished with $\delta^2 n^3$ flops. The ith iteration of the GSA involves the multiplication of n Householder matrices of size n times a matrix of size $((i + \delta + 1)n \times n)$. Therefore, since the cost of the multiplication by the hyperbolic rotation is negligible with respect to that of the multiplication by the Householder matrices, the computational cost at iteration i is $4n^3(\delta + i)$. Hence, the computational cost of GSA is $2n^3 n_b(2\delta + n_b^2/2)$.

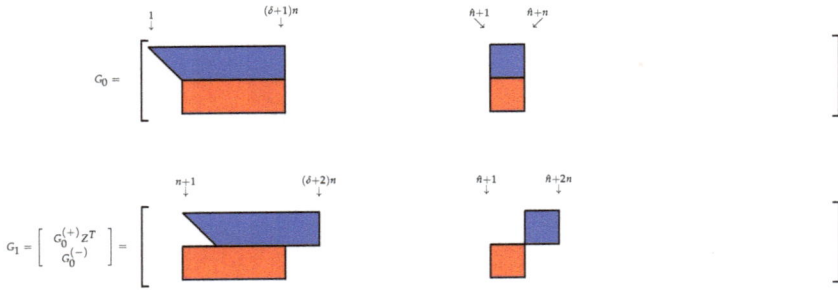

Figure 1. Modification of the sparsity pattern of the generator matrix G after the first iteration.

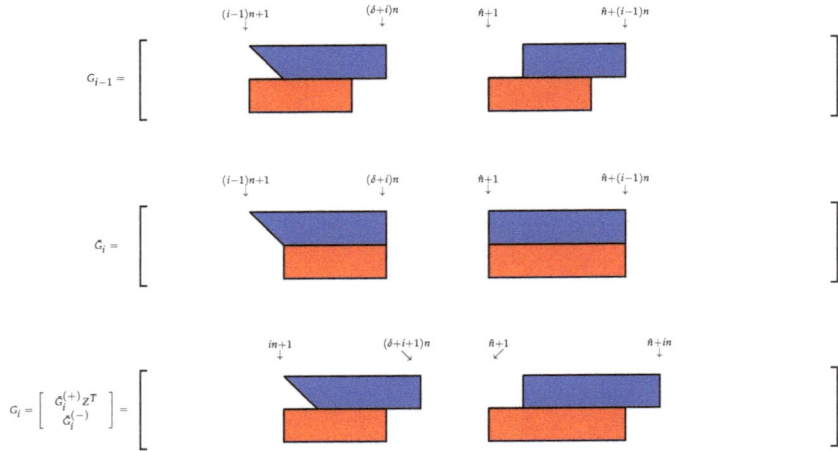

Figure 2. Modification of the sparsity pattern of the generator matrix G after the ith iteration.

5.2. GSA for Computing the Right Null-Space of Semidefinite Block Toeplitz Matrices

As already mentioned in Section 5.1, the number of desired blocks n_b of the matrix T in Equation (13) can be computed as described in [8]. For the sake of simplicity, in the considered examples, we choose n_b large enough to compute the null-space of T.

The structure and the computation via the GSA of the R factor of the QR factorization of the singular block Toeplitz matrix T with rank $\rho < n \leq m$, is considered in [23].

A modification of the GSA for computing the null-space of Toeplitz matrices is described in [25]. In this paper, we extend the latter results to compute the null-space of T by modifying GSA.

Without loss of generality, let us assume that the first $\hat{n} - 1$ columns of T are linear independent and suppose that the $\hat{n}$th column linearly depends on the previous ones. Therefore, the first $\hat{n} - 1$ principal minors of $\hat{T}$ are positive while the $\hat{n}$th one is zero. Let $\hat{T} = Q \Lambda Q^T$ be the spectral decomposition of $\hat{T}$, with $Q = [\mathbf{q}_1, \dots, \mathbf{q}_{\hat{n}}]$ orthogonal and $\Lambda = \operatorname{diag}(\lambda_1, \dots, \lambda_{\hat{n}-1}, \lambda_{\hat{n}})$, with

$$\lambda_1 \geq \lambda_2 \geq \cdots \geq \lambda_{\hat{n}-1} > \lambda_{\hat{n}} = 0,$$

and let $\hat{T}_\varepsilon = Q \Lambda_\varepsilon Q^T$, with $\Lambda_\varepsilon = \operatorname{diag}(\lambda_1, \dots, \lambda_{\hat{n}-1}, \varepsilon^2)$, with $\varepsilon \in \mathbb{R}_+^*$. Hence,

$$\hat{T}_\varepsilon^{-1} = \frac{1}{\varepsilon^2} \left(\sum_{i=1}^{\hat{n}-1} \frac{\varepsilon^2}{\lambda_i} \mathbf{q}_i \mathbf{q}_i^T + \mathbf{q}_{\hat{n}} \mathbf{q}_{\hat{n}}^T \right).$$

Let R_ε be the Cholesky factor of $\hat{T}_\varepsilon$, with R_ε upper triangular, i.e., $\hat{T}_\varepsilon = R_\varepsilon^T R_\varepsilon$. Then,

$$\varepsilon^2 \hat{T}_\varepsilon^{-1} e_{\hat{n}}^{(2\hat{n})} = \left(\sum_{i=1}^{\hat{n}-1} \frac{\varepsilon^2 \mathbf{q}_i^T e_{\hat{n}}^{(2\hat{n})}}{\lambda_i} \mathbf{q}_i + (\mathbf{q}_{\hat{n}}^T e_{\hat{n}}^{(2\hat{n})}) \mathbf{q}_{\hat{n}} \right).$$

On the other hand,

$$\varepsilon^2 \hat{T}_\varepsilon^{-1} e_{\hat{n}}^{(2\hat{n})} = \varepsilon^2 R_\varepsilon^{-1} R_\varepsilon^{-T} e_{\hat{n}}^{(2\hat{n})} = \varepsilon^2 r_{\hat{n},\hat{n}}^{-1} R_\varepsilon^{-1} e_{\hat{n}}^{(2\hat{n})},$$

where $r_{\hat{n},\hat{n}} = e_{\hat{n}}^{(2\hat{n})^T} R_\varepsilon e_{\hat{n}}^{(2\hat{n})}$. Hence, as $\varepsilon \to 0^+$, the last column of R_ε^{-1} becomes closer and closer to a multiple of $\mathbf{q}_{\hat{n}}$, the eigenvector corresponding to the 0 eigenvalue of $\hat{T}$.

Therefore, given

$$W_\varepsilon = \left[\begin{array}{c|c} \hat{T}_\varepsilon & I_{\hat{n}} \\ \hline I_{\hat{n}} & 0_{\hat{n}} \end{array} \right],$$

we have that

$$\nabla_Z(W_\varepsilon) = G^{(+)^T} G^{(+)} - G^{(-)^T} G^{(-)} + \varepsilon^2 e_{\hat{n}}^{(2\hat{n})} e_{\hat{n}}^{(2\hat{n})^T}.$$

Let

$$G_{0,\varepsilon} = \left[\begin{array}{c} G^{(+)} \\ G^{(-)} \\ \varepsilon e_{\hat{n}}^{(2\hat{n})^T} \end{array} \right].$$

Define

$$J_\varepsilon = \mathrm{diag}(\underbrace{1,1,\ldots,1}_{\hat{n}}, \underbrace{-1,-1,\ldots,-1}_{\hat{n}}, 1).$$

Hence,

$$\nabla_Z(W_\varepsilon) = G_{0,\varepsilon}^T J_\varepsilon G_{0,\varepsilon}.$$

We observe that column $\hat{n} + 1$ of the generator matrix is not involved in the GSA until the very last iteration, since only its $\hat{n}$th entry is different from 0. At the very last iteration, the hyperbolic rotation

$$Y = \left[\begin{array}{cc} c & -s \\ -s & c \end{array} \right],$$

with

$$c = \frac{\sqrt{G^{(+)^2}(n,\hat{n}) + \varepsilon^2}}{\sqrt{G^{(+)^2}(n,\hat{n}) + \varepsilon^2 - G^{(-)^2}(1,\hat{n})}}, \qquad s = \frac{G^{(-)}(1,\hat{n})}{\sqrt{G^{(+)^2}(n,\hat{n}) + \varepsilon^2 - G^{(-)^2}(1,\hat{n})}}$$

is applied to the $\hat{n}$th and $(\hat{n}+1)$st rows of G, i.e., to the nth row of $G^{(+)}$ and the first one of $G^{(-)}$. Since $\hat{T}$ is singular, it turns out that $|G^{(+)}(n,\hat{n})| = |G^{(-)}(1,\hat{n})|$ (see [23,25]). Thus,

$$
\begin{aligned}
Y &= \frac{1}{\sqrt{G^{(+)^2}(n,\hat{n}) + \varepsilon^2 - G^{(-)^2}(1,\hat{n})}} \left[\begin{array}{cc} \sqrt{G^{(+)^2}(n,\hat{n}) + \varepsilon^2} & -G^{(-)}(1,\hat{n}) \\ -G^{(-)}(1,\hat{n}) & \sqrt{G^{(+)^2}(n,\hat{n}) + \varepsilon^2} \end{array} \right] \\
&= \frac{|G^{(+)}(n,\hat{n})|}{\varepsilon} \left[\begin{array}{cc} \sqrt{1 + \left(\frac{\varepsilon}{G^{(+)}(n,\hat{n})} \right)^2} & -\theta \\ -\theta & \sqrt{1 + \left(\frac{\varepsilon}{G^{(+)}(n,\hat{n})} \right)^2} \end{array} \right],
\end{aligned}
$$

where

$$\theta = \frac{G^{(-)}(1, \hat{n})}{|G^{(+)}(n, \hat{n})|} = \text{sign}(G^{(-)}(1, \hat{n})).$$

We observe that, as $\varepsilon \to 0^+$,

$$\frac{|G^{(+)}(n, \hat{n})|}{\varepsilon} \to \infty, \quad \sqrt{1 + \left(\frac{\varepsilon}{G^{(+)}(n, \hat{n})}\right)^2} \to 1.$$

Since a vector of the right null-space of T is determined except for the multiplication by a constant, neglecting the term $|G^{(+)}(n, \hat{n})|/\varepsilon$, such a vector can be computed at the last iteration as the first column of the product

$$\begin{bmatrix} 1 & -\theta \\ -\theta & 1 \end{bmatrix} \begin{bmatrix} G^{(+)}(n, \hat{n}+1:2\hat{n}) \\ G^{(-)}(1, \hat{n}+1:2\hat{n}) \end{bmatrix}.$$

When detecting a vector of the null-space as a linear combination of row n of $G^{(+)}$ and row one of $G^{(-)}$, the new generator matrix G for the GSA is obtained removing the latter columns from G [23,25].

The implementation of the modified GSA for computing the null-space of band block-Toeplitz matrices in Equation (13) is rather technical and can be obtained from the authors upon request.

The stability properties of the GSA have been studied in [4,5,24]. The proposed algorithm inherits the stability properties of the GSA, which means that it is weakly stable.

6. Numerical Examples

All the numerical experiments were carried out in `matlab` with machine precision $\varepsilon \approx 2.22 \times 10^{-16}$. Example 1 concerns the computation of the rank of a Sylvester matrix, while Examples 2 and 3 concern the computation of the null-space of polynomial matrices.

Example 1. *Let $x_i, i = 1, \ldots, 12, y_i, i = 1, \ldots, 15,$ and $z_i, i = 1, \ldots, 3,$ be random numbers generated by the* `matlab` *function* `randn`*. Let $w(x)$and $y(x)$ be the two polynomials of degree 15 and 18, constructed by the matlab function* `poly`*, whose roots are, respectively, x_i and z_j, $i = 1, \ldots, 12,$ $j = 1, \ldots, 3,$ and y_i and z_j, $i = 1, \ldots, 15,$ $j = 1, \ldots, 3.$*

The greatest common divisor of w and y has degree 3 and, therefore, the Sylvester matrix $S \in \mathbb{R}^{33 \times 33}$ constructed from $w(x)$ anf $y(x)$ has rank 30. The diagonal entries of the R factor computed by the GSA implementation described in Section 4 are displayed in Figure 3. Observe that the rank of the matrix can be retrieved by the number of entries of R above a certain tolerance. Moreover, it can be noticed that the first $m = 15$ diagonal entries monotonically decrease.

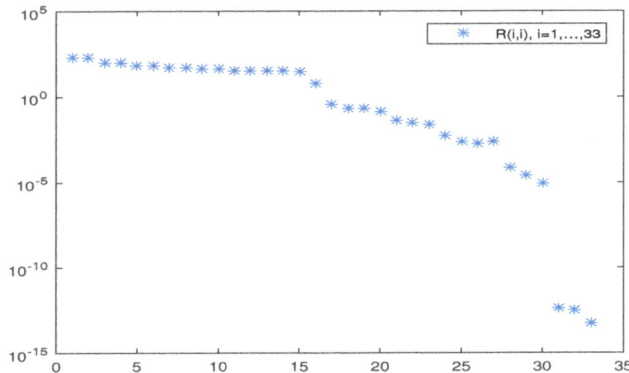

Figure 3. Diagonal entries of R.

Example 2. *As second example, we consider the computation of the coprime factorization of a transfer function matrix, described in [9,26]. The results obtained by the proposed GSA-based algorithm were compared with those obtained computing the null-space of the considered matrix by the function* svd *of* matlab.
   Let $H(s) = N_r(s)D_r^{-1}(s)$ *be the transfer function with*

$$D_r(s) = \begin{bmatrix} 1-s & 0 & 0 & 0 \\ 0 & 1-s & 0 & 0 \\ 0 & -s & 1-s & 0 \\ 0 & 0 & 0 & 1-s \end{bmatrix}, \quad N_r(s) = \begin{bmatrix} s^2 & 0 & 0 & 0 \\ 0 & 0 & 0 & 0 \\ 0 & 0 & 0 & 0 \\ 0 & 0 & s & 0 \\ 0 & 0 & 0 & s \end{bmatrix}.$$

Let

$$M(s) = \begin{bmatrix} N_r^T(s) & -D_r^T(s) \end{bmatrix} = \begin{bmatrix} s^2 & 0 & 0 & 0 & 0 & s-1 & 0 & 0 & 0 \\ 0 & 0 & 0 & 0 & 0 & 0 & s-1 & s & 0 \\ 0 & 0 & 0 & s & 0 & 0 & 0 & s-1 & 0 \\ 0 & 0 & 0 & 0 & s & 0 & 0 & 0 & s-1 \end{bmatrix}.$$

As reported in [9,26], a minimal polynomial basis for the right null-space of $M(s)$ is

$$N(s) = \begin{bmatrix} 1-s & 0 & 0 \\ 0 & 0 & 0 \\ 0 & 0 & 0 \\ 0 & (1-s)^2 & 0 \\ 0 & 0 & 1-s \\ s^2 & 0 & 0 \\ 0 & s^2 & 0 \\ 0 & s-s^2 & 0 \\ 0 & 0 & s \end{bmatrix}.$$

Let us consider $n_b = 3$. Then, $T \in \mathbb{R}^{20 \times 21}$ is the block-Toeplitz matrix constructed from $M(s)$ as described in Section 5.1. Let $\text{rank}(M) = 17$ and $U\Sigma V^T$ be the rank and the singular value decomposition of T computed by matlab, *respectively, and let us define $V_1 = V(:, 1:17)$ and $V_2 = V(:, 18:21)$ the matrices of the right singular vectors corresponding to the nonzero and zero singular values of T, respectively. The modified GSA applied to T yields four vectors $\mathbf{v}_1, Z_{n_b}\mathbf{v}_1, \mathbf{v}_2, \mathbf{v}_3 \in \mathbb{R}^{21}$ belonging to the right null-space of $M(s)$, with $Z_{n_b} = \text{diag}(\text{ones}(14,1), -7)$. Let $X = [\mathbf{v}_1 \; Z_{n_b}\mathbf{v}_1 \; \mathbf{v}_2 \; \mathbf{v}_3]$. In Table 1, the relative norm of TV_2, the relative norm of TX, the norm of $V_1^T V_2$ and the norm of $V_1^T X$, are reported in Columns 1–4, respectively.*

Table 1. Relative norm of TV_2, relative norm of TX, norm of $V_1^T V_2$ and norm of $V_1^T X$, for Example 2.

$\frac{\|TV_2\|_2}{\|T\|_2}$	$\frac{\|TX\|_2}{\|T\|_2}$	$\|V_1^T V_2\|_2$	$\|V_1^T X\|_2$
4.23×10^{-16}	2.89×10^{-16}	9.66×10^{-16}	2.59×10^{-15}

Such values show that the results provided by svd *of* matlab *and by the algorithm based on a modification of GSA are comparable in terms of accuracy.*

Example 3. *This example can be found in [9,26]. Let $H(s) = D_l^{-1}(s)N_r(s)$ be the transfer function with*

$$D_l(s) = (s+2)^2(s+3)\begin{bmatrix} 1 & 0 \\ 0 & 1 \end{bmatrix}, \quad N_l(s) = \begin{bmatrix} 3s+8 & 2s^2+6s+2 \\ s^2+6s+2 & 3s^2+7s+8 \end{bmatrix}.$$

Let $M(s) = [D_l(s), -N_L(s)]$. A right coprime pair for $M(s)$ is given by

$$N_r = \begin{bmatrix} 3 & 2 \\ s+2 & 3 \end{bmatrix}, \quad D_r = \begin{bmatrix} s^2+3s+4 & 2 \\ 2 & s+4 \end{bmatrix},$$

Let us choose $n_b = 4$. Then, $T \in \mathbb{R}^{14 \times 16}$ is the block-Toeplitz matrix constructed from $M(s)$ as described in Section 5.1. Let $\text{rank}(M) = 11$ and $U\Sigma V^T$ be the rank and the singular value decomposition of T computed by matlab, respectively, and let define $V_1 = V(:, 1:11)$ and $V_2 = V(:, 12:16)$ the matrices of the right singular vectors corresponding to the nonzero and zero singular values of T, respectively. The modified GSA applied to T yields the vectors $\mathbf{v}_1, Z_{n_b}\mathbf{v}_1, Z_{n_b}^2\mathbf{v}_1, \mathbf{v}_2, Z_{n_b}\mathbf{v}_2, \mathbf{v}_3 \in \mathbb{R}^{16}$ of the right null-space, with $Z_{n_b} = \text{diag}(\text{ones}(12,1), -4)$. Let $X = [\mathbf{v}_1, Z_{n_b}\mathbf{v}_1, Z_{n_b}^2\mathbf{v}_1, \mathbf{v}_2, Z_{n_b}\mathbf{v}_2, \mathbf{v}_3]$. In Table 2, the relative norm of TV_2, the relative norm of TX, the norm of $V_1^T V_2$ and the norm of $V_1^T X$, are reported in Columns 1–4, respectively.

Table 2. Relative norm of TV_2, relative norm of TX, norm of $V_1^T V_2$ and norm of $V_1^T X$, for Example 3.

$\dfrac{\lVert TV_2 \rVert_2}{\lVert T \rVert_2}$	$\dfrac{\lVert TX \rVert_2}{\lVert T \rVert_2}$	$\lVert V_1^T V_2 \rVert_2$	$\lVert V_1^T X \rVert_2$
1.33×10^{-16}	2.04×10^{-16}	5.56×10^{-16}	4.91×10^{-15}

As in Example 2, the results yielded by the considered algorithms are comparable in accuracy.

7. Conclusions

The Generalized Schur Algorithm is a powerful tool allowing to compute classical decompositions of matrices, such as the QR and LU factorizations. If the involved matrices have a particular structure, such as Toeplitz or Sylvester, the GSA computes the latter factorizations with a complexity of one order of magnitude less than that of classical algorithms based on Householder or elementary transformations.

After having emphasized the main features of the GSA, we have shown in this manuscript that the GSA helps to prove some theoretical properties of the R factor of the QR factorization of some structured matrices. Moreover, a fast implementation of the GSA for computing the rank of Sylvester matrices and the null-space of polynomial matrices is proposed, which relies on a modification of the GSA for computing the R factor and its inverse of the QR factorization of band block-Toeplitz matrices with full column rank. The numerical examples show that the proposed approach yields reliable results comparable to those ones provided by the function svd of matlab.

Author Contributions: All authors contributed equally to this work.

Funding: This research was partly funded by INdAM-GNCS and by CNR under the Short Term Mobility Program.

Conflicts of Interest: The authors declare no conflict of interest.

References

1. Kailath, T.; Sayed, A.H. *Fast Reliable Algorithms for Matrices with Structure*; SIAM: Philadelphia, PA, USA, 1999.
2. Kailath, T.; Sayed, A. Displacement Structure: Theory and Applications. *SIAM Rev.* **1995**, *32*, 297–386. [CrossRef]
3. Chandrasekaran, S.; Sayed, A. Stabilizing the Generalized Schur Algorithm. *SIAM J. Matrix Anal. Appl.* **1996**, *17*, 950–983. [CrossRef]
4. Stewart, M.; Van Dooren, P. Stability issues in the factorization of structured matrices. *SIAM J. Matrix Anal. Appl.* **1997**, *18*, 104–118. [CrossRef]
5. Mastronardi, N.; Van Dooren, P.; Van Huffel, S. On the stability of the generalized Schur algorithm. *Lect. Notes Comput. Sci.* **2001**, *1988*, 560–567.

6. Li, B.; Liu, Z.; Zhi, L. A structured rank-revealing method for Sylvester matrix. *J. Comput. Appl. Math.* **2008**, *213*, 212–223. [CrossRef]
7. Forney, G. Minimal bases of rational vector spaces with applications to multivariable linear systems. *SIAM J. Control Optim.* **1975**, *13*, 493–520. [CrossRef]
8. Zúñiga Anaya, J.; Henrion, D. An improved Toeplitz algorithm for polynomial matrix null-space computation. *Appl. Math. Comput.* **2009**, *207*, 256–272. [CrossRef]
9. Beelen, T.; van der Hurk, G.; Praagman, C. A new method for computing a column reduced polynomial matrix. *Syst. Control Lett.* **1988**, *10*, 217–224. [CrossRef]
10. Neven, W.; Praagman, C. Column reduction of polynomial matrices. *Linear Algebra Its Appl.* **1993**, *188*, 569–589. [CrossRef]
11. Bojańczyk, A.; Brent, R.; Van Dooren, P.; de Hoog, F. A note on downdating the Cholesky factorization. *SIAM J. Sci. Stat. Comput.* **1987**, *8*, 210–221. [CrossRef]
12. Higham, N.J. J-orthogonal matrices: properties and generation. *SIAM Rev.* **2003**, *45*, 504–519. [CrossRef]
13. Lemmerling, P.; Mastronardi, N.; Van Huffel, S. Fast algorithm for solving the Hankel/Toeplitz Structured Total Least Squares Problem. *Numer. Algorithm.* **2000**, *23*, 371–392. [CrossRef]
14. Mastronardi, N.; Lemmerling, P.; Van Huffel, S. Fast structured total least squares algorithm for solving the basic deconvolution problem. *SIAM J. Matrix Anal. Appl.* **2000**, *22*, 533–553.[CrossRef]
15. Golub, G.H.; Van Loan, C.F. *Matrix Computations*, 4th ed.; Johns Hopkins University Press: Baltimore, MD, USA, 2013.
16. Boito, P. *Structured Matrix Based Methods for Approximate Polynomial GCD*; Edizioni Della Normale: Pisa, Italy, 2011.
17. Boito, P.; Bini, D.A. A Fast Algorithm for Approximate Polynomial GCD Based on Structured Matrix Computations. In *Numerical Methods for Structured Matrices and Applications*; Bini, D., Mehrmann, V., Olshevsky, V., Tyrtyshnikov, E., Van Barel, M., Eds.; Birkhauser: Basel, Switzerland, 2010; Volume 199, pp. 155–173.
18. Mastronardi, N.; Lernmerling, P.; Kalsi, A.; O'Leary, D.; Van Huffel, S. Implementation of the regularized structured total least squares algorithms for blind image deblurring. *Linear Algebra Its Appl.* **2004**, *391*, 203–221. [CrossRef]
19. Basilio, J.; Moreira, M. A robust solution of the generalized polynomial Bezout identity. *Linear Algebra Its Appl.* **2004**, *385*, 287–303. [CrossRef]
20. Kailath, T. *Linear Systems*; Prentice Hall: Englewood Cliffs, NJ, USA, 1980.
21. Bueno, M.; De Terán, F.; Dopico, F. Recovery of Eigenvectors and Minimal Bases of Matrix Polynomials from Generalized Fiedler Linearizations. *SIAM J. Matrix Anal. Appl.* **2011**, *32*, 463–483. [CrossRef]
22. De Terán, F.; Dopico, F.; Mackey, D. Fiedler companion linearizations and the recovery of minimal indices. *SIAM J. Matrix Anal. Appl.* **2010**, *31*, 2181–2204. [CrossRef]
23. Gallivan, K.; Thirumalai, S.; Van Dooren, P.; Vermaut, V. High performance algorithms for Toeplitz and block Toeplitz matrices. *Linear Algebra Its Appl.* **1996**, *241–243*, 343–388. [CrossRef]
24. Stewart, M. Cholesky Factorization of Semi-definite Toeplitz Matrices. *Linear Algebra Its Appl.* **1997**, *254*, 497–526. [CrossRef]
25. Mastronardi, N.; Van Barel, M.; Vandebril, R. On the computation of the null space of Toeplitz-like matrices. *Electron. Trans. Numer. Anal.* **2009**, *33*, 151–162.
26. Antoniou, E.; Vardulakis, A.; Vologiannidis, S. Numerical computation of minimal polynomial bases: A generalized resultant approach. *Linear Algebra Its Appl.* **2005**, *405*, 264–278. [CrossRef]

⊞ axioms

MDPI

Article

On Partial Cholesky Factorization and a Variant of Quasi-Newton Preconditioners for Symmetric Positive Definite Matrices

Benedetta Morini † (iD)

Dipartimento di Ingegneria Industriale, Università degli Studi di Firenze, 50134 Florence, Italy;
benedetta.morini@unifi.it; Tel.: +39-055-275-8684
† Member of the INdAM Research Group GNCS.

Received: 23 April 2018; Accepted: 20 June 2018; Published: 1 July 2018

Abstract: This work studies limited memory preconditioners for linear symmetric positive definite systems of equations. Connections are established between a partial Cholesky factorization from the literature and a variant of Quasi-Newton type preconditioners. Then, a strategy for enhancing the Quasi-Newton preconditioner via available information is proposed. Numerical experiments show the behaviour of the resulting preconditioner.

Keywords: linear systems; preconditioners; Cholesky factorization; limited memory

1. Introduction

The numerical solution of linear algebraic systems with symmetric positive definite (SPD) matrix is required in a broad range of applications, see e.g., [1–6]. We consider the case where the linear systems are large and investigate their iterative solution by preconditioned Krylov subspace methods [4,7,8]. Our problem takes the form

$$Hx = b, \tag{1}$$

where $H \in \mathcal{R}^{m \times m}$ is SPD. As a particular case of interest we also consider the case where $H = A\Theta A^T$, $A \in \mathcal{R}^{m \times n}$ is a sparse full row-rank matrix and $\Theta \in \mathcal{R}^{n \times n}$ is SPD. Systems of this kind arise in many contexts, such as the solution of linear and nonlinear least-squares problems and the solution of linear programming problems, see e.g., [4,6,9]. The iterative solver employed is the Conjugate Gradient (CG) method or its variants [4] and we propose its use in combination with limited memory preconditioners.

A preconditioner is denoted as limited memory if it can be stored compactly in a few vectors of length m, and its product by a vector calls for scalar products and, possibly, sums of vectors [10]. The limited memory preconditioners studied in this work belong to both the class of Incomplete Cholesky factorizations and to the class of Quasi-Newton preconditioners. Interestingly, they are approximate inverse preconditioners, i.e., they are approximations for H^{-1}. We point out that the preconditioners proposed can also be used for solving symmetric saddle point linear systems iteratively. In fact, the application of constraint or augmented preconditioners involves the factorization of SPD matrices and a cheap approximation of such matrices or their inverses can be convenient [11,12].

Incomplete Cholesky factorizations use the entries of H and may fail for a general SPD matrix, thus requiring strategies for recovering breakdowns. Further, memory requirements are difficult to predict if a drop tolerance is used to reduce the fill-in. For this reason, some factorizations allow a limited number of fill-ins to be created and the bound on the number of nonzero entries can be prefixed column-wise, typically taking into account the number of nonzero entries of the original matrix. Clearly, Incomplete Cholesky factorizations are not suitable when matrix H is dense, which may be the case if $H = A\Theta A^T$ even though A is sparse, see e.g., [3]. A limited memory and breakdown-free "partial" Cholesky factorization was proposed in [13,14] and used in the solution of compressed sensing, linear and quadratic programming, Lasso problems, maximum cut problems [3,14,15]. This preconditioner is built by computing a trapezoidal partial Cholesky factorization limited to a prefixed and small number of columns and by approximating the resulting Schur complement via its diagonal.

Limited memory Quasi-Newton preconditioners are a class of matrices built drawing inspiration from Quasi-Newton schemes for convex quadratic programming [10]. Given a preconditioner (first-level preconditioner), Quasi-Newton preconditioners provide its update (second-level preconditioner) by exploiting a few vectors of dimension m, i.e., information belonging to a low-dimensional subspace of $\mathcal{R}^m$. Variants of the original Quasi-Newton scheme [10] have been proposed in the literature. They differ in the choice of the low-dimensional subspace [10,16–19] and several instances convey information from the iterative solver, possibly as approximate invariant subspaces.

In this paper, we analyze the connection between the partial Cholesky factorization [13,14] and a variant of the Quasi-Newton preconditioners. We show that the partial Cholesky factorization coincides with a Quasi-Newton preconditioner where the first-level preconditioner is diagonal and the low-dimensional subspace is constituted by a subset of columns of the identity matrix of dimension m. This observation provides a way for building the partial Cholesky factorization which is alternative to the procedures in [13,14] and can offer some advantages in terms of computational effort. Due to the specific form of the low-dimensional subspace spanned by coordinate vectors in $\mathbb{R}^m$, we denote the resulting preconditioner as the Coordinate Limited Memory Preconditioner (Coordinate-LMP). Successively, we propose a strategy for enriching the low-dimensional space that generates the partial Cholesky factorization, and thus enhancing the performance of the preconditioner; such a strategy is guided by the spectral analysis of H preconditioned by the partial Cholesky factorization and it is analyzed from both the theoretical and practical point of view.

The paper is organized as follows. In Section 2 we introduce the partial Cholesky factorization. In Section 3 we show how the partial Cholesky factorization can be formulated as a Quasi-Newton type preconditioner and discuss the application of the two formulations in terms of computational effort. In Section 4 we propose a strategy for enlarging the subspace in the Quasi-Newton formulation and analyze the spectral properties of the preconditioned matrix; the numerical performance of the resulting preconditioner are shown in Section 5.

In the following, for any square matrix B, diag(B) is the diagonal matrix with the same diagonal entries as B. For a SPD matrix $B \in \mathbb{R}^{m \times m}$, an eigenvalue is denoted either as $\lambda(B)$ or as $\lambda_i(B)$, $1 \le i \le m$; the minimum and maximum eigenvalues are denoted as $\lambda_{\min}(B)$ and $\lambda_{\max}(B)$. The identity matrix of dimension q is denoted as I_q. For indicating submatrices we borrow the MATLAB notation. Preconditioned CG method is denoted as PCG.

2. A Limited Memory Partial Cholesky Preconditioner

The convergence behaviour of CG depends on the eigenvalue distribution of H and the condition number of H determines the worst-case behaviour of CG [4,8,20,21]. Further characterizations, possibly sharp, of the convergence behaviour can be gained from additional information on the eigenvalues. More specifically, it is known that if A has t distinct eigenvalues, then the CG method will converge in at most t iterations, while CG applied to a matrix with t tight eigenvalue clusters may not behave similarly as though it were applied to a matrix with t distinct eigenvalues representing the individual clusters ([21], [§5.6.5]).

A proposal for building a partial Cholesky factorization P of H was given by Gondzio in [14] and Bellavia et al., in [13]. It aims at clustering the largest eigenvalues of H at one and reducing both the value of $\lambda_{\max}(P^{-1}H)$ with respect to $\lambda_{\max}(H)$ and the condition number of $P^{-1}H$ with respect to H. The first step is based on the observation that the trace $tr(H)$ of H is such that $tr(H) = \sum_{i=1}^{m} \lambda_i(H)$ and $\lambda_{\max}(H) \le tr(H)$ since H is symmetric and positive definite. Then, it is possible to handle the largest eigenvalues of H by an heuristic technique where the largest k, $k \ll m$, diagonal elements of H are identified and the rank-k partial Cholesky factorization of the corresponding columns of H is performed. More specifically, suppose that H can be partitioned as

$$H = \begin{bmatrix} H_{11} & H_{21}^T \\ H_{21} & H_{22} \end{bmatrix},$$

where $H_{11} \in \mathcal{R}^{k \times k}$, $H_{22} \in \mathcal{R}^{(m-k) \times (m-k)}$ and H_{11} contains the k largest diagonal elements of H (throughout the paper, the symmetric row and column permutations required to move the k largest diagonal elements of H to the $(1,1)$ block are ignored to make the presentation simpler). To handle the large eigenvalues of H, the Cholesky factorization limited to the first k columns $\begin{bmatrix} H_{11} \\ H_{21} \end{bmatrix}$ is computed.

We denote such factors as $\begin{bmatrix} L_{11} \\ L_{21} \end{bmatrix} \in \mathbb{R}^{m \times k}$, with $L_{11} \in \mathbb{R}^{k \times k}$ unit lower triangular, $L_{21} \in \mathbb{R}^{(m-k) \times k}$, and $D_1 \in \mathbb{R}^{k \times k}$ diagonal positive definite and observe that H can be factorized as

$$H = LD_H L^T \stackrel{\text{def}}{=} \begin{bmatrix} L_{11} & \\ L_{21} & I_{m-k} \end{bmatrix} \begin{bmatrix} D_1 & \\ & S \end{bmatrix} \begin{bmatrix} L_{11}^T & L_{21}^T \\ & I_{m-k} \end{bmatrix}, \tag{2}$$

being S the Schur complement of H_{11} in H

$$S = H_{22} - H_{21} H_{11}^{-1} H_{21}^T. \tag{3}$$

Finally, the limited memory Partial Cholesky preconditioner P is obtained by approximating S with its diagonal and setting

$$P = LD_P L^T \stackrel{\text{def}}{=} \begin{bmatrix} L_{11} & \\ L_{21} & I_{m-k} \end{bmatrix} \begin{bmatrix} D_1 & \\ & D_2 \end{bmatrix} \begin{bmatrix} L_{11}^T & L_{21}^T \\ & I_{m-k} \end{bmatrix}, \qquad D_2 = \text{diag}(S). \tag{4}$$

The construction of the preconditioner P is summarized in Algorithm 1 where we use the equalities

$$H_{11} = L_{11} D_1 L_{11}^T, \tag{5}$$
$$H_{21}^T = L_{11} D_1 L_{21}^T \quad \text{i.e.,} \quad L_{21} = H_{21} L_{11}^{-T} D_1^{-1}, \tag{6}$$

derived from Equation (2).

Algorithm 1 Limited Memory Partial Cholesky Preconditioner.

Given the matrix-vector operators $u \to Hu$, $k > 0$.
    1. Form the first k columns of H, i.e. H_{11}, H_{21}.
    2. Compute the diagonal entries of H_{22}.
    3. Compute L_{11}, D_1, L_{21} as in Equations (5) and (6). Discard H_{11} and H_{21}.
    4. Set $D_2 = \text{diag}(H_{22}) - \text{diag}(L_{21} D_1 L_{21}^T)$.
    5. Let P take the form Equation (4).

This procedure is breakdown-free in exact arithmetic while Incomplete Cholesky factorizations employing a drop tolerance to reduce fill-in may fail for a general SPD matrix. The maximum storage requirement is known in advance and the upper bound on the number of nonzero entries in L is $m + k(m - k/2 - 1/2)$.

Forming the preconditioner calls for the complete diagonal of H. If H has the special form $H = A\Theta A^T$, its main diagonal can be constructed by performing m matrix-vector products $r_i = A^T e_i$, $i = 1, \ldots, m$, and then computing $(H)_{ii} = r_i^T \Theta r_i$. The products $A^T e_i$ are cheap if A is sparse and involve no extra effort at all if A can be accessed row-wise and then retrieving the ith row comes at no extra cost. Moreover, the k products $A\Theta A^T e_i$ in Step 1 are expected to be cheaper than the products $A\Theta A^T v$ required by a CG-like method because the unit vectors e_i are typically sparser than v.

The cost to perform the factorization Equation (5) is negligible because matrix H_{11} has small dimension k, while, using the first equation in Equation (6), the computation of L_{21} in Step 4 requires solving $m - k$ triangular linear systems of dimension k. Finally, in Step 4 computing $\text{diag}(L_{21}D_1 L_{21}^T)$ amounts to scaling the rows of L_{21}^T by the entries of D_1 and performing $m - k$ scalar products between vectors of dimension k.

The spectral properties of $P^{-1}H$ are analyzed in [13] and reported below for completeness.

Theorem 1. *Let k be a positive integer and P be as in Equation (4). Then, k eigenvalues of $P^{-1}H$ are equal to 1 and the remaining are equal to the eigenvalues of $D_2^{-1}S$. Moreover, any eigenvalue $\lambda(D_2^{-1}S)$ lies in the interval*
$$\left[\frac{\lambda_{\min}(S)}{\lambda_{\max}(D_2)}, \frac{\lambda_{\max}(S)}{\lambda_{\min}(D_2)} \right] \subseteq \left[\frac{\lambda_{\min}(H)}{\lambda_{\max}(D_2)}, \frac{\lambda_{\max}(H_{22})}{\lambda_{\min}(D_2)} \right].$$

Proof of Theorem 1. Theorem 2.1 in [13] proves that k eigenvalues of $P^{-1}H$ are equal to 1 and the remaining are equal to the eigenvalues of $D_2^{-1}S$. As for the bounds on the eigenvalues $\lambda(D_2^{-1}S)$, let $v \in \mathcal{R}^{m-k}$ be an eigenvector of $D_2^{-1}S$. Then, by $D_2^{-1}Sv = \lambda v$ we get $\lambda(D_2^{-1}S) = \dfrac{v^T S v}{v^T D_2 v}$ and

$$\lambda(D_2^{-1}S) \geq \frac{\lambda_{\min}(S)}{\lambda_{\max}(D_2)} \geq \frac{\lambda_{\min}(H)}{\lambda_{\max}(D_2)}, \tag{7}$$

$$\lambda(D_2^{-1}S) \leq \frac{\lambda_{\max}(S)}{\lambda_{\min}(D_2)} \leq \frac{\lambda_{\max}(H_{22})}{\lambda_{\min}(D_2)}, \tag{8}$$

where we used the bounds $\lambda_{\min}(H) \leq \lambda(S) \leq \lambda_{\max}(H_{22})$, see [6]. $\square$

We point out that the above preconditioner was used in [13] in conjunction with Deflated-CG [22] in order to handle also the smallest eigenvalues of $P^{-1}H$.

We conclude this section observing that the Partial Cholesky preconditioner can be formulated as an Approximate Inverse preconditioner. In fact, from (4) matrix P^{-1} can be factorized as the product of sparse matrices and takes the form

$$P^{-1} = L^{-T} D_P^{-1} L^{-1} = \begin{bmatrix} L_{11}^{-T} & -L_{11}^{-T} L_{21}^T \\ & I_{m-k} \end{bmatrix} \begin{bmatrix} D_1^{-1} & \\ & D_2^{-1} \end{bmatrix} \begin{bmatrix} L_{11}^{-1} & \\ -L_{21}L_{11}^{-1} & I_{m-k} \end{bmatrix}. \tag{9}$$

3. Limited Memory Quasi-Newton Type Preconditioners

Limited memory Quasi-Newton type preconditioners for SPD matrices were proposed in several works, see e.g., [10,17–19]. These preconditioners are generated using a small number k of linear independent vectors in $\mathbb{R}^m$.

Let us consider the formulation by Gratton et al. in [19]. Suppose that a first preconditioner M (called first-level preconditioner) is available. To improve the efficiency of the first-level preconditioner, a class of limited memory preconditioners (called second-level preconditioners) is defined on the base of the explicit knowledge of an m by k, $k \ll m$, full rank matrix Z. The aim of the second-level preconditioner is to capture directions lying in the range of HZ which have been left out by the first-level preconditioner and are slowing down the convergence of the CG solver; e.g., this is the case when the first-level preconditioner is able to cluster many eigenvalues at 1 with relatively few outliers.

Let $M \in \mathbb{R}^{m \times m}$ be symmetric and positive definite, $Z \in \mathbb{R}^{m \times k}$, $k \ll m$, be a full column-rank matrix. The symmetric second-level preconditioner, say Π, takes the form

$$\Pi = (I - TH)M(I - HT) + T, \qquad T = Z(Z^T HZ)^{-1}Z^T. \tag{10}$$

The spectral properties of ΠH established in ([19], [Lemma 3.3, Theorem 3.4]) are summarized below.

Theorem 2. *Let H and M be symmetric and positive definite matrices of order m, $Z \in \mathbb{R}^{m \times k}$, be a full column-rank matrix and Π given by Equation (10). Then the matrix Π is positive definite.*

Let the positive eigenvalues $\lambda_1(MH), \ldots \lambda_m(MH)$ of MH be sorted in nondecreasing order. Then the set of eigenvalues $\lambda_1(\Pi H), \ldots \lambda_m(\Pi H)$ of ΠH can be split in two subsets:

$$\lambda_i(MH) \leq \lambda_i(\Pi H) \leq \lambda_{i+k}(MH) \quad \text{for } i = 1, \ldots, m-k, \tag{11}$$

and

$$\lambda_i(\Pi H) = 1 \quad \text{for } i = m-k+1, \ldots, m.$$

Equation (10) provides a general formulation for designing second-level preconditioners and was inspired by the BFGS inverse Hessian approximation in Quasi-Newton algorithms [9]. In fact, if the BFGS method with exact linesearch is applied to a quadratic function, then the inverse Hessian approximation generated has the form of Π in Equation (10); we refer to ([19], [§2]) for details on this interpretation. In the general setting, any set of linearly independent vectors can provide candidates for the columns of Z and gives rise to a preconditioner of form Equation (10); k eigenvalues of ΠH equal to 1 are obtained while the remaining eigenvalues satisfy the relevant interlacing property Equation (11). On the other hand, specific choices for Z guided by information on the problem at hand are preferable.

The preconditioner Π has been specialized to the case of: *spectral-LMP* where the columns of Z consist of eigenvectors of MH, *Ritz-LMP* where the columns of Z consist of Ritz vectors generated by the iterative linear solver, *Quasi-Newton-LMP* where the columns of Z consist of descent directions from optimization methods applied to continuous optimization problems, see e.g., [10,16–19]. All these vectors are often available when systems with multiple right-hand sides of slowly varying sequence of systems are considered.

In this work, we propose and analyze preconditioners of the form Equation (10) where the first-level preconditioner is the diagonal matrix D_P^{-1} given in Equation (4) and Z is chosen as a suitable submatrix of the identity matrix I_m, i.e., HZ consists of k properly chosen columns of H. Due to the fact that Z consists of coordinate vectors in $\mathbb{R}^m$ we denote the resulting limited memory preconditioner as Coordinate-LMP. We start analyzing the case where

$$M = D_P^{-1}, \text{ and } Z = I_m(:, 1:k). \tag{12}$$

Forming the preconditioner Π with Equation (12) requires the steps listed in Algorithm 2.

Algorithm 2 Coordinate Limited Memory Preconditioner.

Given the matrix-vector operators $u \rightarrow Hu$, $k > 0$.

    1. Form the first k columns of H, i.e. H_{11}, H_{21}.
    2. Compute the diagonal entries of H_{22}.
    3. Compute L_{11}, D_1 as in Equation (5).
    4. Set $D_2 = \text{diag}(H_{22}) - \text{diag}(H_{21}H_{11}^{-1}H_{21}^T)$, D_P as in Equation (4).
    5. Set M and Z as in Equation (12).
    6. Let Π take the form Equation (10).

In this specific variant, Π coincides with the inverse of the Partial Cholesky preconditioner P. We show this fact in the following theorem.

Theorem 3. *Let k be a positive integer and P be as in Equation (4). If matrix Π has the form Equation (10) with M and Z as in Equation (12), then $\Pi = P^{-1}$. Moreover,*

$$P^{-1}H = \Pi H = \begin{bmatrix} I & H_{11}^{-1}H_{21}^T(I - D_2^{-1}S) \\ 0 & D_2^{-1}S \end{bmatrix}. \tag{13}$$

Proof of Theorem 3. By Equation (9) it follows

$$P^{-1} = \begin{bmatrix} L_{11}^{-T}D_1^{-1}L_{11}^{-1} + L_{11}^{-T}L_{21}^T D_2^{-1}L_{21}L_{11}^{-1} & -L_{11}^{-T}L_{21}^T D_2^{-1} \\ -D_2^{-1}L_{21}L_{11}^{-1} & D_2^{-1} \end{bmatrix}.$$

Using Equations (5) and (6) we obtain $L_{21}L_{11}^{-1} = H_{21}L_{11}^{-T}D_1^{-1}L_{11}^{-1} = H_{21}H_{11}^{-1}$ and we conclude

$$P^{-1} = \begin{bmatrix} H_{11}^{-1} + H_{11}^{-1}H_{21}^T D_2^{-1}H_{21}H_{11}^{-1} & -H_{11}^{-1}H_{21}^T D_2^{-1} \\ -D_2^{-1}H_{21}H_{11}^{-1} & D_2^{-1} \end{bmatrix}.$$

Now consider Π and first observe that the matrices appearing in Equation (10) have the form:

$$HZ = \begin{bmatrix} H_{11} \\ H_{21} \end{bmatrix}, \qquad Z^T HZ = H_{11}, \qquad T = \begin{bmatrix} H_{11}^{-1} & \\ & 0 \end{bmatrix}, \tag{14}$$

$$(I - TH) = \begin{bmatrix} 0 & -H_{11}^{-1}H_{21}^T \\ 0 & I_{m-k} \end{bmatrix}, \qquad (I - HT) = \begin{bmatrix} 0 & 0 \\ -H_{21}H_{11}^{-1} & I_{m-k} \end{bmatrix}. \tag{15}$$

Then,

$$\begin{aligned}
\Pi &= \begin{bmatrix} 0 & -H_{11}^{-1}H_{21}^T \\ & I_{m-k} \end{bmatrix} \begin{bmatrix} D_1^{-1} & \\ & D_2^{-1} \end{bmatrix} \begin{bmatrix} 0 & \\ -H_{21}H_{11}^{-1} & I_{m-k} \end{bmatrix} + \begin{bmatrix} H_{11}^{-1} & \\ & 0 \end{bmatrix} \\
&= \begin{bmatrix} H_{11}^{-1} + H_{11}^{-1}H_{21}^T D_2^{-1}H_{21}H_{11}^{-1} & -H_{11}^{-1}H_{21}^T D_2^{-1} \\ -D_2^{-1}H_{21}H_{11}^{-1} & D_2^{-1} \end{bmatrix},
\end{aligned}$$

i.e., $P^{-1} = \Pi$. Finally, it is trivial to verify that $P^{-1}H$ takes the upper block triangular form Equation (13) which also provides the spectrum of $P^{-1}H$ stated in Theorem 1. $\square$

3.1. Application of the Preconditioners

In the previous section, we have shown that P^{-1} and Π can reduce to the same preconditioner. Clearly, their application as a preconditioner calls for matrix-vector products and this computational cost may depend on the formulation used i.e., either Equation (9) or Equations (10) and (12). Let us analyze the cost for performing matrix-vector products of both P^{-1} and Π times a vector. As stated in Section 2, the symmetric row and column permutations required to move the k largest diagonal elements of H to the $(1,1)$ block are ignored to make the presentation simpler.

If the triangular factors in Equation (9) have been formed, the application of the Partial Cholesky preconditioner P^{-1} to a vector amounts to: two products of L_{11}^{-1} by a vector $\mathbb{R}^k$, one matrix-vector product with D_P^{-1}, $m - k$ scalar products in $\mathbb{R}^k$, k scalar products in $\mathbb{R}^{m-k}$. It is worthy pointing out that the partial Cholesky factorization may be dense. In fact, for sparse Cholesky factorizations, permutation matrices are normally chosen to enhance the sparsity of the triangular factor, see e.g., [4], while here we choose the k columns in advance from the largest diagonals of H.

The application of the Coordinate limited memory preconditioner Π to a vector also calls for matrix-vector and scalar products. The computation of a product, say Πv, can be implemented efficiently using Equations (10) and (14) and performing the following steps

$$
\begin{aligned}
a &= H_{11}^{-1}(Z^T v), \\
b &= D_P^{-1}\left(v - \begin{bmatrix} H_{11} \\ H_{21} \end{bmatrix} a \right), \\
\Pi v &= b - Z\left(H_{11}^{-1} \begin{bmatrix} H_{11} & H_{21}^T \end{bmatrix} b \right) + Za.
\end{aligned}
$$

These steps call for two products of H_{11}^{-1} by a vector in $\mathbb{R}^k$, one matrix-vector product with D_P^{-1}, m scalar products in $\mathbb{R}^k$, k scalar products in $\mathbb{R}^m$. The cost for the product of Z by a vector is negligible due to the form of Z.

The computational cost for applying both P^{-1} and Π to a vector is expected to be comparable if the scalar products performed have similar computational effort; this is the case when the density of the first k columns of L^{-1} is similar to the density of the first k columns of H. On the other hand, if the density of the first k columns of L^{-1} is considerably larger than the density of the first k columns of H, the application of P^{-1} is less convenient than the application of Π. This issue is shown in Table 1 where we report on the numerical solution of four linear systems with matrices of the form $H = AA^T$ and matrix A from the LPnetlib group in the University of Florida Sparse Matrix Collection [23]. Preconditioned Conjugate Gradient [4] is applied with both the Partial Cholesky preconditioner and the Coordinate-LMP , setting $k = 50$. We display the dimension m, n of A, the number of PCG iterations (Itns), the execution time in seconds (Time), the density of first k columns of L^{-1} (dens$_{L,k}$), the density of the first k columns of H (dens$_{H,k}$); the density is computed as the ratio between the number of nonzero entries and the overall number of entries of the mentioned submatrices.

Table 1. Solution of systems with $H = AA^T$, $A \in \mathbb{R}^{m \times n}$, using Partial Cholesky preconditioner and Coordinate-LMP with $k = 50$. Number of PCG iterations (Itns), execution time in seconds (Time), density of first k columns of L^{-1} (dens$_{L,k}$), density of the first k columns of H (dens$_{H,k}$).

Test name	m	n	P^{-1}			Π		
			Itns	Time	dens$_{L,k}$	Itns	Time	dens$_{H,k}$
lp_dfl001	6071	12,230	736	3.87	6.0×10^{-1}	736	1.65	2.8×10^{-2}
lpi_ceria3d	3576	4400	79	0.42	8.4×10^{-1}	80	0.27	3.9×10^{-1}
lp_ken_13	28,632	42,659	186	1.82	1.1×10^{-2}	186	1.70	1.1×10^{-2}
lp_osa_60	10,280	243,246	35	2.92	9.5×10^{-1}	39	3.09	8.0×10^{-1}

We observe that dens$_{L,k}$ is larger than dens$_{H,k}$ in the first two tests and runs with P^{-1} are slower than with Π, while the two densities are similar in the last two runs as well as the timings obtained using P^{-1} and Π.

4. Enlarging the Subspace in the Coordinate-LMP Preconditioner

The Partial Cholesky preconditioner P and the Coordinate-LMP Π with first level preconditioner and subspace as in Equation (12) aim at clustering the largest eigenvalues of H. In this section we investigate how to enlarge the subspace Z by means of information available from Algorithm 2 and the potential impact on the resulting preconditioner.

We consider the Coordinate-LMP Equation (10), suppose to use again $M = D_P^{-1}$ as first level preconditioner, and to select a larger number of columns of I_m for the subspace defined by Z. We let

$$
M = D_P^{-1}, \quad Z = I_m(:, 1:q), \quad q = k + \ell, \tag{16}
$$

with ℓ being a positive integer, i.e., besides the first k columns of I_m used in Equation (12) we employ ℓ more columns (for simplicity suppose the first ℓ subsequent columns). The effect of this choice can be analyzed by considering the block partition of H where the leading block $\tilde{H}_{11}$ has dimension q by q, i.e.,

$$H = \begin{bmatrix} \tilde{H}_{11} & \tilde{H}_{21}^T \\ \tilde{H}_{21} & \tilde{H}_{22} \end{bmatrix}, \tag{17}$$

with $\tilde{H}_{11} \in \mathcal{R}^{q \times q}$, $\tilde{H}_{22} \in \mathcal{R}^{(m-q) \times (m-q)}$. Analogously, let us consider the block partition of D_P in Equation (4) where the leading block $\tilde{D}_{P,1}$ has dimension q by q, i.e.,

$$D_P = \begin{bmatrix} \tilde{D}_{P,1} & \\ & \tilde{D}_{P,2} \end{bmatrix} \stackrel{\text{def}}{=} \begin{bmatrix} D_P(1:q,1:q) & \\ & D_P(q+1:m,q+1:m) \end{bmatrix}, \tag{18}$$

with $\tilde{D}_{P,1} \in \mathcal{R}^{q \times q}$, $\tilde{D}_{P,2} \in \mathcal{R}^{m-q \times m-q}$.

The spectral properties of the resulting Coordinate-LMP preconditioner are given in the following theorem.

Theorem 4. *Let q be a positive integer, H and D_P be symmetric positive definite matrices partitioned as in Equations (17) and (18). If matrix Π has the form Equation (10) with M and Z as in Equation (16), then ΠH has q eigenvalues equal to 1 and the remaining are equal to the eigenvalues of $\tilde{D}_{P,2}^{-1}\tilde{S}$ where*

$$\tilde{S} = \tilde{H}_{22} - \tilde{H}_{21}\tilde{H}_{11}^{-1}\tilde{H}_{21}^T. \tag{19}$$

Moreover any eigenvalue $\lambda(\tilde{D}_{P,2}^{-1}\tilde{S})$ lies in the interval $\left[\frac{\lambda_{\min}(\tilde{S})}{\lambda_{\max}(\tilde{D}_{P,2})}, \frac{\lambda_{\max}(\tilde{S})}{\lambda_{\min}(\tilde{D}_{P,2})} \right] \subseteq$ $\left[\frac{\lambda_{\min}(H)}{\lambda_{\max}(\tilde{D}_{P,2})}, \frac{\lambda_{\max}(\tilde{H}_{22})}{\lambda_{\min}(\tilde{D}_{P,2})} \right]$.

Proof of Theorem 4. Similarly to the proof of Theorem 3 we have

$$HZ = \begin{bmatrix} \tilde{H}_{11} \\ \tilde{H}_{21} \end{bmatrix}, \quad Z^THZ = \tilde{H}_{11}, \quad T = \begin{bmatrix} \tilde{H}_{11}^{-1} & \\ & 0 \end{bmatrix}, \tag{20}$$

$$(I - TH) = \begin{bmatrix} 0 & -\tilde{H}_{11}^{-1}\tilde{H}_{21}^T \\ 0 & I_{m-q} \end{bmatrix}, \quad (I - HT) = \begin{bmatrix} 0 & 0 \\ -\tilde{H}_{21}\tilde{H}_{11}^{-1} & I_{m-q} \end{bmatrix}. \tag{21}$$

Then, by Equation (10) we get

$$\Pi = \begin{bmatrix} 0 & -\tilde{H}_{11}^{-1}\tilde{H}_{21}^T \\ & I_{m-q} \end{bmatrix} \begin{bmatrix} \tilde{D}_{P,1}^{-1} & \\ & \tilde{D}_{P,2}^{-1} \end{bmatrix} \begin{bmatrix} 0 & \\ -\tilde{H}_{21}\tilde{H}_{11}^{-1} & I_{m-q} \end{bmatrix} + \begin{bmatrix} \tilde{H}_{11}^{-1} & \\ & 0 \end{bmatrix}$$

$$= \begin{bmatrix} \tilde{H}_{11}^{-1} + \tilde{H}_{11}^{-1}\tilde{H}_{21}^T\tilde{D}_{P,2}^{-1}\tilde{H}_{21}\tilde{H}_{11}^{-1} & -\tilde{H}_{11}^{-1}\tilde{H}_{21}^T\tilde{D}_{P,2}^{-1} \\ -\tilde{D}_{P,2}^{-1}\tilde{H}_{21}\tilde{H}_{11}^{-1} & \tilde{D}_{P,2}^{-1} \end{bmatrix},$$

and consequently

$$\Pi H = \begin{bmatrix} I_q & (\tilde{H}_{11})^{-1}\tilde{H}_{21}^T(I - \tilde{D}_{P,2}^{-1}\tilde{S}) \\ 0 & \tilde{D}_{P,2}^{-1}\tilde{S} \end{bmatrix}.$$

Bounds on the eigenvalues $\lambda(\widetilde{D}_{P,2}^{-1}\widetilde{S})$ can be derived by fixing an eigenvector $v \in \mathcal{R}^{m-k}$ of $\widetilde{D}_{P,2}^{-1}\widetilde{S}$. Then, by $\widetilde{D}_{P,2}^{-1}\widetilde{S}v = \lambda v$ we get $\lambda(\widetilde{D}_{P,2}^{-1}\widetilde{S}) = \dfrac{v^T\widetilde{S}v}{v^T\widetilde{D}_{P,2}v}$. Thus, similarly to the proof of Theorem 1 it follows

$$\lambda(\widetilde{D}_{P,2}^{-1}\widetilde{S}) \geq \frac{\lambda_{\min}(\widetilde{S})}{\lambda_{\max}(\widetilde{D}_{P,2})} \geq \frac{\lambda_{\min}(H)}{\lambda_{\max}(\widetilde{D}_{P,2})}, \tag{22}$$

$$\lambda(\widetilde{D}_{P,2}^{-1}\widetilde{S}) \leq \frac{\lambda_{\max}(\widetilde{S})}{\lambda_{\min}(\widetilde{D}_{P,2})} \leq \frac{\lambda_{\max}(\widetilde{H}_{22})}{\lambda_{\min}(\widetilde{D}_{P,2})}, \tag{23}$$

since $\lambda_{\min}(\widetilde{S}) \geq \lambda_{\min}(H)$ and $\lambda_{\max}(\widetilde{S}) \leq \lambda_{\max}(\widetilde{H}_{22})$ [6]. □

Remark 1. *Let* $I_1 = \left[\frac{\lambda_{\min}(H)}{\lambda_{\max}(D_2)}, \frac{\lambda_{\max}(\widetilde{H}_{22})}{\lambda_{\min}(D_2)}\right]$ *be the interval in the statement of Theorem 1 and* $I_2 = \left[\frac{\lambda_{\min}(H)}{\lambda_{\max}(\widetilde{D}_{P,2})}, \frac{\lambda_{\max}(\widetilde{H}_{22})}{\lambda_{\min}(\widetilde{D}_{P,2})}\right]$ *be the interval in the statement of Theorem 4. It holds* $\lambda_{\max}(\widetilde{H}_{22}) \leq \lambda_{\max}(H_{22})$ *by the Cauchy Interlace Theorem [24] [p. 396]. Moreover, for any choice of* $\widetilde{D}_{P,2}$ *trivially it holds*

$$\lambda_{\min}(\widetilde{D}_{P,2}) \geq \lambda_{\min}(D_2), \qquad \lambda_{\max}(\widetilde{D}_{P,2}) \leq \lambda_{\max}(D_2).$$

Then, $I_2 \subseteq I_1$.

A comparison between bounds Equations (7) and (8) and Equations (22) and (23) suggests that the choice of the extremal diagonal elements of S can be beneficial for improving, at a low computational cost, the clustering of the eigenvalues. In fact, choosing the extremal diagonal elements of S promotes a reduction of the width of the interval containing the eigenvalues of ΠH and this issue can favorably affect the performance of the iterative solver and the condition number of ΠH.

Accordingly to Remark 1, let I_1 and I_2 be the intervals containing the eigenvalues of ΠH with Π generated by Equation (12) and by Equation (16) respectively. If the ℓ largest diagonal entries of diag(S) are contained in matrix $\widetilde{D}_{P,1}$ in Equation (18) and are separated from the remaining, then we obtain an increase of the lower bound of I_2 with respect to lower bound of I_1; clearly, the better the ℓ largest diagonal entries of S are separated from the remaining elements of diag(S) the larger such increase is. Handling small eigenvalues of ΠH seems to be convenient when enlarging the subspace Z for Π as the Partial Cholesky factorization is intended to take care of the largest eigenvalues of H. Alternatively, if the ℓ smallest diagonal entries of S are contained in matrix $\widetilde{D}_{P,1}$ in Equation (18) and are separated from the remaining, the upper bound of I_2 is expected to be smaller than the upper bound of I_1.

As mentioned in Section 2, in [13] the partial Cholesky preconditioner was used in conjunction with Deflated-CG [22] in order to handle the small eigenvalues of $P^{-1}H$. A rough approximation of the five smallest eigenvalues of $P^{-1}H$ was injected into the Krylov subspace and yielded an improvement in some tests where Deflated-CG performed consistently fewer iterations than the usual CG. Although the deflation strategy brought undeniable benefits in terms of reducing the number of CG iterations, it involved an extra storage requirement and an extra cost which ultimately increased the overall solution time. In fact, the application of such a strategy was convenient when the eigenvalue information was used for a sequence of related linear systems, such as slowly varying systems or systems with multiple right-hand-side vectors. The strategy, presented in this section and based on selecting a prefixed number of the largest diagonal entries of S, can be viewed as a cheap alternative procedure for handling the smallest eigenvalues of $P^{-1}H$.

5. Numerical Results

In this section we present a preliminary numerical validation of the performance of the Coordinate-LMP discussed in Sections 3 and 4. All numerical experiments reported were performed on a Dell Latitude E4200 with a Intel(R) Cote(TM)2 Duo CPU U9600, @1.60 GHz, RAM 3.00 GB, using MATLAB and machine precision 2.2×10^{-16}.

We report results on a set of 18 linear systems where the coefficient matrix has the form $H = A\Theta A^T$ and the right-hand side is chosen as a normally distributed vector. In Table 2 we list the name of matrices $A \in \mathbb{R}^{m \times n}$ used along with their dimensions and the density of both A and H computed as the ratio between the number of nonzero entries and the number of rows.

Table 2. Test problems with $H = A\Theta A^T$: source and name of $A \in \mathbb{R}^{m \times n}$, dimension of A, density of A (dens(A)) and density of H (dens(H)).

Group/Test Name	m	n	dens(A)	dens(H)
LPnetlib/lp_bnl2	2424	4486	6.5	12.6
LPnetlib/lp_d2q06c	2171	5831	15.2	25.9
LPnetlib/lp_dfl001[#]	6071	12,230	5.9	13.5
LPnetlib/lp_degen3[#]	1503	2604	16.9	67.7
LPnetlib/lp_ganges	1309	1706	5.3	12.7
LPnetlib/lp_ken_13	28,632	42,659	3.4	5.7
LPnetlib/lp_ken_18	105,127	154,699	3.4	5.8
LPnetlib/lp_osa_30	4350	104,374	139.0	100.4
LPnetlib/lp_osa_60	10,280	243,246	137.0	98.9
LPnetlib/lp_pds_10[#]	16,558	49,932	6.5	9.0
LPnetlib/lp_pilot	1441	4680	30.8	86.3
LPnetlib/lp_pilot87	2030	6680	36.9	117.5
LPnetlib/ lp_sierra[#]	1227	2735	6.5	4.7
Meszaros/cq9	9278	21,534	10.4	23.9
Meszaros/nl	7039	15,325	6.7	14.9
M5_3000_maxcut	3000	9,000,000	1	3000
M2_K4_5000_maxcut	5000	25,000,000	1	5000
M3_K4_5000_maxcut	5000	25,000,000	1	5000

The first 15 matrices A are taken from the groups LPnetlib and Meszaros in the University of Florida Sparse Matrix Collection [23] and are constraint matrices of linear programming problems; in the associated linear systems we set $\Theta = I_n$. The symbol "#" indicates when matrix A was regularized by a shift 10^{-2} in order to get a numerically nonsingular matrix H. We observe that both A and H are sparse and H can be preconditioned by either Incomplete Cholesky factorizations or by our preconditioner. The last three systems were generated by the dual logarithmic barrier method [3] applied to semidefinite programming relaxations of maximum cut problems. In these problems each row of A is the unrolled representation of a rank-one $m \times m$ matrix and has one nonzero entry while Θ is a full matrix of dimension $m^2 \times m^2$ defined as the Kronecker product of matrices of dimension $m \times m$; consequently H is full. Iterative methods are an option for solving these systems when H cannot be allocated due to memory limitations [3]. Incomplete Cholesky factorizations are not applicable while our preconditioner is viable.

The linear systems have been solved by Preconditioned Conjugate Gradient (PCG) method starting from the null initial guess and using the stopping criterion:

$$\|Hx - b\| \leq 10^{-6}\|b\|. \tag{24}$$

A failure is declared after 1000 iterations. The preconditioner Π was applied as described in Section 3.1.

The preconditioners used in our tests are: the Incomplete Cholesky factorization with zero-fill (IC(0)) computed by the built-in MATLAB function ichol, the Coordinate-LMP Equation (12) with $k = 50$, and the Coordinate-LMP Equation (16) with $(k, \ell) = (50, 25)$. Concerning the preconditioner with enlarged subspace Equation (16), we consider the two strategies for enlarging Z discussed at the end of Section 4. The first strategy consists in selecting the columns of I_m associated to the ℓ

largest diagonal entries of $D_2 = \mathrm{diag}(S)$ and in the following is denoted as D2_LARGE. The second strategy consists in selecting the columns of I_m associated to the ℓ smallest diagonal entries of D_2 and is denoted as D2_SMALL. In the following tables, the symbol "*" indicates a failure of CG solver while the symbol "†" indicates a failure in computing the Incomplete Cholesky factorization due to encountering a nonpositive pivot. The timing in seconds "Time" includes the construction of the preconditioner and the total execution time for PCG.

Our focus is on the reliability of the preconditioners tested and the computational gain provided. Regarding the latter issue, clearly it depends on both the number of PCG iterations and the cost of PCG per iteration.

Table 3 displays the results obtained solving the linear systems with: unpreconditioned CG, CG coupled with IC(0), CG coupled with the Coordinate-LMP Equation (12), CG coupled with the Coordinate-LMP Equation (16) and implemented using the D2_LARGE strategy. We report the number of PCG iterations (Itns) and the timing (Time). We observe that IC(0) factorization cannot be applied to the linear systems deriving from maximum cut problems since the resulting matrices H are full, see Table 2.

We start observing that CG preconditioned by the Coordinate-LMP preconditioners Equation (16) and the D2_LARGE strategy solved all the systems, whereas a breakdown of IC(0) occurred five times out of fifteen as a nonpositive pivot was encountered. In eight systems out of ten, PCG with IC(0) required several iterations considerably smaller than the number of iterations performed with the limited memory preconditioners; correspondingly the execution time was favorable to IC(0) preconditioner. On the other hand, IC(0) preconditioner was less effective than the limited memory preconditioner on problems lp_osa_30 and lp_osa_60; this occurrence is motivated by several linear iterations comparable to that of limited preconditioners and the density of the Cholesky factor, cf. Table 2. Finally, we point out that breakdowns of IC(0) can be recovered using Incomplete Cholesky factorization with very small threshold dropping and, consequently, high fill-in in the Incomplete Cholesky factor and computational overhead.

Comparing the limited memory preconditioners in terms of CG iterations, we observe that enlarging the subspace provides a reduction in such a number. The gain in CG iterations using the enlarged subspace is very limited for Problems lp_ganges, lp_dfl001 and lp_pds_10, while it varies between 3% and 52% for the remaining problems. Savings in time depend on both the reduction in the number of CG iterations performed and the cost of matrix-vector products. Namely, when the application of H is cheap, savings in PCG iterations between 11% and 31% do not yield a significant gain in time, see lp_degen3, lp_ken_13, lp_pilot; on the other hand when matrix-vector products are expensive, saving in time can occur even in the presence of a mild reduction in the number of CG iterations, see lp_pds_10, M2_K4_5000_maxcut, M3_K4_5000_maxcut. Interestingly the cost for forming and applying the preconditioners does not offset the convergence gain in PCG; this feature is evident from the value Time in runs where the reduction in PCG iterations is small, see lp_ganges and M5_3000_maxcut. In fact, we can expect that for matrices having the same eigenvalue distribution as our test matrices, and a substantial number of nonzero elements, significant reductions in computing time can be achieved with the Quasi-Newton preconditioner and enlarged subspace.

The effect of enlarging the subspace in the Coordinate-LMP preconditioner is further analyzed in Table 4 where we report the minimum $\lambda_{\min}$ and maximum $\lambda_{\max}$ eigenvalues of the original matrix H and of the preconditioned matrices ΠH in four problems. We observe that the maximum eigenvalue of the preconditioned matrix is consistently smaller than the eigenvalue of H and this shows the effectiveness of handling the largest eigenvalues by using the trace of H. On the other hand, the smallest eigenvalue of the matrix preconditioned by the Partial Cholesky factorization (Π with $k = 50$) is moved towards the origin. As shown in the table, an increase in the value of the smallest eigenvalue can be obtained with the D2_LARGE implementation for Π while the effect on the largest eigenvalue is marginal, as expected.

Table 3. Solution of systems with $H = A\Theta A^T$: unpreconditioned CG, CG coupled with IC(0), CG coupled with Coordinate-LMP Equation (12) $k = 50$, CG coupled with Coordinate-LMP Equation (16) $(k, \ell) = (50, 25)$ and D2_LARGE implementation. Number of PCG iterations (Itns), execution time in seconds (Time) for building the preconditioner and for PCG.

Test Name	CG		PCG with IC(0)		PCG with Π $k = 50$		PCG with Π $(k, \ell) = (50, 25)$ D2_LARGE Implementation	
	Itns	Time	Itns	Time	Itns	Time	Itns	Time
lp_bnl2	*		51	0.1	353	0.4	295	0.3
lp_d2q06c	*		†		*		844	1.2
lp_dfl001	*		320	0.7	736	1.7	720	1.7
lp_degen3	*		284	0.4	599	0.7	530	0.7
lp_ganges	215	0.1	35	0.1	126	0.1	124	0.1
lp_ken_13	510	3.3	87	0.7	186	1.6	165	1.6
lp_ken_18	*		167	10.5	499	20.7	485	20.2
lp_osa_30	126	3.4	47	2.4	38	1.5	18	0.9
lp_osa_60	127	7.7	26	7.4	39	3.6	26	2.8
lp_pds_10	*		431	5.2	897	8.1	892	7.6
lp_pilot	*		†		369	0.6	252	0.4
lp_pilot87	*		†		559	1.7	505	1.6
lp_sierra	*		241	0.1	*		590	0.3
cq9	*		†		554	2.9	472	2.5
nl	*		†		944	2.6	621	1.7
M5_3000_maxcut	60	2.5			55	2.5	51	2.4
M2_K4_5000_maxcut	497	52.2			427	46.6	413	44.9
M3_K4_5000_maxcut	641	65.8			585	62.5	542	58.5

Table 4. Minimum eigenvalue λ_{min} and maximum eigenvalue λ_{max} of: matrix H, matrix ΠH with Coordinate-LMP Equation (12) $k = 50$, matrix ΠH with Coordinate-LMP Equation (16) $(k, \ell) = (50, 25)$.

Test Name	H		ΠH with $k = 50$		ΠH with $(k, \ell) = (50, 25)$ D2_LARGE Implementation	
	λ_{min}	λ_{max}	λ_{min}	λ_{max}	λ_{min}	λ_{max}
lp_d2q06c	6.3×10^{-4}	1.2×10^{6}	3.3×10^{-5}	6.4×10^{0}	4.8×10^{-5}	5.7×10^{0}
lp_osa_30	1.0×10^{0}	1.8×10^{6}	2.4×10^{-5}	2.8×10^{0}	5.9×10^{-4}	2.2×10^{0}
lp_pilot	1.0×10^{-2}	1.0×10^{6}	2.5×10^{-4}	1.2×10^{1}	1.4×10^{-3}	1.2×10^{1}
nl	7.0×10^{-3}	8.2×10^{4}	1.6×10^{-4}	7.3×10^{0}	5.7×10^{-4}	6.7×10^{0}

We conclude our presentation reporting the performance of the D2_SMALL implementation in Table 5; the number of PCG iterations is displayed and part of the results in Table 3 are repeated for clarity. We recall that the number of unit eigenvalues is $q = k + \ell$ for both the D2_LARGE and D2_SMALL implementations, but the former strategy is more effective than the latter. In fact, the behaviour of the Partial Cholesky factorization and of the D2_SMALL implementation of the Coordinate-LMP preconditioner are similar in terms of PCG iterations, apart for problems lp_ganges, lp_sierra, cq9 where the latter approach is convenient. This confirms that the largest eigenvalues are handled by the Partial Cholesky factorization and a further reduction of the upper bound on the eigenvalues is not useful.

Table 5. Solution of systems with $H = A\Theta A^T$, $A \in \mathbb{R}^{m \times n}$: CG coupled with Coordinate-LMP Equation (12) $k = 50$, CG coupled with Coordinate-LMP Equation (16) $(k, \ell) = (50, 25)$ and D2_LARGE implementation, CG coupled with Coordinate-LMP Equation (16) $(k, \ell) = (50, 25)$ and D2_SMALL implementation. Number of PCG iterations (Itns).

Test Name	PCG with Π $k = 50$	PCG with Π $(k, \ell) = (50, 25)$ D2_LARGE Implementation	PCG with Π $(k, \ell) = (50, 25)$ D2_SMALL Implementation
	Itns	Itns	Itns
lp_bnl2	353	295	353
lp_d2q06c	*	844	*
lp_dfl001	736	720	733
lp_degen3	599	530	595
lp_ganges	126	124	78
lp_ken_13	186	165	186
lp_ken_18	499	485	520
lp_osa_30	38	18	37
lp_osa_60	39	26	39
lp_pds_10	897	892	900
lp_pilot	369	252	361
lp_pilot87	559	505	565
lp_sierra	*	590	706
cq9	554	472	528
nl	944	621	946
M5_3000_maxcut	55	51	56
M2_K4_5000_maxcut	427	413	428
M3_K4_5000_maxcut	585	542	596

Summarizing, the results presented in this section seem to indicate that: enlarging the subspace with columns associated to the largest diagonal entries of the Schur complement S in Equation (3) reduces the number of PCG iterations; the cost for forming and applying the preconditioner with enlarged subspace does not offset the gain from reducing PCG iterations; saving in times are obtained accordingly to the cost of matrix-vector products in PCG. Moreover, the Quasi-Newton preconditioner proposed is suitable for application to dense matrices H of the form $H = A\Theta A^T$, A sparse, where computing the Incomplete Cholesky factor is too expensive in terms of computational cost and/or storage requirement.

Funding: This work was partially supported by INdAM-GNCS under Progetti di Ricerca 2018. Università degli Studi di Firenze covered the cost to publish in open access.

Acknowledgments: The author wish to thank the referees for their helpful comments and suggestions.

Conflicts of Interest: The author declares no conflict of interest.

References

1. Bellavia, S.; De Simone, V.; di Serafino, D.; Morini, B. Efficient preconditioner updates for shifted linear systems. *SIAM J. Sci. Comput.* **2011**, *33*, 1785–1809. [CrossRef]
2. Bellavia, S.; De Simone, V.; di Serafino, D.; Morini, B. A preconditioning framework for sequences of diagonally modified linear systems arising in optimization. *SIAM J. Numer. Anal.* **2012**, *50*, 3280–3302. [CrossRef]
3. Bellavia, S.; Gondzio, J.; Porcelli, M. An inexact dual logarithmic barrier method for solving sparse semidefinite programs. *Math. Program.* **2018**, doi:10.1007/s10107-018-1281-5. [CrossRef]
4. Björck, A. *Numerical Methods for Least Squares Problems*; Society for Industrial and Applied Mathematics: Philadelphia, PA, USA, 1996; ISBN 978-0-89871-360-2.
5. Málek, J.; Strakoš, Z. *Preconditioning and the Conjugate Gradient Method in the Context of Solving PDEs, SIAM Spotlights*; Society for Industrial and Applied Mathematics: Philadelphia, PA, USA, 2014; ISBN 978-1-611973-83-9.

6. Zhang, F. *The Schur Complement and Its Applications*; Series: Numerical Methods and Algorithms; Claude Brezinski; Springer: New York, NY, USA, 2005; ISBN 978-0-387-24273-6.

7. Benzi, M. Preconditioning techniques for large linear systems: A survey. *J. Comput. Phys.* **2002**, *2*, 418–477. [CrossRef]

8. Saad, Y. Iterative Method for Sparse Linear System. *Other Titles in Applied Mathematics*; Society for Industrial and Applied Mathematics: Philadelphia, PA, USA, 2003; ISBN 978-0-89871-534-7.

9. Nocedal, J.; Wright, S.J. *Numerical Optimization*; Springer Series in Operations Research; Springer: Berlin, Germany, 1999; ISBN 978-0-387-40065-5.

10. Morales, J.L.; Nocedal, J. Automatic preconditioning by limited memory Quasi-Newton updating. *SIAM J. Optim.* **2000**, *10*, 1079–1096. [CrossRef]

11. Bellavia, S.; De Simone, V.; di Serafino, D.; Morini, B. Updating constraint preconditioners for KKT systems in quadratic programming via low-rank corrections. *SIAM J. Optim.* **2015**, *25*, 1787–1808. [CrossRef]

12. Morini, B.; Simoncini, V.; Tani, M. A comparison of reduced and unreduced KKT systems arising from interior point methods. *Comput. Optim. Appl.* **2017**, *68*, 1–27. [CrossRef]

13. Bellavia, S.; Gondzio, J.; Morini, B. A matrix-free preconditioner for sparse symmetric positive definite systems and least-squares problems. *SIAM J. Sci. Comput.* **2013**, *35*, A192–A211. [CrossRef]

14. Gondzio, J. Matrix-free interior point Method. *Comput. Optim. Appl.* **2012**, *51*, 457–480, doi:10.1007/s10589-010-9361-3. [CrossRef]

15. Fountoulakis, K.; Gondzio, J. A second-order method for strongly convex L1-regularization problems. *Math. Prog. A* **2016**, *156*, 189–219, doi:0.1007/s10107-015-0875-4. [CrossRef]

16. Bergamaschi, L.; De Simone, V.; di Serafino, A.; Martínez, D. BFGS-like updates of constraint preconditioners for sequences of KKT linear systems in quadratic programming. *Numer. Linear Algebra Appl.* **2018**, doi:10.1002/nla.2144. [CrossRef]

17. Caliciotti, A.; Fasano, G.; Roma, M. Preconditioned Nonlinear Conjugate Gradient methods based on a modified secant equation. *Appl. Math. Comput.* **2018**, *318*, 196–214. [CrossRef]

18. Fasano, G.; Roma, M. A novel class of approximate inverse preconditioners for large positive definite linear systems in optimization. *Comput. Optim. Appl.* **2016**, *65*, 399–429. [CrossRef]

19. Gratton, S.; Sartenaer, A.; Tshimanga, J. On a class of limited memory preconditioners for large scale linear systems with multiple right-hand sides. *SIAM J. Optim.* **2011**, *21*, 912–935. [CrossRef]

20. Greenbaum, A. *Iterative Methods for Solving Linear Systems*; Society for Industrial and Applied Mathematics: Philadelphia, PA, USA, 1996; ISBN 978-0-89871-396-X.

21. Liesen, J.; Strakoš, Z. *Krylov Subspace Methods, Principles and Analysis*; Oxford University Press: Oxford, UK, 2012; ISBN 978-0-19-965541-0.

22. Saad, Y.; Yeung, M.; Erhel, J.; Guyomarc'h, F. A deflated version of the conjugate gradient method. *SIAM J. Sci. Comput.* **2000**, *21*, 1909–1926. [CrossRef]

23. Davis, T.; Hu, Y. The University of Florida Sparse Matrix Collection. *ACM Trans. Math. Softw.* **2011**, *38*, 1–25. [CrossRef]

24. Golub, G.H.; van Loan, C.F. *Matrix Computations*, 3rd ed.; The John Hopkins University Press: Baltimore, MD, USA, 1996; ISBN 0-8018-5414-8.

![axioms logo] *axioms*

MDPI

Article

Block Generalized Locally Toeplitz Sequences: From the Theory to the Applications

Carlo Garoni [1,2,*] ![ORCID], **Mariarosa Mazza** [3] ![ORCID] **and Stefano Serra-Capizzano** [2,4] ![ORCID]

1   Institute of Computational Science, University of Italian Switzerland, 6900 Lugano, Switzerland; carlo.garoni@usi.ch
2   Department of Science and High Technology, University of Insubria, 22100 Como, Italy; carlo.garoni@uninsubria.it (C.G.); stefano.serrac@uninsubria.it (S.S.-C.)
3   Division of Numerical Methods in Plasma Physics, Max Planck Institute for Plasma Physics, 85748 Garching bei München, Germany; mariarosa.mazza@ipp.mpg.de
4   Department of Information Technology, Uppsala University, P.O. Box 337, SE-751 05 Uppsala, Sweden; stefano.serra@it.uu.se
*   Correspondence: carlo.garoni@uninsubria.it; Tel.: +39-335-588-9197

Received: 9 May 2018; Accepted: 16 July 2018; Published: 19 July 2018

Abstract: The theory of generalized locally Toeplitz (GLT) sequences is a powerful apparatus for computing the asymptotic spectral distribution of matrices A_n arising from virtually any kind of numerical discretization of differential equations (DEs). Indeed, when the mesh fineness parameter n tends to infinity, these matrices A_n give rise to a sequence $\{A_n\}_n$, which often turns out to be a GLT sequence or one of its "relatives", i.e., a block GLT sequence or a reduced GLT sequence. In particular, block GLT sequences are encountered in the discretization of systems of DEs as well as in the higher-order finite element or discontinuous Galerkin approximation of scalar DEs. Despite the applicative interest, a solid theory of block GLT sequences has been developed only recently, in 2018. The purpose of the present paper is to illustrate the potential of this theory by presenting a few noteworthy examples of applications in the context of DE discretizations.

Keywords: spectral (eigenvalue) and singular value distributions; generalized locally Toeplitz sequences; discretization of systems of differential equations; higher-order finite element methods; discontinuous Galerkin methods; finite difference methods; isogeometric analysis; B-splines; curl–curl operator; time harmonic Maxwell's equations and magnetostatic problems

MSC: 47B06; 15A18; 15B05; 65N30; 65N06; 65D07

---

1. Introduction

The theory of generalized locally Toeplitz (GLT) sequences stems from Tilli's work on locally Toeplitz (LT) sequences [1] and from the spectral theory of Toeplitz matrices [2–12]. It was then carried forward in [13–16], and was recently extended by Barbarino [17]. This theory is a powerful apparatus for computing the asymptotic spectral distribution of matrices arising from the numerical discretization of continuous problems, such as integral equations (IEs) and, especially, differential equations (DEs). The experience reveals that virtually any kind of numerical methods for the discretization of DEs gives rise to structured matrices A_n whose asymptotic spectral distribution, as the mesh fineness parameter n tends to infinity, can be computed through the theory of GLT sequences. We refer the reader to ([13] Section 10.5), ([14] Section 7.3), and [15,16,18] for applications of the theory of GLT sequences in the context of finite difference (FD) discretizations of DEs; to ([13] Section 10.6), ([14] Section 7.4), and [16,18,19] for the finite element (FE) case; to [20] for the finite volume (FV) case; to ([13] Section 10.7), ([14] Sections 7.5–7.7), and [21–26] for the case of isogeometric analysis (IgA) discretizations, both in the collocation and Galerkin frameworks; and to [27] for a further recent application to fractional DEs. We also refer the reader to ([13] Section 10.4) and [28,29] for a look at the GLT approach for sequences of matrices arising from IE discretizations.

It is worth emphasizing that the asymptotic spectral distribution of DE discretization matrices, whose computation is the main objective of the theory of GLT sequences, is not only interesting from a theoretical viewpoint, but can also be used for practical purposes. For example, it is known that the convergence properties of mainstream iterative solvers, such as multigrid and preconditioned Krylov methods, strongly depend on the spectral features of the matrices to which they are applied. The spectral distribution can then be exploited to design efficient solvers of this kind and to analyze/predict their performance. In this regard, we recall that noteworthy estimates on the superlinear convergence of the conjugate gradient method obtained by Beckermann and Kuijlaars in [30] are closely related to the asymptotic spectral distribution of the considered matrices. Furthermore, in the context of Galerkin and collocation IgA discretizations of elliptic DEs, the spectral distribution computed through the theory of GLT sequences in a series of recent papers [21–25] was exploited in [31–33] to devise and analyze optimal and robust multigrid solvers for IgA linear systems.

In the very recent work [34], starting from the original intuition by the third author ([16] Section 3.3), the theory of block GLT sequences has been developed in a systematic way as an extension of the theory of GLT sequences. Such an extension is of the utmost importance in practical applications. In particular, it provides the necessary tools for computing the spectral distribution of block structured matrices arising from the discretization of systems of DEs ([16] Section 3.3) and from the higher-order finite element or discontinuous Galerkin approximation of scalar/vectorial DEs [35–37]. The purpose of this paper is to illustrate the potential of the theory of block GLT sequences [34] and of its multivariate version—which combines the results of [34] with the "multivariate technicalities" from [14]—by presenting a few noteworthy examples of applications. Actually, the present paper can be seen as a necessary completion of the purely theoretical work [34].

The paper is organized as follows. In Section 2, we report a summary of the theory of block GLT sequences. In Section 3, we focus on the FD discretization of a model system of univariate DEs; through the theory of block GLT sequences, we compute the spectral distribution of the related discretization matrices. In Section 4, we focus on the higher-order FE approximation of the univariate diffusion equation; again, we compute the spectral distribution of the associated discretization matrices through the theory of block GLT sequences. In Section 5, we summarize the multivariate version of the theory of block GLT sequences, also known as the theory of multilevel block GLT sequences. In Section 6, we describe the general GLT approach for computing the spectral distribution of matrices arising from the discretization of systems of partial differential equations (PDEs). In Section 7, we focus on the B-spline IgA approximation of a bivariate variational problem for the curl–curl operator, which is of interest in magnetostatics; through the theory of multilevel block GLT sequences, we compute the spectral distribution of the related discretization matrices. Final considerations are collected in Section 8.

2. The Theory of Block GLT Sequences

In this section, we summarize the theory of block GLT sequences, which was originally introduced in ([16] Section 3.3) and has been recently revised and systematically developed in [34].

Sequences of Matrices and Block Matrix-Sequences. Throughout this paper, a sequence of matrices is any sequence of the form $\{A_n\}_n$, where A_n is a square matrix of size d_n and $d_n \to \infty$ as $n \to \infty$. Let $s \geq 1$ be a fixed positive integer independent of n; an s-block matrix-sequence (or simply a matrix-sequence if s can be inferred from the context or we do not need/want to specify it) is a special sequence of matrices $\{A_n\}_n$ in which the size of A_n is $d_n = sn$.

Singular Value and Eigenvalue Distribution of a Sequence of Matrices. Let μ_k be the Lebesgue measure in $\mathbb{R}^k$. Throughout this paper, all the terminology from measure theory (such as "measurable set", "measurable function", "a.e.", etc.) is referred to the Lebesgue measure. A matrix-valued function $f : D \subseteq \mathbb{R}^k \to \mathbb{C}^{r \times r}$ is said to be measurable (resp., continuous, Riemann-integrable, in $L^p(D)$, etc.) if its components $f_{\alpha\beta} : D \to \mathbb{C}$, $\alpha, \beta = 1, \ldots, r$, are measurable (resp., continuous, Riemann-integrable, in $L^p(D)$, etc.). We denote by $C_c(\mathbb{R})$ (resp., $C_c(\mathbb{C})$) the space of continuous complex-valued functions with bounded support defined on $\mathbb{R}$ (resp., $\mathbb{C}$). If $A \in \mathbb{C}^{m \times m}$, the singular values and the eigenvalues of A are denoted by $\sigma_1(A), \ldots, \sigma_m(A)$ and $\lambda_1(A), \ldots, \lambda_m(A)$, respectively.

Definition 1. *Let $\{A_n\}_n$ be a sequence of matrices, with A_n of size d_n, and let $f : D \subset \mathbb{R}^k \to \mathbb{C}^{r \times r}$ be a measurable function defined on a set D with $0 < \mu_k(D) < \infty$.*

- *We say that $\{A_n\}_n$ has a (asymptotic) singular value distribution described by f, and we write $\{A_n\}_n \sim_\sigma f$, if*

$$\lim_{n \to \infty} \frac{1}{d_n} \sum_{i=1}^{d_n} F(\sigma_i(A_n)) = \frac{1}{\mu_k(D)} \int_D \frac{\sum_{i=1}^r F(\sigma_i(f(\mathbf{x})))}{r} d\mathbf{x}, \qquad \forall F \in C_c(\mathbb{R}). \tag{1}$$

  In this case, f is referred to as a singular value symbol of $\{A_n\}_n$.
- *We say that $\{A_n\}_n$ has a (asymptotic) spectral (or eigenvalue) distribution described by f, and we write $\{A_n\}_n \sim_\lambda f$, if*

$$\lim_{n \to \infty} \frac{1}{d_n} \sum_{i=1}^{d_n} F(\lambda_i(A_n)) = \frac{1}{\mu_k(D)} \int_D \frac{\sum_{i=1}^r F(\lambda_i(f(\mathbf{x})))}{r} d\mathbf{x}, \qquad \forall F \in C_c(\mathbb{C}). \tag{2}$$

  In this case, f is referred to as a spectral (or eigenvalue) symbol of $\{A_n\}_n$.

If $\{A_n\}_n$ has both a singular value and an eigenvalue distribution described by f, we write $\{A_n\}_n \sim_{\sigma,\lambda} f$.

We note that Definition 1 is well-posed because the functions $\mathbf{x} \mapsto \sum_{i=1}^r F(\sigma_i(f(\mathbf{x})))$ and $\mathbf{x} \mapsto \sum_{i=1}^r F(\lambda_i(f(\mathbf{x})))$ are measurable ([34] Lemma 2.1). Whenever we write a relation such as $\{A_n\}_n \sim_\sigma f$ or $\{A_n\}_n \sim_\lambda f$, it is understood that f is as in Definition 1; that is, f is a measurable function defined on a subset D of some $\mathbb{R}^k$ with $0 < \mu_k(D) < \infty$, and f takes values in $\mathbb{C}^{r \times r}$ for some $r \geq 1$.

Remark 1. *The informal meaning behind the spectral distribution (2) is the following: assuming that f possesses r Riemann-integrable eigenvalue functions $\lambda_i(f(\mathbf{x}))$, $i = 1, \ldots, r$, the eigenvalues of A_n, except possibly for $o(d_n)$ outliers, can be subdivided into r different subsets of approximately the same cardinality; and, for n large enough, the eigenvalues belonging to the ith subset are approximately equal to the samples of the ith eigenvalue function $\lambda_i(f(\mathbf{x}))$ over a uniform grid in the domain D. For instance, if $k = 1$, $d_n = nr$, and $D = [a, b]$, then, assuming we have no outliers, the eigenvalues of A_n are approximately equal to*

$$\lambda_i\left(f\left(a + j\frac{b-a}{n}\right)\right), \qquad j = 1, \ldots, n, \qquad i = 1, \ldots, r,$$

for n large enough; similarly, if $k = 2$, $d_n = n^2 r$, and $D = [a_1, b_1] \times [a_2, b_2]$, then, assuming we have no outliers, the eigenvalues of A_n are approximately equal to

$$\lambda_i\left(f\left(a_1 + j_1\frac{b_1 - a_1}{n}, a_2 + j_2\frac{b_2 - a_2}{n}\right)\right), \qquad j_1, j_2 = 1, \ldots, n, \qquad i = 1, \ldots, r,$$

for n large enough; and so on for $k \geq 3$. A completely analogous meaning can also be given for the singular value distribution (1).

Remark 2. *Let* $D = [a_1, b_1] \times \cdots \times [a_k, b_k] \subset \mathbb{R}^k$ *and let* $f : D \to \mathbb{C}^{r \times r}$ *be a measurable function possessing* r *real-valued Riemann-integrable eigenvalue functions* $\lambda_i(f(\mathbf{x}))$, $i = 1, \ldots, r$. *Compute for each* $\rho \in \mathbb{N}$ *the uniform samples*

$$\lambda_i\left(f\left(a_1 + j_1 \frac{b_1 - a_1}{\rho}, \ldots, a_k + j_k \frac{b_k - a_k}{\rho}\right)\right), \qquad j_1, \ldots, j_k = 1, \ldots, \rho, \qquad i = 1, \ldots, r,$$

sort them in non-decreasing order and put them in a vector $(\varsigma_1, \varsigma_2, \ldots, \varsigma_{r\rho^k})$. *Let* $\phi_\rho : [0, 1] \to \mathbb{R}$ *be the piecewise linear non-decreasing function that interpolates the samples* $(\varsigma_0 = \varsigma_1, \varsigma_1, \varsigma_2, \ldots, \varsigma_{r\rho^k})$ *over the nodes* $(0, \frac{1}{r\rho^k}, \frac{2}{r\rho^k}, \ldots, 1)$, *i.e.*,

$$\begin{cases} \phi_\rho\left(\dfrac{i}{r\rho^k}\right) = \varsigma_i, & i = 0, \ldots, r\rho^k, \\[2mm] \phi_\rho \text{ linear on } \left[\dfrac{i}{r\rho^k}, \dfrac{i+1}{r\rho^k}\right] \text{ for } i = 0, \ldots, r\rho^k - 1. \end{cases}$$

Suppose ϕ_ρ *converges in measure over* $[0, 1]$ *to some function* ϕ *as* $\rho \to \infty$ *(this is always the case in real-world applications). Then,*

$$\int_0^1 F(\phi(t)) \mathrm{d}t = \frac{1}{\mu_k(D)} \int_D \frac{\sum_{i=1}^r F(\lambda_i(f(\mathbf{x})))}{r} \mathrm{d}\mathbf{x}, \qquad \forall F \in C_c(\mathbb{C}). \tag{3}$$

This result can be proved by adapting the argument used in ([13] solution of Exercise 3.1). The function ϕ *is referred to as the canonical rearranged version of* f. *What is interesting about* ϕ *is that, by* (3), *if* $\{A_n\}_n \sim_\lambda f$ *then* $\{A_n\}_n \sim_\lambda \phi$, *i.e., if* f *is a spectral symbol of* $\{A_n\}_n$ *then the same is true for* ϕ. *Moreover,* ϕ *is a univariate scalar function and hence it is much easier to handle than* f. *According to Remark 1, assuming that* ϕ *is Riemann-integrable, if we have* $\{A_n\}_n \sim_\lambda f$ *(and hence also* $\{A_n\}_n \sim_\lambda \phi$), *then, for* n *large enough, the eigenvalues of* A_n, *with the possible exception of* $o(d_n)$ *outliers, are approximately equal to the samples of* ϕ *over a uniform grid in* $[0, 1]$.

The next two theorems are useful tools for computing the spectral distribution of sequences formed by Hermitian or perturbed Hermitian matrices. For the related proofs, we refer the reader to ([38] Theorem 4.3) and ([39] Theorem 1.1). In the following, the conjugate transpose of the matrix A is denoted by A^*. If $A \in \mathbb{C}^{m \times m}$ and $1 \leq p \leq \infty$, we denote by $\|A\|_p$ the Schatten p-norm of A, i.e., the p-norm of the vector $(\sigma_1(A), \ldots, \sigma_m(A))$. The Schatten ∞-norm $\|A\|_\infty$ is the largest singular value of A and coincides with the spectral norm $\|A\|$. The Schatten 1-norm $\|A\|_1$ is the sum of the singular values of A and is often referred to as the trace-norm of A. The Schatten 2-norm $\|A\|_2$ coincides with the Frobenius norm of A. For more on Schatten p-norms, see [40].

Theorem 1. *Let* $\{X_n\}_n$ *be a sequence of matrices, with* X_n *Hermitian of size* d_n, *and let* $\{P_n\}_n$ *be a sequence such that* $P_n \in \mathbb{C}^{d_n \times \delta_n}$, $P_n^* P_n = I_{\delta_n}$, $\delta_n \leq d_n$ *and* $\delta_n / d_n \to 1$ *as* $n \to \infty$. *Then,* $\{X_n\}_n \sim_{\sigma, \lambda} \kappa$ *if and only if* $\{P_n^* X_n P_n\}_n \sim_{\sigma, \lambda} \kappa$.

Theorem 2. *Let* $\{X_n\}_n$ *and* $\{Y_n\}_n$ *be sequences of matrices, with* X_n *and* Y_n *of size* d_n. *Assume that:*

- *the matrices* X_n *are Hermitian and* $\{X_n\}_n \sim_\lambda \kappa$;
- $\|Y_n\|_2 = o(\sqrt{d_n})$;

then $\{X_n + Y_n\}_n \sim_\lambda \kappa$.

Block Toeplitz Matrices. Given a function $f : [-\pi, \pi] \to \mathbb{C}^{s \times s}$ in $L^1([-\pi, \pi])$, its Fourier coefficients are denoted by

$$f_k = \frac{1}{2\pi} \int_{-\pi}^\pi f(\theta) \mathrm{e}^{-\mathrm{i}k\theta} \mathrm{d}\theta \in \mathbb{C}^{s \times s}, \qquad k \in \mathbb{Z},$$

where the integrals are computed componentwise. The nth block Toeplitz matrix generated by f is defined as

$$T_n(f) = [f_{i-j}]_{i,j=1}^n \in \mathbb{C}^{sn \times sn}.$$

It is not difficult to see that all the matrices $T_n(f)$ are Hermitian when f is Hermitian a.e.

Block Diagonal Sampling Matrices. For $n \in \mathbb{N}$ and $a : [0,1] \to \mathbb{C}^{s \times s}$, we define the block diagonal sampling matrix $D_n(a)$ as the diagonal matrix

$$D_n(a) = \operatorname*{diag}_{i=1,\dots,n} a\left(\frac{i}{n}\right) = \begin{bmatrix} a\left(\frac{1}{n}\right) & & & \\ & a\left(\frac{2}{n}\right) & & \\ & & \ddots & \\ & & & a(1) \end{bmatrix} \in \mathbb{C}^{sn \times sn}.$$

Zero-Distributed Sequences. A sequence of matrices $\{Z_n\}_n$ such that $\{Z_n\}_n \sim_\sigma 0$ is referred to as a zero-distributed sequence. Note that, for any $r \geq 1$, $\{Z_n\}_n \sim_\sigma 0$ is equivalent to $\{Z_n\}_n \sim_\sigma O_r$ (throughout this paper, O_m and I_m denote the $m \times m$ zero matrix and the $m \times m$ identity matrix, respectively). Proposition 1 provides an important characterization of zero-distributed sequences together with a useful sufficient condition for detecting such sequences. Throughout this paper, we use the natural convention $1/\infty = 0$.

Proposition 1. *Let $\{Z_n\}_n$ be a sequence of matrices, with Z_n of size d_n.*

- *$\{Z_n\}_n$ is zero-distributed if and only if $Z_n = R_n + N_n$ with $\operatorname{rank}(R_n)/d_n \to 0$ and $\|N_n\| \to 0$.*
- *$\{Z_n\}_n$ is zero-distributed if there exists a $p \in [1,\infty]$ such that $\|Z_n\|_p/(d_n)^{1/p} \to 0$.*

Approximating Classes of Sequences. The notion of approximating classes of sequences (a.c.s.) is the fundamental concept on which the theory of block GLT sequences is based.

Definition 2. *Let $\{A_n\}_n$ be a sequence of matrices, with A_n of size d_n, and let $\{\{B_{n,m}\}_n\}_m$ be a sequence of sequences of matrices, with $B_{n,m}$ of size d_n. We say that $\{\{B_{n,m}\}_n\}_m$ is an approximating class of sequences (a.c.s.) for $\{A_n\}_n$ if the following condition is met: for every m there exists n_m such that, for $n \geq n_m$,*

$$A_n = B_{n,m} + R_{n,m} + N_{n,m}, \quad \operatorname{rank}(R_{n,m}) \leq c(m)d_n, \quad \|N_{n,m}\| \leq \omega(m),$$

where n_m, $c(m)$, $\omega(m)$ depend only on m and $\lim_{m \to \infty} c(m) = \lim_{m \to \infty} \omega(m) = 0$.

Roughly speaking, $\{\{B_{n,m}\}_n\}_m$ is an a.c.s. for $\{A_n\}_n$ if, for large m, the sequence $\{B_{n,m}\}_n$ approximates $\{A_n\}_n$ in the sense that A_n is eventually equal to $B_{n,m}$ plus a small-rank matrix (with respect to the matrix size d_n) plus a small-norm matrix. It turns out that, for each fixed sequence of positive integers d_n such that $d_n \to \infty$, the notion of a.c.s. is a notion of convergence in the space

$$\mathscr{E} = \{\{A_n\}_n : A_n \in \mathbb{C}^{d_n \times d_n}\}.$$

More precisely, there exists a pseudometric $d_{\text{a.c.s.}}$ in $\mathscr{E}$ such that $\{\{B_{n,m}\}_n\}_m$ is an a.c.s. for $\{A_n\}_n$ if and only if $d_{\text{a.c.s.}}(\{B_{n,m}\}_n, \{A_n\}_n) \to 0$ as $m \to \infty$. We therefore use the convergence notation $\{B_{n,m}\}_n \xrightarrow{\text{a.c.s.}} \{A_n\}_n$ to indicate that $\{\{B_{n,m}\}_n\}_m$ is an a.c.s. for $\{A_n\}_n$. A useful criterion to identify an a.c.s. is provided in the next proposition ([13] Corollary 5.3).

Proposition 2. *Let $\{A_n\}_n$ be a sequence of matrices, with A_n of size d_n, let $\{\{B_{n,m}\}_n\}_m$ be a sequence of sequences of matrices, with $B_{n,m}$ of size d_n, and let $p \in [1,\infty]$. Suppose that for every m there exists n_m such that, for $n \geq n_m$,*

$$\|A_n - B_{n,m}\|_p \leq \varepsilon(m,n)(d_n)^{1/p},$$

where $\lim\limits_{m\to\infty} \limsup\limits_{n\to\infty} \varepsilon(m,n) = 0$. Then, $\{B_{n,m}\}_n \xrightarrow{\text{a.c.s.}} \{A_n\}_n$.

If $X \in \mathbb{C}^{m_1 \times m_2}$ and $Y \in \mathbb{C}^{\ell_1 \times \ell_2}$ are any two matrices, the tensor (Kronecker) product of X and Y is the $m_1\ell_1 \times m_2\ell_2$ matrix defined as follows:

$$X \otimes Y = [x_{ij}Y]_{\substack{i=1,\ldots,m_1 \\ j=1,\ldots,m_2}} = \begin{bmatrix} x_{11}Y & \cdots & x_{1m_2}Y \\ \vdots & & \vdots \\ x_{m_11}Y & \cdots & x_{m_1m_2}Y \end{bmatrix}.$$

We recall that the tensor product operation $\otimes$ is associative and bilinear. Moreover,

$$\|X \otimes Y\| = \|X\|\,\|Y\|, \tag{4}$$

$$\text{rank}(X \otimes Y) = \text{rank}(X)\text{rank}(Y), \tag{5}$$

$$(X \otimes Y)^T = X^T \otimes Y^T. \tag{6}$$

Finally, if X_1, X_2 can be multiplied and Y_1, Y_2 can be multiplied, then

$$(X_1 \otimes Y_1)(X_2 \otimes Y_2) = (X_1 X_2) \otimes (Y_1 Y_2). \tag{7}$$

Lemma 1. *For $i,j = 1,\ldots,s$, let $\{A_{n,ij}\}_n$ be a sequence of matrices and suppose that $\{B_{n,ij}^{(m)}\}_n \xrightarrow{\text{a.c.s.}} \{A_{n,ij}\}_n$. Then,*

$$[B_{n,ij}^{(m)}]_{i,j=1}^s \xrightarrow{\text{a.c.s.}} [A_{n,ij}]_{i,j=1}^s.$$

Proof. Let E_{ij} be the $s \times s$ matrix having 1 in position (i,j) and 0 elsewhere. Note that

$$[A_{n,ij}]_{i,j=1}^s = \sum_{i,j=1}^s E_{ij} \otimes A_{n,ij}, \qquad [B_{n,ij}^{(m)}]_{i,j=1}^s = \sum_{i,j=1}^s E_{ij} \otimes B_{n,ij}^{(m)}. \tag{8}$$

Since $\{B_{n,ij}^{(m)}\}_n \xrightarrow{\text{a.c.s.}} \{A_{n,ij}\}_n$, it is clear from (4), (5) and the definition of a.c.s. that

$$E_{ij} \otimes B_{n,ij}^{(m)} \xrightarrow{\text{a.c.s.}} E_{ij} \otimes A_{n,ij}, \qquad i,j = 1,\ldots,s. \tag{9}$$

Now, if $\{B_{n,m}^{[k]}\}_n \xrightarrow{\text{a.c.s.}} \{A_n^{[k]}\}_n$ for $k = 1,\ldots,K$ then $\{\sum_{k=1}^K B_{n,m}^{[k]}\}_n \xrightarrow{\text{a.c.s.}} \{\sum_{k=1}^K A_n^{[k]}\}_n$ (this is an obvious consequence of the definition of a.c.s.). Thus, the thesis follows from (8) and (9). $\square$

Block GLT Sequences. Let $s \geq 1$ be a fixed positive integer. An s-block GLT sequence (or simply a GLT sequence if s can be inferred from the context or we do not need/want to specify it) is a special s-block matrix-sequence $\{A_n\}_n$ equipped with a measurable function $\kappa : [0,1] \times [-\pi,\pi] \to \mathbb{C}^{s \times s}$, the so-called symbol. We use the notation $\{A_n\}_n \sim_{\text{GLT}} \kappa$ to indicate that $\{A_n\}_n$ is a GLT sequence with symbol κ. The symbol of a GLT sequence is unique in the sense that if $\{A_n\}_n \sim_{\text{GLT}} \kappa$ and $\{A_n\}_n \sim_{\text{GLT}} \varsigma$ then $\kappa = \varsigma$ a.e. in $[0,1] \times [-\pi,\pi]$. The main properties of s-block GLT sequences proved in [34] are listed below. If A is a matrix, we denote by $A^\dagger$ the Moore–Penrose pseudoinverse of A (recall that $A^\dagger = A^{-1}$ whenever A is invertible). If $f_m, f : D \subseteq \mathbb{R}^k \to \mathbb{C}^{r \times r}$ are measurable matrix-valued functions, we say that f_m converges to f in measure (resp., a.e., in $L^p(D)$, etc.) if $(f_m)_{\alpha\beta}$ converges to $f_{\alpha\beta}$ in measure (resp., a.e., in $L^p(D)$, etc.) for all $\alpha, \beta = 1,\ldots,r$.

GLT 1. If $\{A_n\}_n \sim_{\mathrm{GLT}} \kappa$ then $\{A_n\}_n \sim_\sigma \kappa$. If, moreover, each A_n is Hermitian, then $\{A_n\}_n \sim_\lambda \kappa$.

GLT 2. We have:
- $\{T_n(f)\}_n \sim_{\mathrm{GLT}} \kappa(x,\theta) = f(\theta)$ if $f : [-\pi,\pi] \to \mathbb{C}^{s\times s}$ is in $L^1([-\pi,\pi])$;
- $\{D_n(a)\}_n \sim_{\mathrm{GLT}} \kappa(x,\theta) = a(x)$ if $a : [0,1] \to \mathbb{C}^{s\times s}$ is Riemann-integrable;
- $\{Z_n\}_n \sim_{\mathrm{GLT}} \kappa(x,\theta) = O_s$ if and only if $\{Z_n\}_n \sim_\sigma 0$.

GLT 3. If $\{A_n\}_n \sim_{\mathrm{GLT}} \kappa$ and $\{B_n\}_n \sim_{\mathrm{GLT}} \varsigma$, then:
- $\{A_n^*\}_n \sim_{\mathrm{GLT}} \kappa^*$;
- $\{\alpha A_n + \beta B_n\}_n \sim_{\mathrm{GLT}} \alpha\kappa + \beta\varsigma$ for all $\alpha, \beta \in \mathbb{C}$;
- $\{A_n B_n\}_n \sim_{\mathrm{GLT}} \kappa\varsigma$;
- $\{A_n^\dagger\}_n \sim_{\mathrm{GLT}} \kappa^{-1}$ provided that κ is invertible a.e.

GLT 4. $\{A_n\}_n \sim_{\mathrm{GLT}} \kappa$ if and only if there exist s-block GLT sequences $\{B_{n,m}\}_n \sim_{\mathrm{GLT}} \kappa_m$ such that $\{B_{n,m}\}_n \overset{\text{a.c.s.}}{\longrightarrow} \{A_n\}_n$ and $\kappa_m \to \kappa$ in measure.

Remark 3. *The reader might be astonished by the fact that we have talked so far about block GLT sequences without defining them. Actually, we intentionally avoided giving a definition for two reasons. First, the definition is rather cumbersome as it requires introducing other related (and complicated) concepts such as "block LT operators" and "block LT sequences". Second, from a practical viewpoint, the definition is completely useless because everything that can be derived from it can also be derived from* **GLT 1**–**GLT 4** *(and in a much easier way). The reader who is interested in the formal definition of block GLT sequences can find it in ([34] Section 5) along with the proofs of properties* **GLT 1**–**GLT 4**.

3. FD Discretization of a System of DEs

Consider the following system of DEs:

$$
\begin{cases}
-a_{11}(x)u_1''(x) + a_{12}(x)u_2'(x) = f_1(x), & x \in (0,1), \\
a_{21}(x)u_1'(x) + a_{22}(x)u_2(x) = f_2(x), & x \in (0,1), \\
u_1(0) = 0, \quad u_1(1) = 0, \\
u_2(0) = 0, \quad u_2(1) = 0.
\end{cases}
\tag{10}
$$

In this section, we consider the classical central FD discretization of (10). Through the theory of block GLT sequences, we show that the corresponding sequence of (normalized) FD discretization matrices enjoys a spectral distribution described by a 2×2 matrix-valued function. We remark that the number 2, which identifies the matrix space $\mathbb{C}^{2\times 2}$ where the spectral symbol takes values, coincides with the number of equations that compose the system (10). In what follows, we use the following notation:

$$
\underset{j=1,\ldots,n}{\mathrm{tridiag}}\left[\, \beta_j \,\middle|\, \alpha_j \,\middle|\, \gamma_j \,\right] =
\begin{bmatrix}
\alpha_1 & \gamma_1 & & & \\
\beta_2 & \alpha_2 & \gamma_2 & & \\
& \ddots & \ddots & \ddots & \\
& & \beta_{n-1} & \alpha_{n-1} & \gamma_{n-1} \\
& & & \beta_n & \alpha_n
\end{bmatrix}.
$$

The parameters $\alpha_j, \beta_j, \gamma_j$ may be either scalars or $s \times s$ blocks for some $s > 1$, in which case the previous matrix is a block tridiagonal matrix.

3.1. FD Discretization

Let $n \geq 1$, and set $h = \frac{1}{n+1}$ and $x_j = jh$ for $j = 0, \ldots, n+1$. Using the classical central FD schemes $(-1, 2, -1)$ and $\frac{1}{2}(-1, 0, 1)$ for the discretization of, respectively, the (negative) second derivative and the first derivative, for each $j = 1, \ldots, n$ we obtain the following approximations:

$$
[-a_{11}(x)u_1''(x) + a_{12}(x)u_2'(x)]\big|_{x=x_j} \approx a_{11}(x_j)\frac{-u_1(x_{j+1}) + 2u_1(x_j) - u_1(x_{j-1})}{h^2}
$$
$$
+ a_{12}(x_j)\frac{u_2(x_{j+1}) - u_2(x_{j-1})}{2h},
$$
$$
[a_{21}(x)u_1'(x) + a_{22}(x)u_2(x)]\big|_{x=x_j} \approx a_{21}(x_j)\frac{u_1(x_{j+1}) - u_1(x_{j-1})}{2h} + a_{22}(x_j)u_2(x_j).
$$

This means that the nodal values of the solutions u_1, u_2 of (10) satisfy approximately the equations

$$
a_{11}(x_j)\left[-u_1(x_{j+1}) + 2u_1(x_j) - u_1(x_{j-1})\right] + \frac{h}{2}a_{12}(x_j)\left[u_2(x_{j+1}) - u_2(x_{j-1})\right] = h^2 f_1(x_j),
$$
$$
\frac{1}{2}a_{21}(x_j)\left[u_1(x_{j+1}) - u_1(x_{j-1})\right] + h a_{22}(x_j)u_2(x_j) = h f_2(x_j),
$$

for $j = 1, \ldots, n$. We then approximate the solution u_1 (resp., u_2) by the piecewise linear function that takes the value $u_{1,j}$ (resp., $u_{2,j}$) at x_j for all $j = 0, \ldots, n+1$, where $u_{1,0} = u_{1,n+1} = u_{2,0} = u_{2,n+1} = 0$ and the vectors $\mathbf{u}_1 = (u_{1,1}, \ldots, u_{1,n})^T$ and $\mathbf{u}_2 = (u_{2,1}, \ldots, u_{2,n})^T$ solve the linear system

$$
a_{11}(x_j)\left[-u_{1,j+1} + 2u_{1,j} - u_{1,j-1}\right] + \frac{h}{2}a_{12}(x_j)\left[u_{2,j+1} - u_{2,j-1}\right] = h^2 f_1(x_j), \qquad j = 1, \ldots, n,
$$
$$
\frac{1}{2}a_{21}(x_j)\left[u_{1,j+1} - u_{1,j-1}\right] + h a_{22}(x_j)u_{2,j} = h f_2(x_j), \qquad j = 1, \ldots, n. \tag{11}
$$

This linear system can be rewritten in matrix form as follows:

$$
A_n \begin{bmatrix} \mathbf{u}_1 \\ \mathbf{u}_2 \end{bmatrix} = \begin{bmatrix} h^2\mathbf{f}_1 \\ h\mathbf{f}_2 \end{bmatrix}, \tag{12}
$$

where $\mathbf{f}_1 = [f_1(x_j)]_{j=1}^n$, $\mathbf{f}_2 = [f_2(x_j)]_{j=1}^n$,

$$
A_n = \begin{bmatrix} K_n(a_{11}) & hH_n(a_{12}) \\ H_n(a_{21}) & hM_n(a_{22}) \end{bmatrix} = \begin{bmatrix} K_n(a_{11}) & H_n(a_{12}) \\ H_n(a_{21}) & M_n(a_{22}) \end{bmatrix} \begin{bmatrix} I_n & O_n \\ O_n & hI_n \end{bmatrix}, \tag{13}
$$

and

$$
K_n(a_{11}) = \operatorname*{tridiag}_{j=1,\ldots,n}\left[\, -a_{11}(x_j) \,\big|\, 2a_{11}(x_j) \,\big|\, -a_{11}(x_j) \,\right] = \left(\operatorname*{diag}_{j=1,\ldots,n} a_{11}(x_j)\right)T_n(2 - 2\cos\theta),
$$
$$
H_n(a_{12}) = \operatorname*{tridiag}_{j=1,\ldots,n}\left[\, -\tfrac{1}{2}a_{12}(x_j) \,\big|\, 0 \,\big|\, \tfrac{1}{2}a_{12}(x_j) \,\right] = \left(\operatorname*{diag}_{j=1,\ldots,n} a_{12}(x_j)\right)T_n(-i\sin\theta),
$$
$$
H_n(a_{21}) = \operatorname*{tridiag}_{j=1,\ldots,n}\left[\, -\tfrac{1}{2}a_{21}(x_j) \,\big|\, 0 \,\big|\, \tfrac{1}{2}a_{21}(x_j) \,\right] = \left(\operatorname*{diag}_{j=1,\ldots,n} a_{21}(x_j)\right)T_n(-i\sin\theta),
$$
$$
M_n(a_{22}) = \operatorname*{diag}_{j=1,\ldots,n} a_{22}(x_j).
$$

In view of (13), the linear system (12) is equivalent to

$$
B_n \begin{bmatrix} \mathbf{v}_1 \\ \mathbf{v}_2 \end{bmatrix} = \begin{bmatrix} h^2\mathbf{f}_1 \\ h\mathbf{f}_2 \end{bmatrix}, \tag{14}
$$

where $\mathbf{v}_1 = \mathbf{u}_1$, $\mathbf{v}_2 = h\mathbf{u}_2$, and

$$B_n = \begin{bmatrix} K_n(a_{11}) & H_n(a_{12}) \\ H_n(a_{21}) & M_n(a_{22}) \end{bmatrix}. \tag{15}$$

Let $v_{1,1}, \ldots, v_{1,n}$ and $v_{2,1}, \ldots, v_{2,n}$ be the components of $\mathbf{v}_1$ and $\mathbf{v}_2$, respectively. When writing the linear system (11) in the form (14), we are implicitly assuming the following.

- The unknowns are sorted as follows:

$$\begin{bmatrix} [v_{1,j}]_{j=1,\ldots,n} \\ \hline [v_{2,j}]_{j=1,\ldots,n} \end{bmatrix} = \begin{bmatrix} v_{1,1} \\ v_{1,2} \\ \vdots \\ v_{1,n} \\ \hline v_{2,1} \\ v_{2,2} \\ \vdots \\ v_{2,n} \end{bmatrix}. \tag{16}$$

- The equations are sorted as follows, in accordance with the ordering (16) for the unknowns:

$$\begin{bmatrix} \left[a_{11}(x_j) \left[-v_{1,j+1} + 2v_{1,j} - v_{1,j-1} \right] + \frac{1}{2}a_{12}(x_j) \left[v_{2,j+1} - v_{2,j-1} \right] = h^2 f_1(x_j) \right]_{j=1,\ldots,n} \\ \hline \left[\frac{1}{2}a_{21}(x_j) \left[u_{1,j+1} - u_{1,j-1} \right] + a_{22}(x_j)v_{2,j} = hf_2(x_j) \right]_{j=1,\ldots,n} \end{bmatrix}. \tag{17}$$

Suppose we decide to change the ordering for both the unknowns and the equations. More precisely, suppose we opt for the following orderings.

- The unknowns are sorted as follows:

$$\begin{bmatrix} v_{1,j} \\ v_{2,j} \end{bmatrix}_{j=1,\ldots,n} = \begin{bmatrix} v_{1,1} \\ v_{2,1} \\ v_{1,2} \\ v_{2,2} \\ \vdots \\ v_{1,n} \\ v_{2,n} \end{bmatrix}. \tag{18}$$

- The equations are sorted as follows, in accordance with the ordering (18) for the unknowns:

$$\begin{bmatrix} a_{11}(x_j) \left[-v_{1,j+1} + 2v_{1,j} - v_{1,j-1} \right] + \frac{1}{2}a_{12}(x_j) \left[v_{2,j+1} - v_{2,j-1} \right] = h^2 f_1(x_j) \\ \frac{1}{2}a_{21}(x_j) \left[v_{1,j+1} - v_{1,j-1} \right] + a_{22}(x_j)v_{2,j} = hf_2(x_j) \end{bmatrix}_{j=1,\ldots,n}. \tag{19}$$

The matrix C_n associated with the linear system (11) assuming the new orderings (18) and (19) is the 2×2 block tridiagonal matrix given by

$$C_n = \operatorname*{tridiag}_{j=1,\ldots,n} \begin{bmatrix} -a_{11}(x_j) & -\frac{1}{2}a_{12}(x_j) & 2a_{11}(x_j) & 0 & -a_{11}(x_j) & \frac{1}{2}a_{12}(x_j) \\ -\frac{1}{2}a_{21}(x_j) & 0 & 0 & a_{22}(x_j) & \frac{1}{2}a_{21}(x_j) & 0 \end{bmatrix}. \tag{20}$$

The matrix C_n is similar to B_n. Indeed, by permuting both rows and columns of B_n according to the permutation $1, n+1, 2, n+2, \ldots, n, 2n$ we obtain C_n. More precisely, let $\mathbf{e}_1, \ldots, \mathbf{e}_n$ and $\tilde{\mathbf{e}}_1, \ldots, \tilde{\mathbf{e}}_{2n}$ be the vectors of the canonical basis of $\mathbb{R}^n$ and $\mathbb{R}^{2n}$, respectively, and let Π_n be the permutation matrix associated with the permutation $1, n+1, 2, n+2, \ldots, n, 2n$, that is,

$$
\Pi_n = \begin{bmatrix} \tilde{\mathbf{e}}_1^T \\ \tilde{\mathbf{e}}_{n+1}^T \\ \tilde{\mathbf{e}}_2^T \\ \tilde{\mathbf{e}}_{n+2}^T \\ \vdots \\ \tilde{\mathbf{e}}_n^T \\ \tilde{\mathbf{e}}_{2n}^T \end{bmatrix} = \begin{bmatrix} I_2 \otimes \mathbf{e}_1^T \\ I_2 \otimes \mathbf{e}_2^T \\ \vdots \\ I_2 \otimes \mathbf{e}_n^T \end{bmatrix}. \tag{21}
$$

Then, $C_n = \Pi_n B_n \Pi_n^T$.

3.2. GLT Analysis of the FD Discretization Matrices

The main result of this section (Theorem 3) shows that $\{C_n\}_n$ is a block GLT sequence whose spectral distribution is described by a 2×2 matrix-valued symbol, which is obtained by replacing the matrix-sequences $\{K_n(a_{11})\}_n$, $\{H_n(a_{12})\}_n$, $\{H_n(a_{21})\}_n$, $\{M_n(a_{22})\}_n$ appearing in the expression (15) of B_n with the corresponding symbols $a_{11}(x)(2 - 2\cos\theta)$, $-ia_{12}(x)\sin\theta$, $-ia_{12}(x)\sin\theta$, $a_{22}(x)$. In this regard, we note that, assuming for instance $a_{11}, a_{12}, a_{21}, a_{22} \in C([0,1])$, we have

$$
\{K_n(a_{11})\}_n \sim_{\text{GLT}} a_{11}(x)(2 - 2\cos\theta), \tag{22}
$$

$$
\{H_n(a_{12})\}_n \sim_{\text{GLT}} -ia_{12}(x)\sin\theta, \tag{23}
$$

$$
\{H_n(a_{21})\}_n \sim_{\text{GLT}} -ia_{21}(x)\sin\theta, \tag{24}
$$

$$
\{M_n(a_{22})\}_n \sim_{\text{GLT}} a_{22}(x). \tag{25}
$$

To prove (22), it suffices to observe that

$$
\|K_n(a_{11}) - D_n(a_{11})T_n(2 - 2\cos\theta)\| \leq \left\| \operatorname*{diag}_{j=1,\ldots,n} a_{11}(x_j) - D_n(a_{11}) \right\| \|T_n(2 - 2\cos\theta)\|
$$

$$
= \max_{j=1,\ldots,n} \left| a_{11}(x_j) - a_{11}\left(\frac{j}{n}\right) \right| \|T_n(2 - 2\cos\theta)\| \leq 4\omega_{a_{11}}(h),
$$

where $\omega_{a_{11}}(\cdot)$ is the modulus of continuity of a_{11}. Since $\omega_{a_{11}}(h) \to 0$ as $n \to \infty$, it follows from Proposition 1 that $\{K_n(a_{11}) - D_n(a_{11})T_n(2 - 2\cos\theta)\}_n \sim_\sigma 0$, and so **GLT 2** and **GLT 3** immediately yield (22). The relations (23)–(25) are proved in the same way.

Theorem 3. *Suppose that $a_{11}, a_{12}, a_{21}, a_{22} \in C([0,1])$. Then,*

$$
\{C_n\}_n \sim_{\text{GLT}} \kappa(x, \theta) = \begin{bmatrix} a_{11}(x)(2 - 2\cos\theta) & -ia_{12}(x)\sin\theta \\ -ia_{21}(x)\sin\theta & a_{22}(x) \end{bmatrix} \tag{26}
$$

and

$$
\{C_n\}_n \sim_\sigma \kappa(x, \theta). \tag{27}
$$

If, moreover, $a_{21} = -a_{12}$, then we also have

$$
\{C_n\}_n \sim_\lambda \kappa(x, \theta). \tag{28}
$$

Proof. From (20), we have

$$
\begin{aligned}
C_n &= \operatorname*{tridiag}_{j=1,\ldots,n}\left[\begin{array}{cc|cc|cc} -a_{11}(x_j) & -\tfrac{1}{2}a_{12}(x_j) & 2a_{11}(x_j) & 0 & -a_{11}(x_j) & \tfrac{1}{2}a_{12}(x_j) \\ -\tfrac{1}{2}a_{21}(x_j) & 0 & 0 & a_{22}(x_j) & \tfrac{1}{2}a_{21}(x_j) & 0 \end{array}\right] \\
&= \operatorname*{tridiag}_{j=1,\ldots,n}\left[\begin{array}{cc|cc|cc} -a_{11}(x_j) & 0 & 2a_{11}(x_j) & 0 & -a_{11}(x_j) & 0 \\ 0 & 0 & 0 & 0 & 0 & 0 \end{array}\right] \\
&\quad + \operatorname*{tridiag}_{j=1,\ldots,n}\left[\begin{array}{cc|cc|cc} 0 & -\tfrac{1}{2}a_{12}(x_j) & 0 & 0 & 0 & \tfrac{1}{2}a_{12}(x_j) \\ 0 & 0 & 0 & 0 & 0 & 0 \end{array}\right] \\
&\quad + \operatorname*{tridiag}_{j=1,\ldots,n}\left[\begin{array}{cc|cc|cc} 0 & 0 & 0 & 0 & 0 & 0 \\ -\tfrac{1}{2}a_{21}(x_j) & 0 & 0 & 0 & \tfrac{1}{2}a_{21}(x_j) & 0 \end{array}\right] \\
&\quad + \operatorname*{tridiag}_{j=1,\ldots,n}\left[\begin{array}{cc|cc|cc} 0 & 0 & 0 & 0 & 0 & 0 \\ 0 & 0 & 0 & a_{22}(x_j) & 0 & 0 \end{array}\right] \\
&= \operatorname*{diag}_{j=1,\ldots,n} a_{11}(x_j)I_2 \cdot \operatorname*{tridiag}_{j=1,\ldots,n}\left[\begin{array}{cc|cc|cc} -1 & 0 & 2 & 0 & -1 & 0 \\ 0 & 0 & 0 & 0 & 0 & 0 \end{array}\right] \\
&\quad + \operatorname*{diag}_{j=1,\ldots,n} a_{12}(x_j)I_2 \cdot \operatorname*{tridiag}_{j=1,\ldots,n}\left[\begin{array}{cc|cc|cc} 0 & -\tfrac{1}{2} & 0 & 0 & 0 & \tfrac{1}{2} \\ 0 & 0 & 0 & 0 & 0 & 0 \end{array}\right] \\
&\quad + \operatorname*{diag}_{j=1,\ldots,n} a_{21}(x_j)I_2 \cdot \operatorname*{tridiag}_{j=1,\ldots,n}\left[\begin{array}{cc|cc|cc} 0 & 0 & 0 & 0 & 0 & 0 \\ -\tfrac{1}{2} & 0 & 0 & 0 & \tfrac{1}{2} & 0 \end{array}\right] \\
&\quad + \operatorname*{diag}_{j=1,\ldots,n} a_{22}(x_j)I_2 \cdot \operatorname*{tridiag}_{j=1,\ldots,n}\left[\begin{array}{cc|cc|cc} 0 & 0 & 0 & 0 & 0 & 0 \\ 0 & 0 & 0 & 1 & 0 & 0 \end{array}\right] \\
&= \operatorname*{diag}_{j=1,\ldots,n} a_{11}(x_j)I_2 \cdot T_n((2-2\cos\theta)E_{11}) \\
&\quad + \operatorname*{diag}_{j=1,\ldots,n} a_{12}(x_j)I_2 \cdot T_n((-i\sin\theta)E_{12}) \\
&\quad + \operatorname*{diag}_{j=1,\ldots,n} a_{21}(x_j)I_2 \cdot T_n((-i\sin\theta)E_{21}) \\
&\quad + \operatorname*{diag}_{j=1,\ldots,n} a_{22}(x_j)I_2 \cdot T_n(E_{22}),
\end{aligned}
\tag{29}
$$

where E_{pq} is the 2×2 matrix having 1 in position (p,q) and 0 elsewhere. It is clear that, for every $p,q = 1,2$,

$$
\left\|\operatorname*{diag}_{j=1,\ldots,n} a_{pq}(x_j)I_2 - D_n(a_{pq}I_2)\right\| \le \omega_{a_{pq}}(h) \to 0
$$

as $n \to \infty$; hence, by Proposition 1, **GLT 2** and **GLT 3**,

$$
\left\{\operatorname*{diag}_{j=1,\ldots,n} a_{pq}(x_j)I_2\right\}_n \sim_{\mathrm{GLT}} a(x)I_2, \qquad p,q = 1,2.
$$

Consequently, the decomposition (29), **GLT 2** and **GLT 3** imply (26), which in turn implies (27) by **GLT 1**. It only remains to prove (28) in the case where $a_{21} = -a_{12}$. In this case, we have

$$
C_n = \operatorname*{tridiag}_{j=1,\ldots,n}\left[\begin{array}{cc|cc|cc} -a_{11}(x_j) & -\tfrac{1}{2}a_{12}(x_j) & 2a_{11}(x_j) & 0 & -a_{11}(x_j) & \tfrac{1}{2}a_{12}(x_j) \\ \tfrac{1}{2}a_{12}(x_j) & 0 & 0 & a_{22}(x_j) & -\tfrac{1}{2}a_{12}(x_j) & 0 \end{array}\right].
$$

Consider the symmetric approximation of C_n given by

$$\widetilde{C}_n = \operatorname*{tridiag}_{j=1,\ldots,n} \left[\begin{array}{cc|cc|cc} -a_{11}(x_{j-1}) & -\frac{1}{2}a_{12}(x_{j-1}) & 2a_{11}(x_j) & 0 & -a_{11}(x_j) & \frac{1}{2}a_{12}(x_j) \\ \frac{1}{2}a_{12}(x_{j-1}) & 0 & 0 & a_{22}(x_j) & -\frac{1}{2}a_{12}(x_j) & 0 \end{array} \right].$$

It is not difficult to see that $\|C_n - \widetilde{C}_n\| \to 0$ as $n \to \infty$ by invoking the inequality

$$\|X\| \leq \sqrt{\left(\max_{i=1,\ldots,n} \sum_{j=1}^{n} |x_{ij}| \right) \left(\max_{j=1,\ldots,n} \sum_{i=1}^{n} |x_{ij}| \right)}, \qquad X \in \mathbb{C}^{n \times n}; \tag{30}$$

see, e.g., ([13] Section 2.4.1). Therefore:

- in view of the decomposition $\widetilde{C}_n = C_n + (\widetilde{C}_n - C_n)$, we have $\{\widetilde{C}_n\}_n \sim_{\text{GLT}} \kappa(x, \theta)$ by (26), Proposition 1, **GLT 2** and **GLT 3**, so in particular $\{\widetilde{C}_n\}_n \sim_\lambda \kappa(x, \theta)$ by **GLT 1** as $\widetilde{C}_n$ is symmetric;
- $\|C_n - \widetilde{C}_n\|_2 \leq \sqrt{n} \|C_n - \widetilde{C}_n\| = o(\sqrt{2n})$ as $n \to \infty$.

Thus, (28) follows from Theorem 2. $\square$

Example 1. *Suppose that $a_{11}, a_{12}, a_{21}, a_{22} \in C([0, 1])$ and $a_{21} = -a_{12}$, so that $\{C_n\}_n \sim_\lambda \kappa(x, \theta)$ by Theorem 3. The eigenvalue functions of $\kappa(x, \theta)$ are given by*

$$\lambda_{1,2}(\kappa(x, \theta)) = \frac{a_{11}(x)(2 - 2\cos\theta) + a_{22}(x) \pm \sqrt{(a_{11}(x)(2 - 2\cos\theta) - a_{22}(x))^2 + 4(a_{12}(x)\sin\theta)^2}}{2}$$

and are continuous on $[0, 1] \times [-\pi, \pi]$. Let ϕ be the canonical rearranged version of $\kappa(x, \theta)$ obtained as the limit of the piecewise linear functions ϕ_ρ, according to the construction in Remark 2. Figure 1 shows the graph of ϕ and the eigenvalues $\lambda_1, \ldots, \lambda_{2n}$ of C_n for $a_{11}(x) = 2 + \cos(\pi x)$, $a_{12}(x) = -a_{21}(x) = e^{-x}\sin(\pi x)$, $a_{22}(x) = 2x + \sin(\pi x)$ and $n = 40$. The graph of ϕ has been obtained by plotting the graph of ϕ_ρ corresponding to a large value of ρ. The eigenvalues of C_n, which turn out to be real, although C_n is not symmetric, have been sorted in non-decreasing order and placed at the points (t_q, λ_q) with $t_q = \frac{q}{2n}$, $q = 1, \ldots, 2n$. We clearly see from the figure an excellent agreement between ϕ and the eigenvalues of C_n, as predicted by Remark 2. In particular, we observe no outliers in this case.

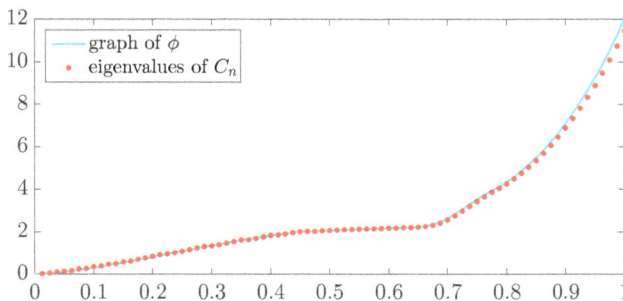

Figure 1. Comparison between the spectrum of C_n and the rearranged version ϕ of the symbol $\kappa(x, \theta)$ for $a_{11}(x) = 2 + \cos(\pi x)$, $a_{12}(x) = -a_{21}(x) = e^{-x}\sin(\pi x)$, $a_{22}(x) = 2x + \sin(\pi x)$ and $n = 40$.

4. Higher-Order FE Discretization of the Diffusion Equation

Consider the diffusion equation

$$\begin{cases} -(a(x)u'(x))' = f(x), & x \in (0,1), \\ u(0) = u(1) = 0. \end{cases} \tag{31}$$

In this section, we consider the higher-order FE discretization of (31). Through the theory of block GLT sequences, we show that the corresponding sequence of (normalized) FE discretization matrices enjoys a spectral distribution described by a $(p - k) \times (p - k)$ matrix-valued function, where p and k represent, respectively, the degree and the smoothness of the piecewise polynomial functions involved in the FE approximation. Note that this result represents a remarkable argument in support of ([35] Conjecture 2).

4.1. FE Discretization

The weak form of (31) reads as follows: find $u \in H_0^1([0,1])$ such that

$$\int_0^1 a(x)u'(x)w'(x)\mathrm{d}x = \int_0^1 f(x)w(x)\mathrm{d}x, \qquad \forall w \in H_0^1([0,1]).$$

In the FE method, we fix a set of basis functions $\{\varphi_1, \ldots, \varphi_N\} \subset H_0^1([0,1])$ and we look for an approximation of the exact solution in the space $\mathscr{W} = \mathrm{span}(\varphi_1, \ldots, \varphi_N)$ by solving the following discrete problem: find $u_{\mathscr{W}} \in \mathscr{W}$ such that

$$\int_0^1 a(x)u'_{\mathscr{W}}(x)w'(x)\mathrm{d}x = \int_0^1 f(x)w(x)\mathrm{d}x, \qquad \forall w \in \mathscr{W}.$$

Since $\{\varphi_1, \ldots, \varphi_N\}$ is a basis of $\mathscr{W}$, we can write $u_{\mathscr{W}} = \sum_{j=1}^N u_j \varphi_j$ for a unique vector $\mathbf{u} = (u_1, \ldots, u_N)^T$. By linearity, the computation of $u_{\mathscr{W}}$ (i.e., of $\mathbf{u}$) reduces to solving the linear system

$$A\mathbf{u} = \mathbf{f},$$

where $\mathbf{f} = \left(\int_0^1 f(x)\varphi_1(x)\mathrm{d}x, \ldots, \int_0^1 f(x)\varphi_N(x)\mathrm{d}x \right)^T$ and A is the stiffness matrix,

$$A = \left[\int_0^1 a(x)\varphi'_j(x)\varphi'_i(x)\mathrm{d}x \right]_{i,j=1}^N. \tag{32}$$

4.2. p-Degree C^k B-spline Basis Functions

Following the higher-order FE approach, the basis functions $\varphi_1, \ldots, \varphi_N$ will be chosen as piecewise polynomials of degree $p \geq 1$. More precisely, for $p, n \geq 1$ and $0 \leq k \leq p - 1$, let $B_{1,[p,k]}, \ldots, B_{n(p-k)+k+1,[p,k]} : \mathbb{R} \to \mathbb{R}$ be the B-splines of degree p and smoothness C^k defined on the knot sequence

$$\{\tau_1, \ldots, \tau_{n(p-k)+p+k+2}\} = \Big\{ \underbrace{0, \ldots, 0}_{p+1}, \underbrace{\frac{1}{n}, \ldots, \frac{1}{n}}_{p-k}, \underbrace{\frac{2}{n}, \ldots, \frac{2}{n}}_{p-k}, \ldots, \underbrace{\frac{n-1}{n}, \ldots, \frac{n-1}{n}}_{p-k}, \underbrace{1, \ldots, 1}_{p+1} \Big\}. \tag{33}$$

We collect here a few properties of $B_{1,[p,k]}, \ldots, B_{n(p-k)+k+1,[p,k]}$ that we shall use in this paper. For the formal definition of B-splines, as well as for the proof of the properties listed below, see [41,42].

- The support of the ith B-spline is given by

$$\mathrm{supp}(B_{i,[p,k]}) = [\tau_i, \tau_{i+p+1}], \qquad i = 1, \ldots, n(p-k)+k+1. \tag{34}$$

- Except for the first and the last one, all the other B-splines vanish on the boundary of $[0, 1]$, i.e.,

$$B_{i,[p,k]}(0) = B_{i,[p,k]}(1) = 0, \qquad i = 2, \dots, n(p-k) + k. \tag{35}$$

- $\{B_{1,[p,k]}, \dots, B_{n(p-k)+k+1,[p,k]}\}$ is a basis for the space of piecewise polynomial functions of degree p and smoothness C^k, that is,

$$\mathscr{V}_{n,[p,k]} = \left\{ v \in C^k([0,1]) : v|_{\left[\frac{i}{n}, \frac{i+1}{n}\right]} \in \mathbb{P}_p \text{ for all } i = 0, \dots, n-1 \right\},$$

where $\mathbb{P}_p$ is the space of polynomials of degree $\leq p$. Moreover, $\{B_{2,[p,k]}, \dots, B_{n(p-k)+k,[p,k]}\}$ is a basis for the space

$$\mathscr{W}_{n,[p,k]} = \{ w \in \mathscr{V}_{n,[p,k]} : w(0) = w(1) = 0 \}.$$

- The B-splines form a non-negative partition of unity over $[0, 1]$:

$$B_{i,[p,k]} \geq 0 \text{ over } \mathbb{R}, \qquad i = 1, \dots, n(p-k) + k + 1, \tag{36}$$

$$\sum_{i=1}^{n(p-k)+k+1} B_{i,[p,k]} = 1 \text{ over } [0,1]. \tag{37}$$

- The derivatives of the B-splines satisfy

$$\sum_{i=1}^{n(p-k)+k+1} |B'_{i,[p,k]}| \leq c_p n \text{ over } [0,1], \tag{38}$$

where c_p is a constant depending only on p. Note that the derivatives $B'_{i,[p,k]}$ may not be defined at some of the grid points $0, \frac{1}{n}, \frac{2}{n}, \dots, \frac{n-1}{n}, 1$ in the case of C^0 smoothness ($k = 0$). In (38), it is assumed that the undefined values are excluded from the summation.

- All the B-splines, except for the first $k+1$ and the last $k+1$, are uniformly shifted-scaled versions of $p-k$ fixed reference functions $\beta_{1,[p,k]}, \dots, \beta_{p-k,[p,k]}$, namely the first $p-k$ B-splines defined on the reference knot sequence

$$\underbrace{0, \dots, 0}_{p-k}, \underbrace{1, \dots, 1}_{p-k}, \dots, \underbrace{\eta, \dots, \eta}_{p-k}, \qquad \eta = \left\lceil \frac{p+1}{p-k} \right\rceil.$$

In formulas, setting

$$\nu = \left\lceil \frac{k+1}{p-k} \right\rceil, \tag{39}$$

for the B-splines $B_{k+2,[p,k]}, \dots, B_{k+1+(n-\nu)(p-k),[p,k]}$, we have

$$B_{k+1+(p-k)(r-1)+q,[p,k]}(x) = \beta_{q,[p,k]}(nx - r + 1), \qquad r = 1, \dots, n-\nu, \qquad q = 1, \dots, p-k. \tag{40}$$

We point out that the supports of the reference B-splines $\beta_{q,[p,k]}$ satisfy

$$\operatorname{supp}(\beta_{1,[p,k]}) \subseteq \operatorname{supp}(\beta_{2,[p,k]}) \subseteq \dots \subseteq \operatorname{supp}(\beta_{p-k,[p,k]}) = [0, \eta].$$

Figures 2 and 3 show the graphs of the B-splines $B_{1,[p,k]}, \dots, B_{n(p-k)+k+1,[p,k]}$ for the degree $p = 3$ and the smoothness $k = 1$, and the graphs of the associated reference B-splines $\beta_{1,[p,k]}, \beta_{2,[p,k]}$.

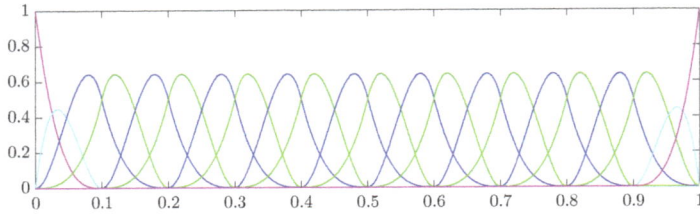

Figure 2. B-splines $B_{1,[p,k]}, \ldots, B_{n(p-k)+k+1,[p,k]}$ for $p = 3$ and $k = 1$, with $n = 10$.

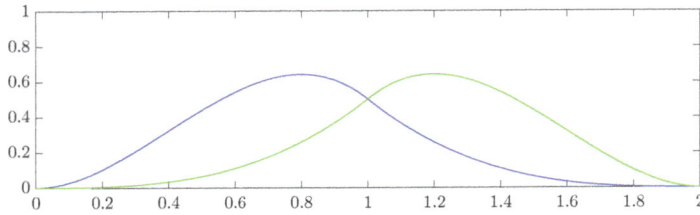

Figure 3. Reference B-splines $\beta_{1,[p,k]}, \beta_{2,[p,k]}$ for $p = 3$ and $k = 1$.

The basis functions $\varphi_1, \ldots, \varphi_N$ are defined as follows:

$$\varphi_i = B_{i+1,[p,k]}, \qquad i = 1, \ldots, n(p - k) + k - 1. \tag{41}$$

In particular, with the notations of Section 4.1, we have $N = n(p - k) + k - 1$ and $\mathscr{W} = \mathscr{W}_{n,[p,k]}$.

4.3. GLT Analysis of the Higher-Order FE Discretization Matrices

The stiffness matrix (32) resulting from the choice of the basis functions as in (41) will be denoted by $A_{n,[p,k]}(a)$,

$$A_{n,[p,k]}(a) = \left[\int_0^1 a(x) B'_{j+1,[p,k]}(x) B'_{i+1,[p,k]}(x) dx \right]_{i,j=1}^{n(p-k)+k-1}. \tag{42}$$

The main result of this section (Theorem 4) gives the spectral distribution of the normalized sequence $\{n^{-1}A_{n,[p,k]}(a)\}_n$. The proof of Theorem 4 is entirely based on the theory of block GLT sequences and it is therefore referred to as "GLT analysis". It also requires the following lemma, which provides an approximate construction of the matrix $A_{n,[p,k]}(1)$ corresponding to the constant-coefficient case where $a(x) = 1$ identically. In view of what follows, define the $(p - k) \times (p - k)$ blocks

$$K_{[p,k]}^{[\ell]} = \left[\int_{\mathbb{R}} \beta'_{j,[p,k]}(t) \beta'_{i,[p,k]}(t - \ell) dt \right]_{i,j=1}^{p-k}, \qquad \ell \in \mathbb{Z}, \tag{43}$$

and the $(p - k) \times (p - k)$ matrix-valued function $\kappa_{[p,k]} : [-\pi, \pi] \to \mathbb{C}^{(p-k) \times (p-k)}$,

$$\kappa_{[p,k]}(\theta) = \sum_{\ell \in \mathbb{Z}} K_{[p,k]}^{[\ell]} e^{i\ell\theta} = K_{[p,k]}^{[0]} + \sum_{\ell > 0} \left(K_{[p,k]}^{[\ell]} e^{i\ell\theta} + (K_{[p,k]}^{[\ell]})^T e^{-i\ell\theta} \right). \tag{44}$$

Due to the compact support of the reference functions $\beta_{1,[p,k]}, \ldots, \beta_{p-k,[p,k]}$, there is only a finite number of nonzero blocks $K_{[p,k]}^{[\ell]}$ and, consequently, the series in (44) is actually a finite sum.

Lemma 2. *Let* $p, n \geq 1$ *and* $0 \leq k \leq p - 1$. *Define* $\widetilde{A}_{n,[p,k]}(1)$ *as the principal submatrix of* $A_{n,[p,k]}(1)$ *of size* $(n - \nu)(p - k)$ *corresponding to the indices* $k + 1, \ldots, k + (n - \nu)(p - k)$, *where* $\nu = \lceil (k+1)/(p-k) \rceil$ *as in* (39). *Then,* $\widetilde{A}_{n,[p,k]}(1) = n T_{n-\nu}(\kappa_{[p,k]})$.

Proof. By (34) and (40), for all $r, R = 1, \ldots, n - \nu$ and $q, Q = 1, \ldots, p - k$ we have

$$
\begin{aligned}
(\widetilde{A}_{n,[p,k]}(1))_{(p-k)(r-1)+q,(p-k)(R-1)+Q} &= \int_0^1 B'_{k+1+(p-k)(R-1)+Q,[p,k]}(x) B'_{k+1+(p-k)(r-1)+q,[p,k]}(x) \mathrm{d}x \\
&= \int_\mathbb{R} B'_{k+1+(p-k)(R-1)+Q,[p,k]}(x) B'_{k+1+(p-k)(r-1)+q,[p,k]}(x) \mathrm{d}x \\
&= n^2 \int_\mathbb{R} \beta'_{Q,[p,k]}(nx - R + 1) \beta'_{q,[p,k]}(nx - r + 1) \mathrm{d}x \\
&= n \int_\mathbb{R} \beta'_{Q,[p,k]}(y) \beta'_{q,[p,k]}(y - r + R) \mathrm{d}y
\end{aligned}
$$

and

$$
(T_{n-\nu}(\kappa_{[p,k]}))_{(p-k)(r-1)+q,(p-k)(R-1)+Q} = (K^{[r-R]}_{[p,k]})_{q,Q} = \int_\mathbb{R} \beta'_{Q,[p,k]}(y) \beta'_{q,[p,k]}(y - r + R) \mathrm{d}y,
$$

which completes the proof. $\square$

Theorem 4. *Let* $a \in L^1([0,1])$, $p \geq 1$ *and* $0 \leq k \leq p - 1$. *Then,* $\{n^{-1} A_{n,[p,k]}(a)\}_n \sim_{\sigma,\lambda} a(x) \kappa_{[p,k]}(\theta)$.

Proof. The proof consists of four steps. Throughout this proof, we use the following notation.

- $\nu = \lceil (k+1)/(p-k) \rceil$ as in (39).
- For every square matrix A of size $n(p - k) + k - 1$, we denote by $\widetilde{A}$ the principal submatrix of A corresponding to the row and column indices $i, j = k + 1, \ldots, k + (n - \nu)(p - k)$.
- $P_{n,[p,k]}$ is the $(n(p - k) + k - 1) \times (n - \nu)(p - k)$ matrix having $I_{(n-\nu)(p-k)}$ as the principal submatrix corresponding to the row and column indices $i, j = k + 1, \ldots, k + (n - \nu)(p - k)$ and zeros elsewhere. Note that $P^T_{n,[p,k]} P_{n,[p,k]} = I_{(n-\nu)(p-k)}$ and $P^T_{n,[p,k]} A P_{n,[p,k]} = \widetilde{A}$ for every square matrix A of size $n(p - k) + k - 1$.

Step 1. Consider the linear operator $A_{n,[p,k]}(\cdot) : L^1([0,1]) \rightarrow \mathbb{R}^{(n(p-k)+k-1) \times (n(p-k)+k-1)}$,

$$
A_{n,[p,k]}(g) = \left[\int_0^1 g(x) B'_{j+1,[p,k]}(x) B'_{i+1,[p,k]}(x) \mathrm{d}x \right]_{i,j=1}^{n(p-k)+k-1}.
$$

The next three steps are devoted to show that

$$
\{P^T_{n,[p,k]}(n^{-1} A_{n,[p,k]}(g)) P_{n,[p,k]}\}_n = \{n^{-1} \widetilde{A}_{n,[p,k]}(g)\}_n \sim_{\text{GLT}} g(x) \kappa_{[p,k]}(\theta), \qquad \forall g \in L^1([0,1]). \quad (45)
$$

Once this is done, the theorem is proven. Indeed, from (45), we immediately obtain the relation $\{P^T_{n,[p,k]}(n^{-1} A_{n,[p,k]}(a)) P_{n,[p,k]}\}_n \sim_{\text{GLT}} a(x) \kappa_{[p,k]}(\theta)$. We infer that $\{P^T_{n,[p,k]}(n^{-1} A_{n,[p,k]}(a)) P_{n,[p,k]}\}_n \sim_{\sigma,\lambda} a(x) \kappa_{[p,k]}(\theta)$ by **GLT 1** and $\{n^{-1} A_{n,[p,k]}(a)\}_n \sim_{\sigma,\lambda} a(x) \kappa_{[p,k]}(\theta)$ by Theorem 1.

Step 2. We first prove (45) in the constant-coefficient case where $g(x) = 1$ identically. In this case, by Lemma 2, we have $n^{-1} \widetilde{A}_{n,[p,k]}(1) = T_{n-\nu}(\kappa_{[p,k]})$. Hence, the desired relation $\{n^{-1} \widetilde{A}_{n,[p,k]}(1)\}_n \sim_{\text{GLT}} \kappa_{[p,k]}(\theta)$ follows from **GLT 2**.

Step 3. Now we prove (45) in the case where $g \in C([0,1])$. Let

$$
Z_{n,[p,k]}(g) = n^{-1} \widetilde{A}_{n,[p,k]}(g) - n^{-1} D_{n-\nu}(g I_{p-k}) \widetilde{A}_{n,[p,k]}(1).
$$

By (33), (34) and (38), for all $r, R = 1, \ldots, n - \nu$ and $q, Q = 1, \ldots, p - k$, we have

$$
\begin{aligned}
&|(n Z_{n,[p,k]}(g))_{(p-k)(r-1)+q,(p-k)(R-1)+Q}| \\
&= \left| (\widetilde{A}_{n,[p,k]}(g))_{(p-k)(r-1)+q,(p-k)(R-1)+Q} - (D_{n-\nu}(g I_{p-k}) \widetilde{A}_{n,[p,k]}(1))_{(p-k)(r-1)+q,(p-k)(R-1)+Q} \right| \\
&= \left| \int_0^1 \left[g(x) - g\left(\frac{r}{n-\nu} \right) \right] B'_{k+1+(p-k)(R-1)+Q,[p,k]}(x) B'_{k+1+(p-k)(r-1)+q,[p,k]}(x) \mathrm{d}x \right| \\
&= \left| \int_{\tau_{k+1+(p-k)(r-1)+q}}^{\tau_{k+1+(p-k)(r-1)+q+p+1}} \left[g(x) - g\left(\frac{r}{n-\nu} \right) \right] B'_{k+1+(p-k)(R-1)+Q,[p,k]}(x) B'_{k+1+(p-k)(r-1)+q,[p,k]}(x) \mathrm{d}x \right| \\
&\le c_p^2 n^2 \int_{(r-1)/n}^{(r+p)/n} \left| g(x) - g\left(\frac{r}{n-\nu} \right) \right| \mathrm{d}x \le c_p^2 (p+1) n \omega_g \left(\frac{\nu + p}{n} \right),
\end{aligned}
$$

where $\omega_g(\cdot)$ is the modulus of continuity of g and the last inequality is justified by the fact that the distance of the point $r/(n-\nu)$ from the interval $[(r-1)/n, (r+p)/n]$ is not larger than $(\nu + p)/n$. It follows that each entry of $Z_{n,[p,k]}(g)$ is bounded in modulus by $C_p \omega_g(1/n)$, where C_p is a constant depending only on p. Moreover, by (34), the matrix $Z_{n,[p,k]}(g)$ is banded with bandwidth bounded by a constant w_p depending only on p. Thus, by (30), $\|Z_{n,[p,k]}(g)\| \le w_p C_p \omega_g(1/n) \to 0$ as $n \to \infty$, and so $\{Z_{n,[p,k]}(g)\}_n$ is zero-distributed by Proposition 1. Since

$$
n^{-1} \widetilde{A}_{n,[p,k]}(g) = n^{-1} D_{n-\nu}(g I_{p-k}) \widetilde{A}_{n,[p,k]}(1) + Z_{n,[p,k]}(g),
$$

we conclude that $\{n^{-1} \widetilde{A}_{n,[p,k]}(g)\}_n \sim_{\mathrm{GLT}} g(x) \kappa_{[p,k]}(\theta)$ by **GLT 2**, **GLT 3** and Step 2.

Step 4. Finally, we prove (45) in the general case where $g \in L^1([0,1])$. By the density of $C([0,1])$ in $L^1([0,1])$, there exist functions $g_m \in C([0,1])$ such that $g_m \to g$ in $L^1([0,1])$. By Step 3,

$$
\{n^{-1} \widetilde{A}_{n,[p,k]}(g_m)\}_n \sim_{\mathrm{GLT}} g_m(x) \kappa_{[p,k]}(\theta). \tag{46}
$$

Moreover,

$$
g_m(x) \kappa_{[p,k]}(\theta) \to g(x) \kappa_{[p,k]}(\theta) \quad \text{in measure.} \tag{47}
$$

We show that

$$
\{n^{-1} \widetilde{A}_{n,[p,k]}(g_m)\}_n \xrightarrow{\text{a.c.s.}} \{n^{-1} \widetilde{A}_{n,[p,k]}(g)\}_n. \tag{48}
$$

Once this is done, the thesis (45) follows immediately from **GLT 4**. To prove (48), we recall that

$$
\|X\|_1 \le \sum_{i,j=1}^N |x_{ij}|, \qquad X \in \mathbb{C}^{N \times N}; \tag{49}
$$

see, e.g., ([13] Section 2.4.3). By (38), we obtain

$$
\begin{aligned}
\|\widetilde{A}_{n,[p,k]}(g) - \widetilde{A}_{n,[p,k]}(g_m)\|_1 &\le \sum_{i,j=1}^{n(p-k)+k-1} \left| \int_0^1 [g(x) - g_m(x)] B'_{j+1,[p,k]}(x) B'_{i+1,[p,k]}(x) \mathrm{d}x \right| \\
&\le \int_0^1 |g(x) - g_m(x)| \sum_{i,j=1}^{n(p-k)+k-1} |B'_{j+1,[p,k]}(x)| \, |B'_{i+1,[p,k]}(x)| \mathrm{d}x \\
&\le c_p^2 n^2 \|g - g_m\|_{L^1}.
\end{aligned}
$$

Thus, the a.c.s. convergence (48) follows from Proposition 2. □

Remark 4. *By following step by step the proof of Theorem 4, we can give an alternative (much simpler) proof of ([36] Theorem A.6) based on the theory of block GLT sequences.*

5. The Theory of Multilevel Block GLT Sequences

As illustrated in Sections 3 and 4, the theory of block GLT sequences allows the computation of the singular value and eigenvalue distribution of block structured matrices arising from the discretization of univariate DEs. In order to cope with multivariate DEs, i.e., PDEs, we need the multivariate version of the theory of block GLT sequences, also known as the theory of multilevel block GLT sequences. The present section is devoted to a careful presentation of this theory, which is obtained by combining the results of [34] with the necessary technicalities for tackling multidimensional problems [14].

Multi-Index Notation. The multi-index notation is an essential tool for dealing with sequences of matrices arising from the discretization of PDEs. A multi-index $i \in \mathbb{Z}^d$, also called a d-index, is simply a (row) vector in $\mathbb{Z}^d$; its components are denoted by $i_1, \ldots, i_d$.

- $\mathbf{0}, \mathbf{1}, \mathbf{2}, \ldots$ are the vectors of all zeros, all ones, all twos, etc. (their size will be clear from the context).
- For any d-index m, we set $N(m) = \prod_{j=1}^{d} m_j$ and we write $m \to \infty$ to indicate that $\min(m) \to \infty$.
- If h, k are d-indices, $h \le k$ means that $h_r \le k_r$ for all $r = 1, \ldots, d$.
- If h, k are d-indices such that $h \le k$, the multi-index range $h, \ldots, k$ is the set $\{j \in \mathbb{Z}^d : h \le j \le k\}$. We assume for this set the standard lexicographic ordering:

$$\left[\cdots \left[\left[(j_1, \ldots, j_d) \right]_{j_d = h_d, \ldots, k_d} \right]_{j_{d-1} = h_{d-1}, \ldots, k_{d-1}} \cdots \right]_{j_1 = h_1, \ldots, k_1}. \tag{50}$$

For instance, in the case $d = 2$, the ordering is the following: (h_1, h_2), $(h_1, h_2 + 1)$, $\ldots$, (h_1, k_2), $(h_1 + 1, h_2)$, $(h_1 + 1, h_2 + 1)$, $\ldots$, $(h_1 + 1, k_2)$, $\ldots \ldots \ldots$, (k_1, h_2), $(k_1, h_2 + 1)$, $\ldots$, (k_1, k_2).
- When a d-index j varies over a multi-index range $h, \ldots, k$ (this is sometimes written as $j = h, \ldots, k$), it is understood that j varies from h to k following the specific ordering (50). For instance, if $m \in \mathbb{N}^d$ and if we write $\mathbf{x} = [x_i]_{i=1}^{m}$, then $\mathbf{x}$ is a vector of size $N(m)$ whose components x_i, $i = 1, \ldots, m$, are ordered in accordance with (50): the first component is $x_1 = x_{(1, \ldots, 1, 1)}$, the second component is $x_{(1, \ldots, 1, 2)}$, and so on until the last component, which is $x_m = x_{(m_1, \ldots, m_d)}$. Similarly, if $X = [x_{ij}]_{i,j=1}^{m}$, then X is a $N(m) \times N(m)$ matrix whose components are indexed by two d-indices i, j, both varying from $\mathbf{1}$ to m according to the lexicographic ordering (50).
- Given $h, k \in \mathbb{Z}^d$ with $h \le k$, the notation $\sum_{j=h}^{k}$ indicates the summation over all j in $h, \ldots, k$.
- Operations involving d-indices that have no meaning in the vector space $\mathbb{Z}^d$ must be interpreted in the componentwise sense. For instance, $ij = (i_1 j_1, \ldots, i_d j_d)$, $i/j = (i_1/j_1, \ldots, i_d/j_d)$, etc.

Multilevel Block Matrix-Sequences. Given $d, s \ge 1$, a d-level s-block matrix-sequence (or simply a matrix-sequence if d and s can be inferred from the context or we do not need/want to specify them) is a sequence of matrices of the form $\{A_n\}_n$, where:

- n varies in some infinite subset of $\mathbb{N}$;
- $n = n(n)$ is a d-index in $\mathbb{N}^d$ which depends on n and satisfies $n \to \infty$ as $n \to \infty$;
- A_n is a square matrix of size $N(n)s$.

Multilevel Block Toeplitz Matrices. Given a function $f : [-\pi, \pi]^d \to \mathbb{C}^{s \times s}$ in $L^1([-\pi, \pi]^d)$, its Fourier coefficients are denoted by

$$f_k = \frac{1}{(2\pi)^d} \int_{[-\pi,\pi]^d} f(\boldsymbol{\theta}) e^{-ik \cdot \boldsymbol{\theta}} d\boldsymbol{\theta} \in \mathbb{C}^{s \times s}, \qquad k \in \mathbb{Z}^d,$$

where $k \cdot \boldsymbol{\theta} = k_1 \theta_1 + \ldots + k_d \theta_d$ and the integrals are computed componentwise. For $n \in \mathbb{N}^d$, the nth multilevel block Toeplitz matrix generated by f is defined as

$$T_n(f) = [f_{i-j}]_{i,j=1}^{n} \in \mathbb{C}^{N(n)s \times N(n)s}.$$

It is not difficult to see that the map $f \mapsto T_n(f)$ is linear. Moreover, it can be shown that

$$T_n(f^*) = (T_n(f))^*, \tag{51}$$

where the transpose conjugate function f^* is defined by $f^*(\theta) = (f(\theta))^*$; in particular, all the matrices $T_n(f)$ are Hermitian whenever f is Hermitian a.e. We also recall that, if $n \in \mathbb{N}^d$ and $f_1, f_2, \ldots, f_d : [-\pi, \pi] \to \mathbb{C}$ belong to $L^1([-\pi, \pi])$, then

$$T_{n_1}(f_1) \otimes T_{n_2}(f_2) \otimes \cdots \otimes T_{n_d}(f_d) = T_n(f), \tag{52}$$

where $f : [-\pi, \pi]^d \to \mathbb{C}$ is defined by $f(\theta) = f(\theta_1)f(\theta_2) \cdots f(\theta_d)$; see, e.g., ([14] Lemma 3.3).

Multilevel Block Diagonal Sampling Matrices. For $n \in \mathbb{N}^d$ and $a : [0,1]^d \to \mathbb{C}^{s \times s}$, we define the multilevel block diagonal sampling matrix $D_n(a)$ as the block diagonal matrix

$$D_n(a) = \operatorname*{diag}_{i=1,\ldots,n} a\left(\frac{i}{n}\right) \in \mathbb{C}^{N(n)s \times N(n)s}.$$

Multilevel Block GLT Sequences. Let $d, s \geq 1$ be fixed positive integers. A d-level s-block GLT sequence (or simply a GLT sequence if d and s can be inferred from the context or we do not need/want to specify them) is a special d-level s-block matrix-sequence $\{A_n\}_n$ equipped with a measurable function $\kappa : [0,1]^d \times [-\pi, \pi]^d \to \mathbb{C}^{s \times s}$, the so-called symbol. We use the notation $\{A_n\}_n \sim_{\mathrm{GLT}} \kappa$ to indicate that $\{A_n\}_n$ is a GLT sequence with symbol κ. The symbol of a GLT sequence is unique in the sense that if $\{A_n\}_n \sim_{\mathrm{GLT}} \kappa$ and $\{A_n\}_n \sim_{\mathrm{GLT}} \varsigma$ then $\kappa = \varsigma$ a.e. in $[0,1]^d \times [-\pi, \pi]^d$. The main properties of d-level s-block GLT sequences are listed below.

GLT 1. If $\{A_n\}_n \sim_{\mathrm{GLT}} \kappa$ then $\{A_n\}_n \sim_\sigma \kappa$. If, moreover, each A_n is Hermitian then $\{A_n\}_n \sim_\lambda \kappa$.

GLT 2. We have:
- $\{T_n(f)\}_n \sim_{\mathrm{GLT}} \kappa(\mathbf{x}, \boldsymbol{\theta}) = f(\boldsymbol{\theta})$ if $f : [-\pi, \pi]^d \to \mathbb{C}^{s \times s}$ is in $L^1([-\pi, \pi]^d)$;
- $\{D_n(a)\}_n \sim_{\mathrm{GLT}} \kappa(\mathbf{x}, \boldsymbol{\theta}) = a(\mathbf{x})$ if $a : [0,1]^d \to \mathbb{C}^{s \times s}$ is Riemann-integrable;
- $\{Z_n\}_n \sim_{\mathrm{GLT}} \kappa(\mathbf{x}, \boldsymbol{\theta}) = O_s$ if and only if $\{Z_n\}_n \sim_\sigma 0$.

GLT 3. If $\{A_n\}_n \sim_{\mathrm{GLT}} \kappa$ and $\{B_n\}_n \sim_{\mathrm{GLT}} \varsigma$ then:
- $\{A_n^*\}_n \sim_{\mathrm{GLT}} \kappa^*$;
- $\{\alpha A_n + \beta B_n\}_n \sim_{\mathrm{GLT}} \alpha\kappa + \beta\varsigma$ for all $\alpha, \beta \in \mathbb{C}$;
- $\{A_n B_n\}_n \sim_{\mathrm{GLT}} \kappa\varsigma$;
- $\{A_n^\dagger\}_n \sim_{\mathrm{GLT}} \kappa^{-1}$ provided that κ is invertible a.e.

GLT 4. $\{A_n\}_n \sim_{\mathrm{GLT}} \kappa$ if and only if there exist GLT sequences $\{B_{n,m}\}_n \sim_{\mathrm{GLT}} \kappa_m$ such that $\{B_{n,m}\}_n \overset{\text{a.c.s.}}{\longrightarrow} \{A_n\}_n$ and $\kappa_m \to \kappa$ in measure.

6. Discretizations of Systems of PDEs: The General GLT Approach

In this section, we outline the main ideas of a multidimensional block GLT analysis for general discretizations of PDE systems. What we are going to present here is then a generalization of what is shown in Section 3. We begin by proving a series of auxiliary results. In the following, given $n \in \mathbb{N}^d$ and $s \geq 1$, we denote by $\Pi_{n,s}$ the permutation matrix given by

$$\Pi_{n,s} = \begin{bmatrix} I_s \otimes \mathbf{e}_1^T \\ I_s \otimes \mathbf{e}_2^T \\ \vdots \\ I_s \otimes \mathbf{e}_n^T \end{bmatrix} = \sum_{k=1}^n \mathbf{e}_k \otimes I_s \otimes \mathbf{e}_k^T, \tag{53}$$

where $\mathbf{e}_i$, $i = 1, \ldots, n$, are the vectors of the canonical basis of $\mathbb{R}^{N(n)}$, which, for convenience, are indexed by a d-index $i = 1, \ldots, n$ instead of a linear index $i = 1, \ldots, N(n)$. Note that $\Pi_{n,2}$ coincides with the matrix Π_n in (21).

Lemma 3. *Let $n \in \mathbb{N}^d$, let $f_{ij} : [-\pi, \pi]^d \to \mathbb{C}$ be in $L^1([-\pi, \pi]^d)$ for $i, j = 1, \ldots, s$, and set $f = [f_{ij}]_{i,j=1}^s$. The block matrix $T_n = [T_n(f_{ij})]_{i,j=1}^s$ is similar via the permutation (53) to the multilevel block Toeplitz matrix $T_n(f)$, that is, $\Pi_{n,s} T_n \Pi_{n,s}^T = T_n(f)$.*

Proof. Let E_{ij} be the $s \times s$ matrix having 1 in position (i, j) and 0 elsewhere. Since $T_n = \sum_{i,j=1}^s E_{ij} \otimes T_n(f_{ij})$ and $T_n(f) = \sum_{i,j=1}^s T_n(f_{ij} E_{ij})$ by the linearity of the map $T_n(\cdot)$, it is enough to show that

$$\Pi_{n,s}(E \otimes T_n(g))\Pi_{n,s}^T = T_n(gE), \qquad \forall g \in L^1([-\pi, \pi]^d), \qquad \forall E \in \mathbb{C}^{s \times s}.$$

By (6) and (7),

$$\Pi_{n,s}(E \otimes T_n(g))\Pi_{n,s}^T = \left[\sum_{k=1}^n \mathbf{e}_k \otimes I_s \otimes \mathbf{e}_k^T\right](E \otimes T_n(g))\left[\sum_{\ell=1}^n \mathbf{e}_\ell^T \otimes I_s \otimes \mathbf{e}_\ell\right]$$

$$= \sum_{k,\ell=1}^n (\mathbf{e}_k \otimes I_s \otimes \mathbf{e}_k^T)(E \otimes T_n(g))(\mathbf{e}_\ell^T \otimes I_s \otimes \mathbf{e}_\ell)$$

$$= \sum_{k,\ell=1}^n \mathbf{e}_k \mathbf{e}_\ell^T \otimes E \otimes \mathbf{e}_k^T T_n(g)\mathbf{e}_\ell = \sum_{k,\ell=1}^n \mathbf{e}_k \mathbf{e}_\ell^T \otimes (T_n(g))_{k\ell} E = T_n(gE),$$

as required. □

Lemma 4. *Let $n \in \mathbb{N}^d$, let $a_{ij} : [0, 1]^d \to \mathbb{C}$ for $i, j = 1, \ldots, s$, and set $a = [a_{ij}]_{i,j=1}^s$. The block matrix $D_n = [D_n(a_{ij})]_{i,j=1}^s$ is similar via the permutation (53) to the multilevel block diagonal sampling matrix $D_n(a)$, that is, $\Pi_{n,s} D_n \Pi_{n,s}^T = D_n(a)$.*

Proof. With obvious adaptations, it is the same as the proof of Lemma 3. □

We recall that a d-variate trigonometric polynomial is a finite linear combination of the d-variate Fourier frequencies $e^{ik \cdot \theta}$, $k \in \mathbb{Z}^d$.

Theorem 5. *For $i, j = 1, \ldots, s$, let $\{A_{n,ij}\}_n$ be a d-level 1-block GLT sequence with symbol $\kappa_{ij} : [0, 1]^d \times [-\pi, \pi]^d \to \mathbb{C}$. Set $A_n = [A_{n,ij}]_{i,j=1}^s$ and $\kappa = [\kappa_{ij}]_{i,j=1}^s$. Then, the matrix-sequence $\{\Pi_{n,s} A_n \Pi_{n,s}^T\}_n$ is a d-level s-block GLT sequence with symbol κ.*

Proof. The proof consists of the following two steps.

Step 1. We first prove the theorem under the additional assumption that $A_{n,ij}$ is of the form

$$A_{n,ij} = \sum_{\ell=1}^{L_{ij}} D_n(a_{\ell,ij}) T_n(f_{\ell,ij}), \qquad (54)$$

where $L_{ij} \in \mathbb{N}$, $a_{\ell,ij} : [0, 1]^d \to \mathbb{C}$ is Riemann-integrable, and $f_{\ell,ij} : [-\pi, \pi]^d \to \mathbb{C}$ belongs to $L^1([-\pi, \pi]^d)$. Note that the symbol of $\{A_{n,ij}\}_n$ is

$$\kappa_{ij}(\mathbf{x}, \boldsymbol{\theta}) = \sum_{\ell=1}^{L_{ij}} a_{\ell,ij}(\mathbf{x}) f_{\ell,ij}(\boldsymbol{\theta}).$$

By setting $L = \max_{i,j=1,\ldots,s} L_{ij}$ and by adding zero matrices of the form $D_n(0)T_n(0)$ in the summation (54) whenever $L_{ij} < L$, we can assume, without loss of generality, that

$$A_{n,ij} = \sum_{\ell=1}^{L} D_n(a_{\ell,ij})T_n(f_{\ell,ij}),$$

$$\kappa_{ij}(\mathbf{x}, \boldsymbol{\theta}) = \sum_{\ell=1}^{L} a_{\ell,ij}(\mathbf{x})f_{\ell,ij}(\boldsymbol{\theta}),$$

with L independent of i, j. Let E_{ij} be the $s \times s$ matrix having 1 in position (i,j) and 0 elsewhere. Then,

$$\begin{aligned}
\Pi_{n,s} A_n \Pi_{n,s}^T &= \sum_{\ell=1}^{L} \Pi_{n,s} \left[D_n(a_{\ell,ij})T_n(f_{\ell,ij}) \right]_{i,j=1}^{s} \Pi_{n,s}^T \\
&= \sum_{\ell=1}^{L} \Pi_{n,s} \left[\sum_{i,j=1}^{s} (E_{ij} \otimes D_n(a_{\ell,ij}))(E_{ij} \otimes T_n(f_{\ell,ij})) \right] \Pi_{n,s}^T \\
&= \sum_{\ell=1}^{L} \sum_{i,j=1}^{s} \Pi_{n,s}(E_{ij} \otimes D_n(a_{\ell,ij}))\Pi_{n,s}^T \Pi_{n,s}(E_{ij} \otimes T_n(f_{\ell,ij}))\Pi_{n,s}^T.
\end{aligned}$$

By Lemmas 3 and 4,

$$\Pi_{n,s}(E_{ij} \otimes D_n(a_{\ell,ij}))\Pi_{n,s}^T = D_n(a_{\ell,ij}E_{ij}),$$
$$\Pi_{n,s}(E_{ij} \otimes T_n(f_{\ell,ij}))\Pi_{n,s}^T = T_n(f_{\ell,ij}E_{ij}).$$

It follows that

$$\Pi_{n,s} A_n \Pi_{n,s}^T = \sum_{\ell=1}^{L} \sum_{i,j=1}^{s} D_n(a_{\ell,ij}E_{ij})T_n(f_{\ell,ij}E_{ij}).$$

Thus, by **GLT 2** and **GLT 3**, $\{\Pi_{n,s} A_n \Pi_{n,s}^T\}_n$ is a d-level s-block GLT sequence with symbol

$$\kappa(\mathbf{x}, \boldsymbol{\theta}) = \sum_{\ell=1}^{L} \sum_{i,j=1}^{s} a_{\ell,ij}(\mathbf{x})f_{\ell,ij}(\boldsymbol{\theta})E_{ij} = [\kappa_{ij}(\mathbf{x}, \boldsymbol{\theta})]_{i,j=1}^{s}.$$

Step 2. We now prove the theorem in its full generality. Since $\{A_{n,ij}\}_n \sim_{\text{GLT}} \kappa_{ij}$, by ([14] Theorem 5.6) there exist functions $a_{\ell,ij}^{(m)}$, $f_{\ell,ij}^{(m)}$, $\ell = 1, \ldots, L_{ij}^{(m)}$, such that

- $a_{\ell,ij}^{(m)} \in C^{\infty}([0,1]^d)$ and $f_{\ell,ij}^{(m)}$ is a d-variate trigonometric polynomial,
- $\kappa_{ij}^{(m)}(\mathbf{x}, \boldsymbol{\theta}) = \sum_{\ell=1}^{L_{ij}^{(m)}} a_{\ell,ij}^{(m)}(\mathbf{x})f_{\ell,ij}^{(m)}(\boldsymbol{\theta}) \to \kappa_{ij}(\mathbf{x}, \boldsymbol{\theta})$ a.e.;
- $\{A_{n,ij}^{(m)} = \sum_{\ell=1}^{L_{ij}^{(m)}} D_n(a_{\ell,ij}^{(m)})T_n(f_{\ell,ij}^{(m)})\}_n \xrightarrow{\text{a.c.s.}} \{A_{n,ij}\}_n$.

Set $A_n^{(m)} = [A_{n,ij}^{(m)}]_{i,j=1}^{s}$ and $\kappa^{(m)}(\mathbf{x}, \boldsymbol{\theta}) = [\kappa_{ij}^{(m)}(\mathbf{x}, \boldsymbol{\theta})]_{i,j=1}^{s}$. We have:

- $\{\Pi_{n,s} A_n^{(m)} \Pi_{n,s}^T\}_n \sim_{\text{GLT}} \kappa^{(m)}$ by Step 1;
- $\kappa^{(m)} \to \kappa$ a.e. (and hence also in measure);
- $\{\Pi_{n,s} A_n^{(m)} \Pi_{n,s}^T\}_n \xrightarrow{\text{a.c.s.}} \{\Pi_{n,s} A_n \Pi_{n,s}^T\}_n$ because $\{A_n^{(m)}\}_n \xrightarrow{\text{a.c.s.}} \{A_n\}_n$ by Lemma 1.

We conclude that $\{\Pi_{n,s} A_n \Pi_{n,s}^T\}_n \sim_{\text{GLT}} \kappa$ by **GLT 4**. $\quad\square$

Now, suppose we have a system of linear PDEs of the form

$$
\begin{cases}
\sum_{j=1}^{s} \mathscr{L}_{1j} u_j(\mathbf{x}) = f_1(\mathbf{x}), \\
\sum_{j=1}^{s} \mathscr{L}_{2j} u_j(\mathbf{x}) = f_2(\mathbf{x}), \\
\quad \vdots \\
\sum_{j=1}^{s} \mathscr{L}_{sj} u_j(\mathbf{x}) = f_s(\mathbf{x}),
\end{cases}
\tag{55}
$$

where $\mathbf{x} \in (0,1)^d$. The matrices A_n resulting from any standard discretization of (55) are parameterized by a d-index $\boldsymbol{n} = (n_1, \ldots, n_d)$, where n_i is related to the discretization step h_i in the ith direction, and $n_i \to \infty$ if and only if $h_i \to 0$ (usually, $h_i \approx 1/n_i$). By choosing each n_i as a function of a unique discretization parameter $n \in \mathbb{N}$, as it normally happens in practice where the most natural choice is $n_i = n$ for all $i = 1, \ldots, d$, we see that $\boldsymbol{n} = \boldsymbol{n}(n)$ and, consequently, $\{A_n\}_n$ is a (d-level) matrix-sequence. Moreover, it turns out that, after a suitable normalization that we ignore in this discussion—the normalization we are talking about is the analog of the normalization that we have seen in Section 3, which allowed us to pass from the matrix A_n in (13) to the matrix B_n in (15)—, A_n has the following block structure:

$$
A_n = [A_{n,ij}]_{i,j=1}^{s},
$$

where $A_{n,ij}$ is the (normalized) matrix arising from the discretization of the differential operator $\mathscr{L}_{ij}$. In the simplest case where the operators $\mathscr{L}_{ij}$ have constant coefficients and we use equispaced grids in each direction, the matrix $A_{n,ij}$ takes the form

$$
A_{n,ij} = T_n(f_{ij}) + Z_{n,ij},
$$

where f_{ij} is a d-variate trigonometric polynomial, while the perturbation $Z_{n,ij}$ is usually a low-rank correction due to boundary conditions and, in any case, we have $\{Z_{n,ij}\}_n \sim_\sigma 0$. Hence,

$$
\{A_{n,ij}\}_n \sim_{\mathrm{GLT}} f_{ij}
$$

by **GLT 2** and **GLT 3**, and it follows from Theorem 5 that

$$
\{\Pi_{n,s} A_n \Pi_{n,s}^T\}_n \sim_{\mathrm{GLT}} [f_{ij}]_{i,j=1}^{s}.
$$

In the case where the operators $\mathscr{L}_{ij}$ have variable coefficients, the matrix $A_{n,ij}$ usually takes the form

$$
A_{n,ij} = \sum_{\ell=1}^{L_{ij}} D_n(a_{\ell,ij}) T_n(f_{\ell,ij}) + Z_{n,ij},
$$

where $L_{ij} \in \mathbb{N}$, $f_{\ell,ij}$ is a d-variate trigonometric polynomial, $\{Z_{n,ij}\}_n \sim_\sigma 0$, and the functions $a_{\ell,ij} : [0,1]^d \to \mathbb{R}$, $\ell = 1, \ldots, L_{ij}$, are related to the coefficients of $\mathscr{L}_{ij}$ (for example, in Section 3, while proving (22), we have seen that $K_n(a_{11})$, which plays there the same role as the matrix $A_{n,11}$ here, is equal to $D_n(a_{11}) T_n(2 - 2\cos\theta) + Z_n$ for some zero-distributed sequence $\{Z_n\}_n$). Hence,

$$
\{A_{n,ij}\}_n \sim_{\mathrm{GLT}} \kappa_{ij}(\mathbf{x}, \boldsymbol{\theta}) = \sum_{\ell=1}^{L_{ij}} a_{\ell,ij}(\mathbf{x}) f_{\ell,ij}(\boldsymbol{\theta})
$$

by **GLT 2** and **GLT 3**, and it follows from Theorem 5 that

$$
\{\Pi_{n,s} A_n \Pi_{n,s}^T\}_n \sim_{\mathrm{GLT}} [\kappa_{ij}]_{i,j=1}^{s}.
$$

7. B-Spline IgA Discretization of a Variational Problem for the Curl–Curl Operator

For any function $\mathbf{u}(x_1, x_2) = [u_1(x_1, x_2), u_2(x_1, x_2)]^T$, defined over some open set $\Omega \subseteq \mathbb{R}^2$ and taking values in $\mathbb{R}^2$, the curl operator is formally defined as follows:

$$(\nabla \times \mathbf{u})(x_1, x_2) = \frac{\partial u_2}{\partial x_1}(x_1, x_2) - \frac{\partial u_1}{\partial x_2}(x_1, x_2), \qquad (x_1, x_2) \in \Omega.$$

Clearly, this definition has meaning when the components u_1, u_2 belong to $H^1(\Omega)$, so that their partial derivatives exist in the Sobolev sense. Now, let $\Omega = (0, 1)^2$, set

$$(L^2(\Omega))^2 = \{\mathbf{u} : \Omega \to \mathbb{R}^2 : u_1, u_2 \in L^2(\Omega)\},$$
$$H(\text{curl}, \Omega) = \{\mathbf{u} \in (L^2(\Omega))^2 : \nabla \times \mathbf{u} \text{ exists in the Sobolev sense}, \ \nabla \times \mathbf{u} \in L^2(\Omega)\},$$

and consider the following variational problem: find $\mathbf{u} \in H(\text{curl}, \Omega)$ such that

$$(\nabla \times \mathbf{u}, \nabla \times \mathbf{v}) = (\mathbf{f}, \mathbf{v}), \qquad \forall \mathbf{v} \in H(\text{curl}, \Omega), \tag{56}$$

where $\mathbf{f}(x_1, x_2) = [f_1(x_1, x_2), f_2(x_1, x_2)]^T$ is a vector field in $(L^2(\Omega))^2$ and

$$(\nabla \times \mathbf{u}, \nabla \times \mathbf{v}) = \int_\Omega (\nabla \times \mathbf{u})(x_1, x_2) \, (\nabla \times \mathbf{v})(x_1, x_2) \, dx_1 dx_2,$$
$$(\mathbf{f}, \mathbf{v}) = \int_\Omega [f_1(x_1, x_2)v_1(x_1, x_2) + f_2(x_1, x_2)v_2(x_1, x_2)] \, dx_1 dx_2.$$

Variational problems of the form of (56) arise in important applications, such as time harmonic Maxwell's equations and magnetostatic problems. In this section, we consider a so-called compatible B-spline IgA discretization of (56); see [43] for details. We show that the corresponding sequence of discretization matrices enjoys a spectral distribution described by a 2×2 matrix-valued function whose determinant is zero everywhere. The results of this section have already been obtained in [38], but the derivation presented here is entirely based on the theory of multilevel block GLT sequences and turns out to be simpler and more lucid than that in [38]. For simplicity, throughout this section, the B-splines $B_{i,[p,p-1]}$, $i = 1, \ldots, n + p$, and the associated reference B-spline $\beta_{1,[p,p-1]}$, are denoted by $B_{i,[p]}$, $i = 1, \ldots, n + p$, and $\beta_{[p]}$, respectively. The function $\beta_{[p]}$ is the so-called cardinal B-spline of degree p over the knot sequence $\{0, 1, \ldots, p + 1\}$. In view of the following, we recall from [42] and ([23] Lemma 4) that the cardinal B-spline $\beta_{[q]}$ is defined for all degrees $q \geq 0$, belongs to $C^{q-1}(\mathbb{R})$, and satisfies the following properties:

$$\text{supp}(\beta_{[q]}) = [0, q + 1] \tag{57}$$

for $q \geq 1$,

$$\beta'_{[q]}(t) = \beta_{[q-1]}(t) - \beta_{[q-1]}(t - 1), \tag{58}$$

for $t \in \mathbb{R}$ and $q \geq 1$, and

$$\int_\mathbb{R} \beta^{(r_1)}_{[q_1]}(\tau)\beta^{(r_2)}_{[q_2]}(\tau + t)d\tau = (-1)^{r_1}\beta^{(r_1+r_2)}_{[q_1+q_2+1]}(q_1 + 1 + t) = (-1)^{r_2}\beta^{(r_1+r_2)}_{[q_1+q_2+1]}(q_2 + 1 - t) \tag{59}$$

for $t \in \mathbb{R}$ and $q_1, q_2, r_1, r_2 \geq 0$. Moreover, property (40) in the case $k = p - 1$ simplifies to

$$B_{i,[p]}(x) = \beta_{[p]}(nx - i + p + 1), \qquad i = p + 1, \ldots, n. \tag{60}$$

7.1. Compatible B-Spline IgA Discretization

Let $n = (n_1, n_2) \in \mathbb{N}^2$, let $p \geq 2$, and define the space

$$
\mathscr{V}_{n,[p]}(\mathrm{curl}, \Omega) = \mathrm{span} \left\{ \begin{bmatrix} B_{i_1,[p-1]}(x_1) B_{i_2,[p]}(x_2) \\ B_{j_1,[p]}(x_1) B_{j_2,[p-1]}(x_2) \end{bmatrix} : \begin{array}{ll} i_1 = 1, \ldots, n_1 + p - 1, & i_2 = 1, \ldots, n_2 + p, \\[2mm] j_1 = 1, \ldots, n_1 + p, & j_2 = 1, \ldots, n_2 + p - 1 \end{array} \right\}. \tag{61}
$$

Following a compatible B-spline approach [43], we look for an approximation of the solution in the space $\mathscr{V}_{n,[p]}(\mathrm{curl}, \Omega)$ by solving the following discrete problem: find $\mathbf{u}_{\mathscr{V}} \in \mathscr{V}_{n,[p]}(\mathrm{curl}, \Omega)$ such that

$$
(\nabla \times \mathbf{u}_{\mathscr{V}}, \nabla \times \mathbf{v}) = (\mathbf{f}, \mathbf{v}), \qquad \forall \mathbf{v} \in \mathscr{V}_{n,[p]}(\mathrm{curl}, \Omega).
$$

After choosing a suitable ordering on the basis functions of $\mathscr{V}_{n,[p]}(\mathrm{curl}, \Omega)$ displayed in (61), by linearity the computation of $\mathbf{u}_{\mathscr{V}}$ reduces to solving a linear system whose coefficient matrix is given by

$$
A_{n,[p]} = \begin{bmatrix} A_{n,[p],11} & A_{n,[p],12} \\ A_{n,[p],21} & A_{n,[p],22} \end{bmatrix} = \begin{bmatrix} M_{n_1,[p-1]} \otimes K_{n_2,[p]} & -H_{n_1,[p]} \otimes (H_{n_2,[p]})^T \\ -(H_{n_1,[p]})^T \otimes H_{n_2,[p]} & K_{n_1,[p]} \otimes M_{n_2,[p-1]} \end{bmatrix},
$$

where

$$
(M_{n,[p-1]})_{ij} = \int_0^1 B_{j,[p-1]}(x) B_{i,[p-1]}(x)\,\mathrm{d}x, \qquad i, j = 1, \ldots, n + p - 1,
$$

$$
(K_{n,[p]})_{ij} = \int_0^1 B'_{j,[p]}(x) B'_{i,[p]}(x)\,\mathrm{d}x, \qquad i, j = 1, \ldots, n + p,
$$

$$
(H_{n,[p]})_{ij} = \int_0^1 B'_{j,[p]}(x) B_{i,[p-1]}(x)\,\mathrm{d}x, \qquad i = 1, \ldots, n + p - 1, \qquad j = 1, \ldots, n + p.
$$

Note that $M_{n,[p-1]}$ is a square matrix of size $n + p - 1$, $K_{n,[p]}$ is a square matrix of size $n + p$, while $H_{n,[p]}$ is a rectangular matrix of size $(n + p - 1) \times (n + p)$.

7.2. GLT Analysis of the B-Spline IgA Discretization Matrices

In the main result of this section (Theorem 6), assuming that $n = n\nu$ for a fixed vector ν, we show that the spectral distribution of the sequence $\{A_{n,[p]}\}_n$ is described by a 2×2 matrix-valued function whose determinant is zero everywhere (Remark 5). To prove Theorem 6, some preliminary work is necessary. We first note that, in view of the application of Theorem 5, the matrix $A_{n,[p]}$ has an unpleasant feature: the anti-diagonal blocks $A_{n,[p],12}$ and $A_{n,[p],21}$ are not square and the square diagonal blocks $A_{n,[p],11}$ and $A_{n,[p],22}$ do not have the same size whenever $n_1 \neq n_2$. Let us then introduce the nicer matrix

$$
\tilde{A}_{n,[p]} = \begin{bmatrix} \tilde{A}_{n,[p],11} & \tilde{A}_{n,[p],12} \\ \tilde{A}_{n,[p],21} & \tilde{A}_{n,[p],22} \end{bmatrix} = \begin{bmatrix} \tilde{M}_{n_1,[p-1]} \otimes K_{n_2,[p]} & -\tilde{H}_{n_1,[p]} \otimes (\tilde{H}_{n_2,[p]})^T \\ -(\tilde{H}_{n_1,[p]})^T \otimes \tilde{H}_{n_2,[p]} & K_{n_1,[p]} \otimes \tilde{M}_{n_2,[p-1]} \end{bmatrix},
$$

where $\tilde{M}_{n,[p-1]}$ and $\tilde{H}_{n,[p]}$ are square matrices of size $n + p$ given by

$$
\tilde{M}_{n,[p-1]} = \left[\begin{array}{ccc|c} & & & 0 \\ & M_{n,[p-1]} & & \vdots \\ & & & 0 \\ \hline 0 & \cdots & 0 & 0 \end{array} \right], \qquad \tilde{H}_{n,[p]} = \left[\begin{array}{ccc} & H_{n,[p]} & \\ \hline 0 & \cdots & 0 \end{array} \right].
$$

Each block $\tilde{A}_{n,[p],ij}$ of the matrix $\tilde{A}_{n,[p]}$ is now a square block of size $(n_1 + p)(n_2 + p)$. Moreover,

$$M_{n,[p-1]} = P_{n,[p]} \tilde{M}_{n,[p-1]} (P_{n,[p]})^T, \qquad H_{n,[p]} = P_{n,[p]} \tilde{H}_{n,[p]},$$

where the matrix

$$P_{n,[p]} = \left[\begin{array}{c|c} I_{n+p-1} & \begin{matrix} 0 \\ \vdots \\ 0 \end{matrix} \end{array} \right]$$

satisfies $P_{n,[p]}(P_{n,[p]})^T = I_{n+p-1}$. By (6) and (7),

$$A_{n,[p],11} = (P_{n_1,[p]} \otimes I_{n_2+p}) \tilde{A}_{n,[p],11} (P_{n_1,[p]} \otimes I_{n_2+p})^T,$$
$$A_{n,[p],12} = (P_{n_1,[p]} \otimes I_{n_2+p}) \tilde{A}_{n,[p],12} (I_{n_1+p} \otimes P_{n_2,[p]})^T,$$
$$A_{n,[p],21} = (I_{n_1+p} \otimes P_{n_2,[p]}) \tilde{A}_{n,[p],21} (P_{n_1,[p]} \otimes I_{n_2+p})^T,$$
$$A_{n,[p],22} = (I_{n_1+p} \otimes P_{n_2,[p]}) \tilde{A}_{n,[p],22} (I_{n_1+p} \otimes P_{n_2,[p]})^T,$$

and so

$$A_{n,[p]} = P_{n,[p]} \tilde{A}_{n,[p]} (P_{n,[p]})^T, \qquad P_{n,[p]} = \left[\begin{matrix} P_{n_1,[p]} \otimes I_{n_2+p} & \\ & I_{n_1+p} \otimes P_{n_2,[p]} \end{matrix} \right]. \tag{62}$$

In view of the application of Theorem 1, we note that

$$P_{n,[p]} \in \mathbb{R}^{[(n_1+p-1)(n_2+p)+(n_1+p)(n_2+p-1)] \times 2(n_1+p)(n_2+p)}, \tag{63}$$

$$P_{n,[p]}(P_{n,[p]})^T = I_{(n_1+p-1)(n_2+p)+(n_1+p)(n_2+p-1)}, \tag{64}$$

$$\lim_{n\to\infty} \frac{(n_1+p-1)(n_2+p)+(n_1+p)(n_2+p-1)}{2(n_1+p)(n_2+p)} = \lim_{n\to\infty} \left[\frac{n_1+p-1}{2(n_1+p)} + \frac{n_2+p-1}{2(n_2+p)} \right] = 1. \tag{65}$$

Lemma 5. *Let $p \geq 2$ and $n \geq 1$. Then,*

$$n^{-1} K_{n,[p]} = T_{n+p}(f_p) + Q_{n,[p]}, \qquad \mathrm{rank}(Q_{n,[p]}) \leq 4p, \tag{66}$$
$$\tilde{H}_{n,[p]} = T_{n+p}(g_p) + R_{n,[p]}, \qquad \mathrm{rank}(R_{n,[p]}) \leq 4p, \tag{67}$$
$$n \tilde{M}_{n,[p-1]} = T_{n+p}(h_p) + S_{n,[p]}, \qquad \mathrm{rank}(S_{n,[p]}) \leq 4p, \tag{68}$$

where

$$f_p(\theta) = \sum_{k \in \mathbb{Z}} -\beta''_{[2p+1]}(p+1-k) e^{ik\theta}, \tag{69}$$

$$g_p(\theta) = \sum_{k \in \mathbb{Z}} -\beta'_{[2p]}(p-k) e^{ik\theta}, \tag{70}$$

$$h_p(\theta) = \sum_{k \in \mathbb{Z}} \beta_{[2p-1]}(p-k) e^{ik\theta}, \tag{71}$$

and we note that the three series are actually finite sums because of (57).

Proof. For every $i,j = p+1,\ldots,n$, since $[-i+p+1, n-i+p+1] \supseteq [0, p+1] = \mathrm{supp}(\beta_{[p]})$ and $[-i+p, n-i+p] \supseteq [0, p] = \mathrm{supp}(\beta_{[p-1]})$, by (59) and (60) we obtain

$$(K_{n,[p]})_{ij} = \int_0^1 B'_{j,[p]}(x) B'_{i,[p]}(x) dx = n^2 \int_0^1 \beta'_{[p]}(nx-j+p+1) \beta'_{[p]}(nx-i+p+1) dx$$
$$= n \int_{-i+p+1}^{n-i+p+1} \beta'_{[p]}(\tau+i-j) \beta'_{[p]}(\tau) d\tau = n \int_{\mathbb{R}} \beta'_{[p]}(\tau) \beta'_{[p]}(\tau+i-j) d\tau$$

$$= -n\beta''_{[2p+1]}(p+1+i-j) = -n\beta''_{[2p+1]}(p+1-i+j),$$

$$(\widetilde{H}_{n,[p]})_{ij} = \int_0^1 B'_{j,[p]}(x)B_{i,[p-1]}(x)\mathrm{d}x = n\int_0^1 \beta'_{[p]}(nx-j+p+1)\beta_{[p-1]}(nx-i+p)\mathrm{d}x$$

$$= \int_{-i+p}^{n-i+p} \beta'_{[p]}(\tau+i-j+1)\beta_{[p-1]}(\tau)\mathrm{d}\tau = \int_{\mathbb{R}} \beta_{[p-1]}(\tau)\beta'_{[p]}(\tau+i-j+1)\mathrm{d}\tau$$

$$= \beta'_{[2p]}(p+i-j+1) = -\beta'_{[2p]}(p-i+j),$$

$$(\widetilde{M}_{n,[p-1]})_{ij} = \int_0^1 B_{j,[p-1]}(x)B_{i,[p-1]}(x)\mathrm{d}x = \int_0^1 \beta_{[p-1]}(nx-j+p)\beta_{[p-1]}(nx-i+p)\mathrm{d}x$$

$$= n^{-1}\int_{-i+p}^{n-i+p} \beta_{[p-1]}(\tau+i-j)\beta_{[p-1]}(\tau)\mathrm{d}\tau = n^{-1}\int_{\mathbb{R}} \beta_{[p-1]}(\tau)\beta_{[p-1]}(\tau+i-j)\mathrm{d}\tau$$

$$= n^{-1}\beta_{[2p-1]}(p+i-j) = n^{-1}\beta_{[2p-1]}(p-i+j).$$

Thus,

$$[(n^{-1}K_{n,[p]})_{ij}]_{i,j=p+1}^n = [-\beta''_{[2p+1]}(p+1-i+j)]_{i,j=p+1}^n = T_{n-p}(f_p), \tag{72}$$

$$[(\widetilde{H}_{n,[p]})_{ij}]_{i,j=p+1}^n = [-\beta'_{[2p]}(p-i+j)]_{i,j=p+1}^n = T_{n-p}(g_p), \tag{73}$$

$$[(n\widetilde{M}_{n,[p-1]})_{ij}]_{i,j=p+1}^n = [\beta_{[2p-1]}(p-i+j)]_{i,j=p+1}^n = T_{n-p}(h_p). \tag{74}$$

It follows from (72) that the principal submatrix of $n^{-1}K_n^{[p]} - T_{n+p}(f_p)$ corresponding to the row and column indices $i,j = p+1,\ldots,n$ is the zero matrix, which implies (66). Similarly, (73) and (74) imply (67) and (68), respectively. $\square$

Theorem 6. *Let $p \geq 2$, let $\nu = (\nu_1, \nu_2) \in \mathbb{Q}^2$ be a vector with positive components, and assume that $n = n\nu$ (it is understood that n varies in the infinite subset of $\mathbb{N}$ such that $n = n\nu \in \mathbb{N}^2$). Then,*

$$\{A_{n,[p]}\}_n \sim_{\sigma,\lambda} \kappa(\theta) = \begin{bmatrix} \dfrac{\nu_2}{\nu_1}h_p(\theta_1)f_p(\theta_2) & -g_p(\theta_1)\overline{g_p(\theta_2)} \\ -\overline{g_p(\theta_1)}g_p(\theta_2) & \dfrac{\nu_1}{\nu_2}f_p(\theta_1)h_p(\theta_2) \end{bmatrix}.$$

Proof. The thesis follows immediately from Theorem 1 and (62)–(65) as soon as we have proven that

$$\{\widetilde{A}_{n,[p]}\}_n \sim_{\sigma,\lambda} \kappa(\theta). \tag{75}$$

We show that

$$\{\widetilde{A}_{n,[p],ij}\}_n \sim_{\mathrm{GLT}} \kappa_{ij}(\theta), \qquad i,j = 1,2. \tag{76}$$

Once this is done, the thesis (75) follows immediately from Theorem 5 and **GLT 1** as the matrix $\widetilde{A}_{n,[p]}$ is symmetric. Actually, we only prove (76) for $(i,j) = (1,2)$ because the proof for the other pairs of indices (i,j) is conceptually the same. Setting $p = (p,p)$ and keeping in mind the assumption $n = n\nu$, by Lemma 5 and Equations (5), (51) and (52), we have

$$\widetilde{A}_{n,[p],12} = -\widetilde{H}_{n_1,[p]} \otimes (\widetilde{H}_{n_2,[p]})^T = -(T_{n_1+p}(g_p) + R_{n_1,[p]}) \otimes (T_{n_2+p}(g_p) + R_{n_2,[p]})^T$$

$$= -(T_{n_1+p}(g_p) + R_{n_1,[p]}) \otimes (T_{n_2+p}(\overline{g_p}) + (R_{n_2,[p]})^T)$$

$$= -(T_{n+p}(g_p(\theta_1)\overline{g_p(\theta_2)}) + T_{n_1+p}(g_p) \otimes (R_{n_2,[p]})^T + R_{n_1,[p]} \otimes (\widetilde{H}_{n_2,[p]})^T)$$

$$= T_{n+p}(\kappa_{12}) + V_{n,[p]},$$

where $\mathrm{rank}(V_{n,[p]}) \leq 4p(n_1+p) + 4p(n_2+p)$. Thus, $\{V_{n,[p]}\}_n \sim_\sigma 0$ by Proposition 1, and (76) (for $(i,j) = (1,2)$) follows from **GLT 2** and **GLT 3**. $\square$

Remark 5. *Using* (58), *it is not difficult to see that the functions* $f_p(\theta)$ *and* $g_p(\theta)$ *in* (69) *and* (70) *can be expressed in terms of* $h_p(\theta)$ *as follows:*

$$f_p(\theta) = (2 - 2\cos\theta)h_p(\theta), \qquad g_p(\theta) = (e^{-i\theta} - 1)h_p(\theta).$$

Therefore, the 2×2 *matrix-valued function* $\kappa(\boldsymbol{\theta})$ *appearing in Theorem* 6 *can be simplified as follows:*

$$\kappa(\boldsymbol{\theta}) = \frac{1}{\nu_1\nu_2}h_p(\theta_1)h_p(\theta_2)\begin{bmatrix} \nu_2(e^{i\theta_2}-1) \\ -\nu_1(e^{i\theta_1}-1) \end{bmatrix}\begin{bmatrix} \nu_2(e^{-i\theta_2}-1) & -\nu_1(e^{-i\theta_1}-1) \end{bmatrix}.$$

In particular, $\det(\kappa(\boldsymbol{\theta})) = 0$ *for all* $\boldsymbol{\theta}$. *According to the informal meaning behind the spectral distribution* $\{A_{\boldsymbol{n},[p]}\}_n \sim_\lambda \kappa(\boldsymbol{\theta})$ *reported in Remark* 1, *this means that, for large* n, *one half of the eigenvalues of* $A_{\boldsymbol{n},[p]}$ *are approximately zero and one half is given by a uniform sampling over* $[-\pi,\pi]^2$ *of*

$$\mathrm{trace}(\kappa(\boldsymbol{\theta})) = \frac{1}{\nu_1\nu_2}h_p(\theta_1)h_p(\theta_2)\left[\nu_1^2(2 - 2\cos\theta_1) + \nu_2^2(2 - 2\cos\theta_1)\right].$$

8. Conclusions

We have illustrated through specific examples the applicative interest of the theory of block GLT sequences and of its multivariate version, thus bringing to completion the purely theoretical work [34]. It should be said, however, that the theory of GLT sequences is still incomplete. In particular, besides filling in the details of the theory of multilevel block GLT sequences—the results of Section 5 have been obtained as a combination of the results in [14,34], but formal proofs of them are still missing and will be the subject of a future paper—, it will be necessary to develop the theory of the so-called reduced GLT sequences, as explained in ([13] Chapter 11).

Author Contributions: C.G. authored Sections 1–4 and co-authored Section 5. S.S.-C. co-authored Sections 5 and 6; he also conceived several important ideas for the proofs of the results of Sections 3, 4 and 7. M.M. co-authored Section 6 and authored Section 7.

Funding: This research was funded by the Italian INdAM (Istituto Nazionale di Alta Matematica) through the grant PCOFUND-GA-2012-600198 and by the INdAM GNCS (Gruppo Nazionale per il Calcolo Scientifico) through a national grant.

Acknowledgments: C.G. is an INdAM Marie-Curie fellow under grant PCOFUND-GA-2012-600198. All authors are members of the INdAM GNCS, which partially supported this work.

Conflicts of Interest: The authors declare no conflict of interest.

References

1. Tilli, P. Locally Toeplitz sequences: Spectral properties and applications. *Linear Algebra Appl.* **1998**, *278*, 91–120. [CrossRef]

2. Avram, F. On bilinear forms in Gaussian random variables and Toeplitz matrices. *Probab. Theory Relat. Fields* **1988**, *79*, 37–45. [CrossRef]

3. Böttcher, A.; Grudsky, S.M. *Toeplitz Matrices, Asymptotic Linear Algebra, and Functional Analysis*; Birkhäuser Verlag: Basel, Switzerland, 2000.

4. Böttcher, A.; Grudsky, S.M. *Spectral Properties of Banded Toeplitz Matrices*; SIAM: Philadelphia, PA, USA, 2005.

5. Böttcher, A.; Silbermann, B. *Introduction to Large Truncated Toeplitz Matrices*; Springer: New York, NY, USA, 1999.

6. Böttcher, A.; Silbermann, B. *Analysis of Toeplitz Operators*, 2nd ed.; Springer: Berlin, Germany, 2006.

7. Grenander, U.; Szegő, G. *Toeplitz Forms and Their Applications*, 2nd ed.; AMS Chelsea Publishing: New York, NY, USA, 1984.

8. Parter, S.V. On the distribution of the singular values of Toeplitz matrices. *Linear Algebra Appl.* **1986**, *80*, 115–130. [CrossRef]

9. Tilli, P. A note on the spectral distribution of Toeplitz matrices. *Linear Multilinear Algebra* **1998**, *45*, 147–159. [CrossRef]
10. Tyrtyshnikov, E.E. A unifying approach to some old and new theorems on distribution and clustering. *Linear Algebra Appl.* **1996**, *232*, 1–43. [CrossRef]
11. Tyrtyshnikov, E.E.; Zamarashkin, N.L. Spectra of multilevel Toeplitz matrices: Advanced theory via simple matrix relationships. *Linear Algebra Appl.* **1998**, *270*, 15–27. [CrossRef]
12. Zamarashkin, N.L.; Tyrtyshnikov, E.E. Distribution of eigenvalues and singular values of Toeplitz matrices under weakened conditions on the generating function. *Sb. Math.* **1997**, *188*, 1191–1201. [CrossRef]
13. Garoni, C.; Serra-Capizzano, S. *Generalized Locally Toeplitz Sequences: Theory and Applications, Volume I*; Springer: Cham, Switzerland, 2017.
14. Garoni, C.; Serra-Capizzano, S. *Generalized Locally Toeplitz Sequences: Theory and Applications, Volume II*; Springer: To appear.
15. Serra-Capizzano, S. Generalized locally Toeplitz sequences: Spectral analysis and applications to discretized partial differential equations. *Linear Algebra Appl.* **2003**, *366*, 371–402. [CrossRef]
16. Serra-Capizzano, S. The GLT class as a generalized Fourier analysis and applications. *Linear Algebra Appl.* **2006**, *419*, 180–233. [CrossRef]
17. Barbarino, G. Equivalence between GLT sequences and measurable functions. *Linear Algebra Appl.* **2017**, *529*, 397–412. [CrossRef]
18. Böttcher, A.; Garoni, C.; Serra-Capizzano, S. Exploration of Toeplitz-like matrices with unbounded symbols is not a purely academic journey. *Sb. Math.* **2017**, *208*, 1602–1627. [CrossRef]
19. Beckermann, B.; Serra-Capizzano, S. On the asymptotic spectrum of finite element matrix sequences. *SIAM J. Numer. Anal.* **2007**, *45*, 746–769. [CrossRef]
20. Bertaccini, D.; Donatelli, M.; Durastante, F.; Serra-Capizzano, S. Optimizing a multigrid Runge–Kutta smoother for variable-coefficient convection-diffusion equations. *Linear Algebra Appl.* **2017**, *533*, 507–535. [CrossRef]
21. Donatelli, M.; Garoni, C.; Manni, C.; Serra-Capizzano, S.; Speleers, H. Spectral analysis and spectral symbol of matrices in isogeometric collocation methods. *Math. Comput.* **2016**, *85*, 1639–1680. [CrossRef]
22. Garoni, C. Spectral distribution of PDE discretization matrices from isogeometric analysis: The case of L^1 coefficients and non-regular geometry. *J. Spectr. Theory* **2018**, *8*, 297–313. [CrossRef]
23. Garoni, C.; Manni, C.; Pelosi, F.; Serra-Capizzano, S.; Speleers, H. On the spectrum of stiffness matrices arising from isogeometric analysis. *Numer. Math.* **2014**, *127*, 751–799. [CrossRef]
24. Garoni, C.; Manni, C.; Serra-Capizzano, S.; Sesana, D.; Speleers, H. Spectral analysis and spectral symbol of matrices in isogeometric Galerkin methods. *Math. Comput.* **2017**, *86*, 1343–1373. [CrossRef]
25. Garoni, C.; Manni, C.; Serra-Capizzano, S.; Sesana, D.; Speleers H. Lusin theorem, GLT sequences and matrix computations: An application to the spectral analysis of PDE discretization matrices. *J. Math. Anal. Appl.* **2017**, *446*, 365–382. [CrossRef]
26. Roman, F.; Manni, C.; Speleers, H. Spectral analysis of matrices in Galerkin methods based on generalized B-splines with high smoothness. *Numer. Math.* **2017**, *135*, 169–216. [CrossRef]
27. Donatelli, M.; Mazza, M.; Serra-Capizzano, S. Spectral analysis and structure preserving preconditioners for fractional diffusion equations. *J. Comput. Phys.* **2016**, *307*, 262–279. [CrossRef]
28. Salinelli, E.; Serra-Capizzano, S.; Sesana, D. Eigenvalue–eigenvector structure of Schoenmakers–Coffey matrices via Toeplitz technology and applications. *Linear Algebra Appl.* **2016**, *491*, 138–160. [CrossRef]
29. Al-Fhaid, A.S.; Serra-Capizzano, S.; Sesana, D.; Ullah, M.Z. Singular-value (and eigenvalue) distribution and Krylov preconditioning of sequences of sampling matrices approximating integral operators. *Numer. Linear Algebra Appl.* **2014**, *21*, 722–743. [CrossRef]
30. Beckermann, B.; Kuijlaars, A.B.J. Superlinear convergence of conjugate gradients. *SIAM J. Numer. Anal.* **2001**, *39*, 300–329. [CrossRef]
31. Donatelli, M.; Garoni, C.; Manni, C.; Serra-Capizzano, S.; Speleers, H. Robust and optimal multi-iterative techniques for IgA Galerkin linear systems. *Comput. Methods Appl. Mech. Eng.* **2015**, *284*, 230–264. [CrossRef]
32. Donatelli, M.; Garoni, C.; Manni, C.; Serra-Capizzano, S.; Speleers, H. Robust and optimal multi-iterative techniques for IgA collocation linear systems. *Comput. Methods Appl. Mech. Eng.* **2015**, *284*, 1120–1146. [CrossRef]

33. Donatelli, M.; Garoni, C.; Manni, C.; Serra-Capizzano, S.; Speleers, H. Symbol-based multigrid methods for Galerkin B-spline isogeometric analysis. *SIAM J. Numer. Anal.* **2017**, *55*, 31–62. [CrossRef]

34. Garoni, C.; Serra-Capizzano, S.; Sesana, D. Block generalized locally Toeplitz sequences: Topological construction, spectral distribution results, and star-algebra structure. In *Structured Matrices in Numerical Linear Algebra: Analysis, Algorithms, and Applications*; Springer INdAM Series; Springer: To appear.

35. Garoni, C.; Serra-Capizzano, S.; Sesana, D. Spectral analysis and spectral symbol of d-variate $\mathbb{Q}_p$ Lagrangian FEM stiffness matrices. *SIAM J. Matrix Anal. Appl.* **2015**, *36*, 1100–1128. [CrossRef]

36. Benedusi, P.; Garoni, C.; Krause, R.; Li, X.; Serra-Capizzano, S. Space–time FE–DG discretization of the anisotropic diffusion equation in any dimension: the spectral symbol. *SIAM J. Matrix Anal. Appl.* To appear.

37. Dumbser, M.; Fambri, F.; Furci, I.; Mazza, M.; Serra-Capizzano, S.; Tavelli, M. Staggered discontinuous Galerkin methods for the incompressible Navier–Stokes equations: Spectral analysis and computational results. *Numer. Linear Algebra Appl.* **2018**, doi:10.1002/nla.2151. [CrossRef]

38. Mazza, M.; Ratnani, A.; Serra-Capizzano, S. Spectral analysis and spectral symbol for the 2D curl–curl (stabilized) operator with applications to the related iterative solutions. *Math. Comput.* **2018**, doi:10.1090/mcom/3366. [CrossRef]

39. Barbarino, G.; Serra-Capizzano, S. *Non-Hermitian Perturbations of Hermitian Matrix-Sequences and Applications to the Spectral Analysis of Approximated PDEs*; Technical Report 2018-004; Department of Information Technology, Uppsala University: Uppsala, Sweden, 2018.

40. Bhatia, R. *Matrix Analysis*; Springer: New York, NY, USA, 1997.

41. Schumaker, L.L. *Spline Functions: Basic Theory*, 3rd ed.; Cambridge University Press: Cambridge, UK, 2007.

42. De Boor, C. *A Practical Guide to Splines*; Revised edition; Springer: New York, NY, USA, 2001.

43. Buffa, A.; Sangalli, G.; Vázquez, R. Isogeometric analysis in electromagnetics: B-splines approximation. *Comput. Methods Appl. Mech. Eng.* **2010**, *199*, 1143–1152. [CrossRef]

MDPI

Article

Trees, Stumps, and Applications

John C. Butcher

Department of Mathematics, University of Auckland, Auckland 92019, New Zealand;
butcher@math.auckland.ac.nz

Received: 23 May 2018; Accepted: 16 July 2018; Published: 1 August 2018

Abstract: The traditional derivation of Runge–Kutta methods is based on the use of the scalar test problem $y'(x) = f(x, y(x))$. However, above order 4, this gives less restrictive order conditions than those obtained from a vector test problem using a tree-based theory. In this paper, stumps, or incomplete trees, are introduced to explain the discrepancy between the two alternative theories. Atomic stumps can be combined multiplicatively to generate all trees. For the scalar test problem, these quantities commute, and certain sets of trees form isomeric classes. There is a single order condition for each class, whereas for the general vector-based problem, for which commutation of atomic stumps does not occur, there is exactly one order condition for each tree. In the case of order 5, the only nontrivial isomeric class contains two trees, and the number of order conditions reduces from 17 to 16 for scalar problems. A method is derived that satisfies the 16 conditions for scalar problems but not the complete set based on 17 trees. Hence, as a practical numerical method, it has order 4 for a general initial value problem, but this increases to order 5 for a scalar problem.

Keywords: ordinary differential equations; Runge–Kutta; tree; stump; order; elementary differential

MSC: 65L05

1. Introduction

Trees have a well-established role in the analysis of numerical methods for ordinary differential equations. In this paper, the more general concept of a stump is introduced and applied to the analysis of B-series and the composition rule. It is also shown how stumps can be used to analyse the order of nonautonomous scalar problems for which the order conditions for Runge–Kutta methods are slightly different. A new explanation is given for this discrepancy.

In Section 2, a brief survey is given of the theory of Runge–Kutta methods, showing the structure of the elementary differentials on which B-series are based and the relationship of elementary differentials to trees. This is followed by Section 3, in which stumps are introduced. These are a generalisation of trees, but, by restricting to "atomic stumps", they also provide a means of generating all trees. Isomeric classes of trees generated in this way provide a framework for the analysis of order conditions in the scalar case, as shown in Section 4. The paper concludes with the derivation of a method of "ambiguous order". That is, the method has order 4 in general, but this increases to 5 for a scalar problem.

The theory of stumps, isomeric trees, and applications to scalar differential equations appear in greater detail in [1]. The theory of trees and applications to vector-based numerical methods can be found, for example, in [2]. The order of the method in [3] was studied in [4].

2. Trees, Elementary Differentials, and B-Series

Trees are graphs such as $\bullet$, $\mathfrak{t}$, $\mathbf{v}$, $\mathbf{\mathfrak{t}}$, $\mathbf{w}$, $\mathbf{\mathfrak{v}}$, Y, $\mathfrak{t}$. The "root" of a tree is the lowest point in the diagram, and all vertices, except for the root, have a single parent. For a given tree t, the "order of t", written as $|t|$, is the number of vertices in t. If a vertex v is the parent of v', then v' is a child of v. If there exists a path

$$(v_0, v_1, v_2, \ldots, v_n), \qquad \text{where } v_i \text{ is a child of } v_{i-1}, \quad i = 1, 2, \ldots, n,$$

then v_n is a "descendant" of v_0. The product of the number of descendants for every vertex in a tree t is defined to be the "factorial of t" and is written as $t!$.

For the first eight trees, the order and factorial are the following:

$$|\bullet| = 1, \quad |\mathfrak{t}| = 2, \quad |\mathbf{V}| = 3, \quad |\mathfrak{t}| = 3, \quad |\mathbf{V}| = 4, \quad |\mathbf{V}| = 4, \quad |\mathbf{Y}| = 4, \quad |\mathfrak{t}| = 4;$$

$$\bullet! = 1, \quad \mathfrak{t}! = 2, \quad \mathbf{V}! = 3, \quad \mathfrak{t}! = 6, \quad \mathbf{V}! = 4, \quad \mathbf{V}! = 8, \quad \mathbf{Y}! = 12, \quad \mathfrak{t}! = 24.$$

2.1. Notation and Recursions

In this paper, $\tau := \bullet$, and we recall two recursions to build other trees in terms of τ. There are two convenient constructions for building complicated trees in terms of simpler trees. They are the following:

1. Given trees $t_1, t_2, \ldots, t_m$, define $t = [t_1 t_2 \cdots t_m]$ from the diagram

$$t_1 \ t_2 \ t_3 \ \cdots \ t_m$$

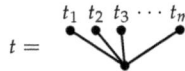

$$t =$$

The notation $\left[t_1^{k_1} t_2^{k_2} \cdots t_m^{k_m}\right]$ is used to show repetitions of $t_1, \ldots$ Assuming the t_i are distinct, then the "symmetry" $\sigma(t)$ is defined recursively by

$$\sigma(\tau) = 1,$$

$$\sigma\left(\left[t_1^{k_1} t_2^{k_2} \cdots t_m^{k_m}\right]\right) = \prod_{i=1}^{m} k_i! \sigma(t_i)^{k_i}.$$

2. Given trees t_1 and t_2, define $t = t_1 * t_2$ from the diagram

$$t = t_1 \nearrow^{t_2}$$

2.2. Polish Notation Tree Construction

Polish notation or prefix (as distinct from infix or postfix) notation is credited to Lukasiewicz. A famous reference to his work is [5]. We generalise the notation so that τ_m acts as a prefix operator on m operands and thus $\tau_m t_1 t_2 \cdots t_m$ has the same meaning as $[t_1 t_2 \cdots t_m]$. This gives a third and bracketless scheme for writing trees. In Table 1, the various notations are given side by side. It is noted that the notation based on $t * t'$ does not always give a unique factorisation.

Table 1. Tree notations.

Tree	Notation 1	Notation 2	Polish Notation
•	τ	τ	τ
‧	$[\tau]$	$\tau * \tau$	$\tau_1 \tau$
∨	$[\tau^2]$	$(\tau * \tau) * \tau$	$\tau_2 \tau \tau$
‧	$[[\tau]]$	$\tau * (\tau * \tau)$	$\tau_1 \tau_1 \tau$
∨	$[\tau^3]$	$((\tau * \tau) * \tau) * \tau$	$\tau_3 \tau \tau \tau$
∨	$[\tau[\tau]]$	$(\tau * \tau) * (\tau * \tau) = (\tau * (\tau * \tau)) * \tau$	$\tau_2 \tau \tau_1 \tau$
Y	$[[\tau^2]]$	$\tau * ((\tau * \tau) * \tau)$	$\tau_1 \tau_2 \tau \tau$
‧	$[[[\tau]]]$	$\tau * (\tau * (\tau * \tau))$	$\tau_1 \tau_1 \tau_1 \tau$

2.3. Elementary Differentials

Given an autonomous initial value problem,

$$y'(x) = f(y(x)), \quad y(x_0) = y_0, \qquad y : \mathbb{R} \to \mathbb{R}^N, f : \mathbb{R}^N \to \mathbb{R}^N, \tag{1}$$

we write $\mathbf{f} = f(y_0)$ and also write the sequence of Fréchet derivatives of f, evaluated at y_0, as $\mathbf{f}'$, $\mathbf{f}''$, $\mathbf{f}^{(3)}$, ... It is noted that, in linear algebra terms, these are linear, bilinear, and multilinear operators. In this paper, we always use Polish notation so that $\mathbf{f}^{(m)}$ acting on the m vectors $v_1, v_2, \ldots, v_m$ is written as $\mathbf{f}^{(m)} v_1 v_2 \cdots v_m$.

Definition 1. *The elementary differential $\mathbf{F}(t)$ associated with the tree t is defined by*

$$\mathbf{F}(\tau) = \mathbf{f},$$
$$\mathbf{F}([t_1 t_2 \cdots t_m]) = \mathbf{f}^{(m)} \mathbf{F}(t_1) \mathbf{F}(t_2) \cdots \mathbf{F}(t_m).$$

It is noted that the recursion formula can also be written as

$$\mathbf{F}(\tau_m t_1 t_2 \cdots t_m) = \mathbf{f}^{(m)} \mathbf{F}(t_1) \mathbf{F}(t_2) \cdots \mathbf{F}(t_m).$$

This makes it possible, in the Polish form of tree notation, to perform a simple substitution. That is, every τ is replaced by $\mathbf{f}$, and every τ_m is replaced by $\mathbf{f}^{(m)}$.

2.4. Application to B-Series

Given a function $a : T \to \mathbb{R}$, the corresponding B-series is a formal Taylor series:

$$y_0 + \sum_{t \in T} \frac{a(t) h^{|t|}}{\sigma(t)} \mathbf{F}(t).$$

Two special cases are the following:

1. $t \mapsto 1/t!$, which gives the Taylor series for the solution to Equation (1) at $x = x_0 + h$. The series is

$$y_0 + \sum_{t \in T} \frac{h^{|t|}}{t! \sigma(t)} \mathbf{F}(t). \tag{2}$$

2. $t \mapsto \Phi(t)$, where $\Phi(t)$ is the corresponding elementary weight for a specific Runge–Kutta method. This gives the Taylor series for the approximation computed by this Runge–Kutta method:

$$y_0 + \sum_{t \in T} \frac{\Phi(t) h^{(|t|)}}{\sigma(t)} \mathbf{F}(t). \tag{3}$$

By comparing Equations (2) and (3), we recover the conditions for a Runge–Kutta method to have order p:

$$\Phi(t) = \frac{1}{t!}, \qquad |t| \le p. \tag{4}$$

3. Trees, Forests, and Stumps

A sequence of items built from $\tau, \tau_1, \tau_2, \ldots$, can be contracted by the rules of Polish operations to form a sequence of trees, together with a final subsequence that might not be a tree but would become one if further operands are appended on the right. The sequence of trees on the left is usually referred to as a forest and can be converted into a single tree by a suitable operator to the left of this subsequence.

Incomplete "trees" are referred to as stumps. Examples are

$$\tau_1, \quad \tau_2, \quad \tau_2 \tau_1 \tau, \quad \tau_1 \tau_2 \tau, \quad \tau_1 \tau_1 \tau_1.$$

The "valency" of a stump is the number of copies of τ, appended to the right, that would be required to convert it into a tree. It is convenient to refer to a tree as a stump with zero valency.

The word "forestump" is introduced to refer to a sequence of items made up from factors τ and τ_m, $m = 1, 2, \ldots$ When a particular forestump is contracted to form as many trees as possible, then the final form will be the formal product of a forest of trees followed by a single stump (possibly the empty stump).

3.1. Bicolour Diagrams to Represent Stumps

We now introduce a generalisation of the way trees are represented diagrammatically to include stumps. We regard stumps as modified trees with some leaves removed but with the edges from these missing leaves to their parents retained.

In the examples given here, a white disc represents the absence of a vertex. The number of white discs is the valency, with the remark that trees are stumps with zero valency.

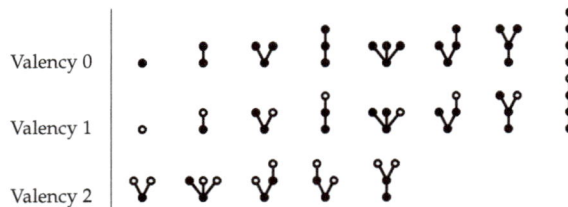

Right multiplication by one or more additional stumps implies grafting to open valency positions. It is noted that the third and fourth examples of valency 2 stumps are mirror images. This is significant in determining the precedence of the operands.

3.1.1. Products of Stumps

Given two stumps s and s', the product ss' has a nontrivial product if s' is not the trivial stump $\circ$ and s has valency of at least 1; that is, if s is not a tree, the product is formed by grafting the root of s' to the rightmost open valency in s.

Two examples of grafting illustrate the significance of stump orientations:

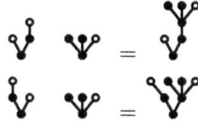

If s is a tree or s' is the trivial stump, no contraction takes place.

3.1.2. Atomic Stumps

An atomic stump is a graph of the following form:

It is noted that no more than two generations can be present.

If m of the children of the root are represented by black discs and n are represented by white discs, then the atomic stump is denoted by s_{mn}. The reason for the designation "atomic" is that every tree can be written as the product of atoms.

This is illustrated for trees of up to order 4:

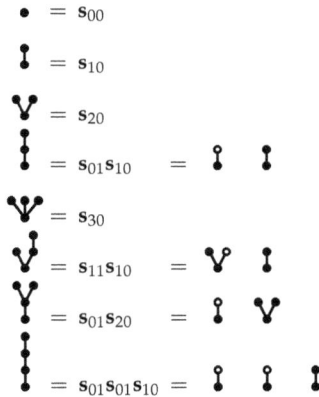

3.1.3. Isomeric Trees

In the factorisation of trees into products of atoms, the factors are written in a specific order, with each factor operating on later factors. However, if we interpret the atoms just as symbols that can commute with each other, we obtain a new equivalence relation, written as $\sim$.

Definition 2. *Two trees are isomeric if their atomic factors are the same.*

Nothing interesting happens up to order 4, but for order 5, we find that

It is a simple exercise to find all isomeric classes of any particular order, but, as far as the author knows, this has not been done above order 6.

For orders 5 and 6, the isomers are, line by line, the following:

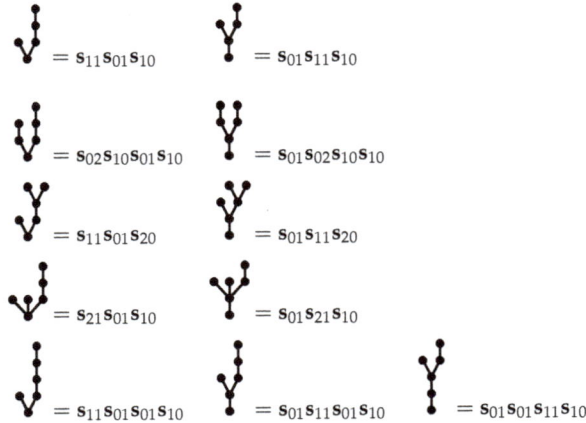

 $= s_{11}s_{01}s_{10}$ $= s_{01}s_{11}s_{10}$

 $= s_{02}s_{10}s_{01}s_{10}$ $= s_{01}s_{02}s_{10}s_{10}$

 $= s_{11}s_{01}s_{20}$ $= s_{01}s_{11}s_{20}$

 $= s_{21}s_{01}s_{10}$ $= s_{01}s_{21}s_{10}$

 $= s_{11}s_{01}s_{01}s_{10}$ $= s_{01}s_{11}s_{01}s_{10}$ $= s_{01}s_{01}s_{11}s_{10}$

We see in Section 4 that isomeric classes for *scalar* differential equations have a similar role to individual trees in the case of differential systems of arbitrarily high dimension. We let a_n denote the number of trees with order n and A_n denote the accumulated total $a_1 + a_2 + \cdots + a_n$. Similarly, we let b_n denote the number of isomeric classes with order n and B_n denote the accumulated total for this quantity. These are shown in Table 2 up to order 6.

Table 2. Trees and isomeric classes for various orders.

n	1	2	3	4	5	6
a_n	1	1	2	4	9	20
A_n	1	2	4	8	17	37
b_n	1	1	2	4	8	15
B_n	1	2	4	8	16	31

4. Scalar Differential Equations

Early studies of Runge–Kutta methods derived order conditions for the scalar initial value problem

$$y'(x) = f(x, y(x)),$$ (5)

instead of using the autonomous test problem (Equation (1)).

The full set of conditions up to some specified order becomes the starting point for finding accurate Runge–Kutta methods. The derivations of these conditions to order 5 were the pioneering contributions of Runge, Heun, and then Kutta [6–8]. We follow their arguments for the same model problem (Equation (5)). In this derivation, $\partial_x f := \partial f / \partial x$ and $\partial_y f := \partial f / \partial y$, with similar notations for higher partial derivatives. First, we find the second derivative of y by the chain rule:

$$y'' = \partial_x f + (\partial_y f) f.$$

Similarly, we find the third derivative:

$$
\begin{aligned}
y^{(3)} &= \left(\partial_x^2 f + (\partial_x \partial_y f) f\right) + \partial_y f \left(\partial_x f + (\partial_y f) f\right) + (\partial_x \partial_y f) f + (\partial_y^2 f) f^2 \\
&= \partial_x^2 f + 2(\partial_x \partial_y f) f + (\partial_y^2 f) f^2 + (\partial_x f \partial_y f) f + (\partial_y f)^2 f
\end{aligned}
$$

and carry on to find fourth and higher derivatives. By evaluating $y^{(n)}$ at $x = x_0$, we can find the Taylor expansions to use in Equation (5). A more complicated calculation leads to the detailed series of Equation (8) in the case of any particular Runge–Kutta method and hence to the determination of its order. We pursue this line of enquiry below.

The greatest achievement in this line of work was given in [3], where sixth order methods involving eight stages were derived. In all the derivations of new methods, up to the publication of this tour de force, a tacit assumption was made. This was that a method derived to have a specific order for a general scalar problem will have this same order for a coupled system of scalar problems; that is, it will have this order for a problem with $N > 1$. This bald assumption is untrue, and it becomes necessary to carry out the order analysis in a multidimensional setting.

4.1. Nonautonomous Vector-Valued Problems

This analysis was carried out in a scalar context, in contrast to later work, for which the application was always to vector-valued problems. To cater for problems that are both nonautonomous and, at the same time, vector-valued, we can use the terminology of the present section but with a multidimensional interpretation.

This is done by regarding factors such as $\partial_y f$ and $\partial_y^2 f$ as linear operators and bilinear operators, respectively, that operate on vector-valued terms to the right, using Polish notation. To maintain this interpretation, when a problem is nonscalar, this requires the strict order of factors to be observed. Of course, in the traditional scalar interpretation, all factors commute, and the order of factors could have been altered.

4.2. Systematic Derivation of Taylor Series

The evaluation of $y^{(n)}$, $n = 1, 2, \ldots, 5$, is now carried out in a systematic manner. We let

$$D_{mn} = \sum_{i=0}^{m} \binom{m}{i} (\partial_x^{m-i} \partial_y^{n+i} f) f^i. \tag{6}$$

We also let D_{mn} denote D_{mn} evaluated at (x_0, y_0).

Lemma 1.
$$\frac{\mathrm{d}}{\mathrm{d}x} D_{mn} = D_{m+1,n} + m D_{m-1,n+1} D_{10}. \tag{7}$$

Proof.

$$\frac{\mathrm{d}}{\mathrm{d}x} \sum_{i=0}^{m} \binom{m}{i} (\partial_x^{m-i} \partial_y^{n+i} f) f^i$$

$$= \left(\sum_{i=0}^{m} \binom{m}{i} (\partial_x^{m-i+1} \partial_y^{n+i} f) f^i + \sum_{i=0}^{m} \binom{m}{i} (\partial_x^{m-i} \partial_y^{n+i+1} f) f^{i+1} \right) + \sum_{i=0}^{m} \binom{m}{i} i (\partial_x f) (\partial_x^{m-i} \partial_y^{n+i} \partial_y f) f^{i-1}$$

$$= \sum_{i=0}^{m+1} \left(\binom{m}{i} + \binom{m}{i-1} \right) \partial_x^{m-i+1} (\partial_y^{n+i} f) f^i + \sum_{i=0}^{m} \left(i \frac{m!}{i!(m-i)!} \right) (\partial_x f) (\partial_x^{m-i} \partial_y^{n+i} \partial_y f) f^{i-1}$$

$$= \sum_{i=0}^{m+1} \binom{m+1}{i} (\partial_x^{m-i+1} \partial_y^{n+i} f) f^i + m \sum_{i=0}^{m-1} \binom{m-1}{i} (\partial_x f) (\partial_x^{m-i-1} \partial_y^{n+1+i} \partial_y f) f^i$$

$$= D_{m+1,n} + m D_{m-1,n+1} D_{10}.$$

$\square$

Using Lemma 1, we find in turn that

$$
\begin{aligned}
y' &= D_{00} \\
y'' &= D_{10} \\
y''' &= D_{20} + D_{01}D_{10} \\
y^{(4)} &= D_{30} + 2D_{11}D_{10} + D_{11}D_{10} + D_{01}(D_{20} + D_{01}D_{10}) \\
&= D_{30} + 3D_{11}D_{10} + D_{01}D_{20} + D_{01}^2 D_{10} \\
y^{(5)} &= D_{40} + 3D_{21}D_{10} + 3(D_{21} + D_{02}D_{10})D_{10} + 3D_{11}(D_{20} + D_{01}D_{10}) \\
&\quad + D_{11}D_{20} + D_{01}(D_{30} + 2D_{11}D_{10}) + 2D_{01}D_{11}D_{10} + D_{01}^2(D_{20} + D_{01}D_{10}) \\
&= D_{40} + 6D_{21}D_{10} + 3D_{02}D_{10}D_{10} + 4D_{11}D_{20} + 7D_{11}D_{01}D_{10} \\
&\quad + D_{01}D_{30} + D_{01}^2 D_{20} + D_{01}^3 D_{10}.
\end{aligned}
\tag{8}
$$

To find the order conditions for a Runge–Kutta method, up to order 5, we need to systematically find the Taylor series for the stages and finally for the output. In this analysis, we assume that $\sum_{j=1}^{s} a_{ij} = c_i$ for all stages. For the stages, it is sufficient to work only to order 4, so that the scaled stage derivatives include h^5 terms.

As a step towards finding the Taylor expansions of the stages and the output, we need to find the series for $hf(Y)$ for a given series $Y = y_0 + \cdots$. In the following result, we use an arbitrary weighted series using the terms in Equation (8).

Lemma 2. *If*

$$
\begin{aligned}
Y &= y_0 + a_1 h D_{00} + a_2 h^2 D_{10} + a_3 h^3 \tfrac{1}{2} D_{20} + a_4 h^3 D_{01}D_{10} \\
&\quad + a_5 h^4 \tfrac{1}{6} D_{30} + a_6 h^4 D_{11}D_{10} + a_7 h^4 \tfrac{1}{2} D_{01}D_{20} + a_8 h^4 D_{01}^2 D_{10} + \mathcal{O}(h^5),
\end{aligned}
$$

then

$$
hf(x_0 + ha_1, Y) = hT_1 + h^2 T_2 + h^3 T_3 + h^4 T_4 + h^5 T_5 + \mathcal{O}(h^6),
$$

where

$$
\begin{aligned}
T_1 &= D_{00}, \\
T_2 &= a_1 D_{10}, \\
T_3 &= \tfrac{1}{2} a_1^2 D_{20} + a_2 D_{01}D_{10} \\
T_4 &= \tfrac{1}{6} a_1^3 D_{30} + a_1 a_2 D_{11}D_{10} + \tfrac{1}{2} a_3 D_{01}D_{20} + a_4 D_{01}^2 D_{10} \\
T_5 &= \tfrac{1}{24} a_1^4 D_{40} + \tfrac{1}{2} a_1^2 a_2 D_{21}D_{10} + a_1 a_3 D_{11}D_{20} + (a_1 a_4 + a_6) D_{11}D_{01}D_{10} \\
&\quad + \tfrac{1}{2} a_2^2 D_{02}D_{10}^2 + \tfrac{1}{6} a_5 D_{30}D_{01} + \tfrac{1}{2} a_7 D_{01}^2 D_{20} + a_8 D_{01}^3 D_{10}.
\end{aligned}
$$

Proof. Throughout this proof, an expression of the form $\partial_x^k \partial_y^m f$ is assumed to have been evaluated at (x_0, y_0). Evaluate T_1, T_2, T_3, and T_4:

$$
T_1 h + T_2 h^2 + T_3 h^3 + T_4 h^4 + \mathcal{O}(h^5),
$$

where

$$T_1 = f(x_0, y_0) = D_{00},$$
$$T_2 = a_1 \partial_x f + a_1 (\partial_y f) f = a_1 D_{10},$$
$$T_3 = \tfrac{1}{2} a_1^2 \partial_x^2 f + a_1^2 (\partial_x \partial_y) D_{00} + \tfrac{1}{2} a_1^2 (\partial_y^2 f) D_{00}^2 + a_2 (\partial_y f) D_{10}$$
$$\quad = \tfrac{1}{2} a_1^2 D_{20} + a_2 D_{01} D_{10},$$
$$T_4 = \tfrac{1}{6} a_1^3 \partial_x^3 f + \tfrac{1}{2} a_1^3 (\partial_x^2 \partial_y f) D_{10} + \tfrac{1}{2} a_1^3 (\partial_x \partial_y^2 f) D_{10}^2 + \tfrac{1}{6} a_1^3 \partial_y^3 f D_{10}^3$$
$$\quad + a_1 a_2 (\partial_x \partial_y f) D_{10} + a_1 a_2 (\partial_y^2 f) D_{10} D_{01} + a_3 (\partial_y f) D_{20} + a_4 (\partial_y f) D_{01} D_{10}$$
$$\quad = \tfrac{1}{6} a_1^3 D_{30} + a_1 a_2 D_{11} D_{10} + a_3 D_{01} D_{20} + a_4 D_{01}^2 D_{10}.$$

The evaluation of T_5 is similar but more complicated and is omitted. $\quad\square$

For the stage values of a Runge–Kutta method, we have

$$Y_i = y_0 + \sum_{j=1}^{s} a_{ij} h f(x_0 + h c_j, Y_j)$$
$$\quad = y_0 + h c_i D_{00} + \mathcal{O}(h^2)$$

and then, to one further order,

$$Y_i = y_0 + \sum_{j=1}^{s} a_{ij} h f(x_0 + h c_j, y_0 + h c_j D_{00}) + \mathcal{O}(h^3)$$
$$\quad = y_0 + h c_i D_{00} + h^2 \sum_j a_{ij} c_j D_{10} + \mathcal{O}(h^3).$$

A similar expression can be written down for the output from a step:

$$y_1 = y_0 + h \sum_i b_i D_{00} + h^2 \sum_i b_i c_i D_{10} + \mathcal{O}(h^3).$$

A comparison with the exact solution, $y_0 + h y'(x_0) + \tfrac{1}{2} h^2 y''(x_0) + \mathcal{O}(h^3)$, evaluated using Equation (8), gives, under second order conditions,

$$\sum_i b_i D_{00} = D_{00},$$
$$\sum_i b_i c_i D_{10} = \tfrac{1}{2} D_{10}.$$

This analysis can be taken further in a straightforward and systematic way and is summarised, as far as order 5, in Theorem 1. This theorem, for which the detailed proof is omitted, has to be read together with Table 3.

Theorem 1. *In the statement of this result, the quantities p, T, σ, and ϕ are given in Table 3.*

1. *The Taylor expansion for the exact solution to the initial value problem*

$$y'(x) = f(x, y), \qquad y(x_0) = y_0 \tag{9}$$

to within $\mathcal{O}(h^6)$ is y_0 plus the sum of terms of the form

$$e h^p \sigma^{-1} T.$$

2. The Taylor expansion for the numerical solution y_1 to Equation (9), using a Runge–Kutta method (A, b^T, c), to within $\mathcal{O}(h^6)$ is y_0 plus the sum of terms of the form

$$\phi h^p \sigma^{-1} T.$$

3. The conditions to order 5, for the solution of Equation (5) using (A, b^T, c), are the equations of the form

$$\phi = e.$$

Table 3. Data for Theorem 1 with reference numbers (O1)–(O11) and (O14)–(O17) shown.

p	σ	T	$\phi = e$	
1	1	D_{00}	$\sum b_i = 1$	(O1)
2	1	D_{10}	$\sum b_i c_i = \frac{1}{2}$	(O2)
3	2	D_{20}	$\sum b_i c_i^2 = \frac{1}{3}$	(O3)
	1	$D_{01}D_{10}$	$\sum b_i a_{ij} c_j = \frac{1}{6}$	(O4)
4	6	D_{30}	$\sum b_i c_i^3 = \frac{1}{4}$	(O5)
	1	$D_{11}D_{10}$	$\sum b_i c_i a_{ij} c_j = \frac{1}{8}$	(O6)
	2	$D_{01}D_{20}$	$\sum b_i a_{ij} c_j^2 = \frac{1}{12}$	(O7)
	1	$D_{01}^2 D_{10}$	$\sum b_i a_{ij} a_{jk} c_k = \frac{1}{24}$	(O8)
5	24	D_{40}	$\sum b_i c_i^4 = \frac{1}{6}$	(O9)
	2	$D_{21}D_{10}$	$\sum b_i c_i^2 a_{ij} c_j = \frac{1}{10}$	(O10)
	2	$D_{11}D_{20}$	$\sum b_i c_i a_{ij} c_j^2 = \frac{1}{15}$	(O11)
	1	$D_{11}D_{01}D_{10}$	$\sum b_i (c_i + c_j) a_{ij} a_{jk} c_k = \frac{7}{120}$	
	2	$D_{02}D_{10}^2$	$\sum b_i a_{ij} c_j a_{ik} c_k = \frac{1}{20}$	(O14)
	6	$D_{01}D_{30}$	$\sum b_i a_{ij} c_j^3 = \frac{1}{20}$	(O15)
	2	$D_{01}^2 D_{20}$	$\sum b_i a_{ij} a_{jk} c_k^2 = \frac{1}{60}$	(O16)
	1	$D_{01}^3 D_{10}$	$\sum b_i a_{ij} a_{jk} a_{k\ell} c_\ell = \frac{1}{120}$	(O17)

4.3. Order Conditions for Vector Problems

The order conditions for the autonomous vector problem, given by Equation (4) for $p = 5$, are identical to (O1)–(O11) and (O14)–(O17) together with the two cases of (4) missing from Table 3:

$$\sum b_i c_i a_{ij} a_{jk} c_k = \frac{1}{30}, \tag{O12}$$
$$\sum b_i a_{ij} c_j a_{jk} c_k = \frac{1}{40}. \tag{O13}$$

Although these do not occur in Table 3, the sum of (O12) and (O13) is equal to

$$\sum b_i (c_i + c_j) a_{ij} a_{jk} c_k = \frac{7}{120}, \tag{10}$$

which does occur as an un-numbered entry in Table 3. Apart from this discrepancy, the order conditions for the scalar and vector problems exactly agree as far as order 5.

4.4. Derivation of Ambiguous Method

We now construct a method that has order 5 for a scalar problem but only order 4 for a vector-based problem. This means that all the conditions $\Phi(t) = 1/t!$ need to be satisfied for the 17 trees such that $|t| \leq 5$, except for (O12) and (O13), which can be replaced by Equation (10). In constructing this method, it is convenient to introduce a vector d^T defined as

$$d^\mathsf{T} = b^\mathsf{T} A + b^\mathsf{T} C - b^\mathsf{T}.$$

This satisfies the property

$$d^\mathsf{T} c^{n-1} = 0, \qquad n = 1, 2, 3, 4, \tag{11}$$

because

$$d^\mathsf{T} c^{n-1} = b^\mathsf{T} A c^{n-1} + b^\mathsf{T} c^n - b^\mathsf{T} c^{n-1} = \frac{1}{n(n+1)} + \frac{1}{n+1} - \frac{1}{n} = 0.$$

In the method to be constructed, some assumptions are made. These are

$$\sum_{j=1}^{i-1} a_{ij} c_j = \tfrac{1}{2} c_i^2, \qquad i \neq 2, 3, \tag{12}$$

$$c_6 = 1, \tag{13}$$

$$b_2 = b_3 = 0. \tag{14}$$

From Equations (13) and (14) and some of the order conditions, it follows that $\sum_{i=1}^{6} b_i c_i (c_i - c_4)(c_i - c_5)(1 - c_i) = 0$, implying that $\frac{1}{120}(20 c_4 c_5 - 10(c_4 + c_5) + 4) = 0$ and hence that $(\frac{1}{2} - c_4)(c_5 - \frac{1}{2}) = \frac{1}{20}$. We choose the convenient values $c_4 = \frac{1}{4}$ and $c_5 = \frac{7}{10}$ together with $c_2 = \frac{1}{2}$ and $c_3 = 1$. The value of b is found from (O1), (O2), (O3), (O5), and (O9), and d is found from Equation (11) with the requirement that $d_6 = 0$. The results are

$$b = \begin{bmatrix} \frac{1}{14} & 0 & 0 & \frac{32}{81} & \frac{250}{567} & \frac{5}{54} \end{bmatrix},$$
$$d = \theta \begin{bmatrix} 1 & 7 & \frac{7}{9} & -\frac{112}{27} & \frac{125}{27} & 0 \end{bmatrix},$$

where θ is a parameter, assumed to be nonzero. The third row of A can be found from

$$d_2\left(-\tfrac{1}{2}c_2^2\right) + d_3\left(a_{32}c_2 - \tfrac{1}{2}c_3^2\right) = 0, \tag{15}$$

because, from several order conditions,

$$d^\mathsf{T}\left(Ac - \tfrac{1}{2}c^2\right) = b^\mathsf{T} A^2 c + b^\mathsf{T} C A c - b^\mathsf{T} A c - \tfrac{1}{2} b^\mathsf{T} A c^2 - \tfrac{1}{2} b^\mathsf{T} c^3 + \tfrac{1}{2} b^\mathsf{T} c^2$$
$$= \tfrac{1}{24} \quad + \tfrac{1}{8} \quad - \tfrac{1}{6} \quad - \tfrac{1}{24} \quad - \tfrac{1}{8} \quad + \tfrac{1}{6} \quad = 0.$$

From Equation (15), it is found that $a_{32} = \frac{13}{4}$. The values of a_{42} and a_{52} can be written in terms of the other elements of rows 4 and 5 of A, and row 6 can be found in terms of the other rows. There are now four free parameters remaining (a_{43}, a_{53}, a_{54}, and θ) and four conditions that are not automatically satisfied. These are (O11), (O16), (O17), and Equation (10). The solutions are given in the complete tableau.

$$
\begin{array}{c|cccccc}
0 & & & & & & \\
\frac{1}{2} & \frac{1}{2} & & & & & \\
1 & -\frac{9}{4} & \frac{13}{4} & & & & \\
\frac{1}{4} & \frac{9}{64} & \frac{5}{32} & -\frac{3}{64} & & & \\
\frac{7}{10} & \frac{63}{625} & \frac{259}{2500} & \frac{231}{2500} & \frac{252}{625} & & \\
1 & -\frac{27}{50} & -\frac{139}{50} & -\frac{21}{50} & \frac{56}{25} & \frac{5}{2} & \\
\hline
 & \frac{1}{14} & 0 & 0 & \frac{32}{81} & \frac{250}{567} & \frac{5}{54}
\end{array}
\tag{16}
$$

5. Numerical Test

A suitable single differential equation to test the order of convergence of this method, together with a closely related autonomous system, is

$$
\frac{dy}{dx} = \frac{y - x}{y + x}, \tag{17}
$$

$$
\frac{d}{dt}\begin{bmatrix} x \\ y \end{bmatrix} = \frac{1}{\sqrt{x^2 + y^2}}\begin{bmatrix} y + x \\ y - x \end{bmatrix} \tag{18}
$$

The solution of Equation (17), in parametric coordinates, is

$$
x = \xi(t) := t \sin(\ln(t)),
$$
$$
y = \eta(t) := t \cos(\ln(t)),
$$

and this is also the solution to Equation (18).

Two experiments were carried out:

1. The scalar problem (Equation (17)) was solved using the method of Equation (16) on the interval $[\xi(\pi/6), \xi(5\pi'12)]$.
2. The two-dimensional problem of Equation (18), using the same method, was solved on the interval $[\pi/6, 5\pi'12]$.

In each case, $n = 10 \times 2^i$ for $i = 0, 1, 2, 3, 4$. The errors for the two methods and the various numbers of steps are shown in Table 4. Also shown are the errors for n steps divided by the error for $2n$ steps.

Table 4. Variation of global errors for a range of step sizes.

n	Problem 1 Error	Ratio	Problem 2 Error	Ratio
10	5.3177×10^{-7}	30.956	1.1830×10^{-5}	15.068
10×2	1.7179×10^{-8}	31.402	7.8506×10^{-7}	15.157
10×2^2	5.4705×10^{-10}	31.679	5.1794×10^{-8}	15.485
10×2^3	1.7268×10^{-11}	31.788	3.3448×10^{-9}	15.720
10×2^4	5.4323×10^{-13}	—	2.1278×10^{-10}	—

As expected, the numerical behaviour for experiment 1 was consistent with order 5. In contrast, for experiment 2, the numerical behaviour was consistent only with order 4.

6. Discussion

There is little scientific interest in the solution of scalar initial value problems, and there is no advantage in constructing numerical methods that are suitable only for this special class of problems. Hence, in the search for useful numerical methods, it is an advantage to use tree-based theory. The results presented here emphasise the danger of using scalar theory to derive methods of order higher than 4 because they could be incorrect.

Funding: This research received no external funding.

Conflicts of Interest: The author declares no conflict of interest.

References

1. Butcher, J.C. *B-Series; Algebraic Analysis of Numerical Methods*; Springer: Berlin, Germany, In preparation.
2. Butcher, J.C. *Numerical Methods for Ordinary Differential Equations*, 3rd ed.; John Wiley & Sons: Chichester, UK, 2016.
3. Huťa, A. Une amélioration de la méthode de Runge–Kutta–Nyström pour la résolution numérique des équations différentielles du premier ordre. *Acta Fac. Nat. Univ. Comenian. Math.* **1956**, *1*, 201–224.
4. Butcher, J.C. On the integration processes of A.Huťa. *J. Austral. Math. Soc.* **1963**, *3*, 202–206. [CrossRef]
5. Łukasiewicz, J.; Tarski, J. Investigations into the Sentential Calculus. *Comp. Rend. Soc. Sci. Lett. Vars.* **1930**, *23*, 31–32. (In German)
6. Heun, K. Neue Methode zur approximativen Integration der Differentialgleichungen einer unabhängigen Veränderlichen. *Z. Math. Phys.* **1900**, *45*, 23–38.
7. Kutta, W. Beitrag zur näherungsweisen Integration totaler Differentialgleichungen. *Z. Math. Phys.* **1901**, *46*, 435–453.
8. Runge, C. Über die numerische Auflösung von Differentialgleichungen. *Math. Ann.* **1895**, *46*, 167–178. [CrossRef]

![Axioms logo] **axioms**

![MDPI logo]

Review

Collocation Methods for Volterra Integral and Integro-Differential Equations: A Review

Angelamaria Cardone [1,*] ⓘ, **Dajana Conte** [1] ⓘ, **Raffaele D'Ambrosio** [2] ⓘ
and Beatrice Paternoster [1] ⓘ

1   Department of Mathematics, University of Salerno, 84084 Fisciano, Italy; dajconte@unisa.it (D.C.);
    beapat@unisa.it (B.P.)
2   Department of Engineering and Computer Science and Mathematics, University of L'Aquila,
    Via Vetoio, Loc. Coppito, 67100 L'Aquila, Italy; raffaele.dambrosio@univaq.it
*   Correspondence: ancardone@unisa.it; Tel.: +39-089-96-3342

Received: 27 April 2018; Accepted: 19 June 2018; Published: 1 July 2018

Abstract: We present a collection of recent results on the numerical approximation of Volterra integral equations and integro-differential equations by means of collocation type methods, which are able to provide better balances between accuracy and stability demanding. We consider both exact and discretized one-step and multistep collocation methods, and illustrate main convergence results, making some comparisons in terms of accuracy and efficiency. Some numerical experiments complete the paper.

Keywords: Volterra integral equations; Volterra integro–differential equations; collocation methods; multistep methods; convergence

MSC: 65R20; 65L03; 65D07; 65L20

---

1. Introduction

It is the purpose of this paper to illustrate recent results on collocation methods for Volterra integral equations (VIEs) and Volterra integro-differential equations (VIDEs), mainly due to the authors. Such equations model evolutionary problems with memory in many applications, such as dynamics of viscoelastic materials with memory, electrodynamics with memory, heat conduction in materials with memory [1–6]. The numerical solution of these equations has a high computational cost due both to the nonlinearity of the advancing term and to the evaluation of the lag term, which contains the past history of the solution. Therefore, a crucial point is finding accurate and efficient numerical methods.

Collocation methods have several desirable properties. They provide an approximation over the entire integration interval to the solution of the equation, which reveals to be quite useful in a variable-stepsize implementation: indeed, it is easy to recover the missing past values when the stepsize is changed, by evaluating the collocation polynomial. Other good properties of collocation methods are their high order of convergence, strong stability properties and flexibility. As a matter of fact, if some information is known on the behavior of the exact solution, then it is possible to choose the collocation functions in order to better follow such behavior, so giving rise to mixed collocation methods, see for example [7] in the case of ordinary differential equations (ODEs), and [8] in the case of VIEs. It is also worthwhile mentioning that collocation also has an important theoretical relevance: in fact, many numerical methods are difficult to be analyzed as discrete schemes while, re-casted as collocation-based methods, their analysis is reasonably simplified and can be carried out in a very elegant way. There is, however, a remarkable drawback of one-step collocation methods: they suffer from order reduction phenomenon when applied to stiff problems [9–11], since the order of convergence is not uniform (for instance, in the case of s-stage collocation based Runge-Kutta methods on Gauss-Legendre collocation points, the order is $p = 2s$ in the grid points, but it degenerates to $p = s$ for stiff problems, since the order is s in the internal stages). Such a drawback is successfully solved by two-step collocation methods [12], having high uniform order on the overall integration interval. On the side of computational cost, collocation methods are usually more expensive than other classes of methods. In fact, a collocation method with m collocation parameters requires at each time-step the solution of a nonlinear system of dimension m. To face this drawback, multistep collocation methods can be adopted which increase the order of convergence at the same computational cost of one-step ones. When a collocation method is applied to an integral equation, several integrals must be computed, thus suitable quadrature rules are needed to complete the discretization, with the introduction of an additional error. Lastly, a reliable error estimation for collocation methods for integral equations is still missing: there have been some advances (compare [13] and references therein contained), however considerable work needs to be done.

One-step collocation methods first appeared in the literature and main results are collected in the monographs [2,3]. Recently, we have proposed multistep collocation methods [13–16] and two step almost collocation methods [13,17,18], where the collocation polynomial depends on the approximate solution in a fixed number of previous time steps, with the aim of increasing the order of convergence of classical one–step collocation methods, without additional computational cost at each time step, and at the same time obtaining highly stable methods. This idea has been already proposed for the numerical solution of ODEs [19–21] (see also [11], Section V.3), and afterward modified in [12], by also using the inherent quadratic technique [22–24]. We also underline that they have high uniform order, thus they do not suffer from the order reduction phenomenon, well-known in the ODEs context [9]. Other approaches, based on multistep collocation, have been proposed in [25–32].

Here we briefly introduce one-step collocation methods and illustrate with more detail the construction and analysis of multistep collocation methods for VIEs and VIDEs, with the aim of giving a complete idea on the recent developments in this context. We give practical indications on how to choose the quadrature formulas in the discretized methods for an efficient implementation. In this review, we consider VIEs and VIDEs with smooth kernel and solution. We illustrate methods with a uniform mesh, however they could easily be applied to a non-uniform mesh (compare [2] for one-step collocation methods).

The paper is organized as follows. Sections 2 and 3 deal with one-step and multistep collocation methods for VIEs, respectively. Section 4 illustrates two-step almost collocation methods for VIEs. Sections 5 and 6 focus on one-step and multistep collocation methods for VIDEs, respectively.

2. One Step Collocation Methods for VIES

We consider VIEs of the form

$$y(t) = g(t) + \int_0^t k(t, \tau, y(\tau)) d\tau, \quad t \in I = [0, T], \tag{1}$$

where $k \in C(D \times \mathbb{R})$, with $D := \{(t, \tau) : 0 \leq \tau \leq t \leq T\}$, and $g \in C(I)$. In the literature, many authors (see [2,3] and references therein contained) have analyzed one step collocation methods for VIEs. As it is well known, a collocation method is based on the idea of approximating the exact solution of a given integral equation with a suitable function belonging to a chosen finite dimensional space, usually a piecewise algebraic polynomial which satisfies the integral equation exactly on a certain subset of the integration interval (called the set of collocation points).

Let us discretize the interval I by introducing a uniform mesh

$$I_h = \{t_n := nh, n = 0, ..., N, \ h \geq 0, Nh = T\}.$$

The Equation (1) can be rewritten, by relating it to this mesh, as

$$y(t) = F_n(t) + \Phi_n(t) \qquad\qquad t \in [t_n, t_{n+1}],$$

where

$$F_n(t) := g(t) + \int_0^{t_n} k(t, \tau, y(\tau))d\tau$$

and

$$\Phi_n(t) := \int_{t_n}^t k(t, \tau, y(\tau))d\tau$$

represent respectively the *lag term* and the *increment function*.

Collocation methods provide an approximation $P(t)$ to the solution $y(t)$ of (1) on $[0, T]$, such that its restriction to each interval $(t_n, t_{n+1}]$ is a polynomial:

$$P(t)|_{(t_n, t_{n+1}]} = P_n(t).$$

2.1. Exact One-Step Collocation Methods

Let us fix m collocation parameters $0 \leq c_1 < \ldots < c_m \leq 1$ and denote by $t_{nj} = t_n + c_j h$ the collocation points. The collocation polynomial, restricted to the interval $[t_n, t_{n+1}]$, is of the form:

$$P_n(t_n + sh) = \sum_{j=1}^m L_j(s)Y_{nj} \quad s \in [0, 1] \quad n = 0, \ldots, N-1 \tag{2}$$

where $L_j(s)$ is the j-th Lagrange fundamental polynomial with respect to the collocation parameters and $Y_{nj} := P_n(t_{nj})$. Exact collocation methods are obtained by imposing that the collocation polynomial (2) exactly satisfies the VIE (1) in the collocation points $t_{n,i}$ and by computing $y_{n+1} = P_n(t_{n+1})$:

$$\begin{cases} Y_{ni} = F_{ni} + \Phi_{ni} \\ y_{n+1} = \sum_{j=1}^m L_j(1)Y_{nj} \end{cases} , \tag{3}$$

where

$$F_{ni} = g(t_{ni}) + h\sum_{\nu=0}^{n-1} \int_0^1 k(t_{ni}, t_\nu + sh, P_\nu(t_\nu + sh))ds \tag{4}$$

$$\Phi_{ni} = h\int_0^{c_i} k(t_{ni}, t_n + sh, P_n(t_n + sh))ds, \tag{5}$$

$i = 1, \ldots, m$. Note that the first equation in (3) represents a system of m nonlinear equations in the m unknowns Y_{ni}. We recall that generally $P(t)$ is not continuous in the mesh points, as

$$P(t) \in S_{m-1}^{(-1)}(I_h), \tag{6}$$

where

$$S_\mu^{(d)}(I_h) = \left\{ v \in C^d(I) : v|_{(t_n, t_{n+1}]} \in \Pi_\mu, n = 0, 1, \ldots, N-1 \right\}.$$

Here, Π_μ denotes the space of (real) polynomials of degree not exceeding μ.

The classical collocation methods have uniform order $O(h^m)$ for any choice of the collocation parameters, and can achieve local superconvergence in the mesh points by opportunely choosing the collocation parameters, i.e., order $2m - 2$ with m Lobatto points or $m - 1$ Gauss points with $c_m = 1$ and order $2m - 1$ with m Radau II points. The optimal superconvergence order $O(h^{2m})$ in the mesh points can be achieved with Gauss nodes in the iterated collocation methods [2,3].

2.2. Discretized One-Step Collocation Methods

The collocation Equation (3) is not yet in a form amenable to numerical computation: another discretization step, based on quadrature formulas $\tilde{F}_{ni} \simeq F_{ni}$ and $\tilde{\Phi}_{ni} \simeq \Phi_{ni}$ for the approximation of (4) and (5) are needed in order to obtain the fully discretised collocation schemes, thus leading to *Discretized* collocation methods.

The discretized collocation polynomial is of the form

$$\tilde{P}_n(t_n + sh) = \sum_{j=1}^{m} L_j(s)\tilde{Y}_{nj} \quad s \in [0,1] \quad n = 0, ..., N-1 \tag{7}$$

where $\tilde{Y}_{nj} := \tilde{P}_n(t_{nj})$. The m unknowns $\tilde{Y}_{nj}$ are determined by imposing that the collocation polynomial (7) satisfies exactly the integral equation at the collocation points and by using quadrature formulas of the form

$$\tilde{\Phi}_n(t_{ni}) = h \sum_{l=0}^{\mu_0} w_{il} k(t_{ni}, t_n + d_{il}h, \tilde{P}_n(t_n + d_{il}h)) \tag{8}$$

$$\tilde{F}_n(t_{ni}) = g(t_{ni}) + h \sum_{v=0}^{n-1} \sum_{l=0}^{\mu_1} b_l k(t_{ni}, t_v + \xi_l h, \tilde{P}_v(t_v + \xi_l h)), \tag{9}$$

$i = 1, ..., m$, for approximating the lag term (4) and the increment function (5). The Formulas (8) and (9) are obtained by using quadrature formulas of the form

$$(\xi_l, b_l)_{l=1}^{\mu_1}, \quad (d_{il}, w_{il})_{l=1}^{\mu_0}, \ i = 1, ..., m, \tag{10}$$

where the quadrature nodes ξ_l and d_{il} satisfy $0 \leq \xi_1 < ... < \xi_{\mu_1} \leq 1$ and $0 \leq d_{i1} < ... < d_{i\mu_0} \leq 1$, μ_0 and μ_1 are positive integers and w_{il}, b_l are suitable weights.

The numerical method is then of the form:

$$\begin{cases} \tilde{Y}_{ni} = \tilde{F}_n(t_{ni}) + \tilde{\Phi}_n(t_{ni}) \\ \tilde{y}_{n+1} = \sum_{j=1}^{m} L_j(1)\tilde{Y}_{nj} \end{cases}, \tag{11}$$

where $\tilde{\Phi}_n(t_{ni})$ and $\tilde{F}_n(t_{ni})$ are given by (8) and (9).

Note that the first equation in (9) represents a system of m nonlinear equations in the m unknowns $\tilde{Y}_{ni}$.

Such methods preserve, under suitable hypothesis on the quadrature Formulas (8) and (9), the same order of the exact collocation methods [3].

A collocation method for VIEs is equivalent to an implicit Runge-Kutta method for VIEs (VRK method) if and only if $c_m = 1$ (see Theorem 5.2.2 of [3]). As the lag–term computation is the most expensive part in the numerical solution of VIEs, fast collocation and Runge–Kutta methods have been constructed for convolution VIEs of Hammerstein type [33,34] in order to reduce the computational effort in the lag–term computation. The stability analysis of collocation methods for VIEs can be found in [3,35] and the related bibliography.

3. Multistep Collocation Methods for VIEs

Multistep collocation methods for VIEs have been introduced in [16] by adding interpolation conditions in r previous step points, with the aim of increasing the uniform order of convergence of one step collocation methods without increasing the computational cost. The multistep collocation polynomial, restricted to the interval $[t_n, t_{n+1}]$, is of the form

$$P_n(t_n + sh) = \sum_{k=0}^{r-1} \varphi_k(s)y_{n-k} + \sum_{j=1}^{m} \psi_j(s)Y_{nj} \quad s \in [0,1], \tag{12}$$

$n = r, ..., N - 1$, where again

$$Y_{nj} := P_n(t_{nj}) \tag{13}$$

and $\varphi_k(s)$, $\psi_j(s)$ are the following polynomials of degree $m + r - 1$

$$\varphi_k(s) = \prod_{i=1}^{m} \frac{s - c_i}{-k - c_i} \cdot \prod_{\substack{i=0 \\ i \neq k}}^{r-1} \frac{s + i}{-k + i}, \quad \psi_j(s) = \prod_{i=0}^{r-1} \frac{s + i}{c_j + i} \cdot \prod_{\substack{i=1 \\ i \neq j}}^{m} \frac{s - c_i}{c_j - c_i}. \tag{14}$$

The collocation parameters are assumed to satisfy $c_i \neq c_j$ and $c_1 \neq 0$.

3.1. Exact Multistep Collocation

The exact multistep collocation methods are obtained by imposing that the collocation polynomial (12) exactly satisfies the VIE (1) at the collocation points t_{ni}, and by computing $y_{n+1} = P_n(t_{n+1})$:

$$\begin{cases} Y_{ni} = F_{ni} + \Phi_{ni}, \\ y_{n+1} = \sum_{k=0}^{r-1} \varphi_k(1) y_{n-k} + \sum_{j=1}^{m} \psi_j(1) Y_{nj}, \end{cases} \tag{15}$$

where the lag–term F_{ni} and increment–term Φ_{ni} are given by (4) and (5) respectively. The r-step m-point exact collocation method (12)–(15) has uniform convergence order of at least $p = m + r$, for any choice of distinct collocation abscissas $0 < c_1 < ... < c_m \leq 1$, as stated in the following theorem proved in [16].

Theorem 1. *Let $\varepsilon(t) = y(t) - P(t)$ be the error of the exact collocation method (12)–(15) and $p = m + r$. Suppose that*

i.   *the given functions describing the VIE (1) satisfy $k \in C^{(p)}(D \times \mathbb{R})$, $g \in C^{(p)}(I)$.*
ii. *the starting error is $\|\varepsilon\|_{\infty,[0,t_r]} = O(h^p)$.*
iii. *$\rho(\mathbf{A}) < 1$, where*

$$\mathbf{A} = \left[\begin{array}{c|c} \mathbf{0}_{r-1,1} & \mathbf{I}_{r-1} \\ \hline \varphi_{r-1}(1) & \varphi_{r-2}(1), ..., \varphi_0(1) \end{array} \right] \tag{16}$$

and ρ denotes the spectral radius.

Then

$$\|\varepsilon\|_\infty = O(h^{m+r}).$$

Moreover, a suitable choice of collocation parameters can ensure superconvergence in the mesh points, as pointed out in the following theorem [16].

Theorem 2. *Let us suppose that*

- *the hypothesis of the Theorem 1 hold with $p = 2m + r - 1$.*
- *the collocation parameters $c_1, ..., c_m$ are the solution of the system*

$$\begin{cases} c_m = 1 \\ \frac{1}{i+1} - \sum_{k=0}^{r-1} \beta_k(-k)^i - \sum_{j=1}^{m} \gamma_j(c_j)^i = 0, \quad i = m + r, ..., 2m + r - 2 \end{cases} \tag{17}$$

with

$$\beta_k = \int_0^1 \varphi_k(s) ds, \quad \gamma_j = \int_0^1 \psi_j(s) ds \tag{18}$$

then

$$\max_{n=0,\dots,N} |\varepsilon(t_n)| = O(h^{2m+r-1}).$$

3.2. Discretized Multistep Collocation

The discretized multistep collocation methods are obtained by using quadrature formulas of the form (8) to (9) for approximating the lag term and the increment function. The discretized multistep collocation polynomial, denoted by $\tilde{P}_n(t)$, is then of the form

$$\tilde{P}_n(t_n + sh) = \sum_{k=0}^{r-1} \varphi_k(s)\tilde{y}_{n-k} + \sum_{j=1}^{m} \psi_j(s)\tilde{Y}_{nj}, \quad s \in [0,1] \tag{19}$$

$n = 0, \dots, N-1$, where the functions $\varphi_k(s)$ and $\psi_j(s)$ are given by (14), and $\tilde{Y}_{nj} := \tilde{P}_n(t_{nj})$ are determined by the solution of the following nonlinear system

$$\begin{cases} \tilde{Y}_{ni} = \tilde{F}_{ni} + \tilde{\Phi}_{ni}, \\ \tilde{y}_{n+1} = \sum_{k=0}^{r-1} \varphi_k(1)\tilde{y}_{n-k} + \sum_{j=1}^{m} \psi_j(1)\tilde{Y}_{nj}. \end{cases} \tag{20}$$

The following theorem [16] shows that, as in the exact case, the r-step m-point discretized collocation method (19) and (20) has convergence order of at least $p = m + r$, for any choice of distinct collocation abscissas $0 < c_1 < \dots < c_m \le 1$.

Theorem 3. *Let $\tilde{\varepsilon}(t) := y(t) - \tilde{P}(t)$ be the error of the discretized collocation method (19) and (20) and let $p = m + r$. Suppose that*

i.   *the given functions describing the VIE (1) satisfy $k \in C^{(p)}(D)$, $g \in C^{(p)}(I)$;*
ii. *the lag–term and increment–term quadrature Formulas (10) are of order respectively at least $p + 1$ and p;*
iii. *the starting error is $\|\tilde{\varepsilon}\|_{\infty,[0,t_r]} = O(h^p)$.*
iv. *$\rho(\mathbf{A}) < 1$, where A is given by (16).*

Then

$$\|\tilde{\varepsilon}\|_{\infty} = O(h^{m+r}).$$

An analogous result holds concerning the local superconvergence:

Theorem 4. *Let us suppose that*

- *the hypothesis of the Theorem 3 hold with $p = 2m + r - 1$.*
- *the collocation parameters $c_1, \dots, c_m$ are the solution of the system (17).*

Then

$$\max_{n=0,\dots,N} |\tilde{\varepsilon}(t_n)| = O(h^{2m+r-1}).$$

4. Two Step Almost Collocation Collocation Methods for VIEs

Within the class of multistep collocation methods, although methods with unbounded stability regions exist, no A-stable methods have been found [16]. In order to determine A-stable methods, two step *almost* collocation (TSAC) methods have been introduced in [18] and further analyzed in [13,17].

The collocation polynomial $P_n(t)$ for TSAC methods is computed by employing the information about the equation on two consecutive steps:

$$P_n(t_n + sh) = \varphi_0(s)y_{n-1} + \varphi_1(s)y_n + \sum_{j=1}^{m} \chi_j(s)Y_j^{[n]} + \sum_{j=1}^{m} \psi_j(s)(F_j^{[n]} + \Phi_j^{[n+1]}), \tag{21}$$

where $Y_j^{[n]} = P(t_{n-1,j})$. Then the method assumes the form:

$$
\begin{cases}
Y_i^{[n+1]} = \varphi_0(c_i)y_{n-1} + \varphi_1(c_i)y_n + \sum_{j=1}^{m} \chi_j(c_i)Y_j^{[n]} + \sum_{j=1}^{m} \psi_j(c_i)\left(F_j^{[n]} + \Phi_j^{[n+1]}\right), \\
y_{n+1} = \varphi_0(1)y_{n-1} + \varphi_1(1)y_n + \sum_{j=1}^{m} \chi_j(1)Y_j^{[n]} + \sum_{j=1}^{m} \psi_j(1)\left(F_j^{[n]} + \Phi_j^{[n+1]}\right),
\end{cases}
\tag{22}
$$

where $F_j^{[n]}$ and $\Phi_j^{[n+1]}$ are suitable sufficiently high order quadrature formulae for the discretization of $F^{[n]}(t_{nj})$ and $\Phi^{[n+1]}(t_{nj})$ respectively, assuming the form

$$
F_j^{[n]} = g(t_{nj}) + h \sum_{v=1}^{n} \sum_{l=0}^{m+1} b_{lk}\left(t_{nj}, t_{v-1,l}, Y_l^{[v]}\right),
\tag{23}
$$

and

$$
\Phi_j^{[n+1]} = h \sum_{l=0}^{m+1} w_{jl}k\left(t_{nj}, t_{nl}, Y_l^{[n+1]}\right).
\tag{24}
$$

In the quadrature Formulas (23) and (24) we mean $t_{v-1,0} = t_{v-1}$, $t_{v-1,m+1} = t_v$, $Y_0^{[v]} = P_n(t_{v-1})$, $Y_{m+1}^{[v]} = P_n(t_v)$ and $t_{n0} = t_n$. We observe as the method (22) requires, at each step, the solution of a nonlinear system of $(m+1)d$ equations in the stage values $Y_i^{[n+1]}$ and y_{n+1}.

The basis functions $\varphi_0(s)$, $\varphi_1(s)$, $\chi_j(s)$ and $\psi_j(s)$, $j = 1, 2, \ldots, m$, are polynomials of degree p, determined from the continuous order conditions, according to the following theorem [18]:

Theorem 5. *Assume that the kernel $k(t, \eta, y)$ and the function $g(t)$ in (1) are sufficiently smooth. Then the method (21) and (22) has uniform order p, i.e.,*

$$
\eta(t_n + sh) = O(h^{p+1}), \quad h \to 0,
$$

for $s \in [0, 1]$, if the polynomials $\varphi_0(s)$, $\varphi_1(s)$, $\chi_j(s)$ and $\psi_j(s)$, $j = 1, 2, \ldots, m$ satisfy the system of equations

$$
\begin{cases}
1 - \varphi_0(s) - \varphi_1(s) - \sum_{j=1}^{m} \chi_j(s) - \sum_{j=1}^{m} \psi_j(s) = 0, \\
s^k - (-1)^k \varphi_0(s) - \sum_{j=1}^{m}(c_j - 1)^k \chi_j(s) - \sum_{j=1}^{m} c_j^k \psi_j(s) = 0,
\end{cases}
\tag{25}
$$

$s \in [0, 1]$, $k = 1, 2, \ldots, p$, where

$$
\eta(t_n + sh) = y(t_n + sh) - \varphi_0(s)y(t_n - h) - \varphi_1(s)y(t_n) - \sum_{j=1}^{m}\left(\chi_j(s)y(t_n + (c_j - 1)h) + \psi_j(s)y(t_n + c_jh)\right). \tag{26}
$$

is the local truncation error.

As regards the global error, the method has uniform order of convergence $p^* = \min\{l+1, q, p+1\}$, where l and q are the order of the starting procedure (for the computation of the starting values y_1 and $Y_i^{[1]}$, $i = 1, 2, \ldots, m$) and the order of the quadrature Formulas (23) and (24) respectively (see Theorem 2.5 in [18]). Then we use as starting procedure a one step collocation method having uniform order of convergence $l = p$.

Two-step collocation methods are obtained by solving the system of order conditions up to the maximum uniform attainable order $p = 2m + 1$, and, in this way, all the basis functions are determined as the unique solution of such system. However, as observed in [18], it is not convenient to impose all the order conditions because it is not possible to achieve high stability properties (e.g., *A*-stability) without getting rid of some of them. Therefore, *almost* collocation methods have been introduced by relaxing a specified number r of order conditions, i.e., by a priori opportunely fixing r basis functions, and determining the remaining ones as the unique solution of the system of order conditions up to $p = 2m + 1 - r$. Within the class of TSAC methods, *A*-stable methods have been constructed in [18] by fixing one (case $r = 1$) or both (case $r = 2$) of the polynomials $\varphi_0(s)$ and $\varphi_1(s)$ as

$$
\begin{aligned}
\varphi_0(s) &= \prod_{k=1}^{m}(s - c_k)(q_0 + q_1 s + \ldots + q_{p-m}s^{p-m}), \\
\varphi_1(s) &= \prod_{k=1}^{m}(s - c_k)(p_0 + p_1 s + \ldots + p_{p-m}s^{p-m}),
\end{aligned}
\tag{27}
$$

where α_j and β_j, $j = 0, 1, \ldots, p - m$, are free parameters, which have to be determined in order to obtain desired stability properties.

A error estimation of the local discretization error for TSAC methods has been derived in [13].

Example 1. *Let us consider the methods with two stages* $m = 2$ *and order* $p = 2m = 4$. *Classes of A-stable methods were derived in [13,18] by considering*

$$
\varphi_0(s) = s(s - c_1)(s - c_2)(q_0 + q_1 s),
$$

where c_1, c_2, q_0, q_1 *are free parameters. The weights in* (23) *and* (24) *were computed in [18] as*

$$
\mathbf{b} = \begin{bmatrix} \dfrac{-1 + 2c_1 + 2c_2 - 6c_1 c_2}{12 c_1 c_2} \\[2ex] \dfrac{1 - 2c_2}{12 c_1 (c_1 - 1)(c_1 - c_2)} \\[2ex] \dfrac{2c_1 - 1}{12 c_2 (c_2 - 1)(c_2 - c_1)} \\[2ex] \dfrac{-3 + 4c_1 + 4c_2 - 6c_1 c_2}{12(c_1 - 1)(c_2 - 1)} \end{bmatrix}, \quad \mathbf{W} = \begin{bmatrix} -\dfrac{c_1^2 - 3c_1 c_2}{6c_2} & \dfrac{c_1(2c_1 - 3c_2)}{6(c_1 - c_2)} & \dfrac{c_1^3}{6c_2(c_1 - c_2)} & 0 \\[2ex] -\dfrac{c_2^2 - 3c_1 c_2}{6c_1} & -\dfrac{c_1^3}{6c_1(c_1 - c_2)} & -\dfrac{c_2(2c_2 - 3c_1)}{6(c_1 - c_2)} & 0 \end{bmatrix}.
$$

An *A*-stable method is obtained by choosing for example $q_0 = 15/10$, $q_1 = -1$, $c_1 = 0.9$, $c_2 = 0.95$, see [13].

4.1. Diagonally Implicit TSAC Methods for VIEs

The computational cost associated to the solution of the nonlinear system (22) can be reduced by making the coefficient matrix have a structured shape, e.g., lower triangular or diagonal. This strategy, in the field of Runge–Kutta methods for ODEs, leads to the raise of the famous classes of Diagonally Implicit and Singly Diagonally Implicit Runge-Kutta methods (DIRK and SDIRK), see [10,11] and bibliography therein contained. Moreover, in the field of collocation-based methods for ODEs, an analogous strategy has been applied, obtaining TSAC methods having structured coefficient matrix [12].

In fact, a lower triangular matrix allows to solve the equations in m successive stages, with only a d-dimensional system to be solved at each stage. Moreover, if all the elements on the diagonal are equal, in solving the nonlinear systems by means of Newton-type iterations, one may hope to use repeatedly the stored LU factorization of the Jacobian. If the structure is diagonal, the problem reduces to the solution of m independent systems of dimension d, and can therefore be solved in a parallel environment.

Methods of this type have been derived in [17], where first of all it was assumed $w_{j,m+1} = 0$, $j = 1, \ldots, m$, in such a way that (22) becomes a nonlinear system of dimension md only depending on the stage values $Y_i^{[n+1]}$, $i = 1, \ldots, m$, and assumes the following form

$$\begin{cases} Y_i^{[n+1]} - h \sum_{j=1}^{m} \sum_{l=1}^{m} \psi_j(c_i) w_{jl} k(t_{nj}, t_{nl}, Y_l^{[n+1]}) = B_i^{[n]}, \\ y_{n+1} = P_n(t_{n+1}), \end{cases} \tag{28}$$

where

$$B_i^{[n]} = \varphi_0(c_i) y_{n-1} + \varphi_1(c_i) y_n + \sum_{j=1}^{m} \chi_j(c_i) Y_j^{[n]} + \sum_{j=1}^{m} \psi_j(c_i) F_j^{[n]} + h \sum_{j=1}^{m} \psi_j(c_i) w_{j0} k(t_{nj}, t_n, y_n). \tag{29}$$

By defining

$$Y^{[n+1]} = \left[Y_1^{[n+1]}, Y_2^{[n+1]}, \ldots, Y_m^{[n+1]} \right]^T, \quad B^{[n]} = \left[B_1^{[n]}, B_2^{[n]}, \ldots, B_m^{[n]} \right]^T, \quad \Psi = \left(\psi_j(c_i) \right)_{i,j=1}^{m},$$

$$W = \left(w_{jl} \right)_{j,l=1}^{m}, \quad K(t_{nc}, t_{nc}, Y^{[n+1]}) = \left(K(t_{ni}, t_{nj}, Y_j^{[n+1]}) \right)_{i,j=1}^{m},$$

the nonlinear system in (28) takes the form

$$Y^{[n+1]} - h(\Psi \otimes I) \left((W \otimes I) \cdot K(t_{nc}, t_{nc}, Y^{[n+1]}) \right) e = B^{[n]}, \tag{30}$$

where $\cdot$ denotes the usual Hadamard product, I is the identity matrix of dimension d and e is the unit vector of dimension md. The tensor form (30) clearly shows as the matrices which determine the structure of the nonlinear system (28) are Ψ and W. In [17] a strategy was described to obtain lower triangular or diagonal structures for the matrices Ψ and W: in particular a quadrature formula of the form

$$\int_0^{c_j} f(s) ds \approx w_{j0} f(0) + \sum_{l=1}^{m} \tilde{w}_{jl} f(c_l - 1) + \sum_{l=1}^{j} w_{jl} f(c_l), \tag{31}$$

was proposed for the increment

$$\Phi^{[n+1]}(t_{nj}, P(\cdot)) = h \int_0^{c_j} k(t_{nj}, t_n + sh, P_n(t_n + sh)) ds, \tag{32}$$

in addition to the quadrature formula

$$\int_0^1 f(s) ds \approx b_0 f(0) + \sum_{l=1}^{m} b_l f(c_l) + b_{m+1} f(1), \tag{33}$$

for the approximation of the lag term

$$F^{[n]}(t_{nj}, P(\cdot)) = g(t_{nj}) + h \sum_{\nu=1}^{n} \int_0^1 k(t_{nj}, t_{\nu-1} + sh, P_{\nu-1}(t_{\nu-1} + sh)) ds. \tag{34}$$

We observe that in Formula (31), in case of triangular structure, $\tilde{w}_{jl} = 0$, $l = 1, \ldots, j$ while, in case of diagonal structure, $\tilde{w}_{j1} = 0$ and $w_{jl} = 0$, $l = 1, \ldots, j - 1$. The determination of the weights in Formulas (31) and (33) has been described in [17].

Assuming that Ψ and W are lower triangular, we obtain the diagonally implicit TSAC methods (DITSAC)

$$
\begin{cases}
Y_i^{[n+1]} - h\psi_i(c_i)w_{ii}k(t_{ni}, t_{ni}, Y_i^{[n+1]}) = B_i^{[n]} + \tilde{B}_i^{[n]} + h\sum_{l=1}^{i-1}\sum_{j=l}^{i}\psi_j(c_i)w_{jl}k(t_{nj}, t_{nl}, Y_l^{[n+1]}), \\
y_{n+1} = \varphi_0(1)y_{n-1} + \varphi_1(1)y_n + \sum_{j=1}^{m}\chi_j(1)Y_j^{[n]} + \sum_{j=1}^{m}\psi_j(1)\left(F_j^{[n]} + \Phi_j^{[n+1]}\right),
\end{cases}
\tag{35}
$$

where $B_i^{[n]}$ is given by (29),

$$
\tilde{B}_i^{[n]} = h\sum_{j=1}^{i}\sum_{l=1}^{m}\psi_j(c_i)\tilde{w}_{jl}k(t_{nj}, t_{n-1,l}Y_l^{[n]}),
\tag{36}
$$

and $F_j^{[n]}$, $\Phi_j^{[n+1]}$ are approximations of (34) by means of the quadrature Formulas (31) and (33).

4.2. Numerical Results

We present some numerical results which confirm that, differently from one step collocation methods, the TSAC methods do not suffer form the order reduction in the integration of stiff systems, as we expect from the uniform order of convergence stated in Theorem 5. In order to illustrate this phenomenon, we show the results obtained on both a non stiff and a stiff equation:

- the non stiff VIE

$$
y(t) = 2 - \cos(t) - \int_0^t \sin(ty(\tau) - \tau)d\tau, \quad t \in [0, 3],
\tag{37}
$$

with exact solution $y(t) \equiv 1$;
- the stiff VIE

$$
y(t) = \int_0^t \left(\lambda\left(y(\tau) - \sin(\tau)\right) + \cos(\tau)\right)d\tau, \quad t \in [0, \frac{3}{4}\pi],
\tag{38}
$$

with $\lambda = -10^4$ and exact solution $y(t) = \sin(t)$. This is a stiff problem because it is equivalent to the Prothero-Robinson problem for ODEs.

We compare TSAC methods with superconvergent one step collocation methods of [2,3], where m denotes the number of collocation points and p denotes the order of the method:

- G2: 1 point Gauss collocation, $c_2 = 1, m = 2, p = 2$;
- R2: 2 points Radau collocation, $m = 2, p = 3$;
- TSAC2: 2 points TSAC method, $m = 2, p = 4$.

The method TSAC2 is the two-stage TSAC method described in Example 1. The accuracy is defined by the number of correct significant digits cd at the end point (the maximal absolute end point error is written as 10^{-cd}). For each test we plot in Figure 1 the number of cd versus the number of mesh points N. We observe as for non stiff Problem (37) the effective order of the all methods is coherent with the theoretical order, while for stiff Problem (38) the one step methods show order reduction as the effective order reduces to $p = 2$.

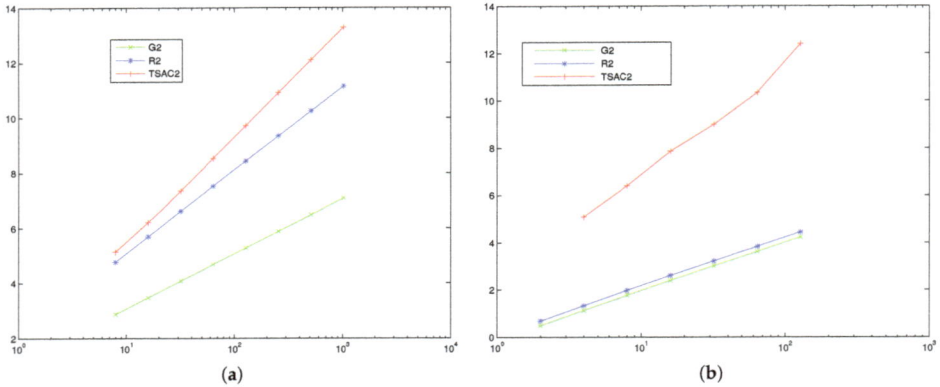

Figure 1. Number of correct significant digits with respect to the number of mesh points. (a) Problem (37); (b) Problem (38).

5. One-Step Collocation Methods for VIDEs

We concentrate on VIDEs of type:

$$y'(t) = g(t, y(t)) + \int_0^t k(t, \tau, y(\tau))d\tau, \ t \in I = [0, T],$$
$$y(0) = y_0,$$

(39)

where $g(t, y) : I \times \mathbf{R}^d \to \mathbf{R}^d$, $k(t, s, y) : S \times \mathbf{R}^d \to \mathbf{R}^d$, $S = \{(t, s) | 0 \leq s \leq t \leq T\}$. For sake of completeness we report the theorem of existence and uniqueness of solution for (39) [3].

Theorem 6. *Let $g(t, y)$ and $k(t, s, y)$ be continuous functions and satisfy a uniform Lipschitz condition with respect to y. Then there exists a unique solution $y \in C^1([0, T])$ of the problem (39).*

Let $I_h = \{t_n : 0 < t_0 < t_1 < \cdots < t_N = T\}$ be a partition of the time interval $[0, T]$ with constant stepsize $h = t_{n+1} - t_n$, $n = 0, \ldots, N - 1$. The integro-differential Equation (39) can be written as follows:

$$y'(t) = g(t, y(t)) + F_n(t, y(\cdot)) + \Phi_n(t, y(\cdot)), \quad t \in [t_n, t_{n+1}],$$

where

$$F_n(t, y(\cdot)) = \int_0^{t_n} k(t, \tau, y(\tau))d\tau, \quad \Phi_n(t, y(\cdot)) = \int_{t_n}^t k(t, \tau, y(\tau))d\tau,$$

represent respectively the *lag term* and the *increment function*.

5.1. Exact One-Step Collocation Methods

Here we briefly expose the classical one-step collocation methods for VIDEs and their main properties [2,3].

A one-step collocation method approximates $y(t)$ by a piecewise polynomial $P(t)$, with $P(t) = P_n(t)$, $t \in [t_n, t_{n+1}]$, $n = 0, \ldots, N - 1$, where

$$P_n(t_n + sh) = y_n + h \sum_{j=1}^m \beta_j(s) U_{nj}, \quad s \in [0, 1],$$

(40)

with $y_n = P_n(t_n)$, $U_{nj} = P_n'(t_n + c_j h)$, $\beta_j(s) = \int_0^s L_j(\tau)ds$, $L_j(\tau)$ being the j-th Lagrange fundamental polynomial with respect to the collocation parameters.

The m unknowns U_{nj} are found by imposing that $P_n(t)$ satisfies (39) at the collocation points $t_{nj} := t_n + c_j h$, $j = 1, \ldots, m$, $n = 0, \ldots, N - 1$, where $0 \le c_1 < \ldots < c_m \le 1$ are the collocation parameters.

The numerical approximation at the point t_{n+1} is then given by $y_{n+1} = P_n(t_{n+1})$. The final form of an exact collocation method is

$$\begin{cases} U_{ni} = & g(t_{ni}, P_n(t_{ni})) + F_n(t_{ni}, P(\cdot)) + \Phi_n(t_{ni}, P(\cdot)), \quad i = 1, \ldots, m, \\ y_{n+1} = & y_n + h \sum_{i=1}^{m} \beta_j(1) U_{ni}, \end{cases} \tag{41}$$

$n = 0, \ldots, N$, where the lag term and the increment function can be written as

$$F_n(t_{ni}, P(\cdot)) = h \sum_{v=0}^{n-1} \int_0^1 k\left(t_{ni}, t_v + \tau h, P_v(t_n + \tau h)\right) d\tau, \tag{42}$$

$$\Phi_n(t_{ni}, P(\cdot)) = h \int_0^{c_i} k\left(t_{ni}, t_n + \tau h, P_n(t_n + \tau h)\right) d\tau. \tag{43}$$

The first equation in (41) requires, at each time step, the solution of an m-dimensional nonlinear system in the unknowns $\{U_{ni}\}_{i=1}^{m}$.

For every choice of the collocation parameters $c_1, \ldots, c_m$, the collocation polynomial $P(t)$ is continuous on $[0, T]$ and provides a uniform approximation of order $O(h^m)$. Moreover, if $c_1, \ldots, c_m$ are suitably chosen, the order of convergence at the mesh points increases (local superconvergence): is $2m - 2$ for the Lobatto points, $2m - 1$ for the Radau points and $2m$ for the Gauss ones [2,3].

5.2. Discretized One-Step Collocation Methods

In the general case, the integrals appearing in (42) and (43) cannot be exactly evaluated, so a further approximation is needed in order to fully discretize the method. Let us suppose to approximate these integrals by quadrature formulae of the type:

$$\tilde{F}_n(t_{ni}, P(\cdot)) = h \sum_{v=0}^{n-1} \sum_{l=1}^{\mu_1} w_l k(t_{ni}, t_v + d_l h, P_v(t_v + d_l h)), \tag{44}$$

$$\tilde{\Phi}_n(t_{ni}, P(\cdot)) = h \sum_{l=1}^{\mu_0} w_{il} k(t_{ni}, t_n + d_{il} h, P_n(t_n + d_{il} h)). \tag{45}$$

These formulae are then used to define the discretized collocation methods as

$$\begin{cases} \tilde{U}_{ni} = g(t_{ni}, \tilde{P}_n(t_{ni})) + \tilde{F}_n(t_{ni}, \tilde{P}(\cdot)) + \tilde{\Phi}_n(t_{ni}, \tilde{P}(\cdot)) \\ \tilde{y}_{n+1} = \tilde{y}_n + h \sum_{i=1}^{m} \beta_j(1) \tilde{U}_{ni}, \end{cases} \tag{46}$$

where the collocation polynomial is now of the form

$$\tilde{P}_n(t_n + sh) = y_n + h \sum_{j=1}^{m} \beta_j(s) \tilde{U}_{nj}, \quad s \in [0, 1]. \tag{47}$$

The discretized collocation methods are a special class of the Runge-Kutta extended methods and preserve the order of convergence and superconvergence of exact collocation methods, if the quadrature Formulae (44) and (45) are sufficiently accurate [3].

Some relevant stability results for one-step collocation methods are derived in [36,37].

6. Multistep Collocation for VIDEs

6.1. Exact Multistep Collocation

Recently, in order to obtain an higher order of convergence at the same computational effort, multistep collocation methods have been introduced: the solution $y(t)$ is approximated by a piecewise algebraic polynomial $P(t)$:

$$P(t_n + sh) = \sum_{k=0}^{r-1} \varphi_k(s) y_{n-k} + h \sum_{j=1}^{m} \psi_j(s) U_{nj}, \quad s \in [0,1], \tag{48}$$

where again

$$U_{nj} := P'(t_{nj}), \quad j = 1, \ldots, m, \tag{49}$$

and the functions $\varphi_k(s)$, $\psi_j(s)$ are polynomials of degree $m + r - 1$ which are determined by imposing that the polynomial (48) satisfies (49) and the interpolation conditions:

$$P(t_{n-k}) = y_{n-k}, \quad k = 0, \ldots, r - 1. \tag{50}$$

For any fixed set of collocation parameters $c_1, \ldots, c_m$, conditions (49) and (50) lead to the following non linear system of $(r + m)^2$ equations, where the $(r + m)^2$ unknowns are the coefficients of the polynomials $\varphi_k(s)$ and $\psi_j(s)$:

$$\begin{aligned} \varphi_l(-k) &= \delta_{lk}, \quad \varphi'_l(c_j) = 0, \\ \psi'_i(c_j) &= \delta_{ij}, \quad \psi_i(-k) = 0, \end{aligned} \tag{51}$$

$l, k = 0, \ldots, r - 1, \ i, j = 1, \ldots, m$.

Exact multistep collocation methods are obtained by imposing that the collocation polynomial (48) satisfies the VIDE at the collocation points t_{ni}, and by computing $y_{n+1} = P_n(t_{n+1})$:

$$\begin{cases} U_{ni} = g(t_{ni}, P(t_{ni})) + F_n(t_{ni}, P(\cdot)) + \Phi_n(t_{ni}, P(\cdot)), \quad i = 1, \ldots, m \\ y_{n+1} = \sum_{k=0}^{r-1} \varphi_k(1) y_{n-k} + h \sum_{i=1}^{m} \psi_j(1) U_{ni}. \end{cases} \tag{52}$$

$n = r - 1, \ldots, N$, where now the lag term and the increment function can be written as

$$F_n(t_{ni}, P(\cdot)) = h \sum_{v=0}^{n-1} \int_0^1 k \left(t_{ni}, t_v + \tau h, \sum_{k=0}^{r-1} \varphi_k(\tau) y_{v-k} + h \sum_{j=1}^{m} \psi_j(\tau) U_{vj} \right) d\tau, \tag{53}$$

$$\Phi_n(t_{ni}, P(\cdot)) = h \int_0^{c_i} k \left(t_{ni}, t_n + \tau h, \sum_{k=0}^{r-1} \varphi_k(\tau) y_{n-k} + h \sum_{j=1}^{m} \psi_j(\tau) U_{nj} \right) d\tau. \tag{54}$$

We note that at each time step, the approximations $y_{n-k}, k = 0, \ldots, r - 1$ are already known, so only the unknowns $\{U_{ni}\}_{i=1}^{m}$ need to be computed, by solving the nonlinear system given by the first equation of (52).

Observe that we are able to give an approximate value $P(t)$ of the solution $y(t)$ at each point t of the integration interval, therefore we have a uniform approximation of the solution on $[0, T]$.

The classical one-step collocation methods described in the previous section can be seen as a particular case of multistep methods with $r = 1$ and

$$\varphi_0(s) \equiv 1, \quad \psi_j(s) = \int_0^s L_j(\tau) d\tau,$$

where $L_j(\tau)$ is the j-th Lagrange fundamental polynomial with respect to the collocation parameters. We observe that, at each time step, both one-step and multistep collocation methods require the solution of a non linear system of dimension m for the stages $U_{ni}, i = 1, \ldots, m$. The multistep methods only need in addition the computation of the starting values $y_1, \ldots, y_{r-1}$.

6.2. Discretized Multistep Collocation

As in the case of one-step collocation methods, it is evident that the exact multistep collocation methods (52) are not directly applicable for the implementation, since approximations of the integrals $F_n(t_{ni}, P(\cdot))$ and $\Phi_n(t_{ni}, P(\cdot))$ are needed. With the aim of fully discretizing the multistep collocation methods we consider the following quadrature formulas

$$\int_0^{c_i} \alpha(x)dx \approx Q_i(\alpha(\cdot)) := \sum_{l=1}^{\mu_0} w_{il}\alpha(d_{il}), \quad \int_0^1 \alpha(x)dx \approx Q(\alpha(\cdot)) := \sum_{l=1}^{\mu_1} w_l\alpha(d_l), \tag{55}$$

where the weights and nodes are suitably chosen, as it will be illustrated later.

The *discretized* multistep collocation method for the problem (39) approximates the solution $y(t)$ with a piecewise polynomial $\tilde{P}(t)$, with

$$\tilde{P}(t_n + sh) = \sum_{k=0}^{r-1} \varphi_k(s)\tilde{y}_{n-k} + h \sum_{j=1}^{m} \psi_j(s)\tilde{U}_{nj}, s \in [0,1], \tag{56}$$

where the polynomials $\{\varphi_k(s)\}_{k=0}^{r-1}, \{\psi_j(s)\}_{j=1}^{m}$ are the same as in the exact collocation, and can be computed by solving the system (51).

We impose that at the collocation points $\tilde{P}(t)$ satisfies the VIDE (39), where the integrals appearing in both the lag term (53) and the increment function (54) are approximated by the quadrature formulae defined in (55), and we set $\tilde{y}_{n+1} = \tilde{P}(t_n + h)$. Thus the discretized multistep method is

$$\begin{cases} \tilde{U}_{ni} = g(t_{ni}, \tilde{P}(t_{ni})) + \tilde{F}_n(t_{ni}, \tilde{P}(\cdot)) + \tilde{\Phi}_n(t_{ni}, \tilde{P}(\cdot)), \quad i = 1, \ldots, m \\ \tilde{y}_{n+1} = \sum_{k=0}^{r-1} \varphi_k(1)\tilde{y}_{n-k} + h\sum_{i=1}^{m} \psi_i(1)\tilde{U}_{ni}. \end{cases} \tag{57}$$

where $\tilde{F}_n(t_{ni}, \tilde{P}(\cdot))$ and $\tilde{\Phi}_n(t_{ni}, \tilde{P}(\cdot))$ are of the form

$$\tilde{F}_n(t_{ni}, \tilde{P}(\cdot)) = h\sum_{v=0}^{n-1}\sum_{l=1}^{\mu_1} w_l k\left(t_{ni}, t_v + d_l h, \sum_{k=0}^{r-1} \varphi_k(d_l)\tilde{y}_{v-k} + h\sum_{j=1}^{m} \psi_j(d_l)\tilde{U}_{vj}\right)$$

$$\tilde{\Phi}_n(t_{ni}, \tilde{P}(\cdot)) = h\sum_{l=1}^{\mu_0} w_{il} k\left(t_{ni}, t_n + d_{il}h, \sum_{k=0}^{r-1} \varphi_k(d_{il})\tilde{y}_{n-k} + h\sum_{j=1}^{m} \psi_j(d_{il})\tilde{U}_{nj}\right).$$

i.e., they are obtained by applying the quadrature Formula (55) to the integrals appearing in (53) and (54).

6.3. Convergence Analysis

The multivalue nature of the multistep methods imposes to analyze first the zero-stability of the methods. When $h \to 0$, second equation of (52) reduces to

$$y_{n+1} = \sum_{k=0}^{r-1} \varphi_k(1)y_{n-k}.$$

Therefore, the method (48) and (52) is said to be zero-stable, if all of the roots of the polynomial

$$p(\lambda) = \lambda^r - \sum_{k=0}^{r-1} \varphi_k(1)\lambda^{r-k-1} \tag{58}$$

have modulus less than or equal to unity, and those of modulus unity are simple.

On this basis, the following theorem studies the convergence of the method.

Theorem 7. *Consider the problem (39) with $d = 1$. Let $p = m + r - 1$ and assume that:*

1. $k \in C^p(S \times \mathbf{R})$ *and* $g \in C^p([0, T] \times \mathbf{R})$ *and have bounded derivatives with respect to y;*
2. *the method (48) and (52) is zero-stable;*
3. *the starting error satisfies* $|e(t)| = O(h^p)$, *for any* $t \in [t_0, t_{r-1}]$.

Then, the global error $e(t) = y(t) - P(t)$ *of the exact MCM (48) and (52) satisfies*

$$\max_{[0,T]} |e(t)| \le Ch^{m+r-1}. \tag{59}$$

By a suitable choice of the collocation parameters, it is possible to increase the order of convergence at the mesh points (local *superconvergence*), following the lines of multistep methods for ODEs (compare [21], Section 3).

Theorem 8. *Assume that hypotheses of Theorem 7 hold with $p = 2m + r - 1$ and that the collocation parameters satisfy these conditions*

$$\sum_{k=-1}^{r-1} \frac{1}{c_i + k} + 2 \sum_{\substack{j=1 \\ j \neq i}}^{m} \frac{1}{c_i - c_j} = 0, \ i = 1, \dots, m. \tag{60}$$

Then the order of the exact MCM (48) and (52) at the mesh points is p, i.e.,:

$$\max_{1 \le n \le N} |e(t_n)| = O(h^{2m+r-1}).$$

Similar convergence and superconvergence results hold also for the discretized MCM (56) and (57). We can summarize them in the following theorem.

Theorem 9. *Assume that hypotheses of Theorem 7 hold. If quadrature formulae Q and Q_i defined in (55) have order $m + r$ and $m + r - 1$ respectively, then the uniform order of the discretized method (56) and (57) is equal to $m + r - 1$.*

Moreover, if hypotheses of Theorem 8 are fulfilled, Q and Q_i defined in (55) have order $2m + r$ and $2m + r - 1$ respectively, then the order of the discretized method (56) and (57), at the mesh points, is $2m + r - 1$.

We observe that, at the same cost of one-step collocation methods with m collocation parameters, multistep collocation methods have an higher computational cost. A further improvement of the efficiency could be obtained by exploiting parallel techniques, as done for example in [38–40].

An extensive analysis of the stability properties on basic test equations is contained in [14]. A possible future development may regard new multistep methods with some relaxing order conditions, which leave some parameters free to perform a numerical search for the methods with optimal stability properties, as done in [22,23,41–43] in the context of ODEs.

6.4. Numerical Results

Now we give a short numerical illustration of discretized MCMs (56) and (57), on the linear test equation

$$y'(t) = g(t, y) - \int_0^t t^2 \exp(-st)y(s)ds, \quad t \in [0, 1], \tag{61}$$
$$y(0) = 1,$$

with $g(t, y)$ such that $y(t) = \exp(-t)$; and on the nonlinear problem

$$y'(t) = g(t, y) - \int_0^t 2t \sin(s) \exp(-y(s))ds, \quad t \in [0, 1],$$
$$y(0) = 1,$$

(62)

with $g(t, y)$ such that $y(t) = \cos(t)$.

We consider three methods

- TS3: superconvergent discretized two-step collocation method, with $r = 2$ and $m = 1$, with order $p = 3$;
- TS3b: two-step discretized collocation method, with $r = 2$ and $m = 2$, $c_1 = 0.9$, $c_2 = 1$, with uniform order 3;
- TS5: superconvergent discretized two-step collocation method, with $r = 2$ and $m = 2$, with order $p = 5$.

Method TS3b has an unbounded stability region, while TS3 and TS5 have a bounded stability region. The exact expression of the methods and their stability region can be found in [14]. To confirm the theoretical order of convergence, in Figure 2 the error (in logarithmic scale) produced by methods TS3, TS3b and TS5 when applied to problems (61) and (62), and the slopes corresponding to order 3 and 5. We see that the effective order is equal to the theoretical one.

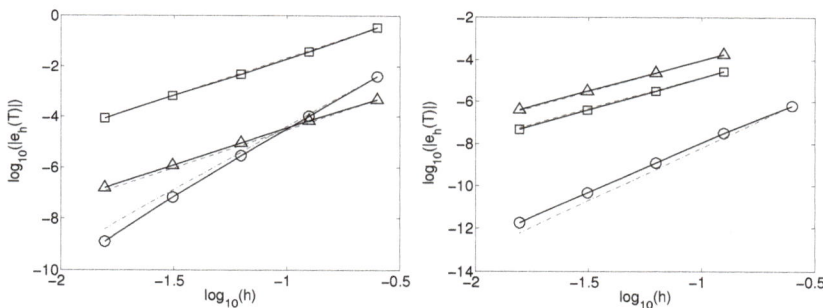

Figure 2. Error of two-step methods TS3 (⊟), TS3b (△) and TS5 (⊝), and slopes of order 3 (dashed line) and of order 5 (dash-dot line), applied to problem (61) (left) and on problem (62) (right).

7. Conclusions

We have illustrated multistep collocation methods for VIEs and VIDEs and gave an overview of their convergence and superconvergence properties. This idea may be exploited to obtain high order methods for solving other types of equations as well. For example, recently two-step collocation methods have been proposed for fractional differential equations [44], and further developments may be achieved for other fractional models, as time fractional differential equations [45]. Further issues of this research will focus on oscillatory problems [46,47] and in particular on the application of multistep collocation methods to periodic integral equations [48,49]. Moreover, it seems reasonable to consider the possibility of employing collocation spaces based on functions other than polynomials, as in [50–52] and similarly as in the case of oscillatory problems [53], and merge into the numerical scheme as many known qualitative properties of the continuous problem as possible, in a structure-preserving perspective [54].

The literature on the numerical treatment of VIEs is quite rich and goes beyond the results considered in this review. Here we would like to mention some other results, at least. In [55] the modified Newton–Kantorovich method combined with collocation were applied non linear and nonlinear VIE with piecewise smooth kernels. Such VIE were introduced in [56] and asymptotic approximations to parametric families of solutions were constructed and the existence of continuous solutions was proved. The review of the numerical methods of optimal accuracy (spline-collocation technique) for multidimensional weakly singular VIEs is given in [57]. Some other interesting papers regard the distance between the approximate and exact solutions of various generalizations of the Volterra equations [58–63]. Lastly, we underline that in the practical applications of VIE based models it is extremely important to have the numerical method to be stable with respect to the measurement errors both in the source function and in the kernel. It is well known that the 1st kind of VIEs enjoy self-regularization property when the mesh step serves as the regularisation parameter. In addition, the Lavrentiev type regularisation is a good option [64,65]. These issues have been discussed in [66].

Funding: This research was supported by GNCS-INdAM.

Conflicts of Interest: The authors declare no conflict of interest. The founding sponsors had no role in the design of the study; in the collection, analyses, or interpretation of data; in the writing of the manuscript, and in the decision to publish the results.

References

1. Bonaccorsi, S.; Fantozzi, M. Volterra Integro-Differential Equations with Accretive Operators and Non-Autonomous Perturbations. *J. Integral Equ. Appl.* **2006**, *18*, 437–470. [CrossRef]
2. Brunner, H. *Collocation Methods for Volterra Integral and Related Functional Equations*; Cambridge University Press: Cambridge, UK, 2004.
3. Brunner, H.; van der Houwen, P.J. *The Numerical Solution of Volterra Equations*; CWI Monographs, 3; Elsevier Science Ltd.: Amsterdam, The Netherlands, 1986.
4. Hoppensteadt, F.C.; Jackiewicz, Z.; Zubik-Kowal, B. Numerical solution of Volterra integral and integro-differential equations with rapidly vanishing convolution kernels. *BIT Numer. Math.* **2007**, *47*, 325–350. [CrossRef]
5. Hrusa, W.J.; Renardy, M. A model problem in one-dimensional viscoelasticity with a singular kernel. In *Volterra Integrodifferential Equations in Banach Spaces and Applications*; Pitman Research Notes in Mathematics; da Prato, G., Iannelli, M., Eds.; Longman Scientific & Technical: Harlow, UK; Wiley: New York, NY, USA, 1989; Volume 190, pp. 221–230.
6. Nohel, J.A.; Rogers, R.C.; Tzavaras, A.E. Hyperbolic conservation laws in viscoelasticity. In *Volterra Integro Differential Equations in Banach Spaces and Applications*; Da Prato, G., Iannelli, M., Eds.; Pitman Research Notes in Mathematics Series, 190; Longman Science Technology: Harlow, UK, 1989; pp. 320–338.
7. Coleman, J.P.; Duxbury, S.C. Mixed collocation methods for $y = f(x, y)$. *J. Comput. Appl. Math.* **2000**, *126*, 47–75. [CrossRef]
8. Brunner, H.; Makroglou, A.; Miller, R.K. Mixed interpolation collocation methods for first and second order Volterra integro-differential equations with periodic solution. *Appl. Numer. Math.* **1997**, *23*, 381–402. [CrossRef]
9. Braś, M.; Cardone, A.; Jackiewicz, Z.; Welfert, B. Order reduction phenomenon for general linear methods. *J. Comput. Appl. Math.* **2015**, *290*, 44–64. [CrossRef]
10. Butcher, J.C. *Numerical Methods for Ordinary Differential Equations*, 2nd ed.; John Wiley & Sons: Chichester, UK, 2008.
11. Hairer, E.; Wanner, G. Solving Ordinary Differential Equations. II. In *Springer Series in Computational Mathematics*; Springer: Berlin, Germany, 1991; Volume 14.
12. D'Ambrosio, R.; Paternoster, B. Two-step modified collocation methods with structured coefficient matrices for ordinary differential equations. *Appl. Numer. Math.* **2012**, *62*, 1325–1334. [CrossRef]
13. Capobianco, G.; Conte, D.; Paternoster, B. Construction and implementation of two-step continuous methods for Volterra Integral Equations. *Appl. Numer. Math* **2017**, *119*, 239–247. [CrossRef]
14. Cardone, A.; Conte, D. Multistep collocation methods for Volterra integro-differential equations. *Appl. Math. Comput.* **2013**, *221*, 770–785. [CrossRef]
15. Cardone, A.; Conte, D.; Paternoster, B. A family of multistep collocation methods for Volterra integro-differential equations. *AIP Conf. Proc.* **2009**, *1168*, 358–361. [CrossRef]

16. Conte, D.; Paternoster, B. Multistep collocation methods for Volterra Integral Equations. *Appl. Numer. Math.* **2009**, *59*, 1721–1736. [CrossRef]

17. Conte, D.; D'Ambrosio, R.; Paternoster, B. Two-step diagonally-implicit collocation-based methods for Volterra Integral Equations. *Appl. Numer. Math.* **2012**, *62*, 1312–1324. [CrossRef]

18. Conte, D.; Jackiewicz, Z.; Paternoster, B. Two-step almost collocation methods for Volterra integral equations. *Appl. Math. Comput.* **2008**, *204*, 839–853. [CrossRef]

19. Guillou, A.; Soulé, J.L. La résolution numérique des problèmes différentiels aux conditions initiales par des méthodes de collocation. *Rev. Fr. Inform. Rech. Opér.* **1969**, *3*, 17–44. [CrossRef]

20. Lie, I. The stability function for multistep collocation methods. *Numer. Math.* **1990**, *57*, 779–787. [CrossRef]

21. Lie, I.; Nørsett, S. Superconvergence for multistep collocation. *Math. Comput.* **1989**, *52*, 65–79. [CrossRef]

22. Braś, M.; Cardone, A. Construction of Efficient General Linear Methods for Non-Stiff Differential Systems. *Math. Model. Anal.* **2012**, *17*, 171–189. [CrossRef]

23. Braś, M.; Cardone, A.; D'Ambrosio, R. Implementation of explicit Nordsieck methods with inherent quadratic stability. *Math. Model. Anal.* **2013**, *18*, 289–307. [CrossRef]

24. Jackiewicz, Z. *General Linear Methods for Ordinary Differential Equations*; John Wiley & Sons: Hoboken, NJ, USA, 2009.

25. Darania, P. Superconvergence analysis of multistep collocation method for delay Volterra integral equations. *Comput. Methods Differ. Equ.* **2016**, *4*, 205–216.

26. Darania, P.; Pishbin, S. High-order collocation methods for nonlinear delay integral equation. *J. Comput. Appl. Math.* **2017**, *326*, 284–295. [CrossRef]

27. Fazeli, S.; Hojjati, G. Numerical solution of Volterra integro-differential equations by superimplicit multistep collocation methods. *Numer. Algorithms* **2015**, *68*, 741–768. [CrossRef]

28. Fazeli, S.; Hojjati, G.; Shahmorad, S. Multistep Hermite collocation methods for solving Volterra integral equations. *Numer. Algorithms* **2012**, *60*, 27–50. [CrossRef]

29. Fazeli, S.; Hojjati, G.; Shahmorad, S. Super implicit multistep collocation methods for nonlinear Volterra integral equations. *Math. Comput. Model.* **2012**, *55*, 590–607. [CrossRef]

30. Fazeli, S.; Hojjati, G.; Shahmorad, S. Multistep collocation and iterated multistep collocation methods for solving two-dimensional Volterra integral equations. *J. Mod. Methods Numer. Math.* **2015**, *6*, 1–21. [CrossRef]

31. Ma, J.; Xiang, S. A collocation boundary value method for linear Volterra integral equations. *J. Sci. Comput.* **2017**, *71*, 1–20. [CrossRef]

32. Sheng, C.; Wang, Z.; Guo, B. A multistep Legendre-Gauss spectral collocation method for nonlinear Volterra integral equations. *SIAM J. Numer. Anal.* **2014**, *52*, 1953–1980. [CrossRef]

33. Lopez-Fernandez, M.; Lubich, C.; Schadle, A. Adaptive, fast, and oblivious convolution in evolution equations with memory. *SIAM J. Sci. Comput.* **2008**, *30*, 1015–1037. [CrossRef]

34. Lopez-Fernandez, M.; Lubich, C.; Schadle, A. Fast and oblivious convolution quadrature. *SIAM J. Sci. Comput.* **2006**, *28*, 421–438.

35. Crisci, M.R.; Russo, E.; Vecchio, A. Stability results for one-step discretized collocation methods in the numerical treatment of Volterra integral equations. *Math. Comput.* **1992**, *58*, 119–134. [CrossRef]

36. Crisci, M.R.; Russo, E.; Jackiewicz, Z.; Vecchio, A. Global stability of exact collocation methods for Volterra integro-differential equations. *Atti Sem. Mat. Fis. Univ. Modena* **1991**, *39*, 527–536.

37. Crisci, M.R.; Russo, E.; Vecchio, A. Stability of Collocation Methods for Volterra Integro-Differential Equations. *J. Integral Equ. Appl.* **1992**, *4*, 491–507. [CrossRef]

38. Cardone, A.; Messina, E.; Vecchio, A. An adaptive method for Volterra–Fredholm integral equations on the half line. *J. Comput. Appl. Math.* **2009**, *228*, 538–547. [CrossRef]

39. Conte, D.; D'Ambrosio, R.; Paternoster, B. GPU acceleration of waveform relaxation methods for large differential systems. *Numer. Algorithms* **2016**, *71*, 293–310. [CrossRef]

40. Conte, D.; Paternoster, B. Parallel methods for weakly singular Volterra Integral Equations on GPUs. *Appl. Numer. Math.* **2017**, *114*, 30–37. [CrossRef]

41. Cardone, A.; Jackiewicz, Z.; Sandu, A.; Zhang, H. Extrapolated Implicit-Explicit Runge-Kutta Methods. *Math. Model. Anal.* **2014**, *19*, 18–43. [CrossRef]

42. Conte, D.; D'Ambrosio, R.; Jackiewicz, Z.; Paternoster, B. A practical approach for the derivation of two-step Runge-Kutta methods. *Math. Model. Anal.* **2012**, *17*, 65–77. [CrossRef]

43. Conte, D.; D'Ambrosio, R.; Jackiewicz, Z.; Paternoster, B. Numerical search for algebraically stable two-step continuous Runge-Kutta methods. *J. Comput. Appl. Math.* **2013**, *239*, 304–321. [CrossRef]

44. Cardone, A.; Conte, D.; Paternoster, B. Two-step collocation methods for fractional differential equations. *Discret. Contin. Dyn. Syst. Ser. B* **2017**, *22*, 1–17. [CrossRef]

45. Burrage, K.; Cardone, A.; D'Ambrosio, R.; Paternoster, B. Numerical solution of time fractional diffusion systems. *Appl. Numer. Math.* **2017**, *116*, 82–94. [CrossRef]

46. Conte, D.; Paternoster, B. Modified Gauss-Laguerre exponential fitting based formulae. *J. Sci. Comput.* **2016**, *69*, 227–243. [CrossRef]

47. Ixaru, L.G.; Paternoster, B. A Gauss quadrature rule for oscillatory integrands. *Comput. Phys. Commun.* **2001**, *133*, 177–188. [CrossRef]

48. Cardone, A.; D'Ambrosio, R.; Paternoster, B. High order exponentially fitted methods for Volterra integral equations with periodic solution. *Appl. Numer. Math.* **2017**, *114*, 18–29. [CrossRef]

49. Cardone, A.; Ixaru, L.G.; Paternoster, B.; Santomauro, G. Ef-Gaussian direct quadrature methods for Volterra integral equations with periodic solution. *Math. Comput. Simul.* **2015**, *110*, 125–143. [CrossRef]

50. Cardone, A.; D'Ambrosio, R.; Paternoster, B. Exponentially fitted IMEX methods for advection—Diffusion problems. *J. Comput. Appl. Math.* **2017**, *316*, 100–108. [CrossRef]

51. D'Ambrosio, R.; Moccaldi, M.; Paternoster, B. Adapted numerical methods for advection-reaction-diffusion problems generating periodic wavefronts. *Comput. Math. Appl.* **2017**, *74*, 1029–1042. [CrossRef]

52. D'Ambrosio, R.; Paternoster, B. Numerical solution of reaction–diffusion systems of $\lambda - \omega$ type by trigonometrically fitted methods. *J. Comput. Appl. Math.* **2016**, *294*, 436–445. [CrossRef]

53. D'Ambrosio, R.; De Martino, G.; Paternoster, B. General Nystrom methods in Nordsieck form: Error analysis. *J. Comput. Appl. Math.* **2016**, *292*, 694–702. [CrossRef]

54. Butcher, J.; D'Ambrosio, R. Partitioned general linear methods for separable Hamiltonian problems. *Appl. Numer. Math.* **2017**, *117*, 69–86. [CrossRef]

55. Muftahov, I.; Tynda, A.; Sidorov, D. Numeric solution of Volterra integral equations of the first kind with discontinuous kernels. *J. Comput. Appl. Math.* **2017**, *313*, 119–128. [CrossRef]

56. Sidorov, D.N. On parametric families of solutions of Volterra integral equations of the first kind with piecewise smooth kernel. *Differ. Equ.* **2013**, *49*, 210–216. [CrossRef]

57. Boykov, I.V.; Tynda, A.N. Numerical methods of optimal accuracy for weakly singular Volterra integral equations. *Ann. Funct. Anal.* **2015**, *6*, 114–133. [CrossRef]

58. Castro, L.P.; Ramos, A. Hyers–Ulam–Rassias stability for a class of nonlinear Volterra integral equations. *Banach J. Math. Anal.* **2009**, *3*, 36–43. [CrossRef]

59. Brzdek, J.; Eghbali, N. On approximate solutions of some delayed fractional differential equations. *Appl. Math. Lett.* **2016**, *54*, 31–35. [CrossRef]

60. Bahyrycz, A.; Brzdek, J.; Lesniak, Z. On approximate solutions of the generalized Volterra integral equation. *Nonlinear Anal. Real World Appl.* **2014**, *20*, 59–66. [CrossRef]

61. Gachpazan, M.; Baghani, O. Hyers-Ulam stability of nonlinear integral equation. *Fixed Point Theory Appl.* **2010**, *2010*, 927640. [CrossRef]

62. Jung, S.M. A fixed point approach to the stability of a Volterra integral equation. *Fixed Point Theory Appl.* **2007**, *2007*, 057064. [CrossRef]

63. Morales, J.R.; Rojas, E.M. Hyers–Ulam and Hyers–Ulam–Rassias stability of nonlinear integral equations with delay. *Int. J. Nonlinear Anal. Appl.* **2011**, *2*, 1–7.

64. Kythy, P.K.; Puri, P. *Computational Methods for Linear Integral Equations*; Birkhauser: Boston, MA, USA, 2002.

65. Muftahov, I.R.; Sidorov, D.N.; Sidorov, N.A. Lavrentiev regularization of integral equations of the first kind in the space of continuous functions. *Izvestiya Irkutskogo Gosudarstvennogo Universiteta* **2016**, *15*, 62–77.

66. Muftahov, I.; Sidorov, D.; Zhukov, A.; Panasetsky, D.; Foley, A.; Li, Y.; Tynda, A. Application of Volterra Equations to Solve Unit Commitment Problem of Optimised Energy Storage and Generation. *arXiv* **2016**, arXiv:1608.05221.

![axioms logo] *axioms*

MDPI

Article

On the Analysis of Mixed-Index Time Fractional Differential Equation Systems

Kevin Burrage [1,2,†] (ID)**, Pamela Burrage** [1,2,*,†] (ID)**, Ian Turner** [1,2,†] (ID) **and Fanhai Zeng** [2,†]

[1] ARC Centre of Excellence for Mathematical and Statistical Frontiers, Queensland University of Technology (QUT), Brisbane 4001, Australia; kevin.burrage@qut.edu.au (K.B.); i.turner@qut.edu.au (I.T.)

[2] School of Mathematical Sciences, Queensland University of Technology (QUT), Brisbane 4001, Australia; f2.zeng@qut.edu.au

* Correspondence: pamela.burrage@qut.edu.au

† These authors contributed equally to this work.

Received: 13 February 2018; Accepted: 11 April 2018; Published: 17 April 2018

Abstract: In this paper, we study the class of mixed-index time fractional differential equations in which different components of the problem have different time fractional derivatives on the left-hand side. We prove a theorem on the solution of the linear system of equations, which collapses to the well-known Mittag–Leffler solution in the case that the indices are the same and also generalises the solution of the so-called linear sequential class of time fractional problems. We also investigate the asymptotic stability properties of this class of problems using Laplace transforms and show how Laplace transforms can be used to write solutions as linear combinations of generalised Mittag–Leffler functions in some cases. Finally, we illustrate our results with some numerical simulations.

Keywords: time fractional differential equations; mixed-index problems; analytical solution; asymptotic stability

1. Introduction

Time fractional and space fractional differential equations are increasingly used as a powerful modelling tool for understanding the role of heterogeneity in the modulating function in such diverse areas as cardiac electrophysiology [1–3], brain dynamics [4], medicine [5], biology [6,7], porous media [8,9] and physics [10]. Time fractional models are typically used to model subdiffusive processes (anomalous diffusion [11,12]), while space fractional models are often associated with modelling processes occurring in complex spatially heterogeneous domains [1].

Time fractional models typically have solutions with heavy tails as described by the Mittag–Leffler matrix function [13] that naturally occurs when solving time fractional linear systems. However, such models are usually only described by a single fractional exponent, α, associated with the fractional derivative. The fractional exponent can allow the coupling of different processes that may be occurring in different spatial domains by using different fractional exponents for the different regimes. One natural application here would be the coupling of models describing anomalous diffusion of proteins on the plasma membrane of the cell with the behaviour of other proteins in the cytosol of the cell. Tian et al. [14] addressed this problem by coupling a stochastic model (based on the stochastic simulation algorithm [15]) for the plasma membrane with systems of ordinary differential equations describing reaction cascades within the cell. It may also be necessary to couple more than two models, and so, in this paper, we introduce a formulation that focuses on coupling an arbitrary number of domains in which dynamical processes are occurring described by different anomalous diffusive processes. This leads us to consider the r index time fractional differential equation problem in Caputo form:

$$D_t^{\alpha_i} y_i = \sum_{j=1}^{r} A_{ij} y_j + F_i(y), \quad y_i(0) = z_i, \quad y_i \in \mathbb{R}^{m_i}, \, i = 1, \cdots, r, \tag{1}$$

or in vector form:

$$D_t^\alpha y = A y + F(y).$$

Here, A_{ij} are $m_i \times m_j$ matrices, while A is the associated block matrix of dimension $\sum_{j=1}^r m_j$, and $\alpha = (\alpha_1, \cdots, \alpha_r)^\top$ has all components $\alpha_i \in (0, 1]$.

We believe that a modelling approach based on this formulation has not been fully developed before. We note that scalar linear sequential fractional problems have been considered whose solution can be described by multi-indexed Mittag–Leffler functions [16], and there are a number of articles on the numerical solution of multi-term fractional differential equations [17–22]. While mixed index problems can, in some cases, be written in the form of linear sequential problems, namely $\sum_{i=1}^{R} D_t^{\beta_i} y = f(y)$, we claim that it is inappropriate to do so in many cases. We note that Diethelm et al. [20] have very recently considered the asymptotic behaviour of certain linear multi-order fractional differential equations from a theoretical viewpoint.

Therefore, in this paper, we develop a new theorem that gives the analytical solution of equations such as (1) that reduces to the Mittag–Leffler expansion in the case that all the indices are the same (Section 3) and generalises the class of linear sequential problems (Section 3.1). We then analyse the asymptotic stability properties of these mixed index problems using Laplace transform techniques (Section 3.2), relating our results with known results that have been developed in control theory. In Section 3.2, we also show that, in the case that the α_i are all rational, the solutions to the linear problem can be written as a linear combination of generalised Mittag–Leffler functions, again using ideas from control theory and transfer functions. In Section 4, we present some numerical simulations illustrating the results in this paper and give some discussion on how these ideas can be used to solve semi-linear problems either by extending the methodology of exponential integrators to Mittag–Leffler functions or by writing the solution as sums of certain Mittag–Leffler expansions.

2. Materials and Methods

2.1. Analytical Solutions

We consider the linear system given in (1) with $r = 2$ and $\alpha_1 = \alpha, \alpha_2 = \beta$. It will be convenient to let:

$$A = \begin{pmatrix} A_1 & A_2 \\ B_1 & B_2 \end{pmatrix}, \quad y^\top = (y_1^\top, y_2^\top), \quad z^\top = (z_1^\top, z_2^\top) \tag{2}$$

where A is $m \times m$, $m = m_1 + m_2$. We will call such a system a time fractional index-2 system. Here, the Caputo time fractional derivative with starting point at $t = 0$ is defined (see Podlubny [23], for example), as:

$$D_t^\alpha y(t) = \frac{1}{\Gamma(1-\alpha)} \int_0^t \frac{y'(s)}{(t-s)^\alpha} ds, \quad 0 < \alpha < 1.$$

Furthermore, given a fixed mesh of size h, then a first order approximation of the Caputo derivative [24] is given by:

$$D_t^\alpha y_n = \frac{1}{\Gamma(2-\alpha) h^\alpha} \sum_{j=1}^{n} (j^{1-\alpha} - (j-1)^{1-\alpha})(y_{n-j-1} - y_{n-j}).$$

If $\beta = \alpha$, then the solution to (1) is given by the Mittag–Leffler expansion:

$$y(t) = E_\alpha(t^\alpha A)\, y(0), \quad E_\alpha(z) = \sum_{j=0}^{\infty} \frac{z^j}{\Gamma(1+j\alpha)} \tag{3}$$

where $\Gamma(x)$ is the Gamma function.

If the problem is completely decoupled, say $A_2 = 0$, then from (3), the solution to (1) and (2) satisfies:

$$\begin{aligned} y_1(t) &= E_\alpha(t^\alpha A_1)\, z_1 \\ D_t^\beta y_2 &= B_2 y_2 + B_1 E_\alpha(t^\alpha A_1)\, z_1. \end{aligned} \tag{4}$$

In order to solve (4), this requires us to solve problems of the form:

$$D_t^\beta y_2 = B_2 y_2 + f(t). \tag{5}$$

Before making further headway, we need some additional background material.

Definition 1. *Generalisations of the Mittag–Leffler functions are given by:*

$$\begin{aligned} E_{\alpha,\beta}(z) &= \sum_{k=0}^{\infty} \frac{z^k}{\Gamma(\alpha k + \beta)}, \quad Re(\alpha) > 0 \\ E_{\alpha,\beta}^\gamma(z) &= \sum_{k=0}^{\infty} \frac{(\gamma)_k}{\Gamma(\alpha k + \beta)} \frac{z^k}{k!}, \quad \gamma \in \mathbb{N}_0, \end{aligned}$$

where $(\gamma)_k$ is the Pochhammer symbol:

$$(\gamma)_0 = 1, \quad (\gamma)_k = \gamma(\gamma + 1)\cdots(\gamma + k - 1);$$

see [22]. We will only consider the case where γ is a positive integer, but it can take on positive real values.

Remark 1. $E_{\alpha,1}(z) = E_\alpha(z), \quad E_{\alpha,\beta}^1(z) = E_{\alpha,\beta}(z), \quad E_1(z) = e^z.$

The following Lemmas are standard results; see [16,23], for example.

Lemma 1.

$$\left(\frac{d}{dz}\right)^n E_{\alpha,\beta}(z) = n! E_{\alpha,\beta+\alpha n}^{n+1}(z), \; n \in \mathbb{N}.$$

Lemma 2. *The Laplace transform of $t^{\beta-1} E_{\alpha,\beta}(\lambda t^\alpha)$ satisfies:*

$$X(s) = \frac{s^\alpha}{s^\beta(s^\alpha - \lambda)}. \tag{6}$$

Lemma 3. *The Caputo derivatives satisfy the following relationships.*

(i) $D_t^\alpha I^\alpha y(t) = y(t)$
(ii) $I^\alpha D_t^\alpha y(t) = y(t) - y(0)$
(iii) $D_t^\alpha y(t) = \frac{1}{\Gamma(1-\alpha)} \int_0^t \frac{y'(s)}{(t-s)^\alpha} ds = I^{1-\alpha} D_t y(t).$

Lemma 4. *The solution of the scalar, linear, non-homogeneous problem:*

$$D_t^\alpha y(t) = \lambda y(t) + f(t), \quad y(0) = y_0 \tag{7}$$

is:

$$y(t) = E_\alpha(\lambda t^\alpha)y_0 + \int_0^t (t-s)^{\alpha-1}E_{\alpha\alpha}(\lambda(t-s)^\alpha)f(s)ds. \tag{8}$$

Note, it is insightful that this can be proven using the integral form from Lemma 3, namely:

$$y(t) = y_0 + \frac{\lambda}{\Gamma(\alpha)}\int_0^t \frac{y(s)}{(t-s)^{1-\alpha}}ds + \frac{1}{\Gamma(\alpha)}\int_0^t \frac{f(s)}{(t-s)^{1-\alpha}}ds.$$

We now apply a Picard-style iteration of the form:

$$y_k(t) = y_0(t) + \frac{\lambda}{\Gamma(\alpha)}\int_0^t \frac{y_{k-1}(s)}{(t-s)^{1-\alpha}}ds + \frac{1}{\Gamma(\alpha)}\int_0^t \frac{f(s)}{(t-s)^{1-\alpha}}ds, \quad k = 1,2,\cdots$$

where $y_0(t) = y_0$, $\forall t$. Then, the iteration will converge to (8).

Lemma 5.

$$1 + \int_0^t \lambda s^{\alpha-1}E_{\alpha\alpha}(\lambda s^\alpha)ds = E_\alpha(\lambda s^\alpha).$$

Proof. Use Definition 1, and integrate the left-hand side term by term. □

Remark 2. *The function multiplying $f(s)$ in the integrand of (8), namely:*

$$G_\alpha(t-s) = (t-s)^{\alpha-1}E_{\alpha\alpha}(\lambda(t-s)^\alpha),$$

can be viewed as a Green function. For example, when $\alpha = 1$, $G_1(t-s) = e^{\lambda(t-s)}$.

The generalisation of the class of problems given by (7) to the systems case takes the form:

$$D_t^\alpha y(t) = Ay(t) + F(t), \quad y(0) = y_0, \quad y \in \mathbb{R}^m. \tag{9}$$

In the case that $F(t) = 0$, the solution of the linear homogeneous system is:

$$y(t) = E_\alpha(t^\alpha A)y_0. \tag{10}$$

We have the following theorem (see Podlubny [23]):

Theorem 1. *The solution of (9) is given by:*

$$y(t) = E_\alpha(t^\alpha A)y_0 + \int_0^t (t-s)^{\alpha-1}E_{\alpha\alpha}((t-s)^\alpha A)F(s)ds. \tag{11}$$

2.2. Asymptotic Stability of Multi-Index Systems

The first contribution to the asymptotic stability analysis of time fractional linear systems was by Matignon [25]. Given the linear system $D_t^\alpha y(t) = Ay(t)$ in Caputo form, then taking the Laplace transform and using the definition of the Caputo derivative give:

$$s^\alpha X(s) - s^{\alpha-1}X(0) = AX(s)$$

or:

$$X(s) = \frac{1}{s}(I - s^{-\alpha}A)^{-1}X(0). \tag{12}$$

Here, $X(s)$ is the Laplace transform of $y(t)$. If we write $w = s^\alpha$, then the matrix $s^\alpha I - A$ will be nonsingular if w is not an eigenvalue of A. In the w-domain, this will happen if $Re(\sigma(A)) \leq 0$, where $\sigma(A)$ denotes the spectrum of A. In the s-domain, this will happen if $|Re(\sigma(A))| \geq \frac{\alpha\pi}{2}$. That is, the eigenvalues of A lie in the complex plane minus the sector subtended by angle $\alpha\pi$ symmetric about the positive real axis; see Figure 1.

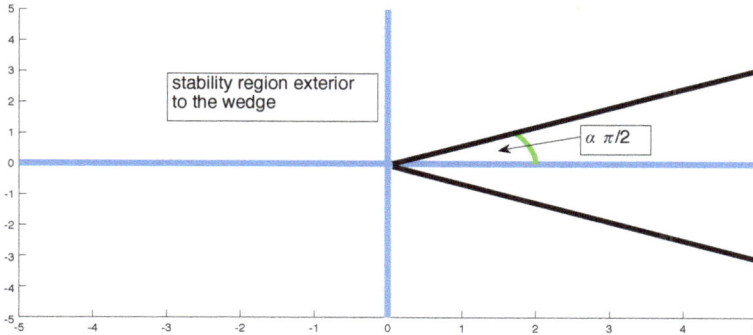

Figure 1. Asymptotic stability region for single index scalar problem, for complex values of λ (imaginary, vertical axis; real, horizontal axis).

In fact, Laplace transforms are a very powerful technique for studying the asymptotic stability of mixed index fractional systems. Deng et al. [26] studied the stability of linear time fractional systems with delays using Laplace transforms. Given the delay system:

$$\frac{d^{\alpha_i} y_i}{dt^{\alpha_i}} = \sum_{j=1}^{m} a_{ij} y_j(t - \tau_{ij}), \quad i = 1, \cdots, m \tag{13}$$

then the Laplace transform results in:

$$\begin{aligned}
\Delta(s)\, X &= b \\
\Delta(s) &= Diag(s^{\alpha_1}, \cdots, s^{\alpha_m}) - L \\
L_{ij} &= a_{ij} e^{-s\tau_{ij}}, \quad i, j = 1, \cdots, m.
\end{aligned} \tag{14}$$

Hence, Deng et al. [26] proved:

Theorem 2. *If all the zeros of the characteristic polynomial of $\Delta(s)$ have a negative real part, then the zero solution of* (13) *is asymptotically stable.*

Deng et al. [26] also proved a very nice result in the case that all the indices $\alpha_1, \cdots, \alpha_m$ are rational.

Theorem 3. *Consider* (13) *with no delays and all the $\alpha_i \in (0, 1)$ and are rational. In particular, let:*

$$\alpha_i = \frac{u_i}{v_i}, \quad gcd(u_i, v_i) = 1$$

Then, let M be the lowest common multiple of all the denominators, and set $\gamma = \frac{1}{M}$. Then, the problem will be asymptotically stable if all the roots, λ, of:

$$p(\lambda) = Det(D - A) = 0, \quad D = diag(\lambda^{Ma_1}, \cdots, \lambda^{Ma_m})$$

satisfy $|\arg(\lambda)| > \gamma\frac{\pi}{2}$.

Remark 3. *If $\alpha_i = \alpha$, $i = 1, \cdots, m$, then Theorem 3 reduces to the result of Matignon. The proof of Theorem 3 comes immediately from (14), where $p(\lambda)$ is the characteristic polynomial of $\Delta(s)$.*

Remark 4. *A nice survey on the stability (both linear and nonlinear) of fractional differential equations is given in Li and Zhang [27], while Saberi Najafi et al. [28] have extended some of these stability results to distributed order fractional differential equations with respect to an order density function. Zhang et al. [29] consider the stability of nonlinear fractional differential equations.*

Remark 5. *Radwan et al. [30] note that the stability analysis of mixed index problems reduces to the study of the roots of the characteristic equation:*

$$\sum_{i=1}^{m} \theta_i s^{\alpha_i} = 0, \quad 0 < \alpha_i \leq 1. \tag{15}$$

In the case that the α_i are arbitrary real numbers, the study of the roots of (15) is difficult. By letting $s = e^z$, we can cast this in the framework of quasi (or exponential) polynomials (Rivero et al. [31]). The zeros of exponential polynomials have been studied by Ritt [32].

The general form of an exponential polynomial with constant coefficients is:

$$f(z) = \sum_{j=0}^{k} a_j e^{\alpha_j z}.$$

An analogue of the fact that a polynomial of degree k can have up to k roots is expressed by a theorem due to Tamarkin, Pólya and Schwengler (see [32]).

Theorem 4. *Let P be the smallest convex polygon containing the values $\alpha_1, \cdots, \alpha_k$, and let the sides of P be $s_1, \cdots, s_k$. Then, there exist k half strips with half rays parallel to the outer normal to b_i that contain all the zeros of f. If $|b_i|$ is the length of b_i, then the number of zeros in the i-th half strip with modulus less than or equal to r is asymptotically $\frac{r|b_i|}{2\pi}$.*

3. Results

3.1. The Solution of Mixed Index Linear Systems

The main focus of this paper is to consider generalisations of (9), where the vector α has different components. In its general form, we will let $y^\top = (y_1^\top, \cdots, y_r^\top) \in \mathbb{R}^m$ where $y_i \in \mathbb{R}^{m_i}$ and $m = \sum_{i=1}^{r} m_i$. We will also assume $F(t)^\top = (F_1(t)^\top, \cdots, F_r(t)^\top)$ and that A can be written in block form $A = (A_{ij})_{i,j=1}^{r}$, $A_{ij} \in \mathbb{R}^{m_i \times m_j}$. We will also let $\alpha = (\alpha_1, \cdots, \alpha_r)$ and consider a class of linear, non-homogeneous multi-indexed systems of FDEs of the form:

$$D_t^\alpha y(t) = Ay(t) + F(t) \tag{16}$$

that we interpret as the system:

$$D_t^{\alpha_i} y_i(t) = \sum_{j=1}^{r} A_{ij} y_j(t) + F_i(t), \quad i = 1, \cdots, r. \tag{17}$$

The index of the system is said to be r.

In the case that $F = 0$, then by letting:

$$E_i = D^{\alpha_i} - A_{1i}$$

we can rewrite (16) as:

$$M y = 0, \tag{18}$$

where M is the block matrix, whose determinant must be zero, with:

$$
\begin{aligned}
M_{ii} &= E_i, \quad i = 1, \cdots, r \\
M_{ij} &= -A_{ij}, \quad i \neq j.
\end{aligned}
$$

Thus, in the case that all $m_i = 1$, so that the individual components are scalar and so $m = r$, (18) implies $\mathrm{Det}(M) y_r = 0$.

For example, when $r = 2$, this becomes:

$$(E_1 E_2 - A_{21} A_{12}) y_2 = 0$$

or

$$(D^{\alpha_1 + \alpha_2} - A_{22} D^{\alpha_1} - A_{11} D^{\alpha_2} + \mathrm{Det}(A)) y_2 = 0;$$

while for $r = 3$ this gives, after some simplification,

$$
\begin{aligned}
D^{\alpha_1 + \alpha_2 + \alpha_3} y_3 \quad - \quad & A_{11} D^{\alpha_2 + \alpha_3} y_3 - A_{22} D^{\alpha_1 + \alpha_3} y_3 - A_{33} D^{\alpha_1 + \alpha_2} y_3 \\
+ \quad & (A_{22} A_{33} - A_{23} A_{32}) D^{\alpha_1} y_3 + (A_{11} A_{33} - A_{13} A_{31}) D^{\alpha_2} y_3 \\
+ \quad & (A_{11} A_{22} - A_{12} A_{21}) D^{\alpha_3} y_3 - \mathrm{Det}(A) = 0.
\end{aligned}
$$

Clearly, there is a general formula for arbitrary r in terms of the cofactors of A. In particular, it can be fitted into the framework of linear sequential FDEs [16,23,24,33,34]. These take the form:

$$D_t^{\beta_0} y_1(t) + \sum_{j=1}^{p} a_j D_t^{\beta_j} y_1(t) = d y_1(t) + f(t), \quad \beta_0 > \beta_1 > \cdots \beta_p. \tag{19}$$

However, this characterisation is not particularly simple, useful or computationally expedient. Furthermore, when the m_i are not one, so that the individual components are not scalar, then there is no simple representation such as (19), and new approaches are needed. Before we consider this new approach, we note the converse, namely that (19) can always be written in the form of (16) for a suitable matrix A with a special structure. In particular, we can write (19) in the form of (16) with $p = r - 1$ as an r dimensional, r index problem with $\alpha = (\beta_0, \beta_1, \cdots, \beta_p)$, and:

$$
A = \begin{pmatrix}
d & -a_1 & -a_2 & \cdots & -a_p \\
0 & 1 & 0 & \cdots & 0 \\
\vdots & & & \ddots & \\
0 & & \cdots & & 1
\end{pmatrix}, \quad F(t) = (f(t), 0, \cdots, 0)^{\top}.
$$

For completeness: we note in the case that $d = 0$ and $f(t) = 0$, an explicit solution to this problem was given in Podlubny [23]. This can be found by considering the transfer function (see Section 3.2) given by:

$$H(s) = \frac{1}{s^{\beta_0} + a_1 s^{\beta_1} + \cdots + a_p s^{\beta_p}}.$$

By finding the poles of this function and converting back to the untransformed domain, Podlubny gives the solution as:

$$y_1(t) = \sum_{m=0}^{\infty} \frac{(-1)^m}{m!} \sum_{\substack{k_0+k_1+\cdots+k_{p-2}=m \\ k_i \geq 0}} \binom{m}{k_0 \cdots k_{p-2}} \prod_{i=0}^{p-2} (a_{p-i})^{k_i} \times$$

$$\epsilon_m \left(t, -a_1; \beta_0 - \beta_1, \beta_0 + \sum_{j=0}^{p-2} (\beta_1 - \beta_{p-j}) k_j + 1 \right)$$

where:

$$\epsilon_k(t, y; \alpha, \beta) = t^{k\alpha + \beta - 1} E_{\alpha,\beta}^k (yt^{\alpha})$$

$$E_{\alpha,\beta}^k(z) = \sum_{i=0}^{\infty} \frac{(i+k)! \, z^i}{i! \, \Gamma(\alpha(i+k) + \beta)}.$$

We now return to the index-2 problem (1) and (2). We first claim that the solution takes the matrix form:

$$y_1 = \alpha_{00} + \sum_{n=1}^{\infty} \sum_{j=0}^{n-1} \alpha_{n,j+1} \frac{t^{n\alpha + j(\beta - \alpha)}}{\Gamma(1 + n\alpha + j(\beta - \alpha))} z$$

$$y_2 = \beta_{00} + \sum_{n=1}^{\infty} \sum_{j=1}^{n} \beta_{n,j} \frac{t^{n\alpha + j(\beta - \alpha)}}{\Gamma(1 + n\alpha + j(\beta - \alpha))} z,$$

$$(20)$$

where the $\alpha_{n,j}$, $\beta_{n,j}$ are appropriate matrices, of size $m_1 \times m$ and $m_2 \times m$, respectively, that are to be determined.

We now use the fact that:

$$D_t^{\alpha} \frac{t^{n\alpha + j(\beta - \alpha)}}{\Gamma(1 + n\alpha + j(\beta - \alpha))} = \frac{1}{\Gamma(1 + (n-1)\alpha + j(\beta - \alpha))} t^{(n-1)\alpha + j(\beta - \alpha)}$$

$$D_t^{\beta} \frac{t^{n\alpha + j(\beta - \alpha)}}{\Gamma(1 + n\alpha + j(\beta - \alpha))} = \frac{1}{\Gamma(1 + (n-1)\alpha + (j-1)(\beta - \alpha))} t^{(n-1)\alpha + (j-1)(\beta - \alpha)}.$$

$$(21)$$

Using (20) and (21), the left-hand side of (1) is:

$$D_t^{\alpha} y_1 = \sum_{n=1}^{\infty} \sum_{j=0}^{n-1} \alpha_{n,j+1} \frac{t^{(n-1)\alpha + j(\beta - \alpha)}}{\Gamma(1 + (n-1)\alpha + j(\beta - \alpha))} z$$

$$D_t^{\beta} y_2 = \sum_{n=1}^{\infty} \sum_{j=0}^{n-1} \beta_{n,j+1} \frac{t^{(n-1)\alpha + j(\beta - \alpha)}}{\Gamma(1 + (n-1)\alpha + j(\beta - \alpha))} z$$

which can be written in matrix form as:

$$\sum_{n=0}^{\infty} \sum_{j=0}^{n} \binom{\alpha_{n+1,j+1}}{\beta_{n+1,j+1}} \frac{t^{n\alpha + j(\beta - \alpha)}}{\Gamma(1 + n\alpha + j(\beta - \alpha))} z.$$

$$(22)$$

If we define:

$$\alpha_{n,n+1} = 0, \quad \beta_{n0} = 0, \quad n = 1, 2, \cdots \tag{23}$$

then the right-hand side of (1) is:

$$A\left(\left(\begin{array}{c} \alpha_{00} \\ \beta_{00} \end{array}\right) + \sum_{n=1}^{\infty} \sum_{j=0}^{n} \left(\begin{array}{c} \alpha_{n,j+1} \\ \beta_{nj} \end{array}\right) \frac{t^{n\alpha + j(\beta - \alpha)}}{\Gamma(1 + n\alpha + j(\beta - \alpha))}\right) z. \tag{24}$$

Equating (22) and (24), we find along with (23) that for $n = 0, 1, 2, \cdots$:

$$\left(\begin{array}{c} \alpha_{00} \\ \beta_{00} \end{array}\right) = I_m, \quad \left(\begin{array}{c} \alpha_{n+1,j+1} \\ \beta_{n+1,j+1} \end{array}\right) = A\left(\begin{array}{c} \alpha_{n,j+1} \\ \beta_{nj} \end{array}\right), \quad j = 0, 1, \cdots, n. \tag{25}$$

In order to get a succinct representation of the solution based on (20) and (25), it will be convenient to write:

$$p_n(t) = \left(\frac{t^{n\alpha}}{\Gamma(1 + n\alpha)}, \frac{t^{(n-1)\alpha + \beta}}{\Gamma(1 + (n-1)\alpha + \beta)}, \cdots, \frac{t^{n\beta}}{\Gamma(1 + n\beta)}\right)^{\top} \otimes I_m, \quad n = 1, 2, \cdots$$

so $p_n(t) \in \mathbb{R}^{m(n+1) \times m}$, and let $p_0(t) = I_m$.

We will also define the matrices:

$$L_n = \left(\begin{array}{ccccc} \alpha_{n1} & \alpha_{n2} & \cdots & \alpha_{nn} & 0 \\ 0 & \beta_{n1} & \cdots & \beta_{n\,n-1} & \beta_{nn} \end{array}\right) \in \mathbb{R}^{m \times m(n+1)}, \quad n = 1, 2, \cdots$$

$$L_0 = I_m$$

where 0 represents appropriately-sized zero matrices. Now, we note that the recursive relation (25) is equivalent to:

$$\left(\begin{array}{ccc} \alpha_{n1} & \cdots & \alpha_{nn} \\ \beta_{n1} & \cdots & \beta_{nn} \end{array}\right) = A\, L_{n-1}, \quad n = 1, 2, \cdots. \tag{26}$$

Thus, we can state the following theorem.

Theorem 5. *The solution of the fractional index-2 system:*

$$D_t^{\alpha,\beta} y(t) = A\, y(t), \; y(0) = z$$

is given by:

$$y(t) = \sum_{n=0}^{\infty} L_n\, p_n(t)\, z, \tag{27}$$

where for $n = 1, 2, \cdots$:

$$L_n = \left(\begin{array}{ccccc} \alpha_{n1} & \alpha_{n2} & \cdots & \alpha_{nn} & 0 \\ 0 & \beta_{n1} & \cdots & \beta_{n\,n-1} & \beta_{nn} \end{array}\right), \quad \left(\begin{array}{ccc} \alpha_{n1} & \cdots & \alpha_{nn} \\ \beta_{n1} & \cdots & \beta_{nn} \end{array}\right) = A\, L_{n-1},$$

$$L_0 = I_m$$

$$p_n(t) = \left(\frac{t^{n\alpha}}{\Gamma(1 + n\alpha)}, \frac{t^{(n-1)\alpha + \beta}}{\Gamma(1 + (n-1)\alpha + \beta)}, \cdots, \frac{t^{n\beta}}{\Gamma(1 + n\beta)}\right)^{\top} \otimes I_m. \tag{28}$$

Remark 6. *In the case $\alpha = \beta$,*

$$p_n(t) = \frac{t^{n\alpha}}{\Gamma(1 + n\alpha)} (1, \cdots, 1)^{\top} \otimes I_m,$$

$$L_n\, p_n(t) = \frac{t^{n\alpha}}{\Gamma(1+n\alpha)} \sum_{j=1}^{n} \begin{pmatrix} \alpha_{nj} \\ \beta_{nj} \end{pmatrix}$$

and with:

$$\sum_{j=1}^{n} \begin{pmatrix} \alpha_{nj} \\ \beta_{nj} \end{pmatrix} = A \sum_{j=1}^{n-1} \begin{pmatrix} \alpha_{n-1,j} \\ \beta_{n-1,j} \end{pmatrix}$$

then (27) reduces, as expected, to:

$$y(t) = E_\alpha(t^\alpha A)z.$$

Remark 7. *It will be convenient to define the matrix:*

$$P_{\alpha,\beta}(t) = \sum_{n=0}^{\infty} L_n p_n(t)$$

so that the solution (27) can be expressed as:

$$y(t) = P_{\alpha,\beta}(t)y_0. \tag{29}$$

Remark 8. *If the fractional index-2 system has initial condition $y(t_0) = z$, then the solution is:*

$$y(t) = P_{\alpha,\beta}(t - t_0)z. \tag{30}$$

We note that in solving (9), an equivalent solution to (11) is:

$$\begin{aligned} y(t) &= E_\alpha(t^\alpha A)y_0 + I_\alpha(G_\alpha(t-s)F(s))ds, \\ G_\alpha(t-s) &= E_\alpha((t-s)^\alpha A), \end{aligned}$$

where G_α is the Green function satisfying:

$$D_t^\alpha G_\alpha(t-s) = AG_\alpha(t-s). \tag{31}$$

This leads us to give a general result on the solution of the mixed index problem (9) (with $r = 2$) with a time-dependent forcing function, but first, we need the following definition.

Definition 2. *Let $y(t) = (y_1^\top(t), y_2^\top(t))^\top$, then define:*

$$I_t^{\alpha,\beta} y(s)ds = \left(I_t^\alpha y_1^\top(s)ds, I_t^\beta y_2^\top(s)ds \right)^\top.$$

Theorem 6. *The solution to the fractional index-2 problem:*

$$D_t^{\alpha,\beta} y(t) = Ay(t) + F(t), \quad y(0) = y_0 \tag{32}$$

is given by:

$$y(t) = P_{\alpha,\beta}(t)y_0 + I_t^{\alpha,\beta}\left(P_{\alpha,\beta}(t-s)F(s)ds \right). \tag{33}$$

Proof. The result follows from $D_t^{\alpha,\beta} P_{\alpha,\beta}(t) = AP_{\alpha,\beta}(t)$, together with the above discussion. □

We now turn to analysing the asymptotic stability of linear fractional index-2 systems.

3.2. Study of Asymptotic Stability

Recalling Theorem 4, we note that if the α_i are rational and with M the lowest common multiple of the denominators, this reduces to the polynomial:

$$\sum_{i=1}^{M} \theta_i W^i = 0, \quad W = s^{\frac{1}{M}}.$$

This leads us to think about stability from a control theory point of view. Thus, given the system:

$$\sum_{j=0}^{n} a_j D^{\alpha_j} y = \sum_{j=0}^{M} b_j D^{\beta_j} y \tag{34}$$

where:

$$\alpha_n > \cdots > \alpha_0, \quad \beta_M > \cdots \beta_0$$

then the solution of (34) can be written in terms of the transfer function:

$$G(s) = \frac{\sum_{j=0}^{M} b_j s^{\beta_j}}{\sum_{j=0}^{n} a_j s^{\alpha_j}} := \frac{Q(s)}{P(s)}, \tag{35}$$

where s is the Laplace variable (see Rivero et al. [31] and Petras [35]).

In the case of the so-called commensurate form in which:

$$\alpha_k = k\alpha, \quad \beta_k = k\beta,$$

then:

$$G(s) = \frac{\sum_{k=0}^{M} b_k (s^{\beta})^k}{\sum_{k=0}^{n} a_k (s^{\alpha})^k} := \frac{Q(s^{\beta})}{P(s^{\alpha})}. \tag{36}$$

Clearly, if $\frac{\beta}{\alpha}$ is rational with $\alpha \geq \beta$ and:

$$\beta = \frac{q}{p}\alpha, \quad q, p \in \mathbb{Z}^+, \quad w = s^{\frac{\alpha}{p}}$$

then (36) can be written as:

$$G(w) := \frac{Q(w^q)}{P(w^p)}, \quad p, q \in \mathbb{Z}^+, \quad q \leq p.$$

Cěrmák and Kisela [36] considered the specific problem:

$$D^{\alpha} y + a D^{\beta} y + b y = 0, \quad y \in \mathbb{R}, \tag{37}$$

where $\alpha = pK$, $\beta = qK$, K real $\in (0, 1)$, $p, q \in \mathbb{Z}^+$, $p \geq q$. In this case, the appropriate stability polynomial is $P(\lambda) := \lambda^p + a\lambda^q + b$, where $\lambda = s^K$. Based on Theorem 3, (37) is asymptotically stable if all the roots of $P(\lambda)$ satisfy $|\arg(\lambda)| > K\frac{\pi}{2}$.

By setting $\lambda = re^{iK\frac{\pi}{2}}$ and substituting into $P(\lambda) = 0$ and equating real and imaginary parts, it is easily seen that:

$$r^p \cos\frac{pK\pi}{2} + a\, r^q \cos\frac{qK\pi}{2} + b \;=\; 0$$
$$r^p \sin\frac{pK\pi}{2} + a\, r^q \sin\frac{qK\pi}{2} \;=\; 0.$$

This leads to the following result, given in Cěrmák and Kisela [36].

Theorem 7. *Equation* (37) *is asymptotically stable with* $\alpha > \beta > 0$ *real and* $\frac{\alpha}{\beta}$ *rational if:*

$$\beta \; < \; 2, \quad \alpha - \beta < 2$$

$$b \; > \; 0, \quad a > \frac{-\sin \frac{\alpha \pi}{2}}{(\sin \frac{\beta \pi}{2})^{\frac{\beta}{\alpha}} (\sin \frac{(\alpha - \beta)\pi}{2})^{\frac{\alpha - \beta}{\alpha}}} \, b^{\frac{\alpha - \beta}{\alpha}}.$$

We now follow this idea, but for arbitrarily-sized systems in our mixed index format, and this leads to slight modifications to (37). We first make a slight simplification and take $m_1 = m_2$, and we also assume that A_2 is nonsingular, then Problem (1) leads to:

$$y_2 = A_2^{-1} (D^\alpha I - A_1) y_1$$

and substituting into the equation for y_1 gives:

$$(D^{\alpha + \beta} I - B_2 D^\alpha I - \bar{A}_1 D^\beta I + B_2 \bar{A}_1 - B_1 A_2) A_2^{-1} y_1 = 0$$
$$\bar{A}_1 = A_2^{-1} A_1 A_2.$$

This leads us to consider the roots of the characteristic function:

$$P(\lambda) := \mathrm{Det}(D^{\alpha + \beta} I - B_2 D^\alpha I - \bar{A}_1 D^\beta I + B_2 \bar{A}_1 - B_1 A_2) = 0. \tag{38}$$

In the scalar case, this gives an extension to (37) where the characteristic equation is:

$$P(\lambda) = \lambda^{\alpha + \beta} - B_2 \lambda^\alpha - A_1 \lambda^\beta + \mathrm{Det}(A). \tag{39}$$

Now, reverting to Laplace transforms of (1) and (2), then:

$$s^\alpha X_1(s) - s^{\alpha - 1} X_1(0) \;=\; A_1 X_1(s) + A_2 X_2(s)$$
$$s^\beta X_2(s) - s^{\beta - 1} X_2(0) \;=\; B_1 X_1(s) + B_2 X_2(s).$$

This can be written in systems form as:

$$(D_1 - A)X(s) = D_2 X(0), \tag{40}$$

where:

$$D_1 = \begin{pmatrix} s^\alpha I & 0 \\ 0 & s^\beta I \end{pmatrix}, \quad D_2 = \begin{pmatrix} s^{\alpha - 1} I & 0 \\ 0 & s^{\beta - 1} I \end{pmatrix}$$

or alternatively as:

$$X(s) = \frac{1}{s}(I - D_1^{-1} A)^{-1} X(0). \tag{41}$$

This can now be considered as a generalised eigenvalue problem. From (40), we require $D_1 - A$ to be nonsingular. That is:

$$\begin{pmatrix} s^\alpha I - A_1 & -A_2 \\ -B_1 & s^\beta I - B_2 \end{pmatrix} v = 0 \implies v = 0.$$

Let us write $v = (v_1^\top, v_2^\top)^\top$ and assume $\alpha \geq \beta$ and that $s^\beta I - B_2$ is nonsingular, so that from the previous analysis, this means:

$$|Re(\sigma(B_2))| \geq \frac{\beta \pi}{2}. \tag{42}$$

Hence:

$$v_2 = (s^\beta I - B_2)^{-1} B_1 v_1$$
$$((s^\alpha I - A_1) - A_2(s^\beta I - B_2)^{-1} B_1) v_1 = 0.$$

Thus, (42) and:

$$\text{Det}((s^\alpha I - A_1) - A_2(s^\beta I - B_2)^{-1} B_1) = 0 \tag{43}$$

define the asymptotic stability boundary; see also (38).

In order to make this more specific, let $m_1 = m_2 = 1$ and:

$$A = \begin{bmatrix} d & b \\ a & d \end{bmatrix}, \quad d < 0. \tag{44}$$

Note that $\sigma(A) = \{d \pm \sqrt{ab}\}$. Then, (43) becomes:

$$(s^\alpha - d)(s^\beta - d) - ab = 0. \tag{45}$$

Furthermore, let $b = -a = \theta$, so that the eigenvalues of A are $d \pm i\theta$ and (45) becomes:

$$(s^\alpha - d)(s^\beta - d) + \theta^2 = 0. \tag{46}$$

If we now assume that:

$$s = r e^{i \frac{\pi}{2}},$$

which defines the asymptotic stability boundary (the imaginary axis) when $\alpha = \beta = 1$, then (46) becomes:

$$\theta^2 = -(r^\alpha e^{i \frac{\pi \alpha}{2}} - d)(r^\beta e^{i \frac{\pi \beta}{2}} - d). \tag{47}$$

Now, since θ and d are real, the imaginary part of the right-hand side of (47) must be zero, so that:

$$r^{\alpha+\beta} \sin \frac{\alpha+\beta}{2} \pi = d(r^\alpha \sin \frac{\alpha\pi}{2} + r^\beta \sin \frac{\beta\pi}{2}). \tag{48}$$

Hence:

$$-\theta^2 = r^{\alpha+\beta} \cos \frac{\alpha+\beta}{2} \pi - d(r^\alpha \cos \frac{\alpha\pi}{2} + r^\beta \cos \frac{\beta\pi}{2}) + d^2. \tag{49}$$

Equations (48) and (49) will define the asymptotic stability boundary with θ as a function of d. Rewriting (48) as:

$$d = \frac{r^{\alpha+\beta} \sin \frac{\alpha+\beta}{2} \pi}{r^\alpha \sin \frac{\alpha\pi}{2} + r^\beta \sin \frac{\beta\pi}{2}}. \tag{50}$$

and substituting (49) leads after simplification to:

$$\frac{\theta^2}{d^2} = \frac{1}{r^{\alpha+\beta}(\sin \frac{\alpha+\beta}{2} \pi)^2} \left[\sin \frac{\alpha+\beta}{2} \pi (\frac{r^{2\alpha}}{2} \sin \alpha\pi + \frac{r^{2\beta}}{2} \sin \beta\pi) \right.$$

$$\left. - \cos \frac{\alpha+\beta}{2} \pi (r^{2\alpha} \sin^2 \frac{\alpha\pi}{2} + r^{2\beta} \sin^2 \frac{\beta\pi}{2} + 2r^{\alpha+\beta} \sin \frac{\alpha\pi}{2} \sin \frac{\beta\pi}{2}) \right].$$

Using the relationships:

$$\sin^2 \theta = \frac{1}{2}(1 - \cos 2\theta)$$
$$\sin A \sin B + \cos A \cos B = \cos(A - B)$$

gives:

$$\frac{\theta^2}{d^2} = \frac{1}{2r^{\alpha+\beta}\sin^2\frac{\alpha+\beta}{2}\pi}\left((r^{2\alpha}+r^{2\beta})\left(\cos\frac{\alpha-\beta}{2}\pi-\cos\frac{\alpha+\beta}{2}\pi\right)\right.$$
$$\left.-4r^{\alpha+\beta}\sin\frac{\alpha\pi}{2}\sin\frac{\beta\pi}{2}\cos\frac{\alpha+\beta}{2}\pi\right). \tag{51}$$

Since:

$$\cos\frac{\alpha-\beta}{2}\pi-\cos\frac{\alpha+\beta}{2}\pi = 2\sin\frac{\alpha\pi}{2}\sin\frac{\beta\pi}{2}$$

and letting $x = r^{\alpha-\beta}$, then we can write (51) as:

$$\left(\frac{\theta}{d}\right)^2 = \frac{\sin\frac{\alpha\pi}{2}\sin\frac{\beta\pi}{2}}{\sin^2\frac{\alpha+\beta}{2}\pi}\left(\frac{x^2+1}{x}-2\cos\frac{\alpha+\beta}{2}\pi\right). \tag{52}$$

Furthermore, we can write (50) as:

$$d = \frac{x^{\frac{\alpha}{\alpha-\beta}}\sin\frac{\alpha+\beta}{2}\pi}{x\sin\frac{\alpha\pi}{2}+\sin\frac{\beta\pi}{2}}. \tag{53}$$

It is easily seen that as a function of x, the minimum of (52) is when $x = 1$. Thus:

$$\frac{\theta}{d} \geq \frac{\sqrt{2\sin\frac{\alpha\pi}{2}\sin\frac{\beta\pi}{2}}}{\sin\frac{\alpha+\beta}{2}\pi}\sqrt{1-\cos\frac{\alpha+\beta}{2}\pi}$$
$$= \frac{2\sqrt{\sin\frac{\alpha\pi}{2}\sin\frac{\beta\pi}{2}}\sin\frac{\alpha+\beta}{4}\pi}{2\sin\frac{\alpha+\beta}{4}\pi\cos\frac{\alpha+\beta}{4}\pi}$$
$$= \frac{\sqrt{\sin\frac{\alpha\pi}{2}\sin\frac{\beta\pi}{2}}}{\cos\frac{\alpha+\beta}{4}\pi}.$$

Thus, we have proven the following result.

Theorem 8. *Given the mixed index problem with A as in (44), the angle for asymptotic stability* $\hat{\theta} = \arctan(\frac{\theta}{d})$ *satisfies:*

$$\tan\hat{\theta} \in \left[\frac{\sqrt{\sin\frac{\alpha\pi}{2}\sin\frac{\beta\pi}{2}}}{\cos\frac{\alpha+\beta}{4}\pi},\infty\right), \tag{54}$$

or in radians with $\tilde{\theta} = \frac{1}{\pi}\arctan(\frac{\theta}{d})$:

$$\tilde{\theta} \in \frac{1}{\pi}\left[\arctan\frac{\sqrt{\sin\frac{\alpha\pi}{2}\sin\frac{\beta\pi}{2}}}{\cos\frac{\alpha+\beta}{4}\pi},\arctan\frac{\pi}{2}\right] \tag{ }$$

with the minimum occurring with:

$$d = \frac{\sin\frac{\alpha+\beta}{2}\pi}{\sin\frac{\alpha\pi}{2}+\sin\frac{\beta\pi}{2}}. \tag{55}$$

Remark 9. *We have the following results for* $\hat{\theta}$ *in three particular cases:*

(i) $\alpha = \beta$: $\hat{\theta} = \alpha\frac{\pi}{2}$, *since in this case,* $(\frac{\theta}{d})^2 = \tan^2\frac{\alpha\pi}{2}$.

(ii) $\alpha + \beta = 1$: $\hat{\theta} \in (\sqrt{\sin \alpha \pi}, \frac{\pi}{2})$, $\alpha \in [\frac{1}{2}, 1]$. *In the case* $\alpha + \beta = 1$, *we see from* (52) *that:*

$$\left(\frac{\theta}{d}\right)^2 = \sin \alpha \pi \left(\frac{x^2 + 1}{2x}\right).$$

Letting $\alpha = \frac{1}{2} + \epsilon$ *with* $\epsilon > 0$ *small, then* $x = r^{2\epsilon}$. *This means that* $\frac{x^2+1}{2x}$, *as a function of* r, *is very shallow apart from when* r *is near the origin or very large. Hence, the asymptotic stability boundary will be almost constant over long periods of* d *when* α *and* β *are close together.*

(iii) $\alpha = 2\beta$: $\hat{\theta} \in [\frac{\sin \frac{\beta \pi}{2} \sqrt{2 \cos \frac{\beta \pi}{2}}}{\cos \frac{3\beta \pi}{4}}, \frac{\pi}{2})$, $\beta \in (0, \frac{1}{2}]$.

Letting:

$$K = \frac{\sin \frac{\alpha \pi}{2} \sin \frac{\beta \pi}{2}}{\sin^2 \frac{\alpha+\beta}{2} \pi}, \quad L = 2 \cos \frac{\alpha + \beta}{2} \pi, \ \phi = \frac{\theta}{d},$$

we can write (52) and (53) as:

$$x^2 - x(L + \frac{\phi^2}{K})x + 1 = 0 \tag{56}$$

$$x^{\frac{\alpha}{\alpha-\beta}} - x d_\alpha - d_\beta = 0, \tag{57}$$

where:

$$d_\alpha = d \frac{\sin \frac{\alpha \pi}{2}}{\sin \frac{\alpha+\beta}{2} \pi}, \quad d_\beta = d \frac{\sin \frac{\beta \pi}{2}}{\sin \frac{\alpha+\beta}{2} \pi}.$$

Due to the nonlinearities in (57), it is hard to determine an explicit simple relation between ϕ and d except if $\alpha = 2\beta$. In this case, we make use of the following Lemma.

Lemma 6. *If* $x^2 - ax + b = 0$ *and* $x^2 - cx + d = 0$, *then there is a solution:*

$$
\begin{array}{ll}
x = 0, & b = d \\
x^2 - ax + b = 0, & a = c, b = d \\
x = \frac{d-b}{c-a}, & c \neq a \ and \ (d-b)^2 = (c-a)(ad-bc).
\end{array}
\tag{58}
$$

Proof. This is by subtraction of the two equations and substitution. □

In the case of (56) and (57), then (58) becomes:

$$(1 + d_\beta)^2 = (P - d_\alpha)(P d_\beta + d_\alpha), \quad P = L + \frac{\phi^2}{K},$$

that is:

$$P^2 d_\beta - P d_\alpha (d_\beta - 1) - (d_\alpha^2 + (1 + d_\beta)^2) = 0.$$

Hence:

$$2 d_\beta P = d_\alpha (d_\beta - 1) \pm (1 + d_\beta) \sqrt{d_\alpha^2 + 4 d_\beta}. \tag{59}$$

Note that:

$$\phi^2 = KP - KL$$

and:

$$d_\alpha d_\beta = d^2 K.$$

Some manipulation of(59) leads to:

$$\phi^2 = \frac{1}{2}\left(\frac{d_\alpha}{d}\right)^2\left(d_\beta - 1 \pm (1+d_\beta)\sqrt{1+4\frac{d_\beta}{d_\alpha^2}} - 2L\frac{d_\beta}{d_\alpha}\right).$$

Now, since $\alpha = 2\beta$, this reduces to:

$$\phi^2 = \frac{1}{2}\left(\frac{\sin\beta\pi}{\sin\frac{3\beta}{2}\pi}\right)^2\left(d_\beta - 1 \pm (1+d_\beta)\sqrt{1+\frac{4}{d}\frac{\sin\frac{\beta}{2}\pi\sin\frac{3\beta}{2}\pi}{(\sin\beta\pi)^2}} - 2\frac{\cos\frac{3\beta}{2}\pi}{\cos\frac{\beta}{2}\pi}\right)$$

$$d_\beta = d\frac{\sin\frac{\beta}{2}\pi}{\sin\frac{3\beta}{2}\pi}. \tag{60}$$

By taking $\tilde{\theta} = \arctan(\phi)$. this gives an explicit relationship between $\tilde{\theta}$ and d for the case $\alpha = 2\beta$.

Remark 10. *Particular solutions are:*

(i) $\beta = \frac{1}{2}$, $\alpha = 1$, $\tan\tilde{\theta} = \sqrt{(1+d)(1+\sqrt{1+\frac{2}{d}})}$

(ii) $\beta = \frac{1}{3}$, $\alpha = \frac{2}{3}$, $\tan\tilde{\theta} = \sqrt{\frac{3}{8}}\sqrt{(1+\frac{d}{2})\sqrt{1+\frac{8}{3d}} + \frac{d}{2} - 1}$.

It is clear from (60) that when $d = 0$ and $d = \infty$, then $\theta = \frac{\pi}{2}$, and then, the angle will make an excursion from $\frac{\pi}{2}$ down to a minimum value and back to $\frac{\pi}{2}$ as d increases. For example, in the case of $\beta = \frac{1}{2}$, $\alpha = 1$, we can see from Remark 10(i) that the minimum value of the angle is when:

$$d = \sqrt{2} - 1, \quad \tan\tilde{\theta} = \sqrt{\sqrt{2}+\sqrt{4+3\sqrt{2}}}.$$

Returning to (40) and taking $m_1 = m_2 = 1$ and:

$$A = \begin{pmatrix} a_1 & a_2 \\ b_1 & b_2 \end{pmatrix}$$

then the Laplace transform in (41) is:

$$X(s) = \frac{1}{Det(s)}\left(s^{\alpha+\beta-1}X(0) + \begin{pmatrix} a_2 \\ -a_1 \end{pmatrix}s^{\beta-1}X_2(0) + \begin{pmatrix} -b_2 \\ b_1 \end{pmatrix}s^{\alpha-1}X_1(0)\right) \tag{61}$$

where:

$$Det(s) = s^{\alpha+\beta} - a_1 s^\beta - b_2 s^\alpha + D_A,$$
$$D_A = a_1 b_2 - a_2 b_1 = Det(A).$$

Now, if α and β are rational ($\alpha \leq \beta$):

$$\alpha = \frac{m}{n}, \quad \beta = \frac{p}{q}, \quad m \leq n, p \leq q, \quad \text{positive integers}$$

and with $z = s^{\frac{1}{nq}}$, then:

$$Det(z) = z^{mq+np} - a_1 z^{np} - b_2 z^{mq} + D_A. \tag{62}$$

Hence, (61) gives:

$$X_1(z) = \frac{1}{z^{(n-m)q}\mathrm{Det}(z)} \left((z^{np} - b_2)X_1(0) + a_2 z^{np-mq}X_2(0) \right) \tag{63}$$

$$X_2(z) = \frac{1}{z^{(n-m)q}\mathrm{Det}(z)} \left(b_1 X_1(0) + (z^{np} - a_1 z^{np-mq})X_2(0) \right). \tag{64}$$

From Descartes' rule of sign, then (62) will have at most four real zeros if $mq + np$ is even, and at most five real zeros if $mq + np$ is odd.

Now, factorise:

$$\mathrm{Det}(z) = \Pi_{j=1}^{N}(z - \lambda_j), \quad N = mq + np,$$

where there are at most four real zeros if N is even and at most five real zeros if N is odd. Then, using (63) and (64), we can write:

$$X_i(s) = \frac{s^{\frac{1}{nq}}}{s^{1-\alpha+\frac{1}{nq}}} \sum_{j=1}^{N} \frac{A_j^{(i)}}{s^{\frac{1}{nq}} - \lambda_j}, \quad i = 1, 2$$

where the $A_j^{(i)}$ can be found by writing:

$$\frac{p_i(z)}{\mathrm{Det}(z)} = \sum_{j=1}^{N} \frac{A_j^{(i)}}{z - \lambda_j}, \quad i = 1, 2$$

where:

$$
\begin{aligned}
p_1(z) &= X_1(0)z^{np} + X_2(0)a_2 z^{np-mq} - b_2 X_1(0) \\
p_2(z) &= X_2(0)z^{np} - X_2(0)a_1 z^{np-mq} + b_1 X_1(0).
\end{aligned}
$$

Using Lemma 2 with:

$$\tilde{\alpha} = \frac{1}{nq}, \quad \tilde{\beta} = 1 - \alpha + \tilde{\alpha}$$

leads to the following result.

Theorem 9. *The solution of the mixed index-2 problem with $\alpha = \frac{m}{n}$, $\beta = \frac{p}{q}$, $m \leq n$, $p \leq q$ all positive integers is, with $N = mq + np$, given by:*

$$
\begin{aligned}
y(t) &= \sum_{j=1}^{N} A_j E_{\frac{1}{nq}, 1-\alpha+\frac{1}{nq}}\left(\lambda_j t^{\frac{1}{nq}}\right) \tag{65} \\
A_j &= (A_j^{(1)}, A_j^{(2)})^{\top},
\end{aligned}
$$

where the λ_j are the zeros of (62) and the A_j are the coefficients in the partial fraction expansion.

Remark 11. *In the case that $\alpha = \beta$, then (65) should collapse to the solution:*

$$y(t) = E_\alpha(t^\alpha A)y(0), \tag{66}$$

and this is not immediately clear. However, in this case, $mq = np$, and so:

$$D(z) = z^{2np} - (a_1 + b_2)z^{np} + D(A)$$

which is a quadratic function in z^{np} while the equivalent p_1 and p_2 numerator functions are linear in z^{np}. Thus, in (65), N is replaced by two, $\frac{1}{nq}$ is replaced by α and $1 - \alpha + \frac{1}{nq}$ becomes one. Thus, (65) reduces to:

$$y(t) = \sum_{j=1}^{2} A_j E_\alpha(\lambda_j t^\alpha)$$

that then becomes (66). On the other hand, if α is rational and $\beta = K\alpha$, K a positive integer, then:

$$Det(s) = (s^\alpha)^{K+1} - a_1(s^\alpha)^K - b_2 s^\alpha + D_A. \tag{67}$$

If we factorise:

$$Det(s) = \Pi_{j=1}^{K+1}(s^\alpha - \lambda_j)$$

and find $A_j^{(1)}$, $A_j^{(2)}$, $j = 1, \cdots, K+1$ by:

$$\sum_{j=1}^{K+1} A_j \frac{1}{s^\alpha - \lambda_j} = \frac{1}{Det(s)} \left((s^\alpha)^K X(0) + \begin{pmatrix} a_2 \\ -a_1 \end{pmatrix} (s^\alpha)^{K-1} X_2(0) + \begin{pmatrix} -b_2 \\ b_1 \end{pmatrix} X_1(0) \right) \tag{68}$$

then we have the following Corollary.

Corollary 1. *The solution of the mixed index-2 problem with α rational and $\beta = K\alpha$, K a positive integer is given by:*

$$y(t) = \sum_{j=1}^{K+1} A_j E_\alpha(\lambda_j t^\alpha),$$

where the vectors A_j and "eigenvalues" λ_j satisfy (68).

As a particular example, take $K = 2$, $\alpha = \frac{p}{q}$, then the λ_j and A_j in Corollary 1 satisfy:

$$D(z) := \Pi_{j=1}^{3}(z - \lambda_j) := z^3 - a_1 z^2 - b_2 z + D_A = 0$$

and:

$$\sum_{j=1}^{3} A_j \frac{1}{z - \lambda_j} = \frac{1}{D(z)} \left(X_0 z^2 + \begin{pmatrix} a_2 \\ -a_1 \end{pmatrix} X_2(0)z + \begin{pmatrix} -b_2 \\ b_1 \end{pmatrix} X_1(0) \right).$$

In other words:

$$A_1(z - \lambda_2)(z - \lambda_3) + A_2(z - \lambda_1)(z - \lambda_3) + A_3(z - \lambda_1)(z - \lambda_2)$$
$$= X_0 z^2 + \begin{pmatrix} a_2 \\ -a_1 \end{pmatrix} X_2(0)z + \begin{pmatrix} -b_2 \\ b_1 \end{pmatrix} X_1(0)$$

or:

$$[A_1 \ A_2 \ A_3] = \left[X_0, \begin{pmatrix} a_2 \\ -a_1 \end{pmatrix} X_2(0), \begin{pmatrix} -b_2 \\ b_1 \end{pmatrix} X_1(0) \right] S^{-1}$$

with:

$$S = \begin{bmatrix} 1 & -(\lambda_2 + \lambda_3) & \lambda_2 \lambda_3 \\ 1 & -(\lambda_1 + \lambda_3) & \lambda_1 \lambda_3 \\ 1 & -(\lambda_1 + \lambda_2) & \lambda_1 \lambda_2 \end{bmatrix}.$$

Clearly, in the case described by Corollary 1, writing the solution as a linear combination of generalised Mittag–Leffler functions makes the evaluation of the solution much more computationally efficient.

4. Simulations and Discussion

In this section, we give a variety of asymptotic stability and dynamics results for different parameter values of the linear mixed index models.

In Figures 2 and 3, we plot the asymptotic stability boundary of the two dimensional, index-2 problem given by (1) where:

$$A = \begin{pmatrix} d & -\theta \\ \theta & d \end{pmatrix}, \quad d > 0 \tag{69}$$

for the two cases considered in Section 3.2, namely $\beta = 1$, $\alpha = \frac{1}{2}$ (Figure 2) and $\beta = \frac{2}{3}$, $\alpha = \frac{1}{3}$ (Figure 3). Since the eigenvalues of A are $d \pm i\theta$, we plot on the vertical axis the angle $\hat{\theta}$ in radians, where $\hat{\theta} = \frac{1}{\pi} \arctan(\frac{\theta}{\lambda})$, as a function of d. In Figure 2, we see that $\hat{\theta} \in (\frac{1}{4}, \frac{1}{2})$ corresponding to an angle lying between 45° and 90°, as expected from the theory. We also plot the angle, in green, corresponding to the midpoint between these two extremes, i.e., $\frac{3}{8}\pi$. We see that for the most part, the asymptotic stability angle lies above this midpoint, except for the values of d, as shown in the right-hand figure.

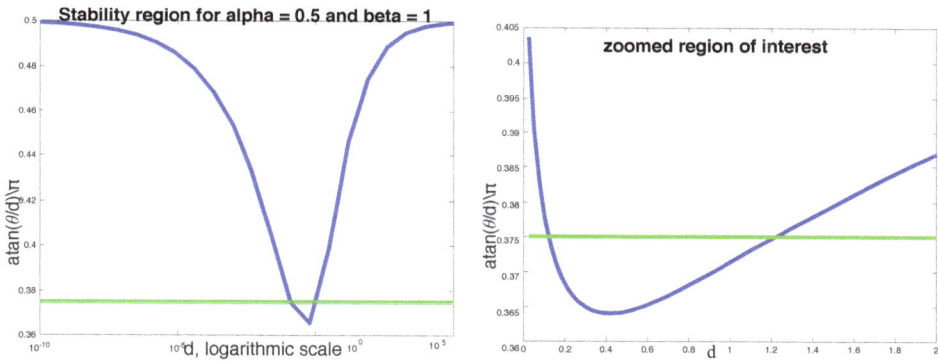

Figure 2. Stability region, above the blue line, for choosing d and θ, when the eigenvalues of A are $d \pm i\theta$, $\alpha = \frac{1}{2}, \beta = 1$. The logarithmic scale is explored in the right-hand figure where the stability boundary dips below the angle $\frac{3\pi}{8}$.

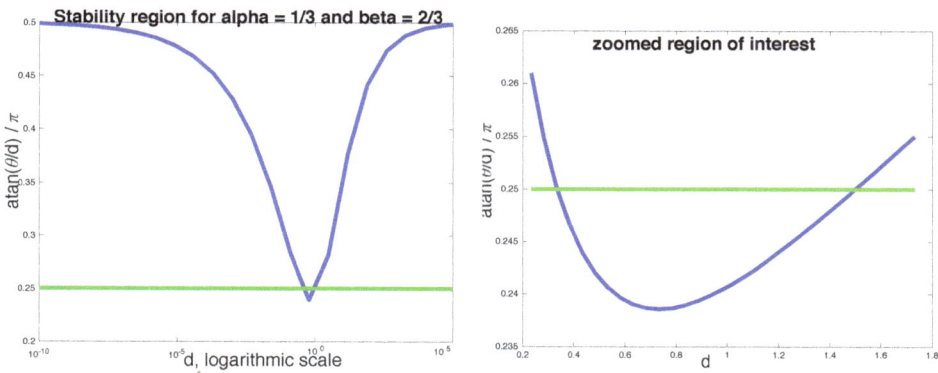

Figure 3. Stability region, above the blue line, for choosing d and θ, when the eigenvalues of A are $d \pm i\theta$, $\alpha = \frac{1}{3}, \beta = \frac{2}{3}$. The logarithmic scale is explored in the right-hand figure where the stability boundary dips below the angle $\frac{\pi}{4}$.

In the case of Figure 3, we give a similar plot as Figure 2. We also plot in green the midpoint between the two lines subtended by angles $\frac{1}{3}\pi$ and $\frac{1}{6}\pi$, namely $\frac{1}{4}\pi$. As with Figure 2, there is a small range of d for which the asymptotic stability angle drops beneath $\frac{1}{4}\pi$. Furthermore, it is clear from Remark 9(ii) that as α and β approach one another, the asymptotic stability boundary will be almost constant over increasingly longer periods of d and will only asymptotically approach the angle $\frac{\pi}{2}$ for very small and very large values of d.

In Figure 4, we confirm the asymptotic stability analysis showing sustained and decaying oscillations with $\alpha = \frac{1}{2}, \beta = 1$ (top panel) and $\alpha = \frac{1}{3}, \beta = \frac{2}{3}$ (bottom panel). In all four cases, $d = 1$, while for the top panel, we take $\theta = \sqrt{2(1+\sqrt{3})}$, $\theta = \sqrt{2(1+\sqrt{3})} + 0.3$, while for the bottom panel we take $\theta = \frac{\sqrt{3}}{4}\sqrt{\sqrt{33}-1}$, $\theta = \frac{\sqrt{3}}{4}\sqrt{\sqrt{33}-1} + 0.3$.

In Figure 5, we present phase plots of y_1 versus y_2 for the two decaying oscillations cases. The figures confirm our theoretical results on the asymptotic stability boundary and also show the effects that the fractional indices have on the period of the solutions. As α approaches β, we expect the oscillatory behaviour to disappear.

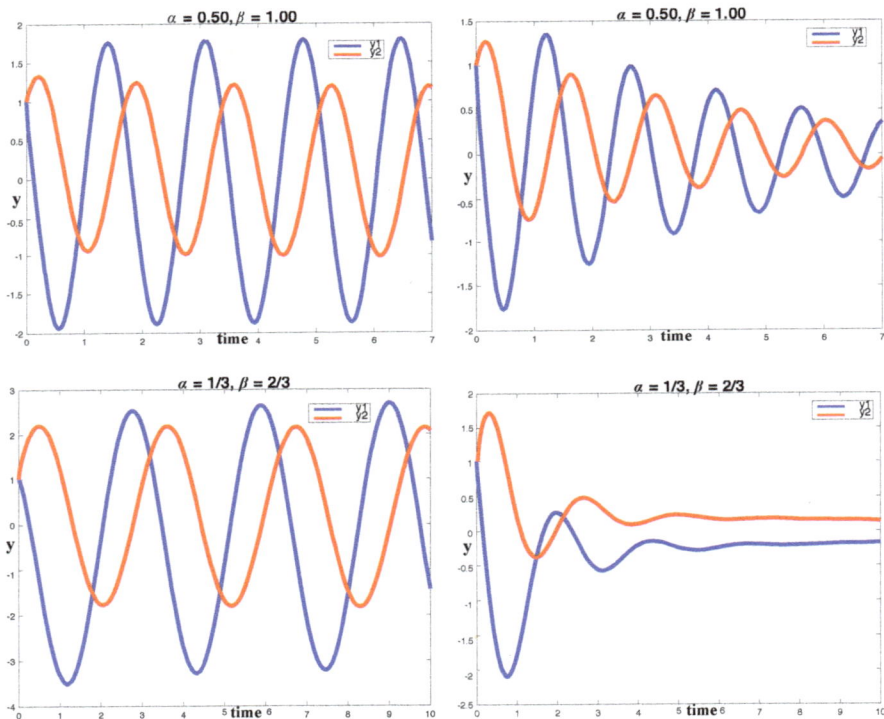

Figure 4. System dynamics with $(\alpha, \beta) = (\frac{1}{2}, 1)$, top, and $(\alpha, \beta) = (\frac{1}{3}, \frac{2}{3})$, bottom. The left-hand column shows sustained dynamics with $d = 1$ and θ chosen so that (d, θ) lies on the stability boundary. The right-hand column corresponds to the same d, but 0.3 has been added to the θ value.

Finally, in Figure 6, we consider the problem:

$$A = \begin{pmatrix} d & \theta \\ \theta & d \end{pmatrix}, \quad d < 0 \tag{70}$$

in which case the eigenvalues of A are $d \pm \theta$. We take $d = -1$, $\theta = \frac{1}{2}$ and present the solutions for four pairs of indices, namely $(\alpha, \beta) = (0.85, 0.95)$, $(0.5, 0.95)$, $(0.2, 0.05)$, $(0.15, 0.95)$. The simulations show that the components of the solution y_1 and y_2 seem to pick up "energy" from one another due to the coupling and that as the distance between α and β grows, there is a greater separation between the two components. Finally, as α gets smaller, the solutions appear to "flat-line" more quickly.

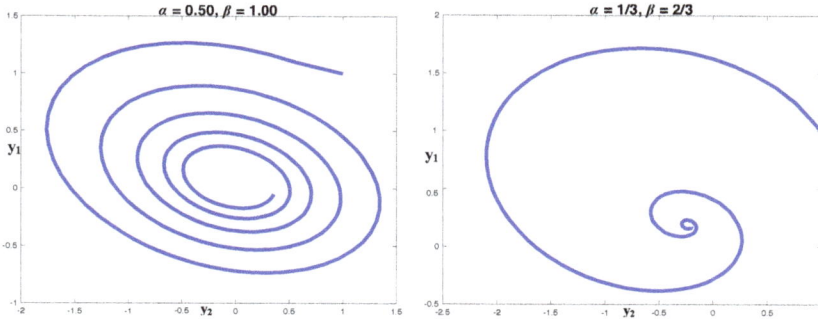

Figure 5. Phase plots of y_1 versus y_2 for the decaying solutions in the right-hand column of Figure 4.

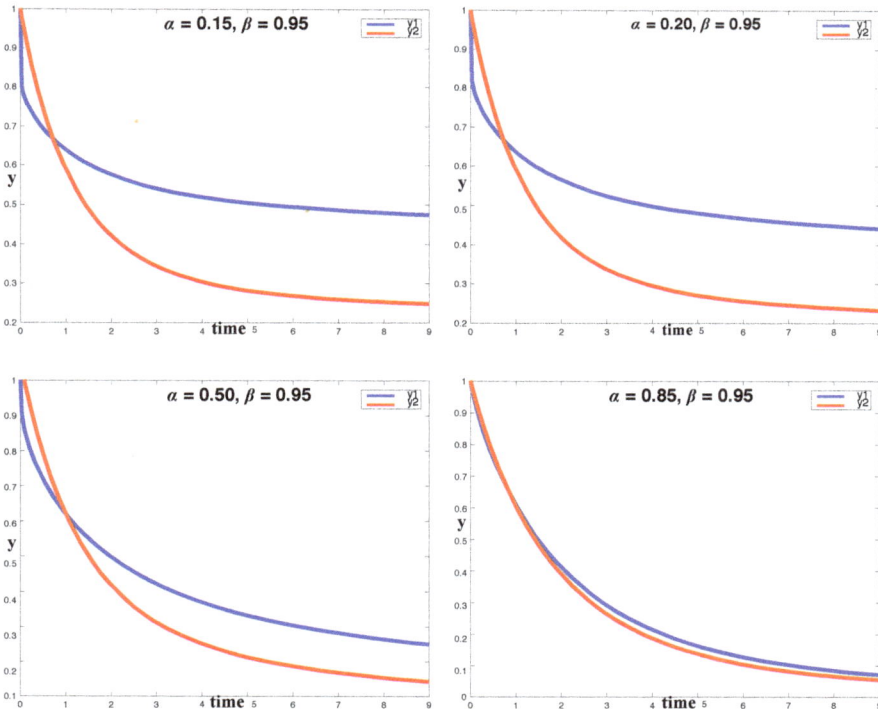

Figure 6. For A given by (70) with $d = -1$, $\theta = \frac{1}{2}$ so that the eigenvalues are $-\frac{3}{2}$, $-\frac{1}{2}$, showing the effect of variation of α with fixed β on the system dynamics.

5. Conclusions

In this paper, we have studied mixed index fractional differential equations with coupling between the different components. We find an analytical expression for the solution of the linear system that generalises the Mittag–Leffler expansion of a matrix and the solution of linear sequential fractional differential equations. We can use this result to derive new numerical methods that generalise the concept of exponential methods used in the approximation of the Mittag–Leffler matrix function (see [37–39], for example) and exponential integrators [40,41]. The second element would deal with developing numerical techniques for the integration component that incorporates the integral of a function times a Green function. We also use Laplace transform techniques to find the asymptotic stability domain in terms of the eigenvalues of the defining linear system. Finally, we have also used Laplace transforms to get analytical expansions of the mixed index problem in terms of a sum of Mittag–Leffler or generalised Mittag–Leffler functions, in the case that the fractional indices are rational.

Acknowledgments: We would like to thank Alfonso Bueno-Orovio in the Department of Computer Science, University of Oxford, for many discussions about fractional differential equations. We also acknowledge the funding support from the ARC Centre of Excellence for Mathematical and Statistical Frontiers.

Author Contributions: All authors contributed equally.

Conflicts of Interest: The authors declare no conflict of interest.

References

1. Bueno-Orovio, A.; Kay, D.; Grau, V.; Rodriguez, B.; Burrage, K. Fractional diffusion models of cardiac electrical propagation: Role of structural heterogeneity in dispersion of repolarization. *J. R. Soc. Interface* **2014**, *11*, 20140352, doi:10.1098/rsif.2014.0352.
2. Cusimano, N.; Bueno-Orovio, A.; Turner, I.; Burrage, K. On the order of the fractional Laplacian in determining the spatio-temporal evolution of a space-fractional model of cardiac electrophysiology. *PLoS ONE* **2015**, *10*, e0143938, doi:10.1371/journal.pone.0143938.
3. Bueno-Orovio, A.; Teh, I.; Schneider, J.E.; Burrage, K.; Grau, V. Anomalous diffusion in cardiac tissue as an index of myocardial microstructure. *IEEE Trans. Med. Imaging* **2016**, *35*, 2200–2207.
4. Henry, B.; Langlands, T. Fractional cable models for spiny neuronal dendrites. *Phys. Rev. Lett.* **2008**, *100*, 128103, doi:10.1103/PhysRevLett.100.128103.
5. Magin, R.; Feng, X.; Baleanu, D. Solving the fractional order Bloch equation. *Concepts Magn. Reson. Part A* **2009**, *34*, 16–23.
6. Klafter, J.; White, B.; Levandowsky, M. Microzooplankton feeding behavior and the Lévy walk. In *Biological Motion*; Lecture Notes in Biomathematics; Springer: Berlin/Heidelberg, Germany, 1990; Volume 89, pp. 281–296.
7. Cusimano, N.; Burrage, K.; Burrage, P. Fractional models for the migration of biological cells in complex spatial domains. *ANZIAM J.* **2013**, *54*, 250–270.
8. Shen, S.; Liu, F.; Liu, Q.; Anh, V. Numerical simulation of anomalous infiltration in porous media. *Numer. Algorithms* **2015**, *68*, 443–454.
9. Carcione, J.; Sanchez-Sesma, F.; Luzon, F.; Gavilan, J.P. Theory and simulation of time-fractional fluid diffusion in porous media. *J. Phys. A* **2013**, *46*, 345501, doi:10.1088/1751-8113/46/34/345501.
10. Metzler, R.; Klafter, J.; Sokolov, I.M. Anomalous transport in external fields: Continuous time random walks and fractional diffusion equations extended. *Phys. Rev. E* **1998**, *58*, 1621–1633.
11. Klages, R.; Radons, G.; Sokolov, I. *Anomalous Transport*; Wiley-VCH Verlag GmbH & Co.: Hoboken, NJ, USA, 2008.
12. Metzler, R.; Klafter, J. The random walk's guide to anomalous diffusion: A fractional dynamics approach. *Phys. Rep.* **2000**, *339*, 1–77.
13. Mittag–Leffler, G.M. Sur la nouvelle function Ea. *C. R. Acad. Sci.* **1903**, *137*, 554–558.
14. Tian, T.; Harding, A.; Inder, K.; Parton, R.G.; Hancock, J.F. Plasma membrane nanoswitches generate high-fidelity Ras signal transduction. *Nat. Cell Biol.* **2007**, *9*, 905–914.
15. Gillespie, D.T. Exact stochastic simulation of coupled chemical reactions. *J. Phys. Chem.* **1977**, *81*, 2340–2361.

16. Kilbas, A.; Srivastava, H.M.; Trujillo, J.J. *Theory and Applications of Fractional Differential Equations, North-Holland Mathematics Studies*; Elsevier: Amsterdam, The Netherlands, 2006; Volume 204.

17. Yu, Q.; Liu, F.; Turner, I.; Burrage, K. Numerical simulation of the fractional Bloch equations. *J. Comp. Appl. Math.* **2014**, *255*, 635–651.

18. Liu, F.; Meerschaert, M.M.; McGough, R.; Zhuang, P.; Liu, Q. Numerical methods for solving the multi-term time fractional wave equations. *Fract. Calc. Appl. Anal.* **2013**, *16*, 9–25.

19. Qin, S.; Liu, F.; Turner, I.; Yu, Q.; Yang, Q.; Vegh, V. Characterization of anomalous relaxation using the time-fractional Bloch equation and multiple echo T2*-weighted magnetic resonance imaging at 7T. *Magn. Reson. Med.* **2017**, *77*, 1485–1494.

20. Diethelm, K.; Siegmund, S.; Tuan, H.T. Asymptotic behavior of soutions of linear multi-order fractional differential systems. *Fract. Calc. Appl. Anal.* **2017**, *20*, 1165–1195.

21. Popolizio, M. Numerical solution of multiterm fractional differential equations using the matrix Mittag–Leffler functions. *Mathematics* **2018**, *6*, 7, doi:10.3390/math6010007.

22. Prabhakar, T.R. A singular integral equation with a generalised Mittag–Leffler function in the kernel. *Yokohama Math. J.* **1971**, *19*, 7–15.

23. Podlubny, I. *Fractional Differential Equations*; Academic Press: New York, NY, USA, 1999.

24. Oldham, K.B.; Spanier, J. *The Fractional Calculus: Theory and Applications of Differentiation and Integration to Arbitrary Order*; Academic Press: New York, NY, USA, 1974.

25. Matignon, D. Stability result on fractional differential equations with applications to control processing. In Proceedings of the July 1996 IMACS-SMC, Lille, France, July 1998; pp. 963–968.

26. Deng, W.; Li, C.; Lu, J. Stability analysis of linear fractional differential system with multiple time delays. *Nonlinear Dyn.* **2007**, *48*, 409–416.

27. Li, C.P.; Zhang, F.R. A survey on the stability of fractional differential equations. *Eur. Phys. J. Spec. Top.* **2011**, *193*, 27–47.

28. Najafi, H.S.; Sheikhani, A.R.; Ansari, A. Stability analysis of distributed order Fractional Differential Equations. *Abstr. Appl. Anal.* **2011**, 175323, doi:/10.1155/2011/175323.

29. Zhang, F.; Li, C.; Chen, Y. Asymptotical stability of nonlinear fractional differential system with Caputo Derivative. *Int. J. Differ. Equ.* **2011**, *2011*, 635165, doi:10.1155/2011/635165.

30. Radwan, A.G.; Soliman, A.M.; Elwakil, A.S.; Sedeek, A. On the stability of linear systems with fractional-order elements. *Chaos Solitons Fractal* **2009**, *40*, 2317–2328.

31. Rivero, M.; Rogosin, S.V.; Machado, J.A.T.; Trujillo, J.J. Stability of Fractional Order Systems. *Math. Probl. Eng.* **2013**, 356215, doi:10.1155/2013/356215.

32. Ritt, J.F. On the zeros of exponential polynomials. *Trans. Am. Math. Soc.* **1929**, *31*, 680–686.

33. Miller, K.S.; Ross, B. Fractional Green's functions. *Indian J. Pure Appl. Math.* **1991**, *22*, 763–767.

34. Vazquez, L. Fractional diffusion equations with internal degrees of freedom. *J. Comp. Math.* **2003**, *21*, 491–494.

35. Petras, I. Stability of fractional order systems with rational orders: A survey. *Fract. Calc. Appl. Anal.* **2009**, *12*, 269–298.

36. Čermák, J.; Kisela, T. Stability properties of two term fractional differential equations. *Nonlinear Dyn.* **2015**, *80*, 1673–1684.

37. Garrappa, R.; Popolizio, M. Evaluation of generalized Mittag–Leffler functions on the real line. *Adv. Comput. Math.* **2013**, *39*, 205–225.

38. Zeng, C.; Chen, Y.Q. Global Padé Approximations of the Generalized Mittag–Leffler Function and its Inverse. *Fract. Calc. Appl. Anal.* **2015**, *18*, 149–156.

39. Garrappa, R. Numerical evaluation of two and three parameters Mittag–Leffler functions. *SIAM J. Numer. Anal.* **2015**, *53*, 1350–1369.

40. Sidje, R.B. Expokit: A software package for computing matrix exponentials. *ACM Trans. Math. Softw.* **1998**, *24*, 130–156.

41. Hochbruck, M.; Ostermann, A. Exponential integrators. *Acta Numer.* **2010**, *19*, 209–286.

![axioms](axioms logo) *axioms*

MDPI

Article

Optimal B-Spline Bases for the Numerical Solution of Fractional Differential Problems

Francesca Pitolli [ORCID]

Dept. SBAI, Università di Roma "La Sapienza", Via Antonio Scarpa 16, 00161 Roma, Italy; francesca.pitolli@sbai.uniroma1.it; Tel.: +39-06-4976-6631

Received: 14 May 2018; Accepted: 22 June 2018; Published: 2 July 2018

Abstract: Efficient numerical methods to solve fractional differential problems are particularly required in order to approximate accurately the nonlocal behavior of the fractional derivative. The aim of the paper is to show how optimal B-spline bases allow us to construct accurate numerical methods that have a low computational cost. First of all, we describe in detail how to construct optimal B-spline bases on bounded intervals and recall their main properties. Then, we give the analytical expression of their derivatives of fractional order and use these bases in the numerical solution of fractional differential problems. Some numerical tests showing the good performances of the bases in solving a time-fractional diffusion problem by a collocation–Galerkin method are also displayed.

Keywords: B-spline; optimal basis; fractional derivative; Galerkin method; collocation method

1. Introduction

In the last few decades, fractional calculus has proved to be a powerful tool for describing real-world phenomena. Differential problems of fractional order, initially introduced to model anomalous diffusion in viscoelastic materials, are now used in several fields, from physics to population dynamics, from signal processing to control theory [1–5]. For an introduction of fractional calculus, we refer to [6–9].

As fractional differential models have become widespread, the development of efficient numerical methods to approximate their solution has become of primary interest. In fact, the nonlocality of the fractional derivative poses a severe challenge in its approximation, and, to face this problem, several numerical methods were proposed in the literature. For a review on numerical methods, see, for instance, [10–15].

In particular, we are interested in the solution of differential problems on bounded intervals. In this case, it is important to have a function basis available that naturally fulfills boundary and/or initial conditions and does not show numerical instabilities at the boundaries. From this point of view, B-spline bases are especially suitable and their use in the numerical solution of classical differential problems is widely diffused (see, for instance, [16–18]). Despite the indubitable success of these methods, the use of B-spline bases for the solution of differential problems of fractional order is not yet very common and limited to few examples. For instance, in [19], the authors used a linear B-spline basis to solve a multi-order fractional differential problem by the operational matrix method, while, in [20], a collocation method based on cubic B-spline wavelets was constructed to solve ordinary differential equations of fractional order.

The aim of this paper is to show how optimal B-spline bases can be profitably used for the solution of fractional differential problems. To this end, we not only describe in detail the construction of optimal B-spline bases of any degree but also give the analytical expression of the basis functions and of their fractional derivatives. These expressions are an efficient tool to construct numerical methods based on spline functions.

The paper is organized as follows. In Section 2, we show how to construct polynomial B-spline bases and recall their main properties. In particular, the construction of cardinal B-splines on the real line through the divided difference operator is described in Sections 2.1–2.2, while, in Section 2.3, we show how to construct optimal B-spline bases on bounded intervals. Their analytical expression in case of equidistant nodes is given in Section 2.4. Section 2.5 is devoted to the evaluation of the fractional derivatives of the B-spline bases. Finally, in Section 2.6, we show how to use optimal bases for the numerical solution of a time-fractional diffusion problem by a collocation–Galerkin method. In Section 3, we use the optimal B-spline basis of degree 3 to solve some test problems; some numerical results are also displayed. Section 4 contains a discussion on the results and highlights possible extensions.

2. Materials and Methods

In this section, we define the B-spline bases through the divided difference operator and recall their main properties. Then, we give the analytical expression of the basis functions and of their fractional derivatives. Finally, B-spline bases are used to solve a fractional differential problem by a collocation–Galerkin method.

2.1. The Divided Difference Operator

Let

$$\mathcal{X} = \{x_1, x_2, \ldots, x_n\}, \qquad \text{with} \qquad x_1 < x_2 < \ldots < x_n \tag{1}$$

be a sequence of *simple, i.e., non coincident, knots* and let

$$D \left(\begin{array}{cccc} u_1(x), & u_2(x), & \cdots & u_n(x) \\ x_1, & x_2, & \cdots & x_n \end{array} \right) := \begin{vmatrix} u_1(x_1) & u_2(x_1) & \ldots & u_n(x_1) \\ u_1(x_2) & u_2(x_2) & \ldots & u_n(x_2) \\ \ldots & \ldots & \ldots & \ldots \\ u_1(x_n) & u_2(x_n) & \ldots & u_n(x_n) \end{vmatrix} \tag{2}$$

be the determinant of the collocation matrix

$$M \left(\begin{array}{cccc} u_1(x), & u_2(x), & \cdots & u_n(x) \\ x_1, & x_2, & \cdots & x_n \end{array} \right) := \begin{pmatrix} u_1(x_1) & u_2(x_1) & \ldots & u_n(x_1) \\ u_1(x_2) & u_2(x_2) & \ldots & u_n(x_2) \\ \ldots & \ldots & \ldots & \ldots \\ u_1(x_n) & u_2(x_n) & \ldots & u_n(x_n) \end{pmatrix} \tag{3}$$

associated with the function system $\mathcal{U} = \{u_1(x), u_2(x), \ldots, u_n(x)\}$ and the knot sequence (1).

The *divided difference operator* $[x_1, x_2, \ldots, x_n]f$ of a function f over the knots $\mathcal{X}$ is defined as

$$[x_1, x_2, \ldots, x_n]f(x) := \frac{D \left(\begin{array}{c} 1, x, x^2, \ldots, x^{n-2}, f(x) \\ x_1, x_2, x_3, \ldots, x_{n-1}, x_n \end{array} \right)}{D \left(\begin{array}{c} 1, x, x^2, \ldots, x^{n-2}, x^{n-1} \\ x_1, x_2, x_3, \ldots, x_{n-1}, x_n \end{array} \right)}. \tag{4}$$

The formula above can be generalized to the case of coincident knots as follows. Let the knots occur more than once, i.e.,

$$x_1 \leq x_2 \leq \ldots \leq x_n, \tag{5}$$

and let

$$\mathcal{M} = \{m_1, m_2, \ldots, m_r\} \qquad \text{with} \qquad \sum_{k=1}^{r} m_k = n \tag{6}$$

be the *multiplicities* of the coincident knots, i.e.,

$$\tilde{\mathcal{X}} = \{x_1, x_2, \ldots, x_n\} = \{\underbrace{\eta_1, \ldots \eta_1}_{m_1}, \underbrace{\eta_2, \ldots \eta_2}_{m_2}, \ldots, \underbrace{\eta_r, \ldots \eta_r}_{m_r}\}, \tag{7}$$

where $\eta_1 < \eta_2 < \ldots < \eta_r$ are the non coincident knots of the sequence $\tilde{\mathcal{X}}$.

Assuming the functions in $\mathcal{U}$ are sufficiently smooth, the collocation matrix associated with the function system $\mathcal{U}$ over the knot sequence (7) is given by

$$\tilde{M}\begin{pmatrix} u_1(x), & u_2(x), & \cdots & u_n(x) \\ x_1, & x_2, & \cdots & x_n \end{pmatrix} := \begin{pmatrix} u_1(\eta_1) & u_2(\eta_1) & \cdots & u_n(\eta_1) \\ u_1'(\eta_1) & u_2'(\eta_1) & \cdots & u_n'(\eta_1) \\ \cdots & \cdots & \cdots & \cdots \\ u_1^{(m_1-1)}(\eta_1) & u_2^{(m_1-1)}(\eta_1) & \cdots & u_n^{(m_1-1)}(\eta_1) \\ \cdots & \cdots & \cdots & \cdots \\ u_1(\eta_r) & u_2(\eta_r) & \cdots & u_n(\eta_r) \\ u_1'(\eta_r) & u_2'(\eta_r) & \cdots & u_n'(\eta_r) \\ \cdots & \cdots & \cdots & \cdots \\ u_1^{(m_r-1)}(\eta_r) & u_2^{(m_r-1)}(\eta_r) & \cdots & u_n^{(m_r-1)}(\eta_r) \end{pmatrix}. \tag{8}$$

Thus, if the function f has sufficient derivatives, the divided difference operator of f over the sequence of multiple knots $\tilde{\mathcal{X}}$ can be defined by using the collocation matrix (8) in Definition (4), i.e.,

$$[x_1, x_2, \ldots, x_n]f(x) := \frac{\tilde{D}\begin{pmatrix} 1, x, x^2, \ldots, x^{n-2}, f(x) \\ x_1, x_2, x_3, \ldots, x_{n-1}, x_n \end{pmatrix}}{\tilde{D}\begin{pmatrix} 1, x, x^2, \ldots, x^{n-2}, x^{n-1} \\ x_1, x_2, x_3, \ldots, x_{n-1}, x_n \end{pmatrix}}, \tag{9}$$

where $\tilde{D}$ denotes the determinant of the collocation matrix as defined in Equation (8).

2.2. The Cardinal B-Splines

The polynomial splines are piecewise polynomials having a certain degree of smoothness which is related to the degree of the polynomial pieces. Spline functions can be represented as a linear combination of B-splines that form a local basis for the spline spaces [21].

The cardinal B-splines, i.e., the polynomial B-splines having break points on integer knots, can be defined by applying the divided difference operator (4) to the *truncated power function* defined as

$$x_+^n := \begin{cases} x^n, & \text{for } x \geq 0, \\ 0, & \text{for } x < 0. \end{cases} \tag{10}$$

Then, the cardinal B-spline B_n of degree n on the integer knots

$$\mathcal{I} = \{0, 1, \ldots, n+1\} \tag{11}$$

is given by

$$\begin{aligned} B_n(x) &:= (n+1)[0, 1, \ldots, n+1](y-x)_+^n \\ &= (n+1)\frac{D\begin{pmatrix} 1, y, y^2, \ldots, y^n, (y-x)_+^n \\ 0, 1, 2, \ldots, n, n+1 \end{pmatrix}}{D\begin{pmatrix} 1, y, y^2, \ldots, y^n, y^{n+1} \\ 0, 1, 2, \ldots, n, n+1 \end{pmatrix}} = \frac{1}{n!}\Delta^{n+1}x_+^n, \end{aligned} \tag{12}$$

where Δ^n is the *finite difference operator*

$$\Delta^n f(x) = \sum_{k=0}^{n} (-1)^k \binom{n}{k} f(x - k).$$ (13)

From the above definition, it follows that the cardinal B-spline B_n is compactly supported on $[0, n+1]$, positive in $(0, n+1)$ and belongs to $C^{n-1}(\mathbb{R})$. Moreover, the system of the integer translates

$$\mathcal{B}_n = \{B_n(x - k), k \in \mathbb{Z}\}$$ (14)

forms a function basis that is a *partition of unity*, i.e.,

$$\sum_{k \in \mathbb{Z}} B_n(x - k) = 1, \qquad \forall x \in \mathbb{R},$$ (15)

and reproduces polynomials up to degree n [21].

The basis $\mathcal{B}_n$ can be generalized to any set of equidistant knots $\mathcal{X}_h = \{x_k = kh, k \in \mathbb{Z}\}$, where $h > 0$ is the space step, by scaling, i.e.,

$$\mathcal{B}_{nh} = \left\{B_n\left(\frac{x}{h} - k\right), k \in \mathbb{Z}\right\}.$$ (16)

For details and further properties on polynomial B-splines, see, for instance, [21].

It is always possible to construct bases on bounded intervals by restricting the B-spline basis $\mathcal{B}_{nh}$ on an interval $[a, b]$. However, in this way, we obtain bases that could be numerically unstable since the functions at the boundaries are obtained by truncation. Moreover, boundary or initial conditions are not easy to satisfy. In the following section, we will show how to construct stable bases on the interval that easily satisfy boundary or initial conditions.

2.3. Optimal B-Spline Bases

Optimal B-spline bases on bounded intervals can be defined by introducing multiple knots at the boundaries of the given interval [21].

Let $L \geq n + 1$ be an integer and let $\mathcal{I} = \{0, 1, \ldots, L - 1, L\}$ be the sequence of the integer knots of the interval $I = [0, L]$. We extend $\mathcal{I}$ to a sequence $\tilde{\mathcal{I}}$ of $L + 1 + 2n$ knots by introducing knots of multiplicity $n + 1$ at the boundaries of the interval, i.e.,

$$\tilde{\mathcal{I}} = \{x_0, x_1, \ldots, x_{2n+L}\} \quad \text{with} \quad \begin{cases} x_0 = x_1 = \cdots = x_n = 0, \\ x_k = k - n, \quad n + 1 \leq k \leq L + n - 1, \\ x_{L+n} = x_{L+n+1} = \cdots = x_{L+2n} = L. \end{cases}$$ (17)

The functions of the optimal B-spline basis

$$\mathcal{N}_n = \{N_{kn}(x), 0 \leq k \leq L - 1 + n\}$$ (18)

of degree n with knots $\tilde{\mathcal{I}}$ are given by the divided difference

$$N_{kn}(x) = (x_{k+n+1} - x_k)[x_k, x_{k+1}, \ldots, x_{k+n+1}](y - x)_+^n, \qquad 0 \leq k \leq L - 1 + n,$$ (19)

so that each function N_{kn} has compact support with

$$\operatorname{supp} N_{kn} = [x_k, x_{k+n+1}] \subset I.$$ (20)

From (19), it follows that the basis $\mathcal{N}_n$ has $L - n$ *interior functions*, i.e., the functions N_{kn}, $n \leq k \leq L - 1$ having all simple knots, and $2n$ *edge functions*, i.e., the functions N_{kn} for $0 \leq k \leq n - 1$ and $L \leq k \leq L - 1 + n$ that have a multiple knot at the left or right boundary, respectively.

The basis $\mathcal{N}_n$ is centrally symmetric, i.e.,

$$N_{kn}(x) = N_{(L-1+n-k)n}(L - x), \qquad 0 \leq k \leq L - 1 + n. \tag{21}$$

Moreover, the functions N_{kn} satisfy the boundary conditions

$$\begin{cases} N_{kn}(0^+) = \delta_{k0}, & 0 \leq k \leq L - 1 + n, \\[2mm] D_x^r N_{kn}(0^+) = 0, & 1 \leq r \leq k - 1, \\[2mm] & \hspace{4em} 1 \leq k \leq n - 1, \\[2mm] (-1)^r D_x^r N_{kn}(0^+) > 0, & k \leq r \leq n, \end{cases} \tag{22}$$

$$\begin{cases} N_{kn}(L^-) = \delta_{k,r-1}, & 0 \leq k \leq L - 1 + n, \\[2mm] D_x^r N_{(L-1+n-k)n}(L^-) = 0, & 1 \leq r \leq k - 1, \\[2mm] & \hspace{4em} 1 \leq k \leq n - 1, \\[2mm] (-1)^r D_x^r N_{(L-1+n-k)n}(L^-) > 0, & k \leq r \leq n, \end{cases} \tag{23}$$

where D_x^r denotes the usual derivative of integer order r.

Finally, the basis $\mathcal{N}_n$ forms a partition of unity, i.e.,

$$\sum_{k=0}^{L-1+n} N_{kn}(x) = 1, \qquad \forall x \in I, \tag{24}$$

and is stable, i.e., for any sequence $c = \{c_0, c_1, \ldots, c_{L-1+n}\}$ with $\|c\|_\infty = \max_{0 \leq k \leq L-1+n} |c_k| < \infty$ it holds

$$\frac{1}{\kappa} \|c\|_\infty \leq \left\| \sum_{k=0}^{L-1+n} c_k N_{kn} \right\|_{L_\infty[0,L]} \leq \|c\|_\infty, \tag{25}$$

where

$$\kappa := \left(\min_{\|c\|_\infty = 1} \left\| \sum_{k=0}^{L-1+n} c_k N_{kn} \right\|_{L_\infty[0,L]} \right)^{-1} \tag{26}$$

is the *condition number* of the basis $\mathcal{N}_n$. From (25), it follows that $\|\sum_k \Delta c_k N_{k3}\|_{L_\infty[0,L]} \leq \|\Delta c\|_\infty$ where Δc denotes a perturbation of the sequence c. As a consequence, optimal B-spline bases are numerically stable so that numerical errors are not amplified when evaluating spline approximations.

The basis $\mathcal{N}_n$ can be generalized to any sequence of equidistant knots on a bounded interval $[a, b]$ by mapping $x \to (x - a)/h$, i.e.,

$$\mathcal{N}_{nh} = \{N_{knh}(x), 0 \leq k \leq L - 1 + n\}, \tag{27}$$

where $L = (b - a)/h$. The interior functions are the $L - n$ functions

$$N_{knh}(x) = B_n \left(\frac{x - a}{h} - (k - n) \right), \qquad n \leq k \leq L - 1. \tag{28}$$

while the 2*n* edge functions are

$$N_{knh}(x) = N_{kn}\left(\frac{x-a}{h}\right), \quad N_{(L-1+n-k)nh}(x) = N_{knh}(b-x), \quad 0 \le k \le n-1. \tag{29}$$

All the properties above make the basis $\mathcal{N}_n$ *optimal* when used in approximation problems [22,23].

2.4. Analytical Expression of the Optimal B-Spline Bases

Let $\mathcal{N}_n$ be the optimal basis (18) associated with the knot sequence (17). The interior functions N_{kn}, $n \le k \le L-1$, are the translates of the B-spline B_n having support $[k-n, k+1]$, i.e.,

$$N_{kn}(x) = B_n(x-k+n), \quad n \le k \le L-1. \tag{30}$$

The left edge functions N_{kn}, $0 \le k \le n-1$, having support on $[0, k+1]$, can be evaluated by using the finite difference operator (9) for coincident knots. For $0 \le k \le n-1$, we get

$$
\begin{aligned}
N_{kn}(x) &= (k+1)\,[x_k, x_{k+1}, \ldots, x_{k+n+1}]\,(y-x)_+^n \\[2mm]
&= (k+1)\,\frac{\tilde{D}\left(\underbrace{1,}_{1}\ \underbrace{y,}_{2}\ \cdots \ \cdots \ , \underbrace{0}_{n+1-k},\ 1, \ \cdots \ y^n, \ (y-x)_+^n \atop \ \ \ 0\ \ \ 0\ 1, \ \ \ k, \ \ k+1 \right)}{\tilde{D}\left(\underbrace{1,}_{1}\ \underbrace{y,}_{2}\ \cdots \ \cdots \ , \underbrace{0}_{n+1-k},\ 1, \ \cdots \ y^n, \ y^{n+1} \atop \ \ \ 0\ \ \ 0\ 1, \ \ \ k, \ \ k+1 \right)}.
\end{aligned} \tag{31}
$$

By Definition (8), we get

$$
\tilde{D}\left(\begin{matrix} 1, & y, & \cdots & & \cdots & \cdots & y^n, & (y-x)_+^n \\ 0 & 0 & \cdots, & 0 & 1, & \cdots & k, & k+1 \end{matrix}\right) = \left|\begin{matrix} D_{kn} & 0 \\ R_{kn} & T_{kn} \end{matrix}\right|, \tag{32}
$$

where D_{kn} is the $(n-k+1)$ order diagonal matrix

$$D_{kn} = \operatorname{diag}\left(1, 1, 2!, \ldots, (n-k)!\right), \tag{33}$$

R_{kn} is the $(k+1) \times (n-k+1)$ dimension matrix

$$R_{kn} = \left(i^{j-1}, 1 \le i \le k+1, 1 \le j \le n-k+1\right), \tag{34}$$

and T_{kn} is the $(k+1)$ order collocation matrix

$$T_{kn} = M\left(\begin{matrix} y^{n-k+1}, & y^{n-k+2}, & \cdots & y^n, & (y-x)_+^n \\ 1, & 2, & \cdots & k, & k+1 \end{matrix}\right). \tag{35}$$

Thus, for the numerator in the last equality of (31), we have

$$
\tilde{D}\left(\begin{matrix} 1, & y, & \cdots & & \cdots & \cdots & y^n, & (y-x)_+^n \\ 0 & 0 & \cdots, & 0 & 1, & \cdots & k, & k+1 \end{matrix}\right) = |T_{nk}|\,|D_{kn}|. \tag{36}
$$

For the denominator, a similar calculation gives

$$
\tilde{D}\left(\begin{array}{cccccccc} 1, & y, & \cdots & \cdots & \cdots & \cdots & y^n, & y^{n+1} \\ \underbrace{0}_{1} & \underbrace{0}_{2} & \cdots, & \underbrace{0}_{n+1-k}, & 1, & \cdots & k, & k+1 \end{array}\right) = |P_{nk}|\,|D_{kn}|,
\tag{37}
$$

where P_{kn} is the $k+1$ order collocation matrix

$$
P_{kn} = M\left(\begin{array}{ccccc} y^{n-k+1}, & y^{n-k+2}, & \cdots & y^n, & y^{n+1} \\ 1, & 2, & \cdots & k, & k+1 \end{array}\right).
\tag{38}
$$

Theorem 1. *Let $\tilde{\mathcal{I}}$ be the sequence of multiple knots (17) on the interval $I = [0, L]$. The analytical expression of the left edge functions is given by*

$$
N_{kn}(x) = (k+1)\,\frac{|T_{nk}|}{|P_{kn}|}, \qquad 0 \le k \le n-1,
\tag{39}
$$

where $|T_{nk}|$ and $|P_{kn}|$ are the determinants of the collocation matrices (35) and (38), respectively.

Proof. The analytical expression (39) follows by using (36) and (37) in the last equality in (31). □

The right edge functions $N_{L-1+n-k,n}, 0 \le k \le n-1$, having support $[L-k-1, L]$, can be easily obtained recalling the symmetry property (21).

In the following corollary, we give the analytical expression of the first three left boundary functions.

Corollary 1. *Let $\tilde{\mathcal{I}}$ be the sequence of multiple knots (17) on the interval $I = [0, L]$. For $0 \le k \le 2$, the left edge functions N_{kn} are given by*

$$
N_{0n}(x) = (1-x)^n_+ = \begin{cases} (1-x)^n, & 0 \le x \le 1, \\ & \qquad\qquad\qquad n \ge 1, \\ 0, & otherwise, \end{cases}
\tag{40}
$$

$$
N_{1n}(x) = 2\left(\frac{1}{2^n}\,(2-x)^n_+ - (1-x)^n_+\right) = \begin{cases} \dfrac{1}{2^{n-1}}\,(2-x)^n - 2\,(1-x)^n, & 0 \le x \le 1, \\[2mm] \dfrac{1}{2^{n-1}}\,(2-x)^n, & 1 < x \le 2, \qquad n \ge 2, \\[2mm] 0, & otherwise, \end{cases}
\tag{41}
$$

$$
N_{2n}(x) = 3\left(\frac{1}{2}\,\frac{1}{3^{n-1}}\,(3-x)^n_+ - \frac{1}{2^{n-1}}\,(2-x)^n_+ + \frac{1}{2}\,(1-x)^n_+\right)
$$

$$
= \begin{cases} \dfrac{1}{2}\,\dfrac{1}{3^{n-2}}\,(3-x)^n - \dfrac{3}{2^{n-1}}\,(2-x)^n + \dfrac{3}{2}\,(1-x)^n, & 0 \le x \le 1, \\[2mm] \dfrac{1}{2}\,\dfrac{1}{3^{n-2}}\,(3-x)^n - \dfrac{3}{2^{n-1}}\,(2-x)^n, & 1 < x \le 2, \\[2mm] \dfrac{1}{2}\,\dfrac{1}{3^{n-2}}\,(3-x)^n, & 2 < x \le 3, \\[2mm] 0, & otherwise, \end{cases} \qquad n \ge 3.
\tag{42}
$$

It is interesting to observe that $N_{0n}(x) = \dfrac{B_n(x+n)}{B_n(n)}$, $0 \le x \le 1$. In particular, when $n = 1$ (linear case), there is just one left edge function $N_{01}(x) = B_1(x+1)$ and the basis $\mathcal{N}_1$ coincides with the basis $\mathcal{B}_1$ restricted to the interval I. We recall that the basis $\mathcal{N}_1$ is the unique interpolatory B-spline basis.

2.5. Fractional Derivatives of the Optimal B-Spline Bases

In this section, we give the explicit expression of the derivatives of fractional order of the functions in the basis $\mathcal{N}_n$.

Let Γ be the Euler's gamma function

$$\Gamma(\beta) = \int_0^\infty z^{\beta-1}\,e^{-z}\,dz\,. \tag{43}$$

Definition 1. *Assuming f is a sufficiently smooth function, the Caputo fractional derivative of f is defined as*

$$^C D_x^\beta f(x) := \frac{1}{\Gamma(m-\beta)} \int_0^x \frac{f^{(m)}(z)}{(x-z)^{\beta-m+1}}\,dz\,, \qquad x \ge 0\,, \qquad m-1 < \beta < m\,, \quad m \in \mathbb{N}. \tag{44}$$

Definition 2. *The Riemann–Liouville fractional derivative of a function f is defined as*

$$^{RL} D_x^\beta f(x) := \frac{1}{\Gamma(m-\beta)} \frac{d^m}{dx^m} \int_0^x \frac{f(z)}{(x-z)^{\beta-m+1}}\,dz\,, \qquad x \ge 0\,, \qquad m-1 < \beta < m\,, \quad m \in \mathbb{N}. \tag{45}$$

We recall that the Caputo derivative and the Riemann–Liouville derivative can be obtained from each other by

$$^{RL} D_x^\beta f(x) = {}^C D_x^\beta f(x) + \sum_{k=0}^{m-1} \frac{x^{k-\beta}}{\Gamma(k-\beta+1)}\,f^{(k)}(0^+)\,. \tag{46}$$

If f satisfies homogeneous initial conditions, i.e., $f^{(k)}(0^+) = 0$, $0 \le k \le m-1$, the Riemann–Liouville derivative coincides with the Caputo derivative.

To use the basis $\mathcal{N}_n$ for the solution of differential problems of fractional order, we need the expression of the fractional derivatives of the functions N_{kn}.

The derivatives of the interior functions can be easily evaluated by the differentiation rule

$$^C D_x^\beta B_n(x) = \frac{\Delta^{n+1} x_+^{n-\beta}}{\Gamma(n-\beta+1)}\,, \qquad x \ge 0\,, \qquad 0 < \beta < n\,, \tag{47}$$

where Δ^n is the finite difference operator (13) [15,24]. Since B_n satisfies homogeneous boundary conditions, the Riemann–Liouville derivative is equal to the Caputo derivative.

Theorem 2. *The Caputo derivative of the left edge functions are given by*

$$^C D_x^\beta N_{kn}(x) = \frac{(k+1)}{|P_{kn}|} \sum_{r=1}^{k+1} (-1)^{r+k+1}\,{}^C D_x^\beta (r-x)_+^n\,|T_{kn}^r|\,, \qquad 0 \le k \le n-1\,, \tag{48}$$

where

$$T_{kn}^r = \begin{pmatrix}
1 & 1 & \cdots & 1 \\
2^{n-k+1} & 2^{n-k+2} & \cdots & 2^n \\
\cdots & \cdots & \cdots & \cdots \\
(r-1)^{n-k+1} & (r-1)^{n-k+2} & \cdots & (r-1)^n \\
(r+1)^{n-k+1} & (r+1)^{n-k+2} & \cdots & (r+1)^n \\
\cdots & \cdots & \cdots & \cdots \\
(k+1)^{n-k+1} & (k+1)^{n-k+2} & \cdots & (k+1)^n
\end{pmatrix}. \tag{49}$$

Proof. First of all, we write the explicit expression of the matrix T_{kn}, i.e.,

$$T_{kn} = \begin{pmatrix} 1 & 1 & \cdots & 1 & (1-x)_+^n \\ 2^{n-k+1} & 2^{n-k+2} & \cdots & 2^n & (2-x)_+^n \\ \cdots & \cdots & \cdots & \cdots & \cdots \\ (k+1)^{n-k+1} & (k+1)^{n-k+2} & \cdots & (k+1)^n & (k+1-x)_+^n \end{pmatrix}.$$

From (39), we get $^C D_x^\beta N_{kn}(x) = \frac{k+1}{|P_{kn}|} {}^C D_x^\beta |T_{kn}|$. The claim follows by expanding the determinant of T_{kn} along the last column. $\square$

Thus, to evaluate the fractional derivatives of the edge functions, we need the fractional derivatives of the translates of the truncated power function. By Definition (44), we get

$$\begin{aligned} {}^C D_x^\beta (r-x)_+^n &= \frac{1}{\Gamma(m-\beta)} \int_0^x \frac{\frac{d^m}{dz^m}(r-z)_+^n}{(x-z)^{\beta-m+1}} dz \\ &= \frac{1}{\Gamma(m-\beta)} \frac{(-1)^m n!}{(n-m)!} \int_0^x \frac{(r-z)_+^{n-m}}{(x-z)^{\beta-m+1}} dz. \end{aligned} \tag{50}$$

Integration rules for rational functions give [25]

$$\int_0^x \frac{(r-z)^{n-m}}{(x-z)^{\beta-m+1}} dz = \frac{1}{m-\beta} \frac{r^{n-m}}{x^{\beta-m}} {}_2F_1\left(1, m-n, m+1-\beta, \frac{x}{r}\right) \tag{51}$$

$$+ \frac{\Gamma(n-m+2)\Gamma(m-\beta)}{\Gamma(n-\beta+1)} \left(-\frac{x}{r}\right)^{\beta-m+1} \Theta(r-x) {}_2F_1\left(\beta-m+1, \beta-n, \beta-m+1, \frac{x}{r}\right),$$

where $\Theta(z)$ is the Heaviside function and $_2F_1(a,b,c,z)$ is the hypergeometric function

$$_2F_1(a,b,c,z) = \sum_{k=0}^\infty \frac{(a)_k (b)_k}{(c)_k} \frac{z^k}{k!}, \tag{52}$$

with $(q)_k$ denoting the rising Pochhammer symbol

$$(q)_k = \begin{cases} 1, & k=0, \\ q(q+1)\cdots(q+k-1), & k>0. \end{cases} \tag{53}$$

Since $m-n<0$, $_2F_1(1, m-n, m+1-\beta, \frac{x}{r})$ is actually a finite sum, i.e.,

$$_2F_1\left(1, m-n, m+1-\beta, \frac{x}{r}\right) = \sum_{k=0}^{n-m} (-1)^k \binom{n-m}{k} \frac{(1)_k}{(m+1-\beta)_k} \left(\frac{x}{r}\right)^k. \tag{54}$$

Moreover, $_2F_1(\beta-m+1, \beta-n, \beta-m+1, \frac{x}{r})$ can be evaluated by a direct calculation, i.e.,

$$_2F_1\left(\beta-m+1, \beta-n, \beta-m+1, \frac{x}{r}\right) = \sum_{k=0}^\infty \frac{(\beta-n)_k}{k!} \left(\frac{x}{r}\right)^k = \left(\frac{r-x}{r}\right)^{n-\beta}. \tag{55}$$

Thus, we get the analytical expression of the Caputo derivative (cf. also [20]).

Theorem 3. *For $0 < \beta < n$ with $m - 1 < \beta < m$ and m a positive integer, the Caputo derivative of the translates of the truncated power function is given by*

$$^{C}D_{x}^{\beta}(r - x)_{+}^{n} = \frac{1}{\Gamma(m + 1 - \beta)} \frac{(-1)^{m} n!}{(n - m)!} \frac{r^{n-m}}{x^{\beta-m}} \sum_{k=0}^{n-m} (-1)^{k} \binom{n-m}{k} \frac{(1)_{k}}{(m + 1 - \beta)_{k}} \left(\frac{x}{r}\right)^{k}$$

$$+ \frac{(-1)^{n+m} n!}{\Gamma(n + 1 - \beta)} (x - r)_{+}^{n-\beta}. \tag{56}$$

The Riemann–Liouville derivatives of the functions N_{kn} can be obtained by the Caputo derivatives using the initial conditions (22) in Equation (46).

2.6. The Collocation-Galerkin Method

In this section, we show how to use the optimal basis introduced in the previous sections to solve a fractional differential problem, i.e., the *time-fractional diffusion problem* (cf. [26–28])

$$\begin{cases} D_{t}^{\beta} u(x,t) - \dfrac{\partial^{2}}{\partial x^{2}} u(x,t) = f(x,t), & x \in [0,\eta], \quad t \in [0,\tau], \\[2mm] u(x,0) = u_{0}, & x \in [0,\eta], \\[2mm] u(0,t) = u(\eta,t) = 0, & t \in [0,\tau], \end{cases} \tag{57}$$

where $D_{t}^{\beta} u$, $0 < \beta < 1$, denotes the partial Caputo derivative with respect to the time t.

Let

$$\mathcal{V}_{nh}^{0} = \overline{\mathrm{span}} \left\{ N_{knh}(x), 1 \le k \le L - 2 + n \right\}, \tag{58}$$

where $L = \eta/h$ is the approximating space generated by the optimal B-spline basis

$$\mathcal{N}_{nh}^{0} = \left\{ N_{knh}(x), 1 \le k \le L - 2 + n \right\}, \tag{59}$$

of the interval $[0, \eta]$ on equidistant knots with space step $h > 0$ (cf. (27)). We observe that the basis $\mathcal{N}_{nh}^{0}$ naturally satisfies the homogeneous boundary conditions

$$N_{knh}(0) = N_{knh}(\eta) = 0, \qquad 1 \le k \le L - 2 + n. \tag{60}$$

For any h held fix, the spline Galerkin method looks for an approximating function

$$u_{h}(x,t) = \sum_{k=1}^{L-2+n} c_{k}(t) N_{knh}(x) \in \mathcal{V}_{nh}^{0} \tag{61}$$

that solves the variational problem

$$\begin{cases} \left(D_{t}^{\beta} u_{h}, N_{knh} \right) - \left(\dfrac{\partial}{\partial x^{2}} u_{h}, N_{knh} \right) = (f, N_{knh}), & 1 \le k \le L - 2 + n, \\[2mm] u_{h}(x,0) = u_{0}, & x \in [0,\eta], \end{cases} \tag{62}$$

where $(f,g) = \int_0^\eta f\,g$. Writing (62) in a weak form and using (61), we get the system of fractional ordinary differential equations

$$Q\,D_t^\beta\,C(t) + L\,C(t) = F(t)\,, \qquad t \in [0,\tau]\,, \tag{63}$$

with initial condition

$$\sum_{k=1}^{L-2+n} c_k(0)\,N_{knh}(x) = u_0\,, \qquad x \in [0,\eta]\,. \tag{64}$$

Here, $C(t) = (c_k(t))_{1\le k\le L-2+n}$ is the unknown function vector, $Q = (q_{kj})_{1\le k,j\le L-2+n}$ is the *mass matrix*, $L = (\ell_{kj})_{1\le k,j\le L-2+n}$ is the *stiffness matrix* and $F = (f_k(t))_{1\le k\le L-2+n}$ is the *load vector* whose entries are

$$q_{kj} = \int_0^\eta N_{knh}\,N_{jnh}\,, \qquad \ell_{kj} = \int_0^\eta N'_{knh}\,N'_{jnh}\,, \qquad f_k(t) = \int_0^\eta f(t,\cdot)\,N_{knh}\,.$$

The entries of Q and L can be evaluated explicitly by using the integration and differentiation rules for B-splines [21]. The entries of F can be evaluated by quadrature formulas especially designed for Galerkin methods [29].

Now, we introduce the sequence of temporal knots. Let

$$\widetilde{\mathcal{T}} = \{t_0, t_1, \dots, t_{T+m}\} \quad \text{with} \quad \begin{cases} t_0 = t_1 = \cdots = t_m = 0\,, \\ t_r = (r-m)s\,, \quad m+1 \le r \le T+m\,, \end{cases} \tag{65}$$

where $T = \tau/s$ with $s > 0$ the time step, be a set of equidistant nodes in the interval $[0,\tau]$ having a knots of multiplicity $m+1$ at the left boundary. We assume the unknown functions $c_k(t)$ belong to the spline space

$$\mathcal{V}_{ms}^1 = \text{span}\,\{N_{rms}(x), 0 \le r \le T-1+m\}\,, \tag{66}$$

so that

$$c_k(t) = u_0\,N_{0ms}(t) + \sum_{r=1}^{T-1+m} \lambda_{rk}\,N_{rms}(t)\,, \qquad 1 \le k \le L-2+m\,. \tag{67}$$

We observe that just the functions N_{rms}, $0 \le r \le n$, are boundary functions while all the other functions in the basis are cardinal B-splines.

To solve the fractional differential system (63), we use the collocation method on *equidistant nodes* introduced in [28,30]. Let $\mathcal{P} = \{t_p, 1 \le p \le P\}$ with $P \ge T-1+m$ be a set of *non coincident collocation points* in the interval $(0,\tau]$. Then, collocating (63) on the nodes t_p, we get the linear system

$$(Q \otimes A + L \otimes B)\,\Lambda = F\,, \tag{68}$$

where $\Lambda = (\lambda_{rk})_{1\le r\le T-1+m, 1\le k\le L-2+n}$ is the unknown vector,

$$A = (a_{pr})_{1\le p\le P, 1\le r\le T-1+m}\,, \qquad a_{pr} = D_t^\beta\,N_{rms}(t_p)\,, \tag{69}$$

$$B = (b_{pr})_{1\le p\le P, 1\le r\le T-1+m}\,, \qquad b_{pr} = N_{rms}(t_p)\,, \tag{70}$$

are collocation matrices, and $F = (f_k(t_p))_{1\le k\le L-2+n, 1\le p\le P}$. When $P = T-1+m$, (68) is a square linear system and the collocation matrix A is non singular if and only if [21],

$$t_p \in \text{supp}\,N_{pms} \tag{71}$$

while B is always non singular [31]. Otherwise, (68) is an overdetermined linear system that can be solved in the least squares sense.

We notice that we must pay a special attention to the evaluation of the entries of A since they involve the values of the fractional derivative $D_t^\beta N_{rms}$ in the collocation points. This can be done by the differentiation rules given in Section 2.5.

3. Results

To test the performance of the optimal B-spline basis when used to solve fractional differential problems, we solved the time-fractional diffusion problem (57) by the collocation–Galerkin method described in Section 2.6. In the tests, we used the optimal basis $\mathcal{N}_3$ having degree 3 since the cubic B-spline has a small support but is sufficiently smooth. In fact, the cubic B-spline belongs to $C^2(I)$ and its support has length 4. In particular, the basis $\mathcal{N}_3$ has 3 right (left) edge functions with support $[0, k]$ ($[L-4+k, L]$), $1 \le k \le 3$, while the interior functions have support $[k, k+4]$, $0 \le k \le L-4$. In the following section, we give the analytical expression of the functions N_{k3} and of their fractional derivatives; then, we show some numerical results.

3.1. The Optimal B-Spline Basis of Degree $n = 3$

When $n = 3$, we get the cubic B-spline basis. The optimal B-spline basis $\mathcal{N}_3$ of the interval $[0, L]$, with L an integer greater than 3, is associated with the sequence of integer knots

$$\widetilde{\mathcal{I}} = \{x_0, x_1, \ldots, x_{L+6}\} \tag{72}$$

with $x_0 = x_1 = x_2 = x_3 = 0$, $x_k = k - 3$, $4 \le k \le L+2$, $x_{L+3} = x_{L+4} = x_{L+5} = x_{L+6} = L$. $\mathcal{N}_3$ has $L - 3$ interior functions and six edge functions, three for each boundary.

The interior functions N_{3k}, $3 \le k \le L - 1$, are the translates of the cubic cardinal B-spline, i.e.,

$$N_{3k}(x) = B_3(x - k + 3), \qquad 3 \le k \le L - 1, \tag{73}$$

where

$$B_3(x) = \begin{cases} \dfrac{1}{6} x^3, & 0 \le x \le 1, \\[2ex] \dfrac{1}{6}\left(-3x^3 + 12x^2 - 12x + 4\right), & 1 \le x \le 2, \\[2ex] \dfrac{1}{6}\left(-3(4-x)^3 + 12(4-x)^2 - 12(4-x) + 4\right), & 2 \le x \le 3, \\[2ex] \dfrac{1}{6}(4-x)^3, & 3 \le x \le 4, \\[2ex] 0, & \text{otherwise}. \end{cases} \tag{74}$$

From Corollary 1, it follows that the three left edge functions N_{kn}, $0 \le k \le 2$, are given by

$$N_{03}(x) = (1-x)_+^3 = \begin{cases} (1-x)^3, & 0 \le x \le 1, \\ 0, & \text{otherwise}, \end{cases} \tag{75}$$

$$N_{13}(x) = \frac{1}{4}(2-x)_+^3 - 2(1-x)_+^3$$

$$= \begin{cases} \frac{1}{4}(2-x)^3 - 2(1-x)^3, & 0 \le x \le 1, \\ \frac{1}{4}(2-x)^3, & 1 < x \le 2, \\ 0, & \text{otherwise}, \end{cases} \tag{76}$$

$$N_{23}(x) = \frac{1}{6}(3-x)_+^3 - \frac{3}{4}(2-x)_+^3 + \frac{3}{2}(1-x)_+^3$$

$$= \begin{cases} \frac{1}{6}(3-x)^3 - \frac{3}{4}(2-x)^3 + \frac{3}{2}(1-x)^3, & 0 \le x \le 1, \\ \frac{1}{6}(3-x)^3 - \frac{3}{4}(2-x)^3, & 1 < x \le 2, \\ \frac{1}{6}(3-x)^3, & 2 < x \le 3, \\ 0, & \text{otherwise}. \end{cases} \tag{77}$$

Now, using Theorems 3, we obtain the analytical expression of the Caputo derivative of fractional order $0 < \beta < 1$ of the translates of the truncated power function

$$^CD_x^\beta(r-x)_+^3 = -\frac{3}{\Gamma(4-\beta)}\left((2x^2 - 2(3-\beta)xr + (2-\beta)_2 r^2)x^{1-\beta} - 2(x-r)_+^{3-\beta}\right). \tag{78}$$

Thus, substituting (78) in (48), we get

$$^CD_x^\beta N_{03}(x) = \frac{3}{\Gamma(4-\beta)}\left(-(2x^2 - 2(3-\beta)x + (2-\beta)_2)\,x^{1-\beta} + 2(x-1)_+^{3-\beta}\right), \tag{79}$$

$$^CD_x^\beta N_{13}(x) = \frac{3}{\Gamma(4-\beta)}\left(((\tfrac{7}{2}x^2 - 3(3-\beta)x + (2-\beta)_2)\,x^{1-\beta} - 4(x-1)_+^{3-\beta} + \tfrac{1}{2}(x-2)_+^{3-\beta}\right), \tag{80}$$

$$^CD_x^\beta N_{23}(x) = \frac{3}{\Gamma(4-\beta)}\left((-\tfrac{11}{6}x^2 + (3-\beta)x)\,x^{1-\beta} + 3(x-1)_+^{3-\beta} - \tfrac{3}{2}(x-2)_+^{3-\beta} + \tfrac{1}{3}(x-3)_+^{3-\beta}\right). \tag{81}$$

The optimal basis and its fractional derivatives in case of equidistant knots on an interval $[a,b]$ can be evaluated by scaling as shown in (27)–(29). The optimal B-spline basis in the interval $[0,2]$ for $h = 0.25$ is shown in Figure 1a. The Caputo derivatives of fractional order $\beta = 0.25, 0.5, 0.75$ are shown in Figure 1b–d.

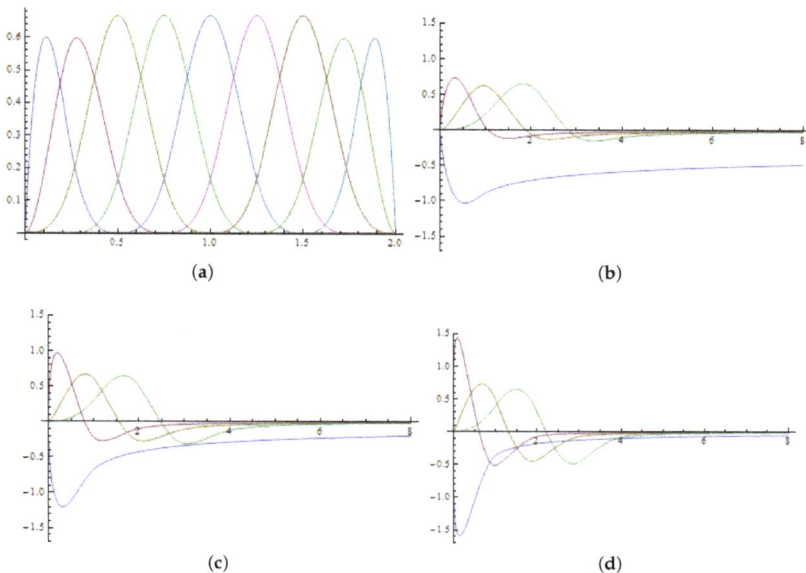

Figure 1. (a) the optimal B-spline basis $\mathcal{N}_3$ in the interval $[0,2]$ for $h = 0.25$; (**b–d**) the fractional derivatives $^{C}D_t^\beta N_{03}$ (blue), $^{C}D_t^\beta N_{13}$ (purple), $^{C}D_t^\beta N_{23}$ (dark yellow), $^{C}D_t^\beta B_3$ (green) of order $\beta = 0.25$ (**b**), $\beta = 0.5$ (**c**), $\beta = 0.75$ (**d**).

To give an idea of the condition number of the discretization matrix $(Q \otimes A + L \otimes B)$ (cf. (68)), in Tables 1–3, we list the condition number of the matrix when using the cubic B-spline basis for different values of the order of the fractional derivative and different values of the space and time steps.

Table 1. The condition number $\mathcal{K}$ (rounded to the nearest integer) and the dimensions $n_{rows} \times n_{cols}$ of the discretization matrix $(Q \otimes A + L \otimes B)$ for $\alpha = 0.25$ as a function of the space step h and the time step δ.

	$\delta = 0.25$		$\delta = 0.125$		$\delta = 0.0625$		$\delta = 0.03125$	
h	$\mathcal{K}$	$n_{rows} \times n_{cols}$	$\mathcal{K}$	$n_{rows} \times n_{cols}$	$\mathcal{K}$	$n_{rows} \times n_{cols}$	$\mathcal{K}$	$n_{rows} \times n_{cols}$
0.25	143	36×36	125	72×54	119	144×90	110	288×162
0.125	543	68×68	472	136×102	450	272×170	415	544×306
0.0625	2144	132×132	1864	264×198	1776	528×330	1639	1056×594
0.03125	8550	260×260	7434	520×390	7083	1040×650	6535	2080×1170

Table 2. The condition number $\mathcal{K}$ (rounded to the nearest integer) and the dimensions $n_{rows} \times n_{cols}$ of the discretization matrix $(Q \otimes A + L \otimes B)$ for $\alpha = 0.5$ as a function of the space step h and the time step δ.

	$\delta = 0.25$		$\delta = 0.125$		$\delta = 0.0625$		$\delta = 0.03125$	
h	$\mathcal{K}$	$n_{rows} \times n_{cols}$	$\mathcal{K}$	$n_{rows} \times n_{cols}$	$\mathcal{K}$	$n_{rows} \times n_{cols}$	$\mathcal{K}$	$n_{rows} \times n_{cols}$
0.25	102	36×36	77	72×54	66	144×90	66	288×162
0.125	386	68×68	291	136×102	234	272×170	179	544×306
0.0625	1525	132×132	1148	264×198	923	528×330	706	1056×594
0.03125	6084	260×260	4579	520×390	3681	1040×650	2817	2080×1170

Table 3. The condition number $\mathcal{K}$ (rounded to the nearest integer) and the dimensions $n_{rows} \times n_{cols}$ of the discretization matrix $(Q \otimes A + L \otimes B)$ for $\alpha = 0.75$ as a function of the space step h and the time step δ.

	$\delta = 0.25$		$\delta = 0.125$		$\delta = 0.0625$		$\delta = 0.03125$	
h	$\mathcal{K}$	$n_{rows} \times n_{cols}$	$\mathcal{K}$	$n_{rows} \times n_{cols}$	$\mathcal{K}$	$n_{rows} \times n_{cols}$	$\mathcal{K}$	$n_{rows} \times n_{cols}$
0.25	69	36×36	62	72×54	63	144×90	60	288×162
0.125	253	68×68	154	136×102	100	272×170	67	544×306
0.0625	1001	132×132	610	264×198	394	528×330	244	1056×594
0.03125	3991	260×260	2434	520×390	1573	1040×650	972	2080×1170

3.2. Numerical Tests

In this section, we show the numerical results we obtained when solving the differential problem (57) for three different expressions of the known term $f(x, t)$ and different values of the fractional derivative. In the tests, we set $\tau = 1$ and $\eta = 2$ and chose as collocation nodes the equidistant points of the interval $[0, 1]$ with time step δ. All the tests were performed in the Mathematica environment [32].

3.2.1. Test 1

To test the accuracy of the method, we first solved the time-fractional diffusion Equation (57) when

$$f(x, t) = \left(\frac{6}{\Gamma(4 - \alpha)} t^{-\alpha} x(L - x) - 2 \right) t^3 . \tag{82}$$

In this case, the exact solution is the bivariate polynomial

$$u(x, t) = x(L - x) t^3 . \tag{83}$$

Since the cubic B-spline basis reproduces polynomials up to degree 3, the approximation error of the collocation–Galerkin method is zero so that we expect the numerical solution coincides with the exact solution.

We evaluated the numerical solution for $\delta = 0.25$, $s = 0.5$ and $h = 0.25$ and different values of the order of the time fractional derivative β.

Let us define the error by

$$e_{h,\delta}(x, t) = u(x, t) - u_{h,\delta}(x, t) , \tag{84}$$

where $u_{h,\delta}(x, t)$ denotes the numerical solution of the fractional differential system (63) obtained by the collocation method.

The tests show that, as expected, the error is on the order of the machine precision even using a large step size. The numerical solution and the error for $\beta = 0.25$ are shown in Figure 2.

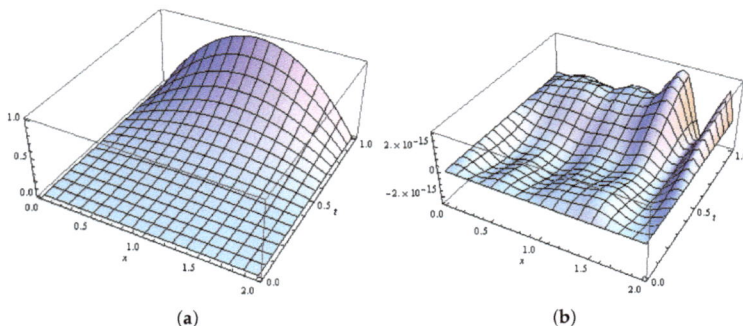

(a)　　　　　　　　　　　　(b)

Figure 2. Test 1: (a) the numerical solution and (b) the error for $\beta = 0.25$ and $\delta = h = 0.25$.

3.2.2. Test 2

To test the accuracy of the method when approximating time fractional derivatives, we solved the time-fractional diffusion Equation (57) for

$$f(x,t) = x(L-x) \frac{\pi t^{1-\beta}}{2\Gamma(2-\beta)} \left({}_1F_1(1,2-\beta,i\pi t) + {}_1F_1(1,2-\beta,-i\pi t) \right) + 2\sin(\pi t), \tag{85}$$

where ${}_1F_1(a,b,z)$ is the Kummer's confluent hypergeometric function

$$ {}_1F_1(a,b,z) = \frac{\Gamma(b)}{\Gamma(a)} \sum_{k=0}^{\infty} \frac{\Gamma(a+k)}{\Gamma(b+k)\, k!} z^k, \qquad a \in \mathbb{R}, \quad -b \notin \mathbb{N}_0, \tag{86}$$

where $\mathbb{N}_0 = \mathbb{N}\backslash\{0\}$ (cf. [9]). In this case, the exact solution is

$$u(x,t) = x(L-x)\,\sin(\pi t) \tag{87}$$

so that the error for the spatial approximation is negligible.

The L_2-norm of the error

$$\|e_{h,\delta}\|_2 = \sqrt{\int_0^\tau dt \int_0^\eta dx \, |e_{h,\delta}(x,t)|^2}, \tag{88}$$

and the numerical convergence order

$$\rho_\delta = \log_2\left(\frac{\|e_{h,\delta}\|_2}{\|e_{h,\delta/2}\|_2} \right), \tag{89}$$

are listed in Table 4. The numerical solution in the case when $\beta = 0.25$ obtained for $\delta = 0.03125$, $s = 2\delta$ and $h = 0.25$ is shown in Figure 3.

Table 4. Test 2: The L_2-norm of the error and the numerical convergence order ρ_δ as a function of the time step δ for different values of the order β of the fractional derivative.

	$\beta = 0.25$		$\beta = 0.5$		$\beta = 0.75$	
δ	$\|e_{h,\delta}\|_2$	ρ_δ	$\|e_{h,\delta}\|_2$	ρ_δ	$\|e_{h,\delta}\|_2$	ρ_δ
0.25	0.42×10^{-2}		0.50×10^{-2}		0.62×10^{-2}	
0.125	0.32×10^{-3}	3.71	0.32×10^{-3}	3.97	0.34×10^{-3}	4.19
0.0625	0.17×10^{-4}	4.23	0.17×10^{-4}	4.23	0.19×10^{-4}	4.16
0.03125	0.10×10^{-5}	4.09	0.11×10^{-5}	3.95	0.12×10^{-5}	3.98

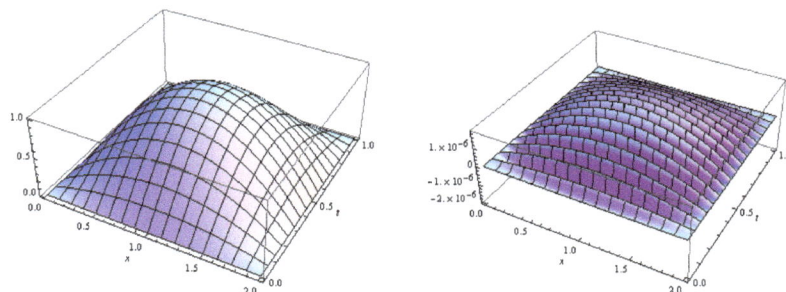

Figure 3. Test 2: (**a**) the numerical solution and (**b**) the error in the case when $\beta = 0.25$ and $\delta = 0.03125$, $s = 2\delta$, $h = 0.25$.

The numerical results show that the method converges with convergence order approximatively equal to 4 and gives a good approximation even using few collocation points. Moreover, the accuracy of the numerical solution does not depend on the order of the fractional derivative.

3.2.3. Test 3

Finally, we solved the time-fractional diffusion equation (57) for

$$f(x,t) = \sin(\pi x) \left(\frac{\pi t^{1-\beta}}{2\Gamma(2-\beta)} \left({}_1F_1(1, 2-\beta, i\pi t) + {}_1F_1(1, 2-\beta, -i\pi t) \right) + \pi^2 \sin(\pi t) \right). \quad (90)$$

In this case, the exact solution is

$$u(x,t) = \sin(\pi x) \sin(\pi t). \quad (91)$$

The L_2-norm of the error and the numerical convergence order

$$\rho_{h,\delta} = \log_2 \left(\frac{\|e_{h,\delta}\|_2}{\|e_{h/2,\delta/2}\|_2} \right) \quad (92)$$

are listed in Table 5. The numerical solution in the case when $\beta = 0.25$ obtained for $\delta = h = 0.03125$ and $s = 2\delta$ is shown in Figure 4.

Also in this case, the numerical solution has a good accuracy that is not affected by the order of the fractional derivative. The numerical convergence order is approximatively 4.

Table 5. Test 3: The L_2-norm of the error and the numerical convergence order $\rho_{h,\delta}$ as a function of the time step δ and the space step h for different values of the order β of the fractional derivative.

	$\beta = 0.25$		$\beta = 0.5$		$\beta = 0.75$	
$h = \delta$	$\|e_{h,\delta}\|_2$	ρ	$\|e_{h,\delta}\|_2$	$\rho_{h,\delta}$	$\|e_{h,\delta}\|_2$	$\rho_{h,\delta}$
0.25	0.38×10^{-2}		0.41×10^{-2}		0.46×10^{-2}	
0.125	0.31×10^{-3}	3.62	0.31×10^{-3}	3.72	0.31×10^{-3}	3.89
0.0625	0.16×10^{-4}	4.28	0.16×10^{-4}	4.28	0.17×10^{-4}	4.19
0.03125	0.98×10^{-6}	4.03	0.99×10^{-6}	4.01	0.10×10^{-5}	4.09

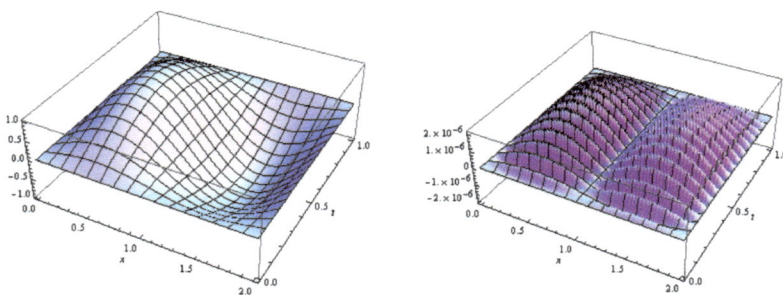

Figure 4. Test 3: (a) the numerical solution and (b) the error for $\beta = 0.25$ and $\delta = h = 0.03125$, $s = 2\delta$.

4. Discussion

In this paper, we showed how optimal B-spline bases have good approximation properties that make them particularly suitable to be used in the solution of fractional differential problems. First of all, boundary and initial conditions can be satisfied easily in view of properties (22)–(23) so that the numerical solution we obtained does not suffer from the numerical instabilities near the boundaries that appear when truncated functions at the boundaries are used (cf. [28]). Moreover, optimal B-spline bases are stable (cf. (25)) meaning that numerical errors are not amplified. Nevertheless, Tables 1–3 show that the condition number of the discretization matrix increases as the dimension of the matrix increases. The conditioning gets worse when the order of the fractional derivative becomes smaller. In this case, the linear system (68) can be accurately solved by Krylov methods [33].

As for the convergence order, we observe that several methods proposed in the literature to solve fractional differential problems have convergence order that depends on the order of the fractional derivative (see, for instance, [26,27]). In the case of the cubic B-spline basis, the numerical tests show that the numerical convergence order does not depend on the order of the fractional derivative being close to 4 for any value of β. This result is in line with the Strang–Fix theory for classical differential problems [34]. We expect that the convergence order for the optimal B-spline basis $\mathcal{N}_n$ is $n + 1$ so that it can be increased easily by increasing the degree of the basis. In fact, optimal bases of high degree and their fractional derivatives can be evaluated by using the analytical expression given in Sections 2.4–2.5. As far as we know, the analytical expressions (39) and (48) are new and, together with (27)–(29), are an easy tool to evaluate optimal B-spline bases on any set of equidistant knots on bounded intervals. Finally, we observe that the accuracy of the numerical solution we obtained is higher than the accuracy obtained in [26] where a quadrature formula was used to approximate the time fractional derivative and a finite element method was used for the spatial approximation.

The optimal bases we described can be used to solve other fractional differential problems, for instance problems involving the Riesz derivative [2], or can be used in other numerical methods, for instance in the operational matrix method or in wavelet methods. We notice that adaptive wavelet methods can be used to solve differential problems whose solutions are non smooth functions. Optimal bases generating multiresolution analyses and wavelet spaces can be obtained starting from a special family of refinable functions [35,36] so that the procedure described above can be generalized to other refinable bases.

Funding: This work was partially supported by grants University of Roma "La Sapienza", Ricerca Scientifica 2016 and INdAM-GNCS, Project 2018.

Conflicts of Interest: The author declares no conflict of interest.

References

1. Hilfer, R. *Applications of Fractional Calculus in Physics*; World Scientific: Singapore, 2000.
2. Kilbas, A.A.; Srivastava, H.M.; Trujillo, J.J. *Theory and Applications of Fractional Differential Equations*; Elsevier Science: North-Holland, The Netherlands, 2006.
3. Sabatier, J.; Agrawal, O.; Tenreiro Machado, J.A. *Advances in Fractional Calculus*; Springer: Berlin, Germany, 2007.
4. Mainardi, F. *Fractional Calculus and Waves in Linear Viscoelasticity: An Introduction to Mathematical Models*; World Scientific: Singapore, 2010.
5. Tarasov, V.E. *Fractional Dynamics*; Nonlinear Physical Science; Springer: Berlin, Germany, 2010.
6. Oldham, K.; Spanier, J. *The Fractional Calculus*; Academic Press: Cambridge, MA, USA, 1974.
7. Samko, S.G.; Kilbas, A.A.; Marichev, O.I. *Fractional Integrals and Derivatives*; Gordon & Breach Science Publishers: Philadelphia, PA, USA, 1993.
8. Podlubny, I. *Fractional Differential Equations*; Academic Press: Cambridge, MA, USA, 1998.
9. Diethelm, K. *The Analysis of Fractional Differential Equations: An Application-Oriented Exposition Using Differential Operators of Caputo Type*; Springer: Berlin, Germany, 2010.
10. Baleanu, D.; Diethelm, K.; Scalas, E.; Trujillo, J.J. *Fractional Calculus: Models and Numerical Methods*; World Scientific: Singapore, 2016.
11. Li, C.; Chen, A.; Ye, J. Numerical approaches to fractional calculus and fractional ordinary differential equation. *J. Comput. Phys.* **2011**, *230*, 3352–3368. [CrossRef]
12. Podlubny, I.; Skovranek, T.; Datsko, B. Recent advances in numerical methods for partial fractional differential equations. In Proceedings of the 2014 15th International Carpathian Control Conference (ICCC), Velke Karlovice, Czech Republic, 28–30 May 2014; pp. 454–457.

13. Li, C.; Zeng, F. *Numerical Methods for Fractional Calculus*; CRC Press: Boca Raton, FL, USA, 2015; Volume 24.
14. Li, C.; Chen, A. Numerical Methods for Fractional Partial Differential Equations. *Int. J. Comput. Math.* **2018**, *95*, 1048–1099. [CrossRef]
15. Pitolli, F. A fractional B-spline collocation method for the numerical solution of fractional predator-prey models. *Fractal Fractional* **2018**, *2*, 13. [CrossRef]
16. lHöllig, K. *Finite Element Methods with B-Splines*; SIAM: Philadelphia, PA, USA, 2003; Volume 26.
17. Buffa, A.; Sangalli, G.; Vázquez, R. Isogeometric analysis in electromagnetics: B-splines approximation. *Comput. Methods Appl. Mech. Eng.* **2010**, *199*, 1143–1152. [CrossRef]
18. Garoni, C.; Manni, C.; Serra-Capizzano, S.; Sesana, D.; Speleers, H. Spectral analysis and spectral symbol of matrices in isogeometric Galerkin methods. *Math. Comput.* **2017**, *86*, 1343–1373. [CrossRef]
19. Lakestani, M.; Dehghan, M.; Irandoust-Pakchin, S. The construction of operational matrix of fractional derivatives using B-spline functions. *Communi. Nonlinear Sci. Numer. Simul.* **2012**, *17*, 1149–1162. [CrossRef]
20. Li, X. Numerical solution of fractional differential equations using cubic B-spline wavelet collocation method. *Communi. Nonlinear Sci. Numer. Simul.* **2012**, *17*, 3934–3946. [CrossRef]
21. Schumaker, L.L. *Spline Functions: Basic Theory*; Cambridge University Press: Cambridge, UK, 2007.
22. Carnicer, J.M.; Peña, J.M. Totally positive bases for shape preserving curve design and optimality of B-splines. *Computer Aided Geometric Design* **1994**, *11*, 633–654. [CrossRef]
23. Carnicer, J.M.; Peña, J.M. Total positivity and optimal bases. In *Total Positivity and its Applications*; Springer: Berlin, Germany, 1996; pp. 133–155.
24. Unser, M.; Blu, T. Fractional splines and wavelets. *SIAM Rev.* **2000**, *42*, 43–67. [CrossRef]
25. Gradshteyn, I.S.; Ryzhik, I.M. *Tables of Integrals, Series and Products: Corrected and Enlarged Edition*; Academic Press: New York, NY, USA, 1980.
26. Ford, N.J.; Xiao, J.; Yan, Y. A finite element method for time fractional partial differential equations. *Fractional Calculus Appl. Anal.* **2011**, *14*, 454–474. [CrossRef]
27. Li, X.; Xu, C. A space-time spectral method for the time fractional diffusion equation. *SIAM J. Numer. Anal.* **2009**, *47*, 2108–2131. [CrossRef]
28. Pezza, L.; Pitolli, F. A fractional spline collocation-Galerkin method for the time-fractional diffusion equation. *Commun. Appl. Ind. Math.* **2018**, *9*, 104–120. [CrossRef]
29. Calabrò, F.; Manni, C.; Pitolli, F. Computation of quadrature rules for integration with respect to refinable functions on assigned nodes. *Appl. Numer. Math.* **2015**, *90*, 168–189. [CrossRef]
30. Pezza, L.; Pitolli, F. A multiscale collocation method for fractional differential problems. *Math. Comput. Simul.* **2018**, *147*, 210–219. [CrossRef]
31. Forster, B.; Massopust, P. Interpolation with fundamental splines of fractional order. In Proceedings of the 9th International Conference on Sampling Theory and Applications (SampTa), Singapore, 2–6 May 2011.
32. Wolfram Research, Inc. *Mathematica, Version 8.0.4.0*; Wolfram Research, Inc.: Champaign, IL, USA, 2011.
33. Calvetti, D.; Pitolli, F.; Somersalo, E.; Vantaggi, B. Bayes meets Krylov: Statistically inspired preconditioners for CGLS. *SIAM Rev.* **2018**, *60*, 429–461. [CrossRef]
34. Strang, G.; Fix, G. A Fourier analysis of the finite element variational method. In *Constructive Aspects of Functional Analysis*; Edizioni Cremonese: Naples, Italy, 1971; pp. 796–830.
35. Gori, L.; Pitolli, F. Refinable functions and positive operators. *Appl. Numer. Math.* **2004**, *49*, 381–393. [CrossRef]
36. Gori, L.; Pezza, L.; Pitolli, F. Recent results on wavelet bases on the interval generated by GP refinable functions. *Appl. Numer. Math.* **2004**, *51*, 549–563. [CrossRef]

axioms

MDPI

Review

Stability Issues for Selected Stochastic Evolutionary Problems: A Review

Angelamaria Cardone [1,†] , **Dajana Conte** [1,†] , **Raffaele D'Ambrosio** [2,†,*]
and Beatrice Paternoster [1,†]

[1] Department of Mathematics, University of Salerno, 84084 Fisciano, Italy; ancardone@unisa.it (A.C.);
    dajconte@unisa.it (D.C.); beapat@unisa.it (B.P.)
[2] Department of Engineering and Computer Science and Mathematics, University of L'Aquila,
    67100 L'Aquila, Italy
[*] Correspondence: raffaele.dambrosio@univaq.it; Tel.: +39-08-6243-4724
[†] The authors contributed equally to this work. The authors are members of the INdAM Research
    group GNCS.

Received: 1 May 2018; Accepted: 21 November 2018; Published: 1 December 2018

Abstract: We review some recent contributions of the authors regarding the numerical approximation of stochastic problems, mostly based on stochastic differential equations modeling random damped oscillators and stochastic Volterra integral equations. The paper focuses on the analysis of selected stability issues, i.e., the preservation of the long-term character of stochastic oscillators over discretized dynamics and the analysis of mean-square and asymptotic stability properties of ϑ-methods for Volterra integral equations.

Keywords: stochastic differential equations; stochastic multistep methods; stochastic Volterra integral equations; mean-square stability; asymptotic stability

1. Introduction

This paper provides a brief review of recent results obtained in the context of stability analysis for stochastic numerical methods. The treatise is essentially twofold: regarding stability properties as the preservation of qualitative features of the continuous problem as well as the numerical preservation of stable behaviors in the solution of the continuous problem.

The first highlighted aspect, i.e., stability as numerical preservation of qualitative properties, is here framed in the context of stochastic differential equations (SDEs), with special emphasis on problems describing stochastic oscillators [1–4].The perspective we follow consists of providing a long-term analysis of numerical methods for SDEs in terms of preserving invariance laws that characterize the dynamics provided by the exact solution. A first contribution in this sense was given in [5], where the author analyzed the invariance of asymptotic laws characterizing linear stochastic systems under given discretizations. The analysis of partitioned methods for linear oscillators in the presence of additive noise has been an object of [6], while the analysis of linear second order SDEs describing damped stochastic oscillators has been provided by Burrage and Lythe in [1,2] and inspired the paper [3] giving a more general two-step framework. More recent contributions have regarded stochastic Hamiltonian problems [7,8] and stochastic oscillators with multiplicative noise [9].

The second highlighted issue, i.e., numerical preservation of stable behaviors in the solution of the continuous problem, the attention is focused on stochastic Volterra integral equations (SVIEs). For such operators, whose numerical discretization has been considered by the recent literature (see, for instance, [10–12] and references therein), researchers have started to provide a parallel with the classical theory of the numerical approximation of SDEs [13,14]. However, as far as the authors are aware, the first contribution in developing a stability analysis in the time-stepping numerics for SVIEs is the recent paper [15], where the analysis is given both in terms of mean-square stability properties as well as on asymptotic ones.

The treatise is divided into two parts: the first one, presented in Section 2, is regarding the approximation of stochastic differential equations modeling the dynamics of damped oscillators subject to both deterministic and random forcing terms; the second part, contained in Section 3, gives a glance on stochastic ϑ-methods for stochastic Volterra integral equations and, in particular, on the analysis of the stability properties with respect to suitable test equations.

2. Damped Linear Stochastic Oscillators: Long-Term Stability Issues

2.1. The Problem

The motion of a particle constrained by a deterministic forcing and a stochastic one, in the presence of damping, can be modeled by the following scalar second order SDE [1,2]

$$\ddot{x} = f(x) - \eta \dot{x} + \varepsilon \xi(t), \tag{1}$$

where $f(x)$ is a deterministic forcing, η is the damping parameter, and $\varepsilon \xi(t)$ is the stochastic forcing of amplitude ε. Clearly, assuming a linear forcing $f(x) = -gX(t)$, Equation (1) admits the following first order Itô formulation

$$dX(t) = V(t)dt,$$
$$dV(t) = -\left(\eta V(t) + gX(t)\right)dt + \varepsilon dW(t), \tag{2}$$

where $X(t)$ and $V(t)$ are, respectively, position and velocity of the particle at time t whose dynamics is also governed by the occurences of the Wiener process $W(t)$ [13,14]. The long-term character of the problem described by Equation (2) is clearly highlighted by its stationary density [1,2]

$$\Pi_\infty(x,v) = \lim_{t \to \infty} \frac{d}{dx}\frac{d}{dv}\mathbb{P}(X(t) < x, V(t) < v) = N_0 \exp\left(-\frac{\eta}{\varepsilon^2}(gx^2 + v^2)\right), \tag{3}$$

where the constant N_0 is the unknown of the equality

$$\int_{-\infty}^{\infty}\int_{-\infty}^{\infty} \Pi_\infty(x,v)dxdv = 1.$$

In other terms, the stochastic dynamics described by Equation (2) has a Gaussian distributed velocity, uncorrelated with the position. An effective representation of such a long-term behavior is given by the following correlation matrix

$$\Sigma = \begin{bmatrix} \sigma_X^2 & \mu \\ \mu & \sigma_V^2 \end{bmatrix} = \frac{\varepsilon^2}{2\eta} \begin{bmatrix} \dfrac{1}{g} & 0 \\ 0 & 1 \end{bmatrix}, \tag{4}$$

where

$$\sigma_X^2 = \lim_{t \to \infty} \mathbb{E}|X(t)|^2 = \frac{\varepsilon^2}{2g\eta}, \quad \sigma_V^2 = \lim_{t \to \infty} \mathbb{E}|V(t)|^2 = \frac{\varepsilon^2}{2\eta}, \quad \mu = \lim_{t \to \infty} \mathbb{E}|X(t)V(t)| = 0.$$

2.2. The Methodology: Indirect Stochastic Linear Two-Step Methods

In [1–3], the authors have analyzed the ability of some of the most common numerical methods for SDEs in preserving the correlation matrix over long times. The analysis given in [1,2] only involves one-step methods, while that provided in [3] is focused on the larger class of indirect two-step methods

(ITS methods), i.e., the family of stochastic two-step methods applied to the first order representation given by Equation (2) of the second order SDE in Equation (1). ITS methods assume the form

$$\begin{bmatrix} X_{i+1} \\ V_{i+1} \end{bmatrix} = R_1(h) \begin{bmatrix} X_i \\ V_i \end{bmatrix} + R_0(h) \begin{bmatrix} X_{i-1} \\ V_{i-1} \end{bmatrix} + r_1(h)\Delta W_i + r_0(h)\Delta W_{i-1}, \tag{5}$$

where the matrices $R_1(h), R_0(h) \in \mathbb{R}^{2\times 2}$ and the vectors $r_1(h), r_0(h) \in \mathbb{R}^2$ are the characteristic coefficients, collected in the Butcher tableau

$$\mathcal{M} = \left[\begin{array}{c|c} R_1(h) & R_0(h) \\ \hline r_1(h) & r_0(h) \end{array} \right]. \tag{6}$$

The corresponding discretized correlation matrix is then given by

$$\widetilde{\Sigma} = \begin{bmatrix} \widetilde{\sigma}_X^2 & \widetilde{\mu} \\ \widetilde{\mu} & \widetilde{\sigma}_V^2 \end{bmatrix}, \tag{7}$$

with

$$\widetilde{\sigma}_X^2 = \lim_{t_n \to \infty} \mathbb{E}|X_n|^2, \quad \widetilde{\sigma}_V^2 = \lim_{t_n \to \infty} \mathbb{E}|V_n|^2, \quad \widetilde{\mu} = \lim_{t_n \to \infty} \mathbb{E}|X_n V_n|,$$

where X_n and V_n are numerical solutions of Equation (2) generated by the ITS method given by Equation (5). As proved in [3], $\widetilde{\Sigma}$ satisfies the matrix equation

$$\widetilde{\Sigma} = R_1(h)\widetilde{\Sigma}R_1(h)^\top + R_0(h)\widetilde{\Sigma}R_0(h)^\top + r_1(h)r_1(h)^\top h + r_0(h)r_0(h)^\top h, \tag{8}$$

which is the most relevant tool in analyzing the long-term dynamics along the solutions computed by the ITS method defined in Equation (5). Indeed, for a fixed ITS method, the matrix in Equation (7) can be a priori computed in order to appreciate how far it is from the exact correlation matrix defined in Equation (4). Alternatively, for those particularly interested in developing new methods, the matrix equality (8) can be used to derive the constraints that the coefficients of an ITS method have to fulfill in order to provide a measure-exact numerical scheme [1,2].

As an example, let us provide the analysis of the famous Euler-Maruyama method [13] that, in the form of the ITS method, assumes the form

$$\mathcal{M} = \left[\begin{array}{cc|cc} 1 & h & 0 & 0 \\ -gh & 1 - \eta h & 0 & 0 \\ 0 & \varepsilon & 0 & 0 \end{array} \right].$$

Solving Equation (8) with respect to the $\widetilde{\Sigma}$ gives

$$\widetilde{\Sigma} = \left[\begin{array}{cc} -\dfrac{\varepsilon^2(2 - h\eta + gh^2)}{g(2\eta^2 h + gh(4 + gh^2) - \eta(4 + 3gh^2))} & \dfrac{\varepsilon^2 h}{2\eta^2 h + gh(4 + gh^2) - \eta(4 + 3gh^2)} \\[4mm] \dfrac{\varepsilon^2 h}{2\eta^2 h + gh(4 + gh^2) - \eta(4 + 3gh^2)} & -\dfrac{2\varepsilon^2}{2\eta^2 h + gh(4 + gh^2) - \eta(4 + 3gh^2)} \end{array} \right].$$

We observe that $\widetilde{\Sigma} = \Sigma + \mathcal{O}(h)$, i.e., the matrix $\widetilde{\Sigma}$ associated with the Euler-Maruyama method is a first order approximation of Σ. In order to appreciate the values of the errors $|\widetilde{\sigma}_X^2 - \sigma_X^2|$, $|\widetilde{\sigma}_V^2 - \sigma_V^2|$ and $|\widetilde{\mu} - \mu|$ over the damping coefficient η, we depict the corresponding graphs in Figure 1. One can recognize that, for increasing values of the damping parameter, position and velocity become less correlated, as happens for the continuous problem defined by Equation (2).

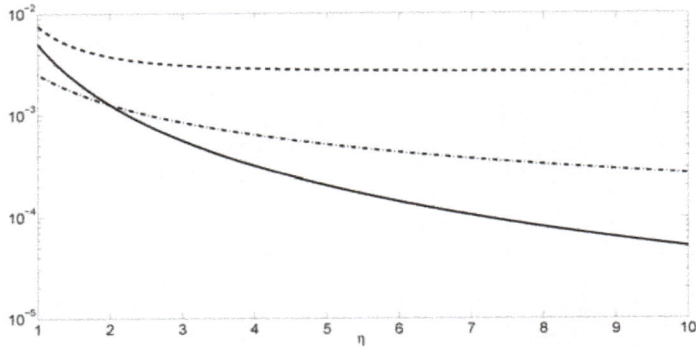

Figure 1. Graphs of $|\tilde{\sigma}_X^2 - \sigma_X^2|$ (continuous line), $|\tilde{\sigma}_V^2 - \sigma_V^2|$ (dashed line) and $|\tilde{\mu} - \mu|$ (dashed-dotted line) over η, for the Euler-Maruyama method applied to Equation (2), with $g = 1$, $h = 10^{-2}$, $\varepsilon = 1$.

Let us now analyze a genuine two-step method, i.e., the following Adams-Moulton method [16]

$$\mathcal{M} = \frac{1}{144 + 60\eta h + 25gh^2}\, \widetilde{\mathcal{M}}, \tag{9}$$

where

$$\widetilde{\mathcal{M}} = \left[\begin{array}{cc|cc} 144 + 60\eta h - 40gh^2 & 156h & 5gh^2 & -12h \\ -156gh & -8(-18 + 12\eta h + 5gh^2) & 12gh & h(12\eta + 5gh) \\ \hline 60\varepsilon h & 144\varepsilon & 0 & 0 \end{array} \right].$$

Solving Equation (8) with respect to the $\widetilde{\Sigma}$ gives

$$\widetilde{\Sigma} = \left[\begin{array}{cc} \frac{90\varepsilon^2 h}{g(468 + 195\eta h - 25gh^2)} & \frac{8640\varepsilon^2 h^2}{625g^2h^4 + 144(12 + 5h\eta)^2 + 120gh^2(264 + 25h\eta)} \\ \frac{8640\varepsilon^2 h^2}{625g^2h^4 + 144(12 + 5h\eta)^2 + 120gh^2(264 + 25h\eta)} & -\frac{2592\varepsilon^2}{125g^2h^3 + 144\eta(-39 + 5h\eta) + 60gh(-39 + 10h\eta)} \end{array} \right].$$

The errors $|\tilde{\sigma}_X^2 - \sigma_X^2|$, $|\tilde{\sigma}_V^2 - \sigma_V^2|$ and $|\tilde{\mu} - \mu|$ over the damping coefficient η, depicted in Figure 2, show that the more the problem is damped, the more the long-term mean-square of the velocity and the position are approximated with slowly decreasing errors. In the long-term, numerical position and velocity of the particle computed by the Adams-Moulton method are only weakly correlated, as desired. A qualitative comparison arising from Figures 1 and 2 shows that the Euler-Maruyama method better approximates the long-term mean-square of the velocity and the position, while the Adams-Moulton method better captures the long term expectation of the product of the velocity and the position.

While the Euler-Maruyama method is able to reproduce the exact correlation matrix Σ as h goes to 0, this is not the case for the Adams-Moulton method. This aspect opens a relevant question on the difference between one-step and genuine multistep methods. The fact that multistep methods do not recover the exact invariants of the continuous problem in the limit as h tends to 0 is typical of the deterministic setting: for instance, multistep methods do not recover the Hamiltonian function of Hamiltonian problems [17]. We conjecture that there are sources of parasitism to be properly addressed also in stochastic linear multistep methods, but this requires a very deep analysis based on the stochastic version of the backward error analysis [17], which will be an object of future investigations.

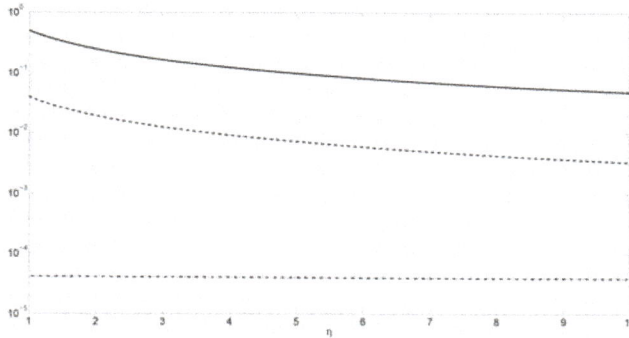

Figure 2. Graphs of $|\tilde{\sigma}_X^2 - \sigma_X^2|$ (continuous line), $|\tilde{\sigma}_V^2 - \sigma_V^2|$ (dashed line) and $|\tilde{\mu} - \mu|$ (dashed-dotted line, the lowest one) over η, for the Adams–Moulton method applied to Problem (2), with $g = 1$, $h = 10^{-2}$, $\varepsilon = 1$.

3. Stability Analysis of ϑ-Methods for Stochastic Volterra Integral Equations

3.1. The Problem

We now focus our attention on stochastic Volterra integral Equations (SVIEs) in Itô form

$$X_t = X_0 + \int_0^t a(t, s, X_s)ds + \int_0^t b(t, s, X_s)dW_s, \quad t \in [0, T]. \tag{10}$$

The next section introduces the general class of stochastic ϑ-methods [15], for which a stability analysis is provided according to the basic linear test equation,

$$X_t = X_0 + \int_0^t \lambda X_s ds + \int_0^t \mu X_s dW_s, \quad t \in [0, T], \tag{11}$$

with $\lambda, \mu \in \mathbb{R}$, and the convolution test equation,

$$X_t = X_0 + \int_0^t (\lambda + \sigma(t - s)) X_s ds + \int_0^t \mu X_s dW_s, \quad t \in [0, T], \tag{12}$$

with $\lambda, \mu, \sigma \in \mathbb{R}$.

3.2. The Methodology: ϑ-Methods for SVIEs

According to the classical theory on the numerical approximation of Volterra integral equations [18–22], we refer to the equidistant grid

$$\mathcal{I}_h = \{t_n = nh, \quad n = 0, ..., N, \ Nh = T\}$$

and, by evaluating Equation (10) in t_n, we obtain

$$X_{t_n} = X_0 + \int_0^{t_n} a(t_n, s, X_s)\, ds + \int_0^{t_n} b(t_n, s, X_s)\, dW_s.$$

The stochastic ϑ-method, introduced in [15], assumes the following form

$$Y_n = Y_0 + h \sum_{i=0}^{n-1} \left(\vartheta a(t_n, t_{i+1}, Y_{i+1}) + (1-\vartheta)a(t_n, t_i, Y_i) \right) + \sqrt{h} \sum_{i=0}^{n-1} b(t_n, t_i, Y_i)G_i, \tag{13}$$

with $Y_0 = X_0$, $h = t_{i+1} - t_i$. G_i is a standard Gaussian random variable, i.e., it is $\mathcal{N}(0,1)$-distributed. Convergence analysis of the stochastic ϑ-method, provided in [15] as well as in [10] for $\vartheta = 0$, relies on the same hypothesis of the existence and uniqueness of the solution to Equation (10) [12,23]. In particular, the authors proved that the stochastic ϑ-method is convergent with strong order $1/2$, i.e., there exists a real constant C such that

$$\mathbb{E}|X_n - Y_n|^2 \le Ch. \tag{14}$$

However, as is usual in the numerical approximation of SDEs, such an order of convergence can be improved by adding further terms in the expansion of the right-hand side [11,24,25], leading to the following improved stochastic ϑ-method

$$
\begin{aligned}
Y_n = Y_0 + h \sum_{i=0}^{n-1} \left(\vartheta a(t_n, t_{i+1}, Y_{i+1}) + (1-\vartheta)a(t_n, t_i, Y_i) \right) + \sqrt{h} \sum_{i=0}^{n-1} b(t_n, t_i, Y_i)G_{i,1} \\
+ \frac{1}{2}h\sqrt{h} \sum_{i=0}^{n-1} \frac{\partial a}{\partial x}(t_n, t_i, Y_i)b(t_i, t_i, Y_i)\left(G_{i,1} + \frac{G_{i,2}}{\sqrt{3}} \right) + \frac{1}{2}h \sum_{i=0}^{n-1} \frac{\partial b}{\partial x}(t_n, t_i, Y_i)b(t_i, t_i, Y_i)\left(G_{i,1}^2 - 1 \right),
\end{aligned}
\tag{15}
$$

where $\frac{\partial}{\partial x}$ denotes partial differentiation with respect to the second argument, while $G_{i,1}$ and $G_{i,2}$ are mutually independent $\mathcal{N}(0,1)$ random variables. Clearly, the numerical scheme can be made free from any derivative by suitable finite difference approximation. For instance, acting as in [11] leads to the following derivative free method

$$
\begin{aligned}
Y_n =\; & Y_0 + h \sum_{i=0}^{n-1} \left(\vartheta a(t_n, t_{i+1}, Y_{i+1}) + (1-\vartheta)a(t_n, t_i, Y_i) \right) + \sqrt{h} \sum_{i=0}^{n-1} b(t_n, t_i, Y_i)G_{i,1} \\
& + \frac{h}{2} \sum_{i=0}^{n-1} \left(a(t_n, t_i, Y_i + a(t_i, t_i, Y_i)h + b(t_i, t_i, Y_i)\sqrt{h}) - a(t_n, t_i, Y_i) \right) \left(G_{i,1} + \frac{G_{i,2}}{\sqrt{3}} \right) \\
& + \frac{\sqrt{h}}{2} \sum_{i=0}^{n-1} \left(b(t_n, t_i, Y_i + a(t_i, t_i, Y_i)h + b(t_i, t_i, Y_i)\sqrt{h}) - b(t_n, t_i, Y_i) \right) \left(G_{i,1}^2 - 1 \right).
\end{aligned}
\tag{16}
$$

One can prove (see [11,15]) that the strong order of convergence of the methods defined by Equations (15) and (16) is equal to 1, since there exists a real constant K such that

$$\mathbb{E}(|X_n - Y_n|^2) \le Kh^2. \tag{17}$$

3.3. Stability Issues

The stability analysis provided in [15] relies on investigating the behavior of the above methods when applied to the linear test Equation (11) and the convolution one (12). As regards the linear test equation, mean-square and asymptotic stability properties of the exact solution respectively occur when

$$\lim_{t \to \infty} \mathbb{E}|X(t)|^2 = 0 \quad \Leftrightarrow \quad \lambda + \frac{1}{2}\mu^2 < 0, \tag{18}$$

$$\lim_{t \to \infty} |X(t)| = 0 \text{ w.p.1} \quad \Leftrightarrow \quad \lambda - \frac{1}{2}\mu^2 < 0. \tag{19}$$

The following result, proved in [15], occurs.

Theorem 1. *Let $x = h\lambda$ and $y = h\mu^2$. The stochastic ϑ-methods defined by Equations (13), (15), (16) are mean-square stable with respect to the basic test Equation (11) if and only if*

$$\left| \alpha^2 + \beta^2 + 3\gamma^2 + \frac{\delta^2}{3} + 2\alpha\gamma + \beta\delta \right| < 1,$$

where

(i) $\alpha = \dfrac{1 + (1 - \vartheta)x}{1 - \vartheta x}$, $\beta = \dfrac{\sqrt{y}}{1 - \vartheta x}$, $\gamma = 0$, $\delta = 0$ *for the method given by Equation (13);*

(ii) $\alpha = \dfrac{1 + (1 - \vartheta)x - \frac{1}{2}y}{1 - \vartheta x}$, $\beta = \dfrac{\sqrt{y}}{1 - \vartheta x}$, $\gamma = \dfrac{y}{2(1 - \vartheta x)}$, $\delta = \dfrac{x\sqrt{y}}{1 - \vartheta x}$ *for the improved method defined by Equation (15);*

(iii) $\alpha = \dfrac{1 + (1 - \vartheta)x - \frac{1}{2}\left(x\sqrt{y} + y\right)}{1 - \vartheta x}$, $\beta = \dfrac{\sqrt{y}}{1 - \vartheta x}$, $\gamma = \dfrac{x\sqrt{y} + y}{2(1 - \vartheta x)}$, $\delta = \dfrac{x\sqrt{y} + x^2}{1 - \vartheta x}$ *for the derivative free method given by Equation (16).*

Moreover, the above methods are asymptotically stable if and only if

$$\mathbb{E}\left(\log \left| \alpha + \beta G_{n,1} + \gamma G_{n,1}^2 + \delta Z_n \right| \right) < 0, \tag{20}$$

with $Z_n = \frac{1}{2}\left(G_{n,1} + \frac{G_{n,2}}{\sqrt{3}} \right)$.

There is a strict analogy between the above presented methods and some similar formulae for SDEs. Indeed, one can see that the stability properties of the ϑ-method in Equation (13) completely parallels those of the Euler-Maruyama ϑ-method for SDEs [26–31]. If we remove from Equation (15) the sum involving the derivative $\frac{\partial a}{\partial x}$, we obtain

$$Y_n = Y_0 + h \sum_{i=0}^{n-1} \left(\vartheta a(t_n, t_{i+1}, Y_{i+1}) + (1 - \vartheta) a(t_n, t_i, Y_i) \right) + \sqrt{h} \sum_{i=0}^{n-1} b(t_n, t_i, Y_i) G_{i,1}$$

$$+ \frac{1}{2} h \sum_{i=0}^{n-1} \frac{\partial b}{\partial x}(t_n, t_i, Y_i) b(t_i, t_i, Y_i) \left(G_{i,1}^2 - 1 \right), \tag{21}$$

i.e., this scheme is analogous to the Milstein ϑ-method for SDEs [26], with

$$\alpha = \frac{1 + (1 - \vartheta)x - \frac{1}{2}y}{1 - \vartheta x}, \quad \beta = \frac{\sqrt{y}}{1 - \vartheta x}, \quad \gamma = \frac{y}{2(1 - \vartheta x)}, \quad \delta = 0.$$

Figures 3 and 4, respectively, show the regions of mean-square stability of the methods defined by Equations (13), (15), (16) and (21) with respect to the basic test Equation (11) for $\vartheta = 1/2$ and $\vartheta = 3/4$. Such methods, introduced in [15], can achieve unbounded stability regions in correspondence with the some values of the parameter $\vartheta \geq 1/2$. As visible from Figure 5, some methods have unbounded regions when $\vartheta > 1$. Regions of asymptotic stability are depicted in Figures 6 and 7, respectively, for $\vartheta = 1/2$ and $\vartheta = 3/4$. We observe that the regions of asymptotic stability are depicted by computing, for each point in the rectangle $[-4, 2] \times [0, 8]$, the value of the expectation contained in the left-hand side of the inequality (20) and checking if it is a negative number. Such expectation has been computed over 500 paths.

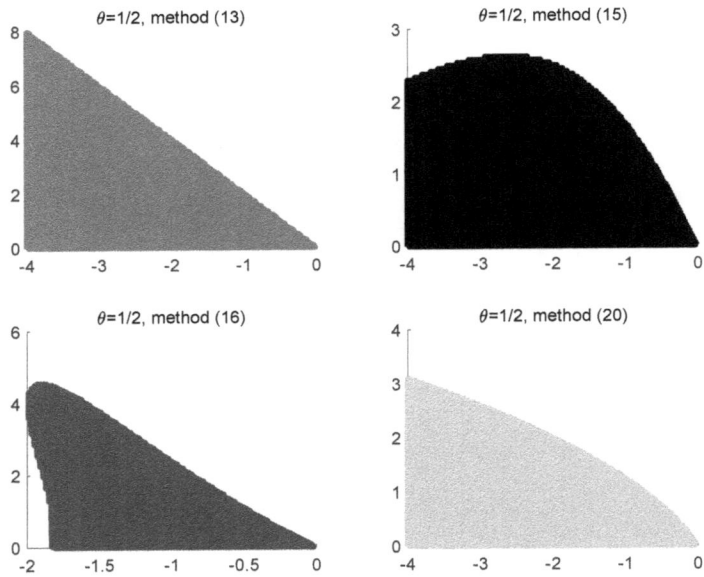

Figure 3. Mean-square stability regions in the (x, y)-plane with respect to the basic test Equation (11), case $\vartheta = 1/2$. The values of x and y are given in Theorem 1.

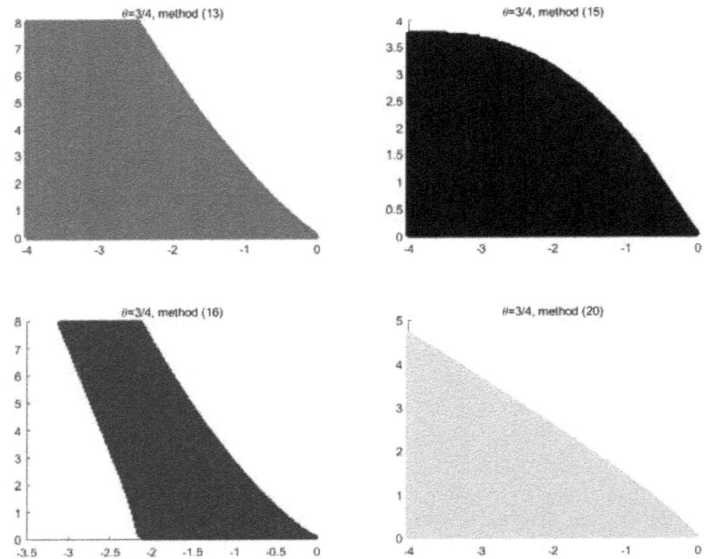

Figure 4. Mean-square stability regions in the (x, y)-plane with respect to the basic test Equation (11), case $\vartheta = 3/4$.

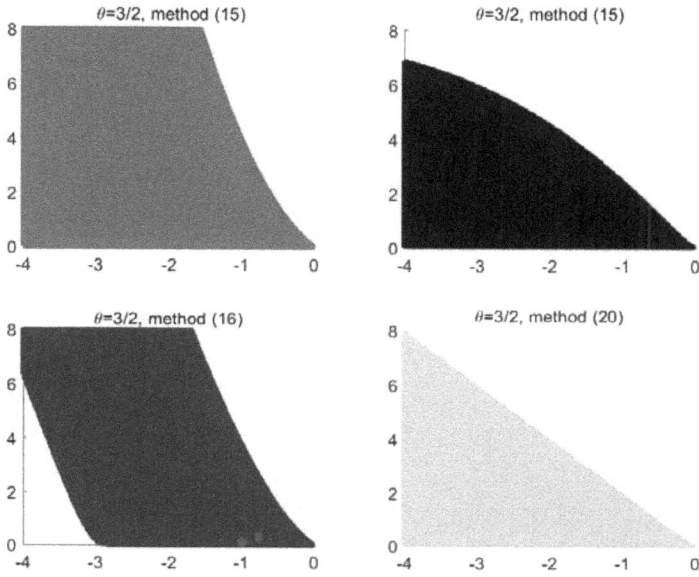

Figure 5. Mean-square stability regions in the (x, y)-plane with respect to the basic test Equation (11), case $\vartheta = 3/2$.

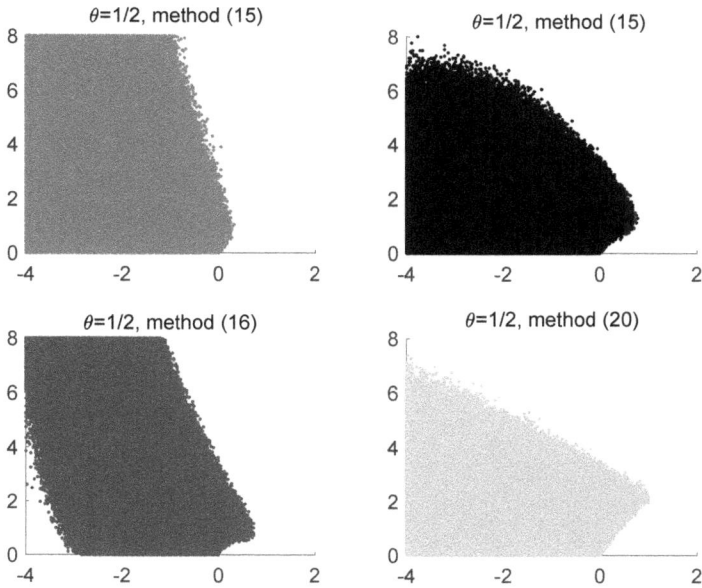

Figure 6. Asymptotic stability regions in the (x, y)-plane with respect to the basic test Equation (11), case $\vartheta = 1/2$.

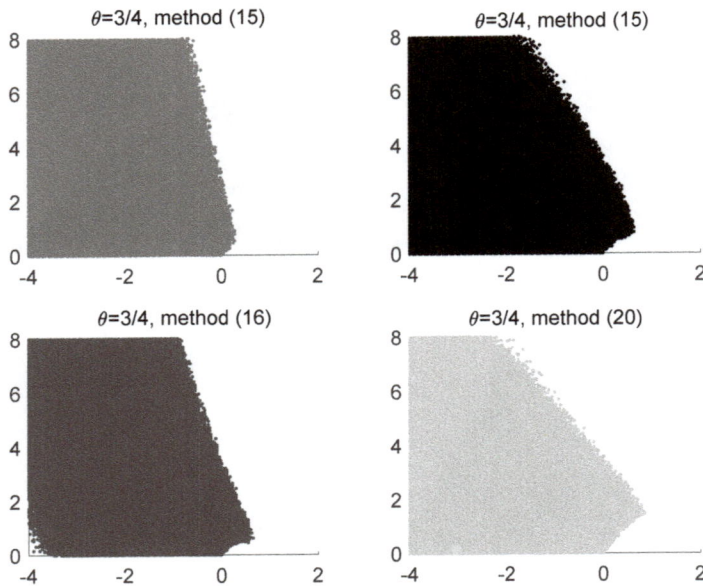

Figure 7. Asymptotic stability regions in the (x, y)-plane with respect to the basic test Equation (11), case $\vartheta = 3/4$.

Let us now move to the analysis with respect to the convolution test Equation (12). The following result occurs.

Theorem 2. *Let $x = h\lambda$, $y = h\mu^2$ and $z = h^2\sigma$. The stochastic ϑ-methods given by Equations (13), (15), (16) and (21) are mean-square stable with respect to the convolution test Equation (12) if the spectral radius $\rho(K)$ of matrix*

$$K = \begin{bmatrix} 0 & 0 & 1 \\ -\dfrac{\mathbb{E}(A_n(A_n + B_n))}{(1-\vartheta x)^2} & -\dfrac{\nu}{1-\vartheta x} & \dfrac{\mu}{1-\vartheta x} \\ \mathbb{E}(\beta_n) - \dfrac{2\mu\mathbb{E}(A_n(A_n + B_n))}{(1-\vartheta x)^3} & -\dfrac{2\nu\mu}{(1-\vartheta x)^2} & \mathbb{E}(\alpha_n) \end{bmatrix} \tag{22}$$

is less than 1, where

$$\mu = 2 + (1 - 2\vartheta)x + z, \quad \nu = 1 + (1 - \vartheta)x \tag{23}$$

and

$$\alpha_n = \left(\frac{\mu + A_{n+1} + B_{n+1}}{1 - \vartheta x}\right)^2, \quad \beta_n = \left(\frac{\nu + A_n}{1 - \vartheta x}\right)^2,$$

where

$$A_n = \sqrt{y}G_{n,1} + \zeta(G_{n,1}^2 - 1) + \eta Z_n, \quad B_n = \psi Z_n$$

and

(i) $\zeta = \eta = \psi = 0$ *for the method given by Equation (13),*

(ii) $\zeta = \frac{1}{2}y$, $\eta = x\sqrt{y}$, $\psi = z\sqrt{y}$ *for for the improved method given by Equation (15),*

(iii) $\zeta = \frac{1}{2}(x\sqrt{y} + y)$, $\eta = x(x + \sqrt{y})$, $\psi = z(x + \sqrt{y})$ *for the derivative free method given by Equation(16), (iv)* $\zeta = \frac{1}{2}y$, $\eta = 0$, $\psi = z\sqrt{y}$ *for the method given by Equation* (21),

with $Z_n = \frac{1}{2}\left(G_{n,1} + \frac{G_{n,2}}{\sqrt{3}}\right)$.

Mean-square and asymptotic stability regions of the above methods with respect to the convolution test Equation (12) are depicted in Figures 8 and 9, respectively. We observe that the region of asymptotic stability is depicted by computing, for each point in the rectangle $[-2, 0] \times [0, 4]$, the absolute value of the solution to Equation (12) and checking if it is smaller than a prescribed threshold, assumed equal to $1e - 4$ in our implementations. We also observe that larger selection of stability regions, in correspondence with different values of z and ϑ, has been provided in [15].

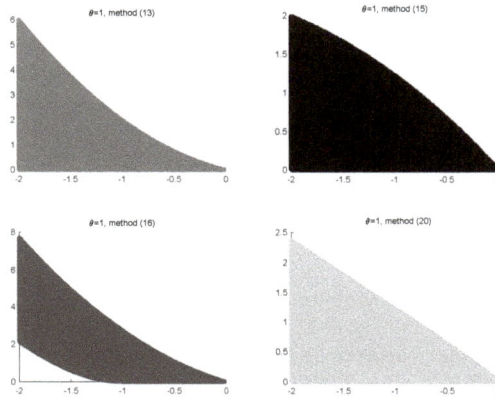

Figure 8. Mean-square stability regions in the (x, y)-plane with respect to the convolution test Equation (12) for $z = -2$ and $\vartheta = 1$.

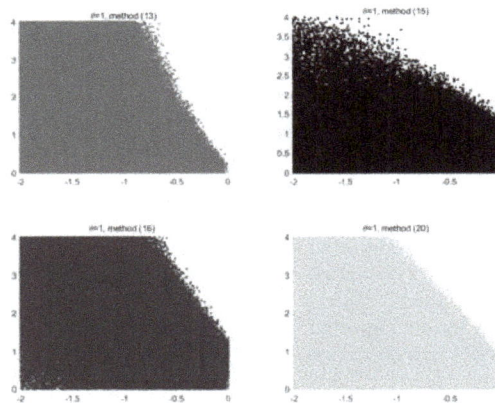

Figure 9. Asymptotic stability regions in the (x, y)-plane with respect to the convolution test Equation (12) for $z = -2$ and $\vartheta = 1$.

4. Conclusions

We have presented a short review of some recent results regarding the analysis of stability issues for stochastic differential Equation (1) and stochastic Volterra integral Equation (10).

As regards SDEs, the attention has been focused on analyzing the long-term behavior of two-step methods when applied to the linear stochastic damped linear oscillator described by Equation (2). We observe that the evidence provided in Section 2 mostly focuses on the case $\eta > 1$, i.e., that of strongly damped oscillators. It is very relevant to focus also on the case $\eta << g$, characterizing weakly damped oscillators, whose relevance is high in many physical problems. This issue will be analyzed in detail in future contributions. The tool introduced in [1–3] gives the possibility to provide a priori analysis relying on simple computations on 2 by 2 matrices. The structure-preserving approach to SDEs will next be devoted to stochastic Hamiltonian problems [7,8] and the stochastic extension of existing deterministic approaches [32–34]. A stochastic version of the non-polynomial fitting for oscillatory problems will also be addressed [35–38].

As regards SVIEs, the investigation has concerned the analysis of mean-square and asymptotic stability properties of ϑ-methods, which is going to be oriented on wider families of methods, such as stochastic Runge-Kutta methods for SVIEs, in future investigations. It is also worth observing that there is a connection between stochastic integral equations and stochastic differential equations. This connection yields, for instance, to the possibility of analyzing the properties of numerical methods for SVIEs through the corresponding SDE theory. Future contributions will focus on the analysis of stability properties of numerical methods for SVIEs that inherit the stability recursion of certain methods for SDEs; this analysis is helpful to assess a stability theory of numerical methods for SVIEs, which has been partially given in Section 3.

A further relevant issue regards the employment of other test equations for the stability analysis of numerical methods for SVIEs. Indeed, it is physically relevant and interesting to consider test equations depending on exponential kernels of the type $\lambda \exp(-\sigma(t - s))$, characteristic of system with fast response, rather than the case of the convolution test Equation (12), typical of slowly responding systems. This issues will also be addressed for stochastic fractional differential equations [35].

Author Contributions: The authors contributed equally to this work.

Funding: The work is supported by GNCS-Indam.

Acknowledgments: The authors are grateful to the anonymous referees for their helpful comments.

Conflicts of Interest: The authors declare no conflict of interest.

References

1. Burrage, K.; Lenane, I.; Lythe, G. Numerical methods for second-order stochastic differential equations. *SIAM J. Sci. Comput.* **2007**, *29*, 245–264. [CrossRef]
2. Burrage, K.; Lythe, G. Accurate stationary densities with partitioned numerical methods for stochastic differential equations. *SIAM J. Numer. Anal.* **2009**, *47*, 1601–1618. [CrossRef]
3. D'Ambrosio, R.; Moccaldi, M.; Paternoster, B. Numerical preservation of long-term dynamics by stochastic two-step methods. *Discret. Cont. Dyn. Syst. B* **2018**, *23*, 2763–2773. [CrossRef]
4. D'Ambrosio, R.; Moccaldi, M.; Paternoster, B.; Rossi, F. Stochastic Numerical Models of Oscillatory Phenomena. In *Artificial Life and Evolutionary Computation*; Pelillo, M., Poli, I., Roli, A., Serra, R., Slanzi, D., Villani, M., Eds.; Springer International Publishing: Berlin/Heidelberg, Germany, 2018; pp. 59–69.
5. Schurz, H. The invariance of asymptotic laws of linear stochastic systems under discretization. *Z. Angew. Math. Mech.* **1999**, *6*, 375–382. [CrossRef]
6. Strömmen Melbö, A.H.; Higham, D.J. Numerical simulation of a linear stochastic oscillator with additive noise. *Appl. Numer. Math.* **2004**, *51*, 89–99. [CrossRef]
7. Burrage, P.M.; Burrage, K. Structure-preserving Runge-Kutta methods for stochastic Hamiltonian equations with additive noise. *Numer. Algor.* **2014**, *65*, 519–532. [CrossRef]

8. Burrage, P.M.; Burrage, K. Low rank Runge-Kutta methods, symplecticity and stochastic Hamiltonian problems with additive noise. *J. Comput. Appl. Math.* **2014**, *236*, 3920–3930. [CrossRef]
9. Vilmart, G. Weak second order multi-revolution composition methods for highly oscillatory stochastic differential equations with additive or multiplicative noise. *SIAM J. Sci. Comput.* **2014**, *36*, 1770–1796. [CrossRef]
10. Wen, C.H.; Zhang, T.S. Rectangular method on stochastic Volterra equations. *Int. J. Appl. Math. Stat.* **2009**, *14*, 12–26. [CrossRef]
11. Wen, C.H.; Zhang, T.S. Improved rectangular method on stochastic Volterra equations. *J. Comput. Appl. Math.* **2011**, *235*, 2492–2501. [CrossRef]
12. Zhang, X. Euler schemes and large deviations for stochastic Volterra equations with singular kernels. *J. Differ. Eq.* **2008**, *244*, 2226–2250. [CrossRef]
13. Higham, D.J. An algorithmic introduction to numerical simulation of stochastic differential equations. *SIAM Rev.* **2011**, *43*, 525–546. [CrossRef]
14. Kloeden, P.E.; Platen, E. *The Numerical Solution of Stochastic Differential Equations*; Springer: Berlin/Heidelberg, Germany, 1992.
15. Conte, D.; D'Ambrosio, R.; Paternoster, B. On the stability of ϑ-methods for stochastic Volterra integral equations. *Discret. Cont. Dyn. Syst. B* **2018**, *23*, 2695–2708.
16. Buckwar, E.; Horvath-Bokor, R.; Winkler, R. Asymptotic mean-square stability of two-step methods for stochastic ordinary differential equations. *BIT Numer. Math.* **2006**, *46*, 261–282. [CrossRef]
17. Hairer, L.; Lubich, C.; Wanner, G. *Geometric Numerical Integration*; Springer: Berlin/Heidelberg, Germany, 2006.
18. Brunner, H.; van der Houwen, P.J. *The Numerical Solution of Volterra Equations*; CWI Monographs 3; North-Holland: Amsterdam, The Netherlands, 1986.
19. Conte, D.; D'Ambrosio, R.; Paternoster, B. Two-step diagonally-implicit collocation based methods for Volterra integral equations. *Appl. Numer. Math.* **2012**, *62*, 1312–1324. [CrossRef]
20. Conte, D.; Jackiewicz, Z.; Paternoster, B. Two-step almost collocation methods for Volterra integral equations. *Appl. Math. Comput.* **2008**, *204*, 839–853. [CrossRef]
21. Conte, D.; Paternoster, B. Multistep collocation methods for Volterra integral equations. *Appl. Numer. Math.* **2009**, *59*, 1721–1736. [CrossRef]
22. Cardone, A.; Conte, D. Multistep collocation methods for Volterra integro-differential equations. *Appl. Math. Comput.* **2013**, *221*, 770—785. [CrossRef]
23. Wang, Z. Existence and uniqueness of solutions to stochastic Volterra equations with singular kernels and non-Lipschitz coefficients. *Stat. Probab. Lett.* **2008**, *78*, 1062–1071. [CrossRef]
24. Buckwar, E.; Sickenberger, T. A comparative linear mean-square stability analysis of Maruyama- and Milstein-type methods. *Math. Comput. Simul.* **2011**, *81*, 1110–1127. [CrossRef]
25. Saito, Y.; Mitsui, T. Stability analysis of numerical schemes for stochastic differential equations. *SIAM J. Numer. Anal.* **1996**, *33*, 2254–2267. [CrossRef]
26. Bryden, A.; Higham, D.J. On the boundedness of asymptotic stability regions for the stochastic theta method. *BIT* **2003**, *43*, 1–6. [CrossRef]
27. Higham, D.J. Mean-square and asymptotic stability of the stochastic theta method. *SIAM J. Numer. Anal.* **2000**, *38*, 753–769 . [CrossRef]
28. Ding, X.; Ma, Q.; Zhang, L. Convergence and stability of the split-step-method for stochastic differential equations. *Comput. Math. Appl.* **2010**, *60*, 1310–1321. [CrossRef]
29. Higham, D.J. A-stability and stochastic mean-square stability. *BIT* **2000**, *40*, 404–409. [CrossRef]
30. Hu, P.; Huang, C. The stochastic ϑ-method for nonlinear stochastic Volterra integro-differential equations. *Abstr. Appl. Anal.* **2014**, *2014*, 583930. [CrossRef]
31. Shi, C.; Xiao, Y.; Zhang, C. The convergence and MS stability of exponential Euler method for semilinear stochastic differential equations. *Abstr. Appl. Anal.* **2012**, *2012*, 350407. [CrossRef]
32. Butcher, J.; D'Ambrosio, R. Partitioned general linear methods for separable Hamiltonian problems. *Appl. Numer. Math.* **2017**, *117*, 69–86. [CrossRef]
33. D'Ambrosio, R.; De Martino, G.; Paternoster, B. Numerical integration of Hamiltonian problems by G-symplectic methods. *Adv. Comput. Math.* **2014**, *40*, 553–575. [CrossRef]

34. D'Ambrosio, R.; Izzo, G.; Jackiewicz, Z. Search for highly stable two-step Runge-Kutta methods. *Appl. Numer. Math.* **2012**, *62*, 1361–1379. [CrossRef]

35. Burrage, K.; Cardone, A.; D'Ambrosio, R.; Paternoster, B. Numerical solution of time fractional diffusion systems. *Appl. Numer. Math.* **2017**, *116*, 82–94. [CrossRef]

36. D'Ambrosio, R.; De Martino, G.; Paternoster, B. General Nystrom methods in Nordsieck form: Error analysis. *J. Comput. Appl. Math.* **2016**, *292*, 694–702. [CrossRef]

37. D'Ambrosio, R.; Paternoster, B.; Santomauro, G. Revised exponentially fitted Runge–Kutta–Nyström methods. *Appl. Math. Lett.* **2014**, *30*, 56–60. [CrossRef]

38. D'Ambrosio, R.; Moccaldi, M.; Paternoster, B. Adapted numerical methods for advection-reaction-diffusion problems generating periodic wavefronts. *Comput. Math. Appl.* **2017**, *74*, 1029–1042. [CrossRef]

axioms

MDPI

Article

Efficient BEM-Based Algorithm for Pricing Floating Strike Asian Barrier Options (with MATLAB® Code)

Alessandra Aimi [1,*,†,‡] ⓘ, **Lorenzo Diazzi** [2,‡] ⓘ **and Chiara Guardasoni** [3,‡] ⓘ

[1] Department of Mathematical Physical and Computer Sciences, University of Parma, Parma 43124, Italy
[2] Department of Physics Informatics and Mathematics, University of Modena and Reggio Emilia, Modena 41125, Italy; lorenzo.diazzi@unimore.it
[3] Department of Mathematical Physical and Computer Sciences, University of Parma, Parma 43124, Italy; chiara.guardasoni@unipr.it
* Correspondence: alessandra.aimi@unipr.it; Tel.: +39-0521-906-944
† Current address: Parco Area delle Scienze 53/A, 43124 Parma, Italy.
‡ Members of the INdAM-GNCS Research Group. These authors contributed equally to this work.

Received: 11 May 2018; Accepted: 12 June 2018; Published: 15 June 2018

Abstract: This paper aims to illustrate how SABO (Semi-Analytical method for Barrier Option pricing) is easily applicable for pricing floating strike Asian barrier options with a continuous geometric average. Recently, this method has been applied in the Black–Scholes framework to European vanilla barrier options with constant and time-dependent parameters or barriers and to geometric Asian barrier options with a fixed strike price. The greater efficiency of SABO with respect to classical finite difference methods is clearly evident in numerical simulations. For the first time, a user-friendly MATLAB® code is made available here.

Keywords: boundary element method; finite difference method; floating strike Asian options; continuous geometric average; barrier options

1. Introduction

The pricing of continuously-monitored Asian options is a relevant task from both a mathematical and a financial point of view.

Asian options are quite common derivatives because they provide protection against strong price fluctuations in volatile markets and reduce the possibilities of price manipulations. The payoff of an Asian option depends on the average price of the underlying asset that is less volatile than the asset price itself. In general, Asian options are hence less valuable than their vanilla European counterparts because an option on a lower volatility asset is worth less.

On the other hand, it is more difficult to deal with Asian options than vanilla options because their price depends on the average value assumed by the underlying asset during the option's life, requiring some mathematical effort in order to describe the dynamics of the average under consideration.

In this paper, Asian options are equipped with a continuously-monitored geometric average [1]. Asian options evaluated with the geometric mean, although not common among practitioners, give some information also about the evaluation of Asian options with the arithmetic mean [2]. From a theoretical point of view, the method illustrated in the paper is extensible to arithmetic Asian options, as well, with slight modification, but from the numerical point of view, there are several problems that we plan to investigate in the near future. Defining the stochastic process

$$A_t := \int_0^t \log(S_\tau) d\tau \tag{1}$$

then the geometric average is defined as $\exp\left(\frac{A_t}{t}\right)$. When A and S are written with subscripts (A_t and S_t), they are intended as stochastic processes; otherwise, they are considered independent variables in the differential analysis context. The differential problem that describes the price evolution of this option is:

$$\frac{\partial V}{\partial t} + \frac{\sigma^2}{2}S^2\frac{\partial^2 V}{\partial S^2} + rS\frac{\partial V}{\partial S} + \log(S)\frac{\partial V}{\partial A} - rV = 0, \qquad S \in \mathbb{R}^+,\ A \in \mathbb{R},\ t \in [0,T] \qquad (2)$$

$$V(S,A,T) \quad \text{assigned}, \qquad S \in \mathbb{R}^+,\ A \in \mathbb{R}. \qquad (3)$$

Wanting to provide a further protection against excessive fluctuations of the strike price, it is possible to apply barriers in the option contract; for example, knock-out barriers make the option cease to exist if the underlying asset reaches a barrier during the life of the option. The model analyzed in this paper concerns an Asian option with an up-and-out barrier at $S = B$ and a floating strike payoff, i.e.,

$$\frac{\partial V}{\partial t} + \frac{\sigma^2}{2}S^2\frac{\partial^2 V}{\partial S^2} + rS\frac{\partial V}{\partial S} + \log(S)\frac{\partial V}{\partial A} - rV = 0, \qquad S \in (0,B),\ A \in \mathbb{R},\ t \in [0,T] \qquad (4)$$

$$V(S,A,T) \quad \text{assigned}, \qquad S \in (0,B),\ A \in \mathbb{R} \qquad (5)$$

$$V(B,A,t) = 0, \qquad A \in \mathbb{R},\ t \in [0,T) \qquad (6)$$

$$\text{asymptotic conditions of vanilla option}, \qquad \{(S,A) : S = 0 \vee A \to -\infty \vee A \to +\infty\}, \quad (7)$$

with the final condition:

$$\text{call} \qquad V(S,A,T) = \max\left(S - \exp\left(\frac{A}{T}\right), 0\right) \qquad \text{or} \qquad (8)$$

$$\text{put} \qquad V(S,A,T) = \max\left(\exp\left(\frac{A}{T}\right) - S, 0\right). \qquad (9)$$

The problem (2)–(3) of pricing a floating strike Asian option with a continuous geometric average and without a barrier has a closed-form solution in the domain $(S,A,t) \in \mathbb{R}^+ \times \mathbb{R} \times [0,T]$ that can be formulated either as the payoff expected value (also known as the Feynman–Kac formula):

$$V(S,A,t) = \int_{-\infty}^{+\infty}\int_0^B V(\tilde{S},\tilde{A},T)G(S,A,t;\tilde{S},\tilde{A},T)\,d\tilde{S}\,d\tilde{A}, \qquad (10)$$

with the transition probability density function G associated with the differential operator defined in Equation (4), which is known to be:

$$\begin{aligned}
G(S,A,t;\tilde{S},\tilde{A},\tilde{t}) = \ & \frac{\sqrt{3}H[\tilde{t}-t]}{\pi\sigma^2(\tilde{t}-t)^2}\exp\Bigg\{-\frac{2}{\sigma^2(\tilde{t}-t)}\log^2\left(\frac{S}{\tilde{S}}\right) + \frac{6}{\sigma^2(\tilde{t}-t)^2}\log\left(\frac{S}{\tilde{S}}\right)(A - \tilde{A} + (\tilde{t}-t)\log(S)) \\
& - \frac{6}{\sigma^2(\tilde{t}-t)^3}(A - \tilde{A} + (\tilde{t}-t)\log(S))^2 - \left(\frac{2r+\sigma^2}{2\sqrt{2}\sigma}\right)^2(\tilde{t}-t)\Bigg\}\left(\frac{\tilde{S}}{S}\right)^{\frac{2r-\sigma^2}{2\sigma^2}}\frac{1}{\tilde{S}},
\end{aligned} \qquad (11)$$

or (see [3]) through the formula:

$$V_{\text{call}}(S,A,t) = S\mathcal{N}[d] - e^{\frac{A}{t}}S^{\frac{T-t}{T}}e^q\mathcal{N}\left[d - \frac{\sigma}{T}\sqrt{\frac{T^3 - t^3}{3}}\right]$$

$$d = \frac{t(\log(S) - \frac{A}{t}) + (r + \frac{\sigma^2}{2})\frac{T^2 - t^2}{2}}{\sigma\sqrt{\frac{T^3 - t^3}{3}}} \qquad (12)$$

$$q = \frac{(t-T)[6Tr(t+T) + (T-t)(2t+T)\sigma^2]}{12T^2}$$

for the call option, where $\mathcal{N}[\cdot]$ is the normal cumulative distribution function, and eventually, the put-call parity:

$$V_{\text{call}}(S, A, t) - V_{\text{put}}(S, A, t) = S - e^{\frac{A}{T}} S^{\frac{T-t}{T}} e^{q}. \tag{13}$$

Instead, when applying barriers, no closed-formulas are available. In this context, SABO is a Semi-Analytical method conceived of for the pricing of Barrier Options, and its milestones are resumed in Section 4.1. It is quite a general method, applicable also to fixed strike payoffs [4,5], put options [6], time-dependent parameters [7] and double barriers [8].

SABO is compared here with two Finite Difference (FD) methods chosen among the wide class of numerical methods at our disposal [9]. Equation (4) is proven to be hypo-elliptic [10–12], a property that guarantees a smooth solution and should benefit from approximations based on Taylor expansions. Anyway, SABO appears to be certainly more efficient looking at the results below.

2. Results

We have performed several simulations related to the pricing of a geometric call Asian option with a floating strike price and with an up-and-out barrier as modeled by the differential problem (4)–(8). Numerical results, some of which are displayed in the following, have been obtained by the MATLAB® codes implementing the algorithms of SABO, FD1 and FD2 described in Sections 4.1, 4.2.1 and 4.2.2, respectively. In Section 5 the SABO code is provided.

Example 1. *In this example, we use the finance parameters displayed in Table 1. The floating strike call option with an up-and-out barrier at $S = 150 =: B$ is evaluated at $t = 0$ and $A = 0$, truncating the A-domain at $A_{min} = 0$ and $A_{max} = 5$, in accordance with (20) and either (19) or (27).*

The approximation by SABO is obtained setting the parameters described in Section 4.1 $N_t = N_A = 20$, and the option value is displayed in Figure 1 as a function of S.

The convergence at $S = 100, 120, 140, 148$ is shown in Table 2 refining the mesh: at each level, parameters N_t and N_A are doubled.

The approximation by FD1 is obtained setting the discretization parameters (defined in (17), (18) and (30)) $\Delta t = 10^{-3}/2^k$, $\Delta A = 10^{-2}/2^k$ and, for the discretization of the asset domain $(0, B)$, either $\Delta S = 2$ or $\Delta S = 1$. The related results are reported in Tables 3 and 4.

The approximation by FD2 is obtained setting the discretization parameters $\Delta t = \Delta A = 0.01/2^k$ and either $\Delta S = 2$ or $\Delta S = 1$. The results are displayed in Tables 5 and 6.

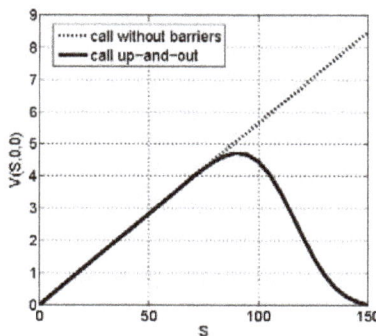

Figure 1. Semi-Analytical method for Barrier Option (SABO) approximation of a call up-and-out geometric Asian option with a floating strike and the data in Table 1.

Table 1. Floating strike up-and-out call option data.

B	T	r	σ
150	1	0.035	0.2

Table 2. $V(S, 0, 0)$ evaluated by SABO at $S = 100, 120, 140, 148$.

$N_t = N_A$	$S = 100$	$S = 120$	$S = 140$	$S = 148$	Elapsed Time (s)
20	4.4059	2.1078	0.2771	0.0441	9.0×10^0
40	4.4339	2.2728	0.3814	0.0632	3.1×10^1
80	4.4335	2.3506	0.4683	0.0807	1.3×10^2
160	4.4334	2.3600	0.4854	0.0843	5.1×10^2
320	4.4333	2.3616	0.4882	0.0849	2.2×10^3
640	4.4333	2.3619	0.4887	0.0850	1.0×10^4

Table 3. $V(S, 0, 0)$ evaluated by FD1at $S = 100, 120, 140, 148$, with $\Delta S = 2$.

k	$S = 100$	$S = 120$	$S = 140$	$S = 148$	Elapsed Time (s)
1	5.8385	3.6788	0.9933	0.1834	6.9×10^0
2	5.2000	3.0951	0.7729	0.1405	3.3×10^1
3	4.8342	2.7511	0.6412	0.1148	1.5×10^2
4	4.6360	2.5610	0.5673	0.1004	5.3×10^2
5	4.5324	2.4605	0.5278	0.0926	2.1×10^3
6	4.4794	2.4087	0.5073	0.0886	8.5×10^3
7	4.4526	2.3824	0.4968	0.0866	4.1×10^4

Table 4. $V(S, 0, 0)$ evaluated by FD1 at $S = 100, 120, 140, 148$, with $\Delta S = 1$.

k	$S = 100$	$S = 120$	$S = 140$	$S = 148$	Elapsed Time (s)
1	5.8427	3.6815	0.9946	0.1837	1.9×10^1
2	5.2047	3.0985	0.7744	0.1408	7.7×10^1
3	4.8392	2.7550	0.6429	0.1151	2.8×10^2
4	4.6413	2.5651	0.5691	0.1007	1.1×10^3
5	4.5380	2.4648	0.5296	0.0930	4.4×10^3
6	4.4551	2.4131	0.5092	0.0890	2.2×10^4
7	4.4583	2.3869	0.4988	0.0870	9.5×10^4

Table 5. $V(S, 0, 0)$ evaluated by FD2 at $S = 100, 120, 140, 148$, with $\Delta S = 2$.

k	$S = 100$	$S = 120$	$S = 140$	$S = 148$	Elapsed Time (s)
1	4.6607	2.5250	0.5016	0.0853	9.8×10^0
2	4.4796	2.3874	0.4806	0.0826	3.7×10^1
3	4.4394	2.3640	0.4845	0.0840	1.5×10^2
4	4.4296	2.3587	0.4861	0.0844	6.0×10^2
5	4.4269	2.3570	0.4864	0.0845	2.4×10^3
6	4.4261	2.3564	0.4864	0.0845	1.0×10^4

Table 6. $V(S, 0, 0)$ evaluated by FD2 at $S = 100, 120, 140, 148$, with $\Delta S = 1$.

k	S = 100	S = 120	S = 140	S = 148	Elapsed Time (s)
1	4.6657	2.5287	0.5034	0.0857	2.8×10^1
2	4.4853	2.3918	0.4826	0.0831	1.2×10^2
3	4.4452	2.3685	0.4865	0.0844	4.5×10^2
4	4.4354	2.3632	0.4881	0.0848	1.9×10^3
5	4.4327	2.3616	0.4884	0.0849	7.6×10^3
6	4.4318	2.3609	0.4884	0.0849	4.8×10^4

Example 2. *In this example, we use the finance parameters displayed in Table 7 and volatility equal to $\sigma = 0.2, 0.3, 0.4$. The floating strike call option with an up-and-out barrier at $S = 115 =: B$ is evaluated at $t = 0$ and $A = 0$, truncating the A-domain at $A_{min} = 0$ and $A_{max} = 5$, in accordance with (20) and (19).*

The approximation by SABO is obtained setting the parameters described in Section 4.1 $N_t = N_A = 20$ and the option value is displayed in Figure 2 as a function of S for the different values of σ (continuous lines), in comparison with the corresponding prices of Asian options without barriers (dotted lines).

Table 7. Floating strike up-and-out call option data.

B	T	r
115	1	0.08

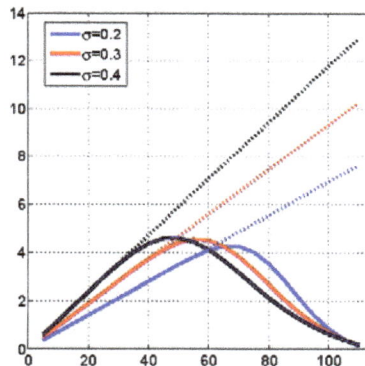

Figure 2. SABO approximation of a call up-and-out geometric Asian option with a floating strike, the data in Table 7 and various values of σ (continuous lines) compared with the corresponding prices of Asian options without barriers (dotted lines).

3. Discussion

Looking at Example 1, the values of a call option with an up-and-out barrier obtained by SABO and displayed in Figure 1 show that the solution, as expected, assumes lower values than the analogous option without barriers whose closed formula is (12) or that can be computed through the evaluation of the payoff expected value (10). The same behavior is recovered by the two proposed FD methods (FD1 and FD2).

Talking about efficiency and convergence, we have to look at the stabilization of the digits in Tables 2–6 where the option values at $S = 100, 120, 140, 148$ are written.

SABO, the results of which are written in Table 2, appears to be faster than the FD methods: doubling parameters N_t and N_A, the CPU time for computation quadruples, but one more digit of accuracy is achieved. The convergence is slower near the barrier because there, the barrier option value is more different from the option value without barriers: the option value without the barrier can be quasi-exactly computed by the Feynman–Kac Formula (10), and therefore, the approximation error introduced by SABO solving the boundary integral Equation (15) related to the barrier case is more involved in representation Formula (14) as the asset nears the barrier (look at (26)).

Comparing Tables 3 and 4, with an analogous computational time, SABO appears much more accurate and therefore efficient than FD1. Furthermore, note that FD1 is still sensitive to the mesh refinement in the S-domain: to halve ΔS means a significant variation in values of V together with a big increase of the computational costs.

Analyzing Tables 5 and 6, we observe that FD2 has a superior accuracy compared to FD1 due to its higher order of consistency: approximations of derivatives in the t and A variables are both of second order. Anyway, FD2 is less efficient than SABO, and the coarseness of the S-grid still significantly affects its results. Refinements in the S-grid would result in much longer computational times, no longer comparable with those of SABO.

Looking at Example 2, SABO maintains its robustness varying the volatility values. The solutions displayed in Figure 2 show the property of smoothness proven in [11]. The increase in volatility causes the expected increase in the value of vanilla options, but on the contrary, it implies a diminishing of barrier option values near the barrier.

4. Methods

4.1. SABO

SABO is a Semi-Analytical method for the pricing of Barrier Options. The foundation on which it is based is the integral representation of the solution of the modeling differential problem based on the knowledge of the transition probability density function.

For a problem like that in (4)–(7), it is proven in [4] that the integral representation formula is: $\forall (S, A, t) \in (0, B) \times \mathbb{R} \times [0, T]$

$$V(S, A, t) = \int_{-\infty}^{+\infty} \int_0^B V(\widetilde{S}, \widetilde{A}, T) G(S, A, t; \widetilde{S}, \widetilde{A}, T) d\widetilde{S} \, d\widetilde{A} + \int_t^T \int_{-\infty}^{+\infty} \frac{\sigma^2}{2} B^2 \frac{\partial V}{\partial \widetilde{S}}(B, \widetilde{A}, \widetilde{t}) G(S, A, t; B, \widetilde{A}, \widetilde{t}) d\widetilde{A} \, d\widetilde{t} \quad (14)$$

where the transition probability density function $G(S, A, t; B, \widetilde{A}, \widetilde{t})$ is defined in (11) and associated with the differential operator defined in Equation (4).

Note that in (14), both $V(S, A, t)$ and $\frac{\partial V}{\partial \widetilde{S}}(B, \widetilde{A}, \widetilde{t})$ are unknown, but, at $S = B$, the Boundary Condition (6) can be applied giving rise to the Volterra integral equation of the first kind:

$$0 = V(B, A, t) = \int_{-\infty}^{+\infty} \int_0^B V(\widetilde{S}, \widetilde{A}, T) G(B, A, t; \widetilde{S}, \widetilde{A}, T) d\widetilde{S} \, d\widetilde{A} + \int_t^T \int_{-\infty}^{+\infty} \frac{\sigma^2}{2} B^2 \frac{\partial V}{\partial \widetilde{S}}(B, \widetilde{A}, \widetilde{t}) G(B, A, t; B, \widetilde{A}, \widetilde{t}) d\widetilde{A} \, d\widetilde{t} \quad (15)$$

in the unknown $\frac{\partial V}{\partial \widetilde{S}}(B, \widetilde{A}, \widetilde{t})$.

The unknown is approximated by:

$$\frac{\partial V}{\partial \widetilde{S}}(B, \widetilde{A}, \widetilde{t}) \approx \sum_{k=1}^{N_t} \sum_{h=1}^{N_A} \alpha_h^{(k)} \psi_h(\widetilde{A}) \varphi_k(\widetilde{t}) \quad (16)$$

having defined piecewise constant basis functions:

$$\varphi_k(\widetilde{t}) := H[\widetilde{t} - t_{k-1}] - H[\widetilde{t} - t_k], \quad k = 1, \ldots, N_t,$$

on a uniform time grid:

$$t_k := k\Delta t, \quad k = 0, \ldots, N_t, \qquad \Delta t := \frac{T}{N_t}, \quad N_t \in \mathbb{N}^+, \quad (17)$$

and piecewise constant basis functions:

$$\psi_h(\widetilde{A}) := H[\widetilde{A} - A_{h-1}] - H[\widetilde{A} - A_h], \quad h = 1, \ldots, N_A,$$

on a uniform A-grid over the truncated A-domain $[A_{\min}, A_{\max}]$:

$$A_h := A_{\min} + h\Delta A, \quad h = 0, \ldots, N_A, \qquad \Delta A := \frac{A_{\max} - A_{\min}}{N_A}, \quad N_A \in \mathbb{N}^+. \tag{18}$$

A careful choice of $A_{\min}$ and $A_{\max}$ has to be performed in such a way that the double integral:

$$\int_t^T \int_{-\infty}^{+\infty} G(B, A, t; B, \widetilde{A}, \widetilde{t}) d\widetilde{A} \, d\widetilde{t}$$

is rightly approximated by:

$$\int_t^T \int_{A_{\min}}^{A_{\max}} G(B, A, t; B, \widetilde{A}, \widetilde{t}) d\widetilde{A} \, d\widetilde{t}.$$

By the analysis developed in [5], this results in seeking for $A_{\max}$ such that:

$$-\int_t^T \frac{\exp\left(-\frac{(\widetilde{t}-t)(2r+\sigma^2)^2}{8\sigma^2}\right)}{2B\sigma\sqrt{2\pi(\widetilde{t}-t)}} \mathrm{Erf}\left[\frac{\sqrt{6}(A - A_{\max} + (\widetilde{t}-t)\log(B))}{(\widetilde{t}-t)^{3/2}\sigma}\right] d\widetilde{t} = \frac{\mathrm{Erf}\left[\frac{(2r+\sigma^2)\sqrt{T-t}}{2\sigma\sqrt{2}}\right]}{B(2r+\sigma^2)} \tag{19}$$

and $A_{\min}$ such that:

$$\int_t^T \frac{\exp\left(-\frac{(\widetilde{t}-t)(2r+\sigma^2)^2}{8\sigma^2}\right)}{2B\sigma\sqrt{2\pi(\widetilde{t}-t)}} \mathrm{Erf}\left[\frac{\sqrt{6}(A - A_{\min} + (\widetilde{t}-t)\log(B))}{(\widetilde{t}-t)^{3/2}\sigma}\right] d\widetilde{t} = \frac{\mathrm{Erf}\left[\frac{(2r+\sigma^2)\sqrt{T-t}}{2\sigma\sqrt{2}}\right]}{B(2r+\sigma^2)} \tag{20}$$

with a suitable tolerance. Otherwise, it is possible to consider the whole unbounded A-domain $\equiv \mathbb{R}$, but with two infinite basis functions, the first and the last, as investigated in [4].

In order to numerically solve the Volterra integral Equation (15), we convert it into a discrete linear system of equations by means of collocation BEM. Hence we collocate (15) at points:

$$\overline{A}_i = \frac{A_i + A_{i-1}}{2}, \quad i = 1, \ldots, N_A \qquad \overline{t}_j = \frac{t_j + t_{j-1}}{2}, \quad j = 1, \ldots, N_t,$$

finally obtaining, for $i = 1, \ldots, N_A, j = 1, \ldots, N_t$:

$$\int_{\overline{t}_j}^T \int_{-\infty}^{+\infty} \frac{\sigma^2}{2} B^2 \sum_{k=1}^{N_t} \sum_{h=1}^{N_A} \alpha_h^{(k)} \psi_h(\widetilde{A}) \varphi_k(\widetilde{t}) G(B, \overline{A}_i, \overline{t}_j; B, \widetilde{A}, \widetilde{t}) d\widetilde{A} \, d\widetilde{t} = -\int_{-\infty}^{+\infty} \int_0^B V(\widetilde{S}, \widetilde{A}, T) G(B, \overline{A}_i, \overline{t}_j; \widetilde{S}, \widetilde{A}, T) d\widetilde{S} \, d\widetilde{A}$$

i.e., in matrix notation,

$$\mathcal{A}\alpha = \mathcal{F}. \tag{21}$$

The unknown vector is:

$$\alpha = \left(\alpha^{(k)}\big|_{k=1,\ldots,N_t}\right) = \left(\left(\alpha_h^{(k)}\big|_{h=1,\ldots,N_A}\right)\big|_{k=1,\ldots,N_t}\right),$$

and the matrix entries are equal to the case of fixed strike Asian options deeply investigated also from a computational point of view in [4,5]: for $i, h = 1, \ldots, N_A$, $j, k = 1, \ldots, N_t$, define $\xi = i - h$, $\xi = -N_A + 1, \ldots, N_A - 1$ and $\ell = k - j$, $\ell = 0, \ldots, N_t - 1$, thus obtaining:

$$
\begin{aligned}
A_{ih}^{(jk)} &= \frac{\sigma B \Delta t}{4\sqrt{2\pi}} \int_{\frac{1}{2} - \frac{1}{2}H[\ell]}^{1} \frac{\exp\left\{-\left(\frac{2r+\sigma^2}{2\sqrt{2}\sigma}\right)^2 \Delta t(\tau + \ell - 1/2)\right\}}{\sqrt{\Delta t(\tau + \ell - 1/2)}} \left\{ \mathrm{Erf}\left[\frac{\sqrt{6}(\Delta A(\xi + \frac{1}{2}) + \Delta t(\tau + \ell - 1/2)\log(B))}{\sigma \Delta t^{3/2}(\tau + \ell - 1/2)^{3/2}}\right] \right. \\
&\left. - \mathrm{Erf}\left[\frac{\sqrt{6}(\Delta A(\xi - \frac{1}{2}) + \Delta t(\tau + \ell - 1/2)\log(B))}{\sigma \Delta t^{3/2}(\tau + \ell - 1/2)^{3/2}}\right] \right\} d\tau =: A_\xi^{(\ell)}.
\end{aligned}
\tag{22}
$$

As we are considering constant coefficients in (4), the fundamental solution (11) depends on $t, \tilde{t}, A, \tilde{A}$ only through the differences $t - \tilde{t}$ and $A - \tilde{A}$ implying that the matrix has a block-Toeplitz structure both in time and A-space:

$$
\mathcal{A} = \begin{bmatrix}
\mathcal{A}^{(0)} & \mathcal{A}^{(1)} & \mathcal{A}^{(2)} & \cdots & \mathcal{A}^{(N_t - 1)} \\
0 & \mathcal{A}^{(0)} & \mathcal{A}^{(1)} & \cdots & \mathcal{A}^{(N_t - 2)} \\
0 & 0 & \mathcal{A}^{(0)} & \ddots & \vdots \\
\vdots & \vdots & \ddots & \ddots & \mathcal{A}^{(1)} \\
0 & 0 & \cdots & 0 & \mathcal{A}^{(0)}
\end{bmatrix}
\tag{23}
$$

$$
\text{with} \quad \mathcal{A}^{(\ell)} = \begin{bmatrix}
\mathcal{A}_0^{(\ell)} & \mathcal{A}_{-1}^{(\ell)} & \mathcal{A}_{-2}^{(\ell)} & \cdots & \mathcal{A}_{-N_A+1}^{(\ell)} \\
\mathcal{A}_1^{(\ell)} & \mathcal{A}_0^{(\ell)} & \mathcal{A}_{-1}^{(\ell)} & \cdots & \mathcal{A}_{-N_A+2}^{(\ell)} \\
\mathcal{A}_2^{(\ell)} & \mathcal{A}_1^{(\ell)} & \mathcal{A}_0^{(\ell)} & \ddots & \vdots \\
\vdots & \vdots & \ddots & \ddots & \mathcal{A}_{-1}^{(\ell)} \\
\mathcal{A}_{N_A-1}^{(\ell)} & \mathcal{A}_{N_A-2}^{(\ell)} & \cdots & \mathcal{A}_1^{(\ell)} & \mathcal{A}_0^{(\ell)}
\end{bmatrix} \quad \text{for } \ell = 0, \ldots, N_t - 1
$$

so it is possible to adopt suitable strategies to save computational costs (as done in [13], even if this feature is not implemented in the code included here) and memory requirements.

A change in the payoff function is caught by the evaluation of the rhs entries: in the case of a floating strike call payoff, for $i = 1, \ldots, N_A$, $j = 1, \ldots, N_t$:

$$
\mathcal{F}_i^{(j)} = -\int_{-\infty}^{+\infty} \int_0^B \max\left(\tilde{S} - e^{\frac{\tilde{A}}{T}}, 0\right) G(B, \overline{A}_i, \overline{t}_j; \tilde{S}, \tilde{A}, T) \, d\tilde{S} \, d\tilde{A}
\tag{24}
$$

$$
\begin{aligned}
&= \frac{B^{-\frac{2r-\sigma^2}{2\sigma^2}}}{2\sigma\sqrt{2\pi(T - \overline{t}_j)}} \int_0^B \tilde{S}^{\frac{2r-3\sigma^2}{2\sigma^2} + \frac{T-\overline{t}_j}{2T}} \exp\left(-\frac{(T - \overline{t}_j)^2(2r + \sigma^2)^2 + 4\log^2\left(\frac{B}{\tilde{S}}\right)}{8\sigma^2(T - \overline{t}_j)}\right) \\
&\quad \left\{ -\tilde{S}^{1 - \frac{T-\overline{t}_j}{2T}} \mathrm{Erfc}\left[\sqrt{\frac{3}{2}} \frac{2\overline{A}_i + 2(T - \overline{t}_j)\log(B) - 2T\log(\tilde{S}) - (T - \overline{t}_j)\log(\frac{B}{\tilde{S}})}{\sigma(T - \overline{t}_j)^{\frac{3}{2}}}\right] \right. \\
&\quad + B^{\frac{T-\overline{t}_j}{2T}} \exp\left(\frac{\overline{A}_i}{T} + \frac{\sigma^2(T - \overline{t}_j)^3}{24 T^2}\right) \\
&\quad \left. \mathrm{Erfc}\left[\frac{12 T\overline{A}_i + \sigma^2(T - \overline{t}_j)^3 - 6T\left(-2(T - \overline{t}_j)\log(B) + 2T\log(\tilde{S}) + (T - \overline{t}_j)\log(\frac{B}{\tilde{S}})\right)}{2\sqrt{6}T\sigma(T - \overline{t}_j)^{\frac{3}{2}}}\right] \right\} d\tilde{S},
\end{aligned}
\tag{25}
$$

with Erfc the complementary of error function Erf.

The approximation of $\frac{\partial V}{\partial S}(B, \widetilde{A}, \widetilde{t})$ through the resolution of (21) is an intermediate step for the final evaluation of the option price by (14):

$$
\begin{aligned}
V(S, A, t) \approx\ & -\frac{S^{-\frac{2r-\sigma^2}{2\sigma^2}}}{2\sigma\sqrt{2\pi(T-t)}} \int_0^B \widetilde{S}^{\frac{2r-3\sigma^2}{2\sigma^2}+\frac{T-t}{2T}} \exp\left(-\frac{(T-t)^2(2r+\sigma^2)^2 + 4\log^2\left(\frac{S}{\widetilde{S}}\right)}{8\sigma^2(T-t)}\right) \\
& \left\{-\widetilde{S}^{1-\frac{T-t}{2T}} \operatorname{Erfc}\left[\sqrt{\frac{3}{2}}\,\frac{2A + 2(T-t)\log(S) - 2T\log(\widetilde{S}) - (T-t)\log(\frac{S}{\widetilde{S}})}{\sigma(T-t)^{\frac{3}{2}}}\right]\right. \\
& +\ S^{\frac{T-t}{2T}} \exp\left(\frac{A}{T} + \frac{\sigma^2(T-t)^3}{24\,T^2}\right) \\
& \left.\operatorname{Erfc}\left[\frac{12\,TA + \sigma^2(T-t)^3 - 6\,T\left(-2(T-t)\log(S) + 2\,T\log(\widetilde{S}) + (T-t)\log(\frac{S}{\widetilde{S}})\right)}{2\sqrt{6}\,T\sigma(T-t)^{\frac{3}{2}}}\right]\right\} d\widetilde{S} \\
& +\ \frac{\sigma B}{4\sqrt{2\pi}}\left(\frac{B}{S}\right)^{\frac{2r-\sigma^2}{2\sigma^2}} \sum_{k=\mathrm{floor}[\frac{t}{\Delta t}]+1}^{N_t} \sum_{h=1}^{N_A} \alpha_h^{(k)} \int_{\max(t,t_{k-1})}^{t_k} \frac{1}{\sqrt{\widetilde{t}-t}} \exp\left(-\frac{(\widetilde{t}-t)^2(2r+\sigma^2)^2 + 4\log^2\left(\frac{S}{B}\right)}{8(\widetilde{t}-t)\sigma^2}\right) \\
& \left\{\operatorname{Erf}\left[\sqrt{\frac{3}{2}}\,\frac{2(A_h - A) - 2(\widetilde{t}-t)\log(S) + (\widetilde{t}-t)\log\left(\frac{S}{B}\right)}{(\widetilde{t}-t)^{3/2}\sigma}\right]\right. \\
& \left.-\ \operatorname{Erf}\left[\sqrt{\frac{3}{2}}\,\frac{2(A_{h-1} - A) - 2(\widetilde{t}-t)\log(S) + (\widetilde{t}-t)\log\left(\frac{S}{B}\right)}{(\widetilde{t}-t)^{3/2}\sigma}\right]\right\} d\widetilde{t}.
\end{aligned}
\tag{26}
$$

4.2. Finite Difference Methods

SABO is compared here (as in [5]) with two Finite Difference (FD) methods chosen among the wide class of FD methods available and deeply analyzed in [14]. This is because Equation (4) is proven to be hypoelliptic [10–12], a property that guarantees a smooth solution and that should benefit from approximations based on Taylor expansions.

The existence of the solution of Problem (4)–(5) and its Feynman–Kac representation (10) are proven involving stochastic arguments without the need for exact boundary conditions. Boundary conditions at $S = 0$ and for $A \to \pm\infty$ have to be empirically deduced analogously to what was done in [5,15]. At $S = B$, Condition (6) holds.

In order to apply FD methods, the first step is to shrink to a bounded domain $[0, B] \times [A_{\min}, A_{\max}]$ enclosing the option evaluation point (S^*, A^*) and then apply boundary conditions at the borders.

- We have chosen $A_{\min}$ less than or equal to the minimum between the value suggested in (20) and A^*. There is not a consistent condition valid at $A = -\infty$ in finance, and as a consequence, it is not easy to conceive of a proper condition at points $(S, A_{\min}, t)$. The use of the upwind method makes this condition unnecessary if $\log(S) > 0$ (as in the herein proposed examples). In fact, $\frac{\partial V}{\partial A}$ is approximated by a forward difference so that the boundary condition at $A_{\min}$ becomes useless; otherwise, if $\log(S) < 0$, a backward difference can be considered together with a condition on the derivative at $A = A_{\min}$, as for example $\frac{\partial^2 V}{\partial A^2}(S, A_{\min}, t) = 0$.

- The upper bound is set at:

$$A_{\max} = T\log(B),\tag{27}$$

the maximum value assumed by the stochastic process A_t defined in (1) if $S_t = B$ during the whole time interval $[0, T]$.

The average $\exp(A/T)$ is a non-decreasing function of time; therefore, if $\exp(A/t) - B > 0$ at time t, then $\exp(A/T) - B > 0$ at maturity, and so, the option is worth nothing as shown by the Payoff Function (8). Hence, if $\log(S) > 0$, $\frac{\partial V}{\partial A}$ is approximated by a forward difference and the boundary condition at $A_{\max}$ is set:

$$V(S, A_{\max}, t) = 0.\tag{28}$$

- At $S = 0$, Equation (4) is degenerate, but if the stochastic process $S_t = 0$ at any time, then the average asset price $\exp(A_t/t) = constant$ and the option value V can be considered as independent of stochastic variables S and A, deducing from Equation (4) that:

$$\lim_{S \to 0} \left[\frac{\partial V}{\partial t} + \frac{\sigma^2}{2} S^2 \frac{\partial^2 V}{\partial S^2} + rS \frac{\partial V}{\partial S} + \log(S) \frac{\partial V}{\partial A} - rV \right] = \lim_{S \to 0} \left[\frac{\partial V}{\partial t} - rV \right] = 0.$$

The ordinary differential equation:

$$\lim_{S \to 0} \left[\frac{\partial V}{\partial t} - rV \right] = 0$$

suggests the condition:

$$\lim_{S \to 0} V(S, A, t) = e^{-r(T-t)} \lim_{S \to 0} V(S, A, T) = 0. \tag{29}$$

The second step, after the definition of the computational domain, is the definition of the grid. Using a (A, t)-grid as defined by (17) and (18) in $[A_{\min}, A_{\max}] \times [0, T]$, let us further introduce the S-grid in $[0, B]$:

$$\Delta S := \frac{B}{N_S}, \quad N_S \in \mathbb{N}^+, \quad S_i := i\Delta S, \quad i = 0, \ldots, N_S \tag{30}$$

and define the approximated option value:

$$V_{i,h}^k :\approx V(S_i, A_h, t_k).$$

- The values $V_{ih}^{N_t}$ can be found from the final condition:

$$V_{i,h}^{N_t} = V(S_i, A_h, T), \quad i = 0, \cdots, N_S - 1, \ h = 0, \cdots, N_A;$$

- in the herein proposed examples, $\log(S_i) > 0$, so everywhere, we apply a forward difference and, at the boundary $A = A_{\max}$, Condition (28):

$$V_{i,N_A}^k = 0 \quad \text{for } k = 0, \cdots, N_t - 1, \ i = 0, \cdots, N_S - 1;$$

- at the boundary $S = 0$, we apply Condition (29):

$$V_{0,h}^k = 0 \quad h = 0, \cdots, N_A - 1, \ k = 0, \cdots, N_t - 1;$$

- at the boundary $S = B$, we apply the Boundary Condition (6), hence:

$$V_{N_S,h}^k = 0, \quad h = 0, \cdots, N_A, \ k = 0, \cdots, N_t.$$

4.2.1. Finite Difference Method FD1

The first scheme FD1 is the very classical FD scheme where the derivatives of V in (4) are approximated by truncations of Taylor expansions:

- first order backward difference for the time derivative approximation:

$$\frac{\partial V}{\partial t}(S_i, A_h, t_k) = \frac{V_{i,h}^k - V_{i,h}^{k-1}}{\Delta t} + O(\Delta t)$$

- second order central difference for the S derivative approximation:

$$\frac{\partial V}{\partial S}(S_i, A_h, t_k) = \frac{V^k_{i+1,h} - V^k_{i-1,h}}{2\Delta S} + O(\Delta S^2)$$

- second order central difference for the S second order derivative approximation:

$$\frac{\partial^2 V}{\partial S^2}(S_i, A_h, t_k) = \frac{V^k_{i+1,h} - 2V^k_{i,h} + V^k_{i-1,h}}{\Delta S^2} + O(\Delta S^2)$$

- if $\log(S_i) > 0$, first order forward difference for the A derivative approximation:

$$\frac{\partial V}{\partial A}(S_i, A_h, t_k) = \frac{V^k_{i,h+1} - V^k_{i,h}}{\Delta A} + O(\Delta A)$$

We can now write down the discrete approximation of (4) at each point (S_i, A_h, t_k):

for $i = 1, \ldots, N_S - 1$, $h = 0, \ldots, N_A - 1$, $k = 0, \ldots, N_t - 1$,

$$\frac{V^k_{i,h} - V^{k-1}_{i,h}}{\Delta t} + \frac{1}{2}\sigma^2 S_i^2 \frac{V^k_{i+1,h} - 2V^k_{i,h} + V^k_{i-1,h}}{\Delta S^2} + rS_i \frac{V^k_{i+1,h} - V^k_{i-1,h}}{2\Delta S} + \log(S_i)\frac{V^k_{i,h+1} - V^k_{i,h}}{\Delta A} - rV^k_{i,h} = 0 \tag{31}$$

and rearrange the scheme in a compact form:

$$V^{k-1}_{i,h} = a_i V^k_{i,h} + b_i V^k_{i+1,h} + c_i V^k_{i-1,h} + d_i V^k_{i,h+1}$$

where:

$$a_i = 1 - \Delta t\left(r + \sigma^2 i^2 + \frac{\log(S_i)}{\Delta A}\right), \quad b_i = \frac{\Delta t}{2}\left(ri + \sigma^2 i^2\right), \quad c_i = \frac{\Delta t}{2}\left(-ri + \sigma^2 i^2\right), \quad d_i = \frac{\Delta t}{\Delta A}\log(S_i).$$

It is easy to see that the scheme, backward computing the new value $V^{k-1}_{i,h}$, is explicit in time.

We want to remark about two more things: first, the weights depend only on i, therefore on the S variable, and second, the values $V^k_{i,0}$ having $A_{\min}$ as the coordinate give no contribution to the scheme.

4.2.2. Finite Difference Method FD2

The method FD2 is suggested in [16]. The PDE (4) is collocated at the points:

$$\left(S_i, A_{h+\frac{1}{2}}, t_{k+\frac{1}{2}}\right) := \left(S_i, A_h + \frac{\Delta A}{2}, t_k + \frac{\Delta t}{2}\right).$$

Then, we use the following approximations, based on suitable Taylor expansions and standard finite difference approximations:

$$
\begin{aligned}
\frac{\partial V}{\partial t}\left(S_i, A_{h+\frac{1}{2}}, t_{k+\frac{1}{2}}\right) &= \frac{1}{2}\left(\frac{\partial V}{\partial t}(S_i, A_h, t_{k+\frac{1}{2}}) + \frac{\partial V}{\partial t}(S_i, A_{h+1}, t_{k+\frac{1}{2}})\right) + O(\Delta A^2) \\
&= \frac{V^{k+1}_{i,h} - V^k_{i,h}}{2\Delta t} + \frac{V^{k+1}_{i,h+1} - V^k_{i,h+1}}{2\Delta t} + O(\Delta t^2 + \Delta A^2),
\end{aligned}
$$

$$
\begin{aligned}
\frac{\partial V}{\partial S}\left(S_i, A_{h+\frac{1}{2}}, t_{k+\frac{1}{2}}\right) &= \frac{1}{2}\left(\frac{\partial V}{\partial S}(S_i, A_{h+1}, t_{k+1}) + \frac{\partial V}{\partial S}(S_i, A_h, t_k)\right) + O(\Delta t^2 + \Delta A^2) \\
&= \frac{V^{k+1}_{i+1,h+1} - V^{k+1}_{i-1,h+1}}{4\Delta S} + \frac{V^k_{i+1,h} - V^k_{i-1,h}}{4\Delta S} + O(\Delta t^2 + \Delta A^2 + \Delta S^2),
\end{aligned}
$$

$$\frac{\partial^2 V}{\partial S^2}\left(S_i, A_{h+\frac{1}{2}}, t_{k+\frac{1}{2}}\right) = \frac{1}{2}\left(\frac{\partial^2 V}{\partial S^2}(S_i, A_{h+1}, t_{k+1}) + \frac{\partial^2 V}{\partial S^2}(S_i, A_h, t_k)\right) + O(\Delta t^2 + \Delta A^2)$$

$$= \frac{V_{i-1,h+1}^{k+1} - 2V_{i,h+1}^{k+1} + V_{i+1,h+1}^{k+1}}{2\Delta S^2} + \frac{V_{i-1,h}^{k} - 2V_{i,h}^{k} + V_{i+1,h}^{k}}{2\Delta S^2} + O(\Delta t^2 + \Delta A^2 + \Delta S^2),$$

$$V\left(S_i, A_{h+\frac{1}{2}}, t_{k+\frac{1}{2}}\right) = \frac{1}{2}\left(V_{i,h}^{k} + V_{i,h+1}^{k+1}\right) + O(\Delta t^2 + \Delta A^2),$$

$$\frac{\partial V}{\partial A}\left(S_i, A_{h+\frac{1}{2}}, t_{k+\frac{1}{2}}\right) = \frac{1}{2}\left(\frac{\partial V}{\partial A}(S_i, A_{h+\frac{1}{2}}, t_{k+1}) + \frac{\partial V}{\partial A}(S_i, A_{h+\frac{1}{2}}, t_k)\right) + O(\Delta t^2)$$

$$= \frac{V_{i,h+1}^{k+1} - V_{i,h}^{k+1}}{2\Delta A} + \frac{V_{i,h+1}^{k} - V_{i,h}^{k}}{2\Delta t} + O(\Delta t^2 + \Delta A^2).$$

After substituting the above approximations into the PDE and discarding the error terms, we get the following equations for the approximate values of the option prices:

$$\overline{a}_i V_{i-1,h}^{k} + \overline{b}_i V_{i,h}^{k} + \overline{c}_i V_{i+1,h}^{k} = \overline{d}_i V_{i-1,h+1}^{k+1} + \overline{e}_i V_{i,h+1}^{k+1} + \overline{f}_i V_{i+1,h+1}^{k+1} + \overline{g}_i(V_{i,h+1}^{k} - V_{i,h}^{k+1}) \tag{32}$$

where, using the notation $\lambda = \frac{\Delta t}{\Delta S^2}$:

$$\overline{a}_i = \frac{\lambda}{2}\left(-S_i^2\sigma^2 + rS_i\Delta S\right), \quad \overline{b}_i = \left(1 + \lambda S_i^2\sigma^2 + \frac{\log(S_i)\Delta t}{\Delta A} + r\Delta t\right), \quad \overline{c}_i = -\frac{\lambda}{2}\left(S_i^2\sigma^2 + rS_i\Delta S\right),$$

$$\overline{d}_i = -a_i, \quad \overline{e}_i = \left(1 - \lambda S_i^2\sigma^2 + \frac{\log(S_i)\Delta t}{\Delta A} - r\Delta t\right), \quad \overline{f}_i = -c_i, \quad \overline{g}_i = \left(-1 + \frac{\log(S_i)\Delta t}{\Delta A}\right).$$

The procedure for solving the option pricing equation is as follows:

1. Fill the values $V_{i,h}^{N_t}$, $i = 0, \cdots, N_S$, $h = 0, \cdots, N_A$ using the payoff function.
2. For each $k = N_t - 1 : -1 : 0$:

   (a) Apply the boundary condition at $A = A_{\max}$ to define V_{i,N_A}^{k}, for $i = 0, \cdots, N_S$.
   (b) For each $h = N_A - 1 : -1 : 0$, solve for the three-diagonal system in the unknowns the values $V_{i,h}^{k}$, for $i = 1, \cdots, N_S - 1$, using the boundary condition at $S = 0$ and $S = B$.

This is a time-explicit difference scheme, as well. If we need the option price only at $t = 0$, then it is not necessary to store the full matrix V of approximate option prices; in fact, we need only two levels, say V_{old} corresponding to $t = t_{k+1}$ and V_{new} corresponding to the current time level $t = t_k$. At the beginning, V_{old} is computed using the final condition, and at the end of each time step, the values of V_{new} are copied to V_{old}. Anyway, it requires the resolution of a linear system: the three-diagonal matrix $M = diag(\overline{a}, \overline{b}, \overline{c})$ can be assembled and factorized at the beginning, outside the cycles, depending only on the S-grid.

5. MATLAB® Code Implementing SABO

All the above-provided numerical results were obtained by codes developed with MATLAB® Release 2007b running on a laptop computer (CPU Intel i5, 4 Gb RAM). The code implementing the SABO algorithm is given below.

```
% Approximation of an up-and-out Asian call option with floating strike
% by SABO

close all
clear
clc


%%%%%%%%%%%%%%%%%%%%%%%%%%%%%%%%%%%%%%%%%%%%%%%%%%%%%%%%%%%%%%%%%%%%%%%%%%%%
```

```
%data set
t0=0; %beginning of the contract
T=1; %expiry of the contract
r=0.035; %free risk interest rate
sigma=0.2; %volatility
B=150; %up barrier value
Sstar=[100,120,140]; %actual spot price
Astar=0; %actual geometric mean value
tstar=t0; %actual time instant of evaluation
prec=10^(-9); % precision required in Matlab "quad" function

%%%%%%%%%%%%%%%%%%%%%%%%%%%%%%%%%%%%%%%%%%%%%
%discretization of time domain
Nt=20; %number of time steps
dt=(T-t0)/Nt; %time step length
t=[0:dt:T]; %time grid
tbar=(t(2:Nt+1)+t(1:Nt))/2; %time collocation points

%%%%%%%%%%%%%%%%%%%%%%%%%%%%%%%%%%%%%%%%%%%%%
%discretization of A-domain
NA=Nt; %number of A-subintervals
AA=-(2*T^(3/2)*sigma)/sqrt(6)+Astar+T*log(B);
Amin=round(fzero(@(a) abs(...
        quad(@(tau) exp(-(tau*(2*r+sigma^2)^2)/(8*sigma^2))./...
        (2*sigma*sqrt(2*pi*tau)).*...
        erf(sqrt(6)*(Astar-a+tau*log(B))./(tau.^(3/2)*sigma)),...
        0,T-t0,10^-11)-...
        erf((2*r+sigma^2)*sqrt(T-t0)/(2*sqrt(2)*sigma))/(2*r+sigma^2))-prec,...
        AA)); %artificial lower bound of A-domain
%Amin=0;
AA=(2*T^(3/2)*sigma)/sqrt(6)+Astar+T*log(B);
Amax=round(fzero(@(a) abs(...
        quad(@(tau) exp(-(tau*(2*r+sigma^2)^2)/(8*sigma^2))./...
        (2*sigma*sqrt(2*pi*tau)).*...
        erf(sqrt(6)*(Astar-a+tau*log(B))./(tau.^(3/2)*sigma)),...
        0,T-t0,10^-11)+...
        erf((2*r+sigma^2)*sqrt(T-t0)/(2*sqrt(2)*sigma))/(2*r+sigma^2))-prec,...
        AA)); %artificial upper bound of A-domain
%Amax=5;
dA=(Amax-Amin)/NA; %A-subinterval length
A=[Amin:dA:Amax]; %A-grid
Abar=(A(2:NA+1)+A(1:NA))/2; %A collocation points
disp('truncated A-domain:')
Amin
Amax
tic
%%%%%%%%%%%%%%%%%%%%%%%%%%%%%%%%%%%%%%%%%%%%%
%computation of matrix entries
M=zeros(NA,NA,Nt);
M_inf=zeros(NA,NA,Nt);
KK=3.5; %bound useful to erf
%--------------- main diagonal ---------------%
for i=1:NA
    for h=1:NA
        % search of integrand function zeros
        if ((sqrt(6)*(Abar(i)-A(h+1)+dt*log(B)*(10^-16)))/...
                (sigma*dt^(3/2)*(10^-16)^(3/2))-KK > 0) || ...
                ((sqrt(6)*(Abar(i)-A(h)+dt*log(B)*(1-0.5)))/...
                (sigma*dt^(3/2)*(1-0.5)^(3/2))+KK < 0)
            M(i,h,1)=0;
        else
            if (sqrt(6)*(Abar(i)-A(h+1)+dt*log(B)*(1-0.5)))/...
                    (sigma*dt^(3/2)*(1-0.5)^(3/2))-KK > 0
```

```
                Upper=fzero(@(s) ...
                    (sqrt(6)*(Abar(i)-A(h+1)+dt*log(B)*(s-0.5)))./...
                    (sigma*dt^(3/2)*(s-0.5).^(3/2))-KK,[0.5+10^-16,1]);
            else
                Upper=1;
            end
            if (sqrt(6)*(Abar(i)-A(h)+dt*log(B)*10^-16))/...
                    (sigma*dt^(3/2)*(10^-16)^(3/2))+KK < 0
                Lower=fzero(@(s)...
                    (sqrt(6)*(Abar(i)-A(h)+dt*log(B)*(s-0.5)))./...
                    (sigma*dt^(3/2)*(s-0.5).^(3/2))+KK,[0.5+10^-16,1]);
            else
                Lower=0.5+10^-16;
            end
            M(i,h,1)=sigma*B*dt/4/sqrt(2*pi)*quad(@(s) ...
                exp(-((2*r+sigma^2)/(2*sqrt(2)*sigma))^2*dt*(s-0.5))./...
                sqrt(dt*(s-0.5)).*...
                (erf((sqrt(6)*(Abar(i)-A(h)+dt*log(B)*(s-0.5)))./...
                (sigma*dt^(3/2)*(s-0.5).^(3/2)))-...
                erf((sqrt(6)*(Abar(i)-A(h+1)+dt*log(B)*(s-0.5)))./...
                (sigma*dt^(3/2)*(s-0.5).^(3/2)))),Lower,Upper,prec);
        end
    end
end
disp('diagonal block: built!')
%--------------- secondary diagonals ---------------%
for ell=1:Nt-1
    for i=1:NA
        for h=1:NA
            % search of integrand function zeros
            if ((sqrt(6)*(Abar(i)-A(h+1)+dt*log(B)*(ell-0.5)))/...
                    (sigma*dt^(3/2)*(ell-0.5)^(3/2))-KK > 0) || ...
                    ((sqrt(6)*(Abar(i)-A(h)+dt*log(B)*(1+ell-0.5)))/...
                    (sigma*dt^(3/2)*(1+ell-0.5)^(3/2))+KK < 0)
                M(i,h,ell+1)=0;
            else
                if (sqrt(6)*(Abar(i)-A(h+1)+dt*log(B)*(1+ell-0.5)))/...
                        (sigma*dt^(3/2)*(1+ell-0.5)^(3/2))-KK > 0
                    Upper=fzero(@(s) ...
                        (sqrt(6)*(Abar(i)-A(h+1)+dt*log(B)*(s+ell-0.5)))./...
                        (sigma*dt^(3/2)*(s+ell-0.5).^(3/2))-KK,[0,1]);
                else
                    Upper=1;
                end
                if (sqrt(6)*(Abar(i)-A(h)+dt*log(B)*(ell-0.5)))/...
                        (sigma*dt^(3/2)*(ell-0.5)^(3/2))+KK < 0
                    Lower=fzero(@(s)...
                        (sqrt(6)*(Abar(i)-A(h)+dt*log(B)*(s+ell-0.5)))./...
                        (sigma*dt^(3/2)*(s+ell-0.5).^(3/2))+KK,[0,1]);
                else
                    Lower=0;
                end
                M(i,h,ell+1)=sigma*B*dt/4/sqrt(2*pi)*quad(@(s) ...
                    exp(-((2*r+sigma^2)/(2*sqrt(2)*sigma))^2*dt*...
                    (s+ell-0.5))./sqrt(dt*(s+ell-0.5)).*...
                    (erf((sqrt(6)*(dA*(i-h+0.5)+dt*log(B)*(s+ell-0.5)))./...
                    (sigma*dt^(3/2)*(s+ell-0.5).^(3/2)))-...
                    erf((sqrt(6)*(dA*(i-h-0.5)+dt*log(B)*(s+ell-0.5)))./...
                    (sigma*dt^(3/2)*(s+ell-0.5).^(3/2)))),...
                    Lower,Upper,prec);
            end
        end
    end
end
```

```
end
disp('matrix: built!')
%%%%%%%%%%%%%%%%%%%%%%%%%%%%%%%%%%%%%%%%%%%%%%%%%%%%%%
%computation of rhs entries
Rhs=zeros(NA,1,Nt);
%————————————————————%
for j=1:Nt
    for i=1:NA
        Rhs(i,1,j)=1/(2*sigma*sqrt(2*pi))*...
            quad(@(S) ...
            exp(-((T-tbar(j))^2*(2*r+sigma^2)^2+4*log(B./S).^2+...
            4*sigma^2*(T-tbar(j))*log(T-tbar(j)))/...
            (8*sigma^2*(T-tbar(j)))+...
            ((2*r-3*sigma^2)/(2*sigma^2)+(T-tbar(j))/(2*T))*log(S)-...
            (2*r-sigma^2)/(2*sigma^2)*log(B)).*...
            (-S.^((T+tbar(j))/(2*T)).*...
            erfc(sqrt(1.5)*(2*Abar(i)+2*(T-tbar(j))*log(B)-2*T*log(S)-...
            (T-tbar(j))*log(B./S))/(sigma*(T-tbar(j))^(3/2)))+...
            B^((T-tbar(j))/(2*T))*exp(Abar(i)/T+(sigma^2*(T-tbar(j))^3)/...
            (24*T^2))*(erfc((12*T*Abar(i)+sigma^2*(T-tbar(j))^3-6*T*...
            (-2*(T-tbar(j))*log(B)+2*T*log(S)+(T-tbar(j))*log(B./S)))/...
            (2*sqrt(6)*T*sigma*(T-tbar(j))^(3/2))))),0,B,prec);
    end
end
disp('rhs: built!')
%%%%%%%%%%%%%%%%%%%%%%%%%%%%%%%%%%%%%%%%%%%%%%%%%%%%%%
%linear system resolution by backward substitution
alpha(:,1,Nt)=M(:,:,1)\Rhs(:,1,Nt);
for i=Nt-1:-1:1
    alpha(:,1,i)=Rhs(:,1,i);
    for j=i+1:Nt
        alpha(:,1,i)=alpha(:,1,i)-M(:,:,j-i+1)*alpha(:,1,j);
    end
    alpha(:,1,i)=M(:,:,1)\alpha(:,1,i);
end
disp('BIE solution: built!')


%%%%%%%%%%%%%%%%%%%%%%%%%%%%%%%%%%%%%%%%%%%%%%%%%%%%%%
%computation of solution in domain nodes (Sstar,Astar) at time tstar
V=zeros(length(Sstar),length(Astar));
V1=zeros(length(Sstar),length(Astar));
F=zeros(NA,Nt);
for jS=1:length(Sstar)
    if Sstar(jS)>=B
        V(jS,:)=zeros(1,length(Astar));
    else
        for jA=1:length(Astar)
            %—————————— first integral ——————————%
            V1(jS,jA)=-1/(2*sigma*sqrt(2*pi))*...
            quad(@(S) ...
            exp(-((T-tstar)^2*(2*r+sigma^2)^2+4*log(Sstar(jS)./S).^2+...
            4*sigma^2*(T-tstar)*log(T-tstar))/(8*sigma^2*(T-tstar))+...
            ((2*r-3*sigma^2)/(2*sigma^2)+(T-tstar)/(2*T))*log(S)-...
            (2*r-sigma^2)/(2*sigma^2)*log(Sstar(jS))).*...
            (-S.^((T+tstar)/(2*T)).*erfc(sqrt(1.5)*(2*Astar(jA)+...
            2*(T-tstar)*log(Sstar(jS))-2*T*log(S)-...
            (T-tstar)*log(Sstar(jS)./S))/(sigma*(T-tstar)^(3/2)))+...
            Sstar(jS)^((T-tstar)/(2*T))*exp(Astar(jA)/T+...
            (sigma^2*(T-tstar)^3)/(24*T^2))*erfc((12*T*Astar(jA)+...
            sigma^2*(T-tstar)^3-6*T*(-2*(T-tstar)*log(Sstar(jS))+...
            2*T*log(S)+(T-tstar)*log(Sstar(jS)./S)))/...
            (2*sqrt(6)*T*sigma*(T-tstar)^(3/2)))),0,B,prec);
            disp('post-pro: first integral!')
```

```
%————————— sum ————————%
         for  kt=floor(tstar/dt)+1:Nt
              Lower=max(tstar,t(kt));
              Upper=t(kt+1);
              for  hA=1:NA
                   F(hA,kt)=(sigma*B)/(4*sqrt(2*pi))*...
                        quad(@(s) exp(-((s-tstar).^2*(2*r+sigma^2)^2+...
                        4*log(Sstar(jS)/B)^2+4*sigma^2*(s-tstar).*...
                        log(s-tstar))./(8*(s-tstar)*sigma^2)+...
                        ((2*r-sigma^2)/(2*sigma^2))*log(B/Sstar(jS))).*...
                        (erf(sqrt(1.5)*(2*(A(hA+1)-Astar(jA))-...
                        2*(s-tstar)*log(Sstar(jS))+(s-tstar)*...
                        log(Sstar(jS)/B))./((s-tstar).^(1.5)*sigma))-...
                        erf(sqrt(1.5)*(2*(A(hA)-Astar(jA))-2*(s-tstar)*...
                        log(Sstar(jS))+(s-tstar)*log(Sstar(jS)/B))./...
                        ((s-tstar).^(1.5)*sigma))),Lower,Upper,prec);
                   V(jS,jA)=V(jS,jA)+alpha(hA,1,kt)*F(hA,kt);
              end
         end
         V(jS,jA)=V(jS,jA)+V1(jS,jA);
    end
  end
end
toc

dataset({[Sstar',V] 'S','V'})
%plot(Sstar',V) %graph of option values as function of asset values
```

Author Contributions: The Authors equally contributed to this work.

Acknowledgments: The dissemination of this research was supported by the INdAM-GNCS group (Italian National Group of Scientific Computing).

Conflicts of Interest: The authors declare no conflicts of interest.

References

1. Hull, J. *Options, Futures, and Other Derivatives*, 8th ed.; Pearson-Prentice Hall: Upper Saddle River, NJ, USA, 2011.

2. Boyle, P.; Potapchik, A. Prices and sensitivities of Asian options: A survey. *Insur. Math. Econ.* **2008**, *42*, 189–211. [CrossRef]

3. Cao, G.; Wang, Y. Risk-neutral pricing for geometric average Asian options with floating strike. *J. Chin. Acad. Sci.* **2015**, *32*, 13–17.

4. Aimi, A.; Guardasoni, C. Collocation Boundary Element Method for the pricing of Geometric Asian Options. *Eng. Anal. Bound. Elem.* **2017**, *92*, 90–100. [CrossRef]

5. Aimi, A.; Diazzi, L.; Guardasoni, C. Numerical Pricing of Geometric Asian Options with Barriers. *Math. Methods Appl. Sci.* **2018**, submitted.

6. Guardasoni, C.; Sanfelici, S. A Boundary Element approach to barrier option pricing in Black-Scholes framework. *Int. J. Comput. Math.* **2016**, *93*, 696–722. [CrossRef]

7. Guardasoni, C. Semi-Analytical method for the pricing of Barrier Options in case of time-dependent parameters (with Matlab codes). *Commun. Appl. Ind. Math.* **2018**, *9*, 42–67. [CrossRef]

8. Ballestra, L.; Pacelli, G. A boundary element method to price time-dependent double barrier options. *Appl. Math. Comput.* **2011**, *218*, 4192–4210. [CrossRef]

9. Wilmott, P. *On Quantitative Finance*; John Wiley and Sons: Hoboken, NJ, USA, 2000.

10. Barucci, E.; Polidoro, S.; Vespri, V. Some results on partial differential equations and Asian options. *Math. Model. Methods Appl. Sci.* **2001**, *11*, 475–497. [CrossRef]

11. Polidoro, S. Uniqueness and representation theorems for solutions of Kolmogorov-Fokker-Planck equations. *Rendiconti di Matematica e delle sue Applicazioni* **1995**, *15*, 535–560.

12. Hörmander, L. Hypoelliptic second order differential equations. *Acta Math.* **1967**, *119*, 147–171. [CrossRef]

13.   Aimi, A.; Diligenti, M.; Guardasoni, C. Energetic BEM for the numerical solution of 2D hard scattering problems of damped waves by open arcs. *Springer INdAM Series* **2018**, accepted.

14.   Hugger, J. A fixed strike Asian option and comments on its numerical solution. *ANZIAM J.* **2004**, *45*, C215–C231. [CrossRef]

15.   Hugger, J. Wellposedness of the boundary value formulation of a fixed strike Asian option. *J. Comput. Appl. Math.* **2006**, *185*, 460–481. [CrossRef]

16.   Kangro, R. *Lecture Notes on Computational Finance*; Technical Report; University of Tartu: Tartu, Estonia, 2011.

![axioms logo] **axioms**

MDPI

Article

Efficient Implementation of ADER Discontinuous Galerkin Schemes for a Scalable Hyperbolic PDE Engine

Michael Dumbser [1,*] , Francesco Fambri [1] , Maurizio Tavelli [1] , Michael Bader [2] and Tobias Weinzierl [3]

[1] Department of Civil, Environmental and Mechanical Engineering, University of Trento, I-38123 Trento, Italy; francesco.fambri@unitn.it (F.F.); m.tavelli@unitn.it (M.T.)
[2] Department of Informatics, Technical University of Munich, D-85748 Munich, Germany; bader@in.tum.de
[3] Department of Computer Science, University of Durham, Durham DH1 3LE, UK; tobias.weinzierl@durham.ac.uk
* Correspondence: michael.dumbser@unitn.it

Received: 21 May 2018; Accepted: 22 August 2018; Published: 1 September 2018

Abstract: In this paper we discuss a new and very efficient implementation of high order accurate arbitrary high order schemes using derivatives discontinuous Galerkin (ADER-DG) finite element schemes on modern massively parallel supercomputers. The numerical methods apply to a very broad class of nonlinear systems of hyperbolic partial differential equations. ADER-DG schemes are by construction communication-avoiding and cache-blocking, and are furthermore very well-suited for vectorization, and so they appear to be a good candidate for the future generation of exascale supercomputers. We introduce the numerical algorithm and show some applications to a set of hyperbolic equations with increasing levels of complexity, ranging from the compressible Euler equations over the equations of linear elasticity and the unified Godunov-Peshkov-Romenski (GPR) model of continuum mechanics to general relativistic magnetohydrodynamics (GRMHD) and the Einstein field equations of general relativity. We present strong scaling results of the new ADER-DG schemes up to 180,000 CPU cores. To our knowledge, these are the largest runs ever carried out with high order ADER-DG schemes for nonlinear hyperbolic PDE systems. We also provide a detailed performance comparison with traditional Runge-Kutta DG schemes.

Keywords: hyperbolic partial differential equations; high order discontinuous Galerkin finite element schemes; shock waves and discontinuities; vectorization and parallelization; high performance computing

---

1. Introduction

Hyperbolic partial differential equations are omnipresent in the mathematical description of time-dependent processes in fluid and solid mechanics, in engineering and geophysics, as well as in plasma physics, and even in general relativity. Among the most widespread applications nowadays are (i) computational fluid mechanics in mechanical and aerospace engineering, in particular compressible gas dynamics at high Mach numbers; (ii) geophysical and environmental free surface flows in oceans, rivers and lakes, describing natural hazards such as tsunami wave propagation, landslides, storm surges and floods; (iii) seismic, acoustic and electromagnetic wave propagation processes in the time domain are described by systems of hyperbolic partial differential equations, namely the equations of linear elasticity, the acoustic wave equation and the well-known Maxwell equations; (iv) high energy density plasma flows in nuclear fusion reactors as well as astrophysical plasma flows in the solar system and the universe, using either the Newtonian limit or the complete equations in full general relativity; (v) the Einstein field equations of general relativity, which govern the dynamics of the spacetime around black holes and neutron stars, can be written under the form of a nonlinear system of hyperbolic partial differential equations.

The main challenge of nonlinear hyperbolic PDE arises from the fact that they can contain at the same time smooth solutions (like sound waves) as well as small scale structures (e.g., turbulent vortices), but they can also develop discontinuous solutions (shock waves) after finite time, even when starting from perfectly smooth initial data. These discontinuities were first discovered by Bernhard Riemann in his ground breaking work on the propagation of waves of finite amplitude in air [1,2], where the term *finite* should actually be understood in the sense of *large*, rather than simple sound waves of *infinitesimal* strength that have been considered in the times before Riemann. In the abstract of his work, Riemann stated that his discovery of the shock waves might probably not be of practical use for applied and experimental science, but should be mainly understood as a contribution to the theory of nonlinear partial differential equations. Several decades later, shock waves were also observed experimentally, thus confirming the new and groundbreaking mathematical concept of Riemann.

The connection between symmetries and conservation laws were established in the work of Emmy Noether [3] at the beginning of the 20th century, while the first methods for the *numerical solution* of hyperbolic conservation laws go back to famous mathematicians such as Courant and Friedrichs and co-workers [4–7]. The connection between hyperbolic conservation laws, symmetric hyperbolic systems in the sense of Friedrichs [8] and thermodynamics was established for the first time by Godunov in 1961 [9], and was rediscovered again by Friedrichs and Lax in 1971 [10]. Within this theoretical framework of symmetric hyperbolic and thermodynamically compatible (SHTC) systems, established by Godunov and Romenski [11,12], it is possible to write down the Euler equations of compressible gas dynamics, the magnetohydrodynamics (MHD) equations [13], the equations of nonlinear elasticity [14], as well as a rather wide class of nonlinear hyperbolic conservation laws [15] with very interesting mathematical properties and structure. Very recently, even a novel and unified formulation of continuum physics, including solid and fluid mechanics only as two particular cases of a more general model, have been cast into the form of a single SHTC system [16–19]. In the 1940ies and 1950ies, major steps forward in numerical methods for hyperbolic PDE have been made in the ground-breaking contributions of von Neumann and Richtmyer [20] and Godunov [21]. While the former introduce an artificial viscosity to stabilize the numerical scheme in the presence of discontinuities, the latter constructs his scheme starting from the most elementary problem in hyperbolic conservation laws for which an exact solution is still available, the so-called Riemann problem. The Riemann problem consists in a particular Cauchy problem where the initial data consist of two piecewise constant states, separated by a discontinuity. In the absence of source terms, its solution is self-similar. While provably robust, these schemes are only first order accurate in space and time and thus only applicable to flows with shock waves, but not to those also involving smooth sound waves and turbulent small scale flow structures. In his paper [21], Godunov has also proven that any linear numerical scheme that is required to be monotone can be at most of order one, which is the well-known Godunov barrier theorem. The main goal in the past decades was to find ways how to circumvent it, since it only applies to linear schemes. The first successful nonlinear monotone and higher order accurate schemes were the method of Kolgan [22] and the schemes of van Leer [23,24]. Subsequently, many other higher order nonlinear schemes have been proposed, such as the ENO [25] and WENO schemes [26], and there is a rapidly growing literature on the subject. In this paper we mainly focus on a rather recent family of schemes, which is of the discontinuous finite element type, namely the so-called discontinuous Galerkin (DG) finite element methods, which were systematically introduced for hyperbolic conservation laws in a well-known series of papers by Cockburn and Shu and collaborators [27–31]. For a review on high order DG methods and WENO schemes, the reader is referred to References [32,33]. In this paper we use a particular variant of the DG scheme that is called the ADER discontinuous Galerkin scheme [34–38], where ADER stands for arbitrary high order schemes using derivatives, first developed by Toro et al. in the context of high order finite volume schemes [39–42]. In comparison to traditional *semi-discrete* DG schemes, which mainly use Runge-Kutta time integration, ADER-DG methods are *fully-discrete* and are based on a predictor-corrector approach that allows the achievement of a naturally cache-blocking and communication-avoiding scheme, which reduces the amount of necessary MPI communications to a minimum. These properties make the method well suitable for high performance computing (HPC).

2. High Order ADER Discontinuous Galerkin Finite Element Schemes

In this paper we consider hyperbolic PDEs with non-conservative products and algebraic source terms of the form (see also References [34,35])

$$\frac{\partial \mathbf{Q}}{\partial t} + \nabla \cdot \mathbf{F}(\mathbf{Q}) + \mathbf{B}(\mathbf{Q}) \cdot \nabla \mathbf{Q} = \mathbf{S}(\mathbf{Q}), \tag{1}$$

where $t \in \mathbb{R}_0^+$ is the time, $\mathbf{x} \in \Omega \subset \mathbb{R}^d$ is the spatial position vector in d space dimensions, $\mathbf{Q} \in \Omega_Q \subset \mathbb{R}^m$ is the state vector, $\mathbf{F}(\mathbf{Q})$ is the nonlinear flux tensor, $\mathbf{B}(\mathbf{Q}) \cdot \nabla \mathbf{Q}$ is a non-conservative product and $\mathbf{S}(\mathbf{Q})$ is a purely algebraic source term. Introducing the system matrix $\mathbf{A}(\mathbf{Q}) = \partial \mathbf{F} / \partial \mathbf{Q} + \mathbf{B}(\mathbf{Q})$ the above system can also be written in quasi-linear form as

$$\frac{\partial \mathbf{Q}}{\partial t} + \mathbf{A}(\mathbf{Q}) \cdot \nabla \mathbf{Q} = \mathbf{S}(\mathbf{Q}). \tag{2}$$

The system is hyperbolic if for all $\mathbf{n} \neq 0$ and for all $\mathbf{Q} \in \Omega_Q$, the matrix $\mathbf{A}(\mathbf{Q}) \cdot \mathbf{n}$ has m real eigenvalues and a full set of m linearly independent right eigenvectors. The system in Equation (1) is provided with an initial condition $\mathbf{Q}(\mathbf{x}, 0) = \mathbf{Q}_0(\mathbf{x})$ and appropriate boundary conditions on $\partial \Omega$. In some parts of the paper we will also make use of the vector of primitive (physical) variables denoted by $\mathbf{V} = \mathbf{V}(\mathbf{Q})$. For very complex PDE systems, such as the general-relativistic MHD equations, it may be much easier to express the flux tensor $\mathbf{F}$ in terms of $\mathbf{V}$ rather than in terms of $\mathbf{Q}$, however the evaluation of $\mathbf{V} = \mathbf{V}(\mathbf{Q})$ can become very complicated.

2.1. Unlimited ADER-DG Scheme and Riemann Solvers

We cover the computational domain Ω with a set of non-overlapping Cartesian control volumes in space $\Omega_i = [x_i - \frac{1}{2}\Delta x_i, x_i + \frac{1}{2}\Delta x_i] \times [y_i - \frac{1}{2}\Delta y_i, y_i + \frac{1}{2}\Delta y_i] \times [z_i - \frac{1}{2}\Delta z_i, z_i + \frac{1}{2}\Delta z_i]$. Here, $\mathbf{x}_i = (x_i, y_i, z_i)$ denotes the barycenter of cell Ω_i and $\Delta \mathbf{x}_i = (\Delta x_i, \Delta y_i, \Delta z_i)$ is the mesh spacing associated with Ω_i in each spatial dimension. The domain $\Omega = \bigcup \Omega_i$ is the union of all spatial control volumes. A key ingredient of the ExaHyPE engine http://exahype.eu is a cell-by-cell adaptive mesh refinement (AMR), which is built upon the space-tree implementation of Peano [43,44]. For further details about cell-by-cell AMR, see Reference [45]. High order finite volume and finite difference schemes for AMR can be found, e.g., in References [46–52]. For high order AMR with better than second order accurate finite volume and DG schemes in combination with time-accurate local time stepping (LTS), the reader is referred to References [37,53–57]. Since the main focus of this paper is not on AMR, at this point we can only give a very brief summary of existing AMR methods and codes for hyperbolic PDE, without pretending to be complete. The starting point of adaptive mesh refinement for hyperbolic conservation laws was of course the pioneering work of Berger et al. [58–60], who were the first to introduce a patched-based block-structured AMR method. Further developments are reported in References [61–63] based on the second order accurate wave-propagation algorithm of LeVeque. We also would like to draw the attention of the reader to the works of Quirk [64], Coirier and Powell [65] and Deiterding et al. [66,67]. For computational astrophysics, relevant AMR techniques have been documented, e.g., in References [68–76], including the RAMSES, PLUTO, NIRVANA, AMRVAC and BHAC codes. For a recent and more complete survey of high level AMR codes, the reader is referred to the review paper Reference [77].

In the following, the discrete solution of the PDE system in Equation (1) is denoted by $\mathbf{u}_h$ and is defined in terms of tensor products of piecewise polynomials of degree N in each spatial direction.

The discrete solution space is denoted by $\mathcal{U}_h$ in the following. Since we adopt a discontinuous Galerkin (DG) finite element method, the numerical solution $\mathbf{u}_h$ is allowed to *jump* across element interfaces, as in the context of finite volume schemes. Within each spatial control volume Ω_i the discrete solution $\mathbf{u}_h$ restricted to that control volume is written at time t^n in terms of some nodal spatial basis functions $\Phi_l(\mathbf{x})$ and some unknown degrees of freedom $\hat{\mathbf{u}}_{i,l}^n$:

$$\mathbf{u}_h(\mathbf{x}, t^n)|_{\Omega_i} = \sum_l \hat{\mathbf{u}}_{i,l} \Phi_l(\mathbf{x}) := \hat{\mathbf{u}}_{i,l}^n \Phi_l(\mathbf{x}), \tag{3}$$

where $l = (l_1, l_2, l_3)$ is a multi-index and the spatial basis functions $\Phi_l(\mathbf{x}) = \varphi_{l_1}(\xi)\varphi_{l_2}(\eta)\varphi_{l_3}(\zeta)$ are generated via tensor products of one-dimensional nodal basis functions $\varphi_k(\xi)$ on the reference interval $[0, 1]$. The transformation from physical coordinates $\mathbf{x} \in \Omega_i$ to reference coordinates $\boldsymbol{\xi} = (\xi, \eta, \zeta) \in [0, 1]^d$ is given by the linear mapping $\mathbf{x} = \mathbf{x}_i - \frac{1}{2}\Delta\mathbf{x}_i + (\xi\Delta x_i, \eta\Delta y_i, \zeta\Delta z_i)^T$. For the one-dimensional basis functions $\varphi_k(\xi)$ we use the Lagrange interpolation polynomials passing through the Gauss-Legendre quadrature nodes ξ_j of an $N + 1$ point Gauss quadrature formula. Therefore, the nodal basis functions satisfy the interpolation property $\varphi_k(\xi_j) = \delta_{kj}$, where δ_{kj} is the usual Kronecker symbol, and the resulting basis is *orthogonal*. Furthermore, due to this particular choice of a *nodal* tensor-product basis, the entire scheme can be written in a dimension-by-dimension fashion, where all integral operators can be decomposed into a sequence of one-dimensional operators acting only on the $N + 1$ degrees of freedom in the respective dimension. For details on multi-dimensional quadrature, see the well-known book of Stroud [78].

In order to derive the ADER-DG method, we first multiply the governing PDE system in Equation (1) with a test function $\Phi_k \in \mathcal{U}_h$ and integrate over the space-time control volume $\Omega_i \times [t^n; t^{n+1}]$. This leads to

$$\int_{t^n}^{t^{n+1}} \int_{\Omega_i} \Phi_k \frac{\partial \mathbf{Q}}{\partial t} \, d\mathbf{x} \, dt + \int_{t^n}^{t^{n+1}} \int_{\Omega_i} \Phi_k \left(\nabla \cdot \mathbf{F}(\mathbf{Q}) + \mathbf{B}(\mathbf{Q}) \cdot \nabla \mathbf{Q} \right) d\mathbf{x} \, dt = \int_{t^n}^{t^{n+1}} \int_{\Omega_i} \Phi_k \mathbf{S}(\mathbf{Q}) \, d\mathbf{x} \, dt, \tag{4}$$

with $d\mathbf{x} = dx \, dy \, dz$. As already mentioned before, the discrete solution is allowed to jump across element interfaces, which means that the resulting jump terms have to be taken properly into account. In our scheme this is achieved via numerical flux functions (approximate Riemann solvers) and via the path-conservative approach that was developed by Castro and Parés in the finite volume context [79,80]. It has later been also extended to the discontinuous Galerkin finite element framework in References [35,81,82]. In classical Runge-Kutta DG schemes, only a weak form in space of the PDE is obtained, while time is still kept continuous, thus reducing the problem to a nonlinear system of ODE, which is subsequently integrated with standard ODE solvers in time. However, this requires MPI communication in each Runge-Kutta stage. Furthermore, each Runge-Kutta stage requires accesses to the entire discrete solution in memory. In the ADER-DG framework, a completely different paradigm is used. Here, higher order in time is achieved with the use of an element-local space-time predictor, denoted by $\mathbf{q}_h(\mathbf{x}, t)$ in the following, and which will be discussed in more detail later. Using Equation (3), integrating the first term by parts in time and integrating the flux divergence term by parts in space, taking into account the jumps between elements and making use of this local space-time predictor solution $\mathbf{q}_h$ instead of $\mathbf{Q}$, the weak formulation of Equation (4) can be rewritten as

$$\left(\int_{\Omega_i} \Phi_k \Phi_l \, d\mathbf{x} \right) \left(\hat{\mathbf{u}}_{i,l}^{n+1} - \hat{\mathbf{u}}_{i,l}^n \right) + \int_{t^n}^{t^{n+1}} \int_{\partial\Omega_i} \Phi_k \mathcal{D}^- \left(\mathbf{q}_h^-, \mathbf{q}_h^+ \right) \cdot \mathbf{n} \, dS \, dt - \int_{t^n}^{t^{n+1}} \int_{\Omega_i^\circ} \left(\nabla\Phi_k \cdot \mathbf{F}(\mathbf{q}_h) \right) d\mathbf{x} \, dt +$$
$$+ \int_{t^n}^{t^{n+1}} \int_{\Omega_i^\circ} \Phi_k \left(\mathbf{B}(\mathbf{q}_h) \cdot \nabla \mathbf{q}_h \right) d\mathbf{x} \, dt = \int_{t^n}^{t^{n+1}} \int_{\Omega_i} \Phi_k \mathbf{S}(\mathbf{q}_h) \, d\mathbf{x} \, dt, \tag{5}$$

where the first integral leads to the element mass matrix, which is diagonal since our basis is orthogonal. The boundary integral contains the approximate Riemann solver and accounts for the jumps across element interfaces, also in the presence of non-conservative products. The third and fourth integral account for the smooth part of the flux and the non-conservative product, while the right hand side takes into account the presence of the algebraic source term. According to the framework of path-conservative schemes [35,79,80,82], the jump terms are defined via a path-integral in phase space between the boundary extrapolated states at the left $\mathbf{q}_h^-$ and at the right $\mathbf{q}_h^+$ of the interface as follows:

$$\mathcal{D}^-\left(\mathbf{q}_h^-,\mathbf{q}_h^+\right)\cdot\mathbf{n} = \frac{1}{2}\left(\mathbf{F}(\mathbf{q}_h^+)+\mathbf{F}(\mathbf{q}_h^-)\right)\cdot\mathbf{n} + \frac{1}{2}\left(\int_0^1 \mathbf{B}(\boldsymbol{\psi})\cdot\mathbf{n}\,ds - \boldsymbol{\Theta}\right)\left(\mathbf{q}_h^+ - \mathbf{q}_h^-\right), \tag{6}$$

with $\mathbf{B}\cdot\mathbf{n} = \mathbf{B}_1 n_1 + \mathbf{B}_2 n_2 + \mathbf{B}_3 n_3$. Throughout this paper, we use the simple straight-line segment path

$$\boldsymbol{\psi} = \boldsymbol{\psi}(\mathbf{q}_h^-,\mathbf{q}_h^+,s) = \mathbf{q}_h^- + s\left(\mathbf{q}_h^+ - \mathbf{q}_h^-\right), \qquad 0 \le s \le 1. \tag{7}$$

In order to achieve exactly well-balanced schemes for certain classes of hyperbolic equations with non-conservative products and source terms, the segment path is not sufficient and a more elaborate choice of the path becomes necessary, see e.g., References [83–86]. In Equation (6) above the symbol $\boldsymbol{\Theta} > 0$ denotes an appropriate numerical dissipation matrix. Following References [35,87,88], the path integral that appears in Equation (6) can be simply evaluated via some sufficiently accurate numerical quadrature formulae. We typically use a three-point Gauss-Legendre rule in order to approximate the path-integral. For a simple path-conservative Rusanov-type method [35,89], the numerical dissipation matrix reads

$$\boldsymbol{\Theta}_{\text{Rus}} = s_{\max}\mathbf{I}, \qquad \text{with} \qquad s_{\max} = \max\left(|\lambda(\mathbf{q}_h^-)|,|\lambda(\mathbf{q}_h^+)|\right), \tag{8}$$

where $\mathbf{I}$ denotes the identity matrix and $s_{\max}$ is the maximum wave speed (eigenvalue λ of matrix $\mathbf{A}\cdot\mathbf{n}$) at the element interface. In order to reduce numerical dissipation, one can use better Riemann solvers, such as the Osher-type schemes proposed in References [88,90], or the recent extension of the original HLLEM method of Einfeldt and Munz [91] to general conservative and non-conservative hyperbolic systems recently put forward in Reference [92]. The choice of the approximate Riemann solver and therefore of the viscosity matrix $\boldsymbol{\Theta}$ completes the numerical scheme in Equation (5). In the next subsection, we shortly discuss the computation of the element-local space-time predictor $\mathbf{q}_h$, which is a key ingredient of our high order accurate and communication-avoiding ADER-DG schemes.

2.2. Space-Time Predictor and Suitable Initial Guess

As already mentioned previously, the element-local space–time predictor is an important *key feature* of ADER-DG schemes and is briefly discussed in this section. The computation of the predictor solution $\mathbf{q}_h(\mathbf{x},t)$ is based on a weak formulation of the governing PDE system in space–time and was first introduced in References [34,93,94]. Starting from the known solution $\mathbf{u}_h(\mathbf{x},t^n)$ at time t^n and following the terminology of Harten et al. [95], we solve a so-called Cauchy problem *in the small*, i.e., without considering the interaction with the neighbor elements. In the ENO scheme of Harten et al. [95] and in the original ADER approach of Toro and Titarev [40–42] the strong differential form of the PDE was used, together with a combination of Taylor series expansions and the so-called Cauchy-Kovalewskaya procedure. The latter is very cumbersome, or becomes even unfeasible for very complicated nonlinear hyperbolic PDE systems, since it requires a lot of analytic manipulations of the governing PDE system in order to replace time derivatives with known space derivatives at time t^n. This is achieved by successively differentiating the governing PDE system with respect to space and time and inserting the resulting terms into the Taylor series. For an explicit example of the Cauchy–Kovalewskaya procedure applied to the three-dimensional Euler equations of compressible gas dynamics and the MHD equations, see References [96,97]. Instead, the local space–time discontinuous Galerkin predictor, introduced in References [34,93,94], requires only pointwise evaluations of the fluxes, source terms and non-conservative products, for element Ω_i the predictor solution $\mathbf{q}_h$ is now expanded in terms of a local space–time basis

$$\mathbf{q}_h(\mathbf{x}, t)|_{\Omega_i^{st}} = \sum_l \theta_l(\mathbf{x}, t) \hat{\mathbf{q}}_l^i := \theta_l(\mathbf{x}, t) \hat{\mathbf{q}}_l^i, \tag{9}$$

with the multi-index $l = (l_0, l_1, l_2, l_3)$ and where the space–time basis functions $\theta_l(\mathbf{x}, t) = \varphi_{l_0}(\tau)\varphi_{l_1}(\xi)\varphi_{l_2}(\eta)\varphi_{l_3}(\zeta)$ are again generated from the same one-dimensional nodal basis functions $\varphi_k(\xi)$ as before, i.e., the Lagrange interpolation polynomials of degree N passing through $N + 1$ Gauss-Legendre quadrature nodes. The spatial mapping $\mathbf{x} = \mathbf{x}(\boldsymbol{\xi})$ is also the same as before and the coordinate time is mapped to the reference time $\tau \in [0, 1]$ via $t = t^n + \tau\Delta t$. Multiplication of the PDE system in Equation (1) with a test function θ_k and integration over the space–time control volume $\Omega_i^{st} = \Omega_i \times [t^n, t^{n+1}]$ yields the following weak form of the governing PDE, which is *different* from Equation (4), because now the test and basis functions are both time dependent:

$$\int_{t^n}^{t^{n+1}}\int_{\Omega_i} \theta_k(\mathbf{x}, t)\frac{\partial \mathbf{q}_h}{\partial t}\, d\mathbf{x}\, dt + \int_{t^n}^{t^{n+1}}\int_{\Omega_i} \theta_k(\mathbf{x}, t)\left(\nabla \cdot \mathbf{F}(\mathbf{Q}) + \mathbf{B}(\mathbf{q}_h) \cdot \nabla\mathbf{q}_h\right) d\mathbf{x}\, dt = \int_{t^n}^{t^{n+1}}\int_{\Omega_i} \theta_k(\mathbf{x}, t)\mathbf{S}(\mathbf{q}_h)\, d\mathbf{x}\, dt. \tag{10}$$

Since we are only interested in an element local predictor solution, i.e., without considering interactions with the neighbor elements we do not yet take into account the jumps in $\mathbf{q}_h$ across the element interfaces, because this will be done in the final corrector step of the ADER-DG scheme in Equation (5). Instead, we introduce the known discrete solution $u_h(\mathbf{x}, t^n)$ at time t^n. For this purpose, the first term is integrated by parts in time. This leads to

$$\int_{\Omega_i} \theta_k(\mathbf{x}, t^{n+1})\mathbf{q}_h(\mathbf{x}, t^{n+1})\, d\mathbf{x} - \int_{t^n}^{t^{n+1}}\int_{\Omega_i} \frac{\partial}{\partial t}\theta_k(\mathbf{x}, t)\mathbf{q}_h(\mathbf{x}, t)\, d\mathbf{x}\, dt - \int_{\Omega_i} \theta_k(\mathbf{x}, t^n)\mathbf{u}_h(\mathbf{x}, t^n)\, d\mathbf{x} = $$
$$\int_{t^n}^{t^{n+1}}\int_{\Omega_i^{\circ}} \theta_k(\mathbf{x}, t)\nabla \cdot \mathbf{F}(\mathbf{q}_h)\, d\mathbf{x}\, dt + \int_{t^n}^{t^{n+1}}\int_{\Omega_i^{\circ}} \theta_k(\mathbf{x}, t)\left(\mathbf{S}(\mathbf{q}_h) - \mathbf{B}(\mathbf{q}_h) \cdot \nabla\mathbf{q}_h\right) d\mathbf{x}\, dt. \tag{11}$$

Using the local space–time ansatz (9), Equation (11) becomes an element-local nonlinear system for the unknown degrees of freedom $\hat{\mathbf{q}}_{i,l}$ of the space–time polynomials $\mathbf{q}_h$. The solution of Equation (11) can be found via a simple and fast converging fixed point iteration (a discrete Picard iteration) as detailed e.g., in References [34,98]. For linear homogeneous systems, the discrete Picard iteration converges in a finite number of at most $N + 1$ steps, since the involved iteration matrix is nilpotent, see Reference [99].

However, we emphasize that the choice of an appropriate *initial guess* $\mathbf{q}_h^0(\mathbf{x}, t)$ for $\mathbf{q}_h(\mathbf{x}, t)$ is of fundamental importance to obtain a faster convergence and thus a computationally more efficient scheme. For this purpose, one can either use an extrapolation of $\mathbf{q}_h$ from the previous time interval $[t^{n-1}, t^n]$, as suggested e.g., in Reference [100], or one can employ a second-order accurate MUSCL-Hancock-type approach, as proposed in Reference [98], which is based on discrete derivatives computed at time t^n. The initial guess is most conveniently written in terms of a Taylor series expansion of the solution in time, where then suitable approximations of the time derivatives are computed. In the following we introduce the operator

$$\mathcal{L}(\mathbf{u}_h(\mathbf{x}, t^n)) = \mathbf{S}(\mathbf{u}_h(\mathbf{x}, t^n)) - \nabla \cdot \mathbf{F}(\mathbf{u}_h(\mathbf{x}, t^n)) - \mathbf{B}(\mathbf{u}_h(\mathbf{x}, t^n)) \cdot \nabla\mathbf{u}_h(\mathbf{x}, t^n), \tag{12}$$

which is an approximation of the time derivative of the solution. The second-order accurate MUSCL-type initial guess [98] then reads

$$\mathbf{q}_h^0(\mathbf{x}, t) = \mathbf{u}_h(\mathbf{x}, t^n) + (t - t^n)\mathcal{L}(\mathbf{u}_h(\mathbf{x}, t^n)), \tag{13}$$

while a third-order accurate initial guess for $\mathbf{q}_h(\mathbf{x}, t)$ is given by

$$\mathbf{q}_h^0(\mathbf{x}, t) = \mathbf{u}_h(\mathbf{x}, t^n) + (t - t^n) \, \mathbf{k}_1 + \frac{1}{2} (t - t^n)^2 \frac{(\mathbf{k}_2 - \mathbf{k}_1)}{\Delta t}. \tag{14}$$

Here, we have used the abbreviations $\mathbf{k}_1 := \mathcal{L} (\mathbf{u}_h(\mathbf{x}, t^n))$ and $\mathbf{k}_2 := \mathcal{L} (\mathbf{u}_h(\mathbf{x}, t^n) + \Delta t \mathbf{k}_1)$. For an initial guess of even higher order of accuracy, it is possible to use the so-called continuous extension Runge-Kutta (CERK) schemes of Owren and Zennaro [101]; see also Reference [102] for the use of CERK time integrators in the context of high-order discontinuous Galerkin finite element methods. If an initial guess with polynomial degree $N - 1$ in time is chosen, it is sufficient to use *one single* Picard iteration to solve Equation (11) to the desired accuracy.

At this point, we make some comments about a suitable data-layout for high order ADER-DG schemes. In order to compute the discrete derivative operators needed in the predictor Equation (11), especially for the computation of the discrete gradient $\nabla \mathbf{q}_h$, it is very convenient to use an array-of-struct (AoS) data structure. In this way, the first or fastest-running unit-stride index is the one associated with the m quantities contained in the vector $\mathbf{Q}$, while the other indices are associated with the space–time degrees of freedom, i.e., we arrange the data contained in the set of degrees of freedom $\hat{\mathbf{q}}_l^i$ as $\hat{\mathbf{q}}_{v, l_1, l_2, l_3, l_0}^i$, with $1 \leq v \leq m$ and $1 \leq l_k \leq N + 1$. The discrete derivatives in space and time direction can then be simply computed by the multiplication of a subset of the degrees of freedom with the transpose of a small $(N + 1) \times (N + 1)$ matrix D_{kl} from the right, which reads

$$D_{kl} = \frac{1}{h} \left(\int_0^1 \phi_k(\xi) \phi_m(\xi) d\xi \right)^{-1} \left(\int_0^1 \phi_m(\xi) \frac{\partial \phi_l(\xi)}{\partial \xi} d\xi \right), \tag{15}$$

where h is the respective spatial or temporal step size in the corresponding coordinate direction, i.e., either Δx_i, Δy_i, Δz_i or Δt. For this purpose, the optimized library for small matrix multiplications `libxsmm` can be employed on Intel machines, see References [103–105] for more details. However, the AoS data layout is *not* convenient for *vectorization* of the PDE evaluation in ADER-DG schemes, since vectorization of the fluxes, source terms and non-conservative products should preferably be done over the integration points l. For this purpose, we convert the AoS data layout *on the fly* into a struct-of-array (SoA) data layout via appropriate transposition of the data and then call the physical flux function $\mathbf{F}(\mathbf{q}_h)$ as well as the combined algebraic source term and non-conservative product contained in the expression $\mathbf{S}(\mathbf{q}_h) - \mathbf{B}(\mathbf{q}_h) \cdot \nabla \mathbf{q}_h$ simultaneously for a subset of VECTORLENGTH space–time degrees of freedom, where VECTORLENGTH is the length of the AVX registers of modern Intel Xeon CPUs, i.e., 4 for those with the old 256 bit AVX and AVX2 instruction sets (Sandy Bridge, Haswell, Broadwell) and 8 for the latest Intel Xeon Scalable CPUs with 512 bit AVX instructions (Skylake). The result of the vectorized evaluation of the PDE, which is still in SoA format, is then converted back to the AoS data layout using appropriate vectorized shuffle commands.

The element-local space–time predictor is arithmetically very intensive, but at the same time it is also by construction cache-blocking. While in traditional RKDG schemes, each Runge-Kutta stage requires touching all spatial degrees of freedom of the entire domain once per Runge-Kutta stage, in our ADER-DG approach the spatial degrees of freedom $\mathbf{u}_h$ need to be loaded only once per element and time step, and from those all space–time degrees of freedom of $\mathbf{q}_h$ are computed. Ideally, this procedure fits entirely into the L3 cache or even into the L2 cache of the CPU, at least up to a certain critical polynomial degree $N_c = N_c(m)$, which is a function of the available L3 or L2 cache size, but also of the number of quantities m to be evolved in the PDE system.

Last but not least, it is important to note that it is possible to hide the entire MPI communication that is inevitably needed on distributed memory supercomputers behind the space–time predictor. For this purpose, the predictor is first invoked on the MPI boundaries of each CPU, which then immediately sends the boundary-extrapolated data $\mathbf{q}_h^-$ and $\mathbf{q}_h^+$ to the neighbor CPUs. While the messages containing the data of these non-blocking MPI send and receive commands are sent around, each CPU can compute the space–time predictor of purely interior elements that do not need any MPI communication.

For an efficient task-based formalism used within ExaHyPE in the context of shared memory parallelism, see Reference [106]. This completes the description of the efficient implementation of the unlimited ADER-DG schemes used within the ExaHyPE engine.

2.3. A Posteriori Subcell Finite Volume Limiter

In regions where the discrete solution is smooth, there is indeed no need for using nonlinear limiters. However, in the presence of shock waves, discontinuities or strong gradients, and taking into account the fact that even a *smooth signal* may become *non-smooth* on the discrete level if it is *underresolved* on the grid, we have to supplement our high order unlimited ADER-DG scheme described above with a nonlinear limiter.

In order to build a simple, robust and accurate limiter, we follow the ideas outlined in References [36–38,107], where a novel *a posteriori* limiting strategy for ADER-DG schemes was developed, based on the ideas of the MOOD paradigm introduced in References [108–111] in the finite volume context. In a first run, the unlimited ADER-DG scheme is used and produces a so-called *candidate solution*, denoted by $\mathbf{u}_h^*(\mathbf{x}, t^{n+1})$ in the following. This candidate solution is then checked *a posteriori* against several physical and numerical detection criteria. For example, we require some relevant physical quantities of the solution to be positive (e.g., pressure and density), we require the absence of floating point errors (NaN) and we impose a relaxed discrete maximum principle (DMP) in the sense of polynomials, see Reference [36]. As soon as one of these detection criteria is not satisfied, a cell is marked as troubled zone and is scheduled for limiting.

A cell Ω_i that has been marked for limiting is now split into $(2N+1)^d$ finite volume subcells, which are denoted by $\Omega_{i,s}$. They satisfy $\Omega_i = \bigcup_s \Omega_{i,s}$. Note that this very fine division of a DG element into finite volume subcells does *not* reduce the time step of the overall ADER-DG scheme, since the CFL number of explicit DG schemes scales with $1/(2N+1)$, while the CFL number of finite volume schemes (used on the subgrid) is of the order of unity. The discrete solution in the subcells $\Omega_{i,s}$ is represented at time t^n in terms of *piecewise constant* subcell averages $\bar{\mathbf{u}}_{i,s}^n$, i.e.,

$$\bar{\mathbf{u}}_{i,s}^n = \frac{1}{|\Omega_{i,s}|} \int_{\Omega_{i,s}} \mathbf{Q}(\mathbf{x}, t^n) \, d\mathbf{x}. \tag{16}$$

These subcell averages are now evolved in time with a second or third order accurate finite volume scheme, which actually looks very similar to the previous ADER-DG scheme in Equation (5), with the difference that now the test function is unity and the spatial control volumes Ω_i are replaced by the sub-volumes $\Omega_{i,s}$:

$$\left(\bar{\mathbf{u}}_{i,s}^{n+1} - \bar{\mathbf{u}}_{i,s}^n \right) + \int_{t^n}^{t^{n+1}} \int_{\partial \Omega_{i,s}} \mathcal{D}^- \left(\mathbf{q}_h^-, \mathbf{q}_h^+ \right) \cdot \mathbf{n} \, dS \, dt + \int_{t^n}^{t^{n+1}} \int_{\Omega_{i,s}^\circ} \left(\mathbf{B}(\mathbf{q}_h) \cdot \nabla \mathbf{q}_h \right) d\mathbf{x} \, dt = \int_{t^n}^{t^{n+1}} \int_{\Omega_{i,s}} \mathbf{S}(\mathbf{q}_h) \, d\mathbf{x} \, dt. \tag{17}$$

Here we use again a space–time predictor solution $\mathbf{q}_h$, but which is now computed from an initial condition given by a second order TVD reconstruction polynomial or from a WENO [26] or CWENO reconstruction [51,112–114] polynomial $\mathbf{w}_h(\mathbf{x}, t^n)$ computed from the cell averages $\bar{\mathbf{u}}_{i,s}^n$ via an appropriate reconstruction operator. The predictor is either computed via a standard second order MUSCL–Hancock-type strategy, or via the space–time DG approach of Equation (11), but where the initial data $\mathbf{u}_h(\mathbf{x}, t^n)$ are now replaced by $\mathbf{w}_h(\mathbf{x}, t^n)$ and the spatial control volumes Ω_i are replaced by the subcells $\Omega_{i,s}$.

Once all subcell averages $\bar{\mathbf{u}}_{i,s}^{n+1}$ inside a cell Ω_i have been computed according to Equation (17), the limited DG polynomial $\mathbf{u}_h'(\mathbf{x}, t^{n+1})$ at the next time level is obtained again via a classical constrained least squares reconstruction procedure requiring

$$\frac{1}{|\Omega_{i,s}|} \int_{\Omega_{i,s}} \mathbf{u}_h'(\mathbf{x}, t^{n+1}) \, dx = \bar{\mathbf{u}}_{i,s}^{n+1} \quad \forall \Omega_{i,s} \in \Omega_i, \quad \text{and} \quad \int_{\Omega_i} \mathbf{u}_h'(\mathbf{x}, t^{n+1}) \, dx = \sum_{\Omega_{i,s} \in \Omega_i} |\Omega_{i,s}| \bar{\mathbf{u}}^{n+1} i, s. \quad (18)$$

Here, the second relation is a constraint and means conservation at the level of the control volume Ω_i. This completes the brief description of the subcell finite volume limiter used here.

3. Some Examples of Typical PDE Systems Solved With the ExaHyPE Engine

The great advantage of ExaHyPE over many existing PDE solvers is its great flexibility and versatility for the solution of a very wide class of hyperbolic PDE systems in Equation (1). The implementation of the numerical method and the definition of the PDE system to be solved are completely independent of each other. The compute kernels are provided either as generic or as an optimized implementation for the general PDE system given by Equation (1), while the user only needs to provide particular implementations of the functions $\mathbf{F}(\mathbf{Q})$, $\mathbf{B}(\mathbf{Q})$ and $\mathbf{S}(\mathbf{Q})$. It is obviously also possible to drop terms that are not needed. This allows to solve all the PDE systems listed below in one single software package. In all numerical examples shown below, we have used a CFL condition of the type

$$\Delta t \leq \frac{\alpha}{\frac{|\lambda_{max}^x|}{\Delta x} + \frac{|\lambda_{max}^y|}{\Delta y} + \frac{|\lambda_{max}^z|}{\Delta z}}, \quad (19)$$

where Δx, Δy and Δz are the mesh spacings and $|\lambda_{max}^x|$, $|\lambda_{max}^x|$ and $|\lambda_{max}^x|$ are the maximal absolute values of the eigenvalues (wave speeds) of the matrix $\mathbf{A} \cdot \mathbf{n}$ in x, y and z direction, respectively. The coefficient $\alpha < 1/(2N+1)$ can be obtained via a numerical von Neumann stability analysis and is reported for some relevant N in Reference [34].

3.1. The Euler Equations of Compressible Gas Dynamics

The Euler equations of compressible gas dynamics are among the simplest nonlinear systems of hyperbolic conservation laws. They only involve a conservative flux $\mathbf{F}(\mathbf{Q})$ and read

$$\frac{\partial}{\partial t} \begin{pmatrix} \rho \\ \rho\mathbf{v} \\ \rho E \end{pmatrix} + \nabla \cdot \begin{pmatrix} \rho\mathbf{v} \\ \rho\mathbf{v} \otimes \mathbf{v} + p\mathbf{I} \\ \mathbf{v}(\rho E + p) \end{pmatrix} = 0. \quad (20)$$

Here, ρ is the mass density, $\mathbf{v}$ is the fluid velocity, ρE is the total energy density and p is the fluid pressure, which is related to ρ, ρE and $\mathbf{v}$ via the so-called equation of state (EOS). In the following we show the computational results for two test problems. The first one is the smooth isentropic vortex test case first proposed in Reference [115] and also used in Reference [36], which has an exact solution and is therefore suitable for a numerical convergence study. Some results of Reference [36] are summarized in Table 1 below, where N_x denotes the number of cells per space dimensions. From the results one can conclude that the high order ADER-DG schemes converge with the designed order of accuracy in both space and time. In order to give a quantitative assessment for the cost of the scheme, we define and provide the TDU metric, which is the cost per degree of freedom update per CPU core, see also Reference [34]. The TDU metric is easily computed by dividing the measured wall clock time (WCT) of a simulation by the number of elements per CPU core and time steps carried out, and by the number of spatial degrees of freedom per element, i.e., $(N+1)^d$. With the appropriate initial guess and AVX 512 vectorization of the code discussed in the previous section, the cost for updating one single degree of freedom for a fourth order ADER-DG scheme ($N = 3$) for the 3D compressible Euler equations is as low as TDU = 0.25 μs when using one single CPU core of a new Intel i9-7900X Skylake test workstation with 3.3 GHz nominal clock speed, 32 GB of RAM and a total number of 10 CPU cores. This cost metric can be directly compared with the cost to update one single point or control volume of existing finite difference and finite volume schemes.

Table 1. L^1, L^2 and L^∞ errors and numerical convergence rates obtained for the two-dimensional isentropic vortex test problem using different unlimited ADER-DG schemes, see Reference [36].

	N_x	L^1 Error	L^2 Error	L^∞ Error	L^1 Order	L^2 Order	L^∞ Order	Theor.
	25	5.77×10^{-4}	9.42×10^{-5}	7.84×10^{-5}	—	—	—	
$N = 3$	50	2.75×10^{-5}	4.52×10^{-6}	4.09×10^{-6}	4.39	4.38	4.26	
	75	4.36×10^{-6}	7.89×10^{-7}	7.55×10^{-7}	4.55	4.30	4.17	4
	100	1.21×10^{-6}	2.37×10^{-7}	2.38×10^{-7}	4.46	4.17	4.01	
	20	1.54×10^{-4}	2.18×10^{-5}	2.20×10^{-5}	—	—	—	
$N = 4$	30	1.79×10^{-5}	2.46×10^{-6}	2.13×10^{-6}	5.32	5.37	5.75	
	40	3.79×10^{-6}	5.35×10^{-7}	5.18×10^{-7}	5.39	5.31	4.92	5
	50	1.11×10^{-6}	1.61×10^{-7}	1.46×10^{-7}	5.50	5.39	5.69	
	10	9.72×10^{-4}	1.59×10^{-4}	2.00×10^{-4}	—	—	—	
$N = 5$	20	1.56×10^{-5}	2.13×10^{-6}	2.14×10^{-6}	5.96	6.22	6.55	
	30	1.14×10^{-6}	1.64×10^{-7}	1.91×10^{-7}	6.45	6.33	5.96	6
	40	2.17×10^{-7}	2.97×10^{-8}	3.59×10^{-8}	5.77	5.93	5.82	

In the following we show the results obtained with an ADER-DG scheme using piecewise polynomials of degree $N = 9$ for a very stringent test case, which is the so-called Sedov blast wave problem detailed in References [100,107,116,117]. It consists in an explosion propagating in a zero pressure gas, leading to an infinitely strong shock wave. In our setup, the outer pressure is set to 10^{-14}, i.e., close to machine zero. In order to get a robust numerical scheme, it is useful to perform the reconstruction step in the subcell finite volume limiter as well as the space–time predictor of the ADER-DG scheme in primitive variables, see Reference [100]. The computational results obtained are shown in Figure 1, where we can observe a very good agreement with the reference solution. One furthermore can see that the discrete solution respects the circular symmetry of the problem and the *a posteriori* subcell limiter is only acting in the vicinity of the shock wave.

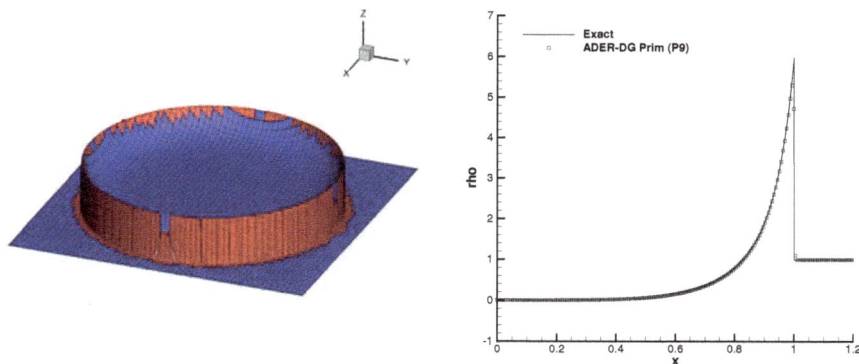

Figure 1. Sedov blast wave problem using an ADER-DG *P9* scheme with *a posteriori* subcell finite volume limiter using predictor and limiter in primitive variables, see Reference [100]. Unlimited cells are depicted in blue, while limited cells are highlighted in red (**left**). 1D cut through the numerical solution and comparison with the exact solution (**right**).

3.2. A Novel Diffuse Interface Approach for Linear Seismic Wave Propagation in Complex Geometries

Seismic wave propagation problems in complex 3D geometries are often very challenging due to the geometric complexity. Standard approaches either use regular curvilinear boundary-fitted meshes, or unstructured tetrahedral or hexahedral meshes. In all cases, a certain amount of user interaction for grid generation is required. Furthermore, the geometric complexity can have a negative impact on the admissible time step size due to the CFL condition, since the mesh generator may create elements with very bad aspect ratio, so-called sliver elements. In the case of regular curvilinear grids, the Jacobian of the mapping may become ill-conditioned and thus reduce the admissible time step size. In Reference [118] a novel diffuse interface approach has been forwarded, where only the definition of a scalar volume fraction function α is required, where $\alpha = 1$ is set inside the solid medium, and $\alpha = 0$ in the surrounding gas or vacuum. The governing PDE system proposed in Reference [118] reads

$$\frac{\partial \sigma}{\partial t} - \mathbf{E}(\lambda, \mu) \cdot \frac{1}{\alpha} \nabla(\alpha \mathbf{v}) + \frac{1}{\alpha} \mathbf{E}(\lambda, \mu) \cdot \mathbf{v} \otimes \nabla \alpha = \mathbf{S}_\sigma, \tag{21}$$

$$\frac{\partial \alpha \mathbf{v}}{\partial t} - \frac{\alpha}{\rho} \nabla \cdot \sigma - \frac{1}{\rho} \sigma \nabla \alpha = \mathbf{S}_v, \tag{22}$$

$$\frac{\partial \alpha}{\partial t} = 0, \qquad \frac{\partial \lambda}{\partial t} = 0, \qquad \frac{\partial \mu}{\partial t} = 0, \qquad \frac{\partial \rho}{\partial t} = 0, \tag{23}$$

and clearly falls into the class of PDE systems described by Equation (1). Here, σ denotes the symmetric stress tensor, $\mathbf{v}$ is the velocity vector, $\alpha \in [0,1]$ is the volume fraction, λ and μ are the Lamé constants and ρ is the density of the solid medium. The elasticity tensor $\mathbf{E}$ is a function of λ and μ and relates stress and strain via the Hooke law. The last four quantities obey trivial evolution equations, which state that these parameters remain constant in time. However, they still need to be properly included in the evolution system, since they have an influence on the solution of the Riemann problem. An analysis of the eigenstructure of Equations (21)–(23) shows that the eigenvalues are all real and are *independent* of the volume fraction function α. Furthermore, the exact solution of a generic Riemann problem with $\alpha = 1$ on the left and $\alpha = 0$ on the right yields the free surface boundary condition $\sigma \cdot \mathbf{n} = 0$ at the interface, see Reference [118] for details. In this new approach, the mesh generation problem can be fully avoided, since all that is needed is the specification of the scalar volume fraction function α, which is set to unity inside the solid and to zero outside. A realistic 3D wave propagation example based on real DTM data of the Mont Blanc region is shown in Figures 2 and 3, where the 3D contour colors of the wave field as well as a set of seismogram recordings in two receiver points are reported. For this simulation, a uniform Cartesian base-grid of 80^3 elements was used, together with one level of AMR refinement close to the free surface boundary determined by the DTM model. A fourth order ADER-DG scheme ($N = 3$) has been used in this simulation. We stress that the entire setup of the computational model in the diffuse interface approach is completely automatic, and no manual user interaction was required. The reference solution was obtained with a high order ADER-DG scheme of the same polynomial degree $N = 3$ using an unstructured boundary-fitted tetrahedral mesh [119] of similar spatial resolution, containing a total of 1,267,717 elements. We observe an excellent agreement between the two simulations, which were obtained with two completely different PDE systems on two different grid topologies.

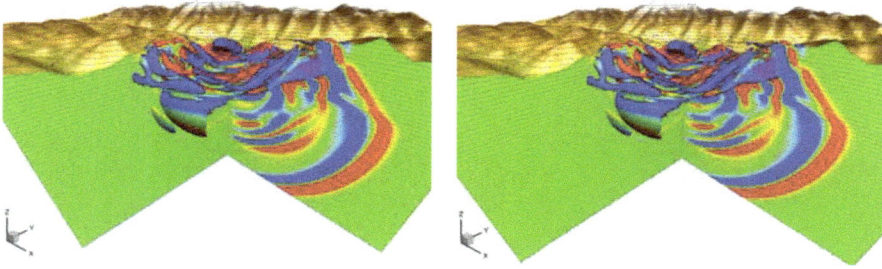

Figure 2. Wave field of a seismic wave propagation problem with the novel diffuse interface approach on adaptive Cartesian grids developed in Reference [118] (**left**) compared with the reference solution obtained on a classical boundary-fitted unstructured tetrahedral mesh [119] (**right**).

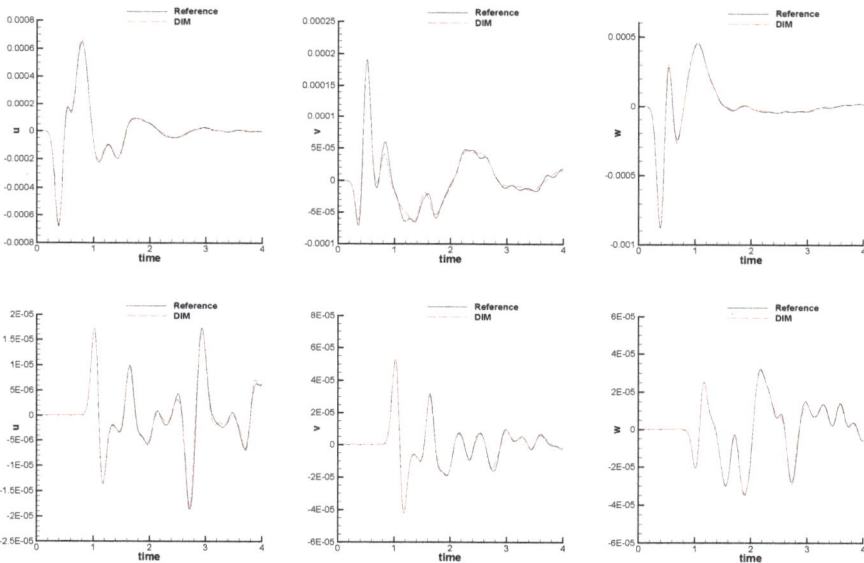

Figure 3. Seismogram recordings in two observation points obtained with the diffuse interface approach on adaptive Cartesian meshes [118] and with a reference solution obtained with high order ADER-DG schemes on boundary-fitted unstructured meshes [119].

3.3. The Unified Godunov-Peshkov-Romenski Model of Continuum Mechanics (GPR)

A major achievement of ExaHyPE was the first successful numerical solution of the unified first order symmetric hyperbolic and thermodynamically compatible Godunov–Peshkov–Romenski (GPR) model of continuum mechanics, see References [17,18]. The GPR model is based on the seminal papers by Godunov and Romenski [14,15,120] on inviscid symmetric hyperbolic systems. The dissipative mechanisms, which allow to model both plastic solids as well as viscous fluids within one single set of equations were added later in the groundbreaking work of Peshkov and Romenski in

Reference [16]. The GPR model is briefly outlined below, while for all details the interested reader is referred to References [16–18]. The governing equations read

$$\frac{\partial \rho}{\partial t} + \frac{\partial}{\partial x_k}(\rho u_k) = 0, \tag{24}$$

$$\frac{\partial \rho u_i}{\partial t} + \frac{\partial}{\partial x_k}(\partial \rho u_i u_k + p\delta_{ik} - \sigma_{ik}) = 0, \tag{25}$$

$$\frac{\partial A_{ik}}{\partial t} + \frac{\partial (A_{im} u_m)}{\partial x_k} + u_j \left(\frac{\partial A_{ik}}{\partial x_j} - \frac{\partial A_{ij}}{\partial x_k} \right) = -\frac{\psi_{ik}}{\theta_1(\tau_1)}, \tag{26}$$

$$\frac{\partial \rho J_i}{\partial t} + \frac{\partial}{\partial x_k}(\rho J_i u_k + T\delta_{ik}) = -\frac{1}{\theta_2(\tau_2)}\rho H_i, \tag{27}$$

$$\frac{\partial \rho E}{\partial t} + \frac{\partial}{\partial x_k}(u_k \rho E + u_i(p\delta_{ik} - \sigma_{ik}) + q_k) = 0. \tag{28}$$

Furthermore, the system is also endowed with an entropy inequality, see Reference [17]. Here, ρ is the mass density, $[u_i] = \mathbf{v} = (u, v, w)$ is the velocity vector, p is a non-equilibrium pressure, $[A_{ik}] = \mathbf{A}$ is the distorsion field, $[J_i] = \mathbf{J}$ is the thermal impulse vector, T is the temperature and ρE is the total energy density that is defined according to Reference [17] as

$$\rho E = \rho e + \frac{1}{2}\rho \mathbf{v}^2 + \frac{1}{4}\rho c_s^2 \operatorname{tr}\left((\operatorname{dev}\mathbf{G})^T(\operatorname{dev}\mathbf{G})\right) + \frac{1}{2}\rho \alpha^2 \mathbf{J}^2 \tag{29}$$

in terms of the specific internal energy $e = e(p, \rho)$ given by the usual equation of state (EOS), the kinetic energy, the energy stored in the medium due to deformations and in the thermal impulse. Furthermore, $\mathbf{G} = \mathbf{A}^T\mathbf{A}$ is a metric tensor induced by the distortion field $\mathbf{A}$, which allows to measure distances and thus deformations in the medium, c_s is the shear sound speed and α is a heat wave propagation speed; the symbol $\operatorname{dev}\mathbf{G} = \mathbf{G} - \frac{1}{3}\operatorname{tr}\mathbf{G}$ indicates the trace-free part of the metric tensor $\mathbf{G}$. From the definition of the total energy Equation (29) and the relations $H_i = E_{J_i}$, $\psi_{ik} = E_{A_{ij}}$, $\sigma_{ik} = -\rho A_{mi} E_{A_{mk}}$, $T = E_S$ and $q_k = E_S E_{J_k}$ the shear stress tensor and the heat flux read $\sigma = -\rho c_s^2 \mathbf{G}\operatorname{dev}\mathbf{G}$ and $\mathbf{q} = \alpha^2 T\mathbf{J}$. It can furthermore be shown via formal asymptotic expansion [17] that via an appropriate choice of θ_1 and θ_2 in the stiff relaxation limit $\tau_1 \to 0$ and $\tau_2 \to 0$, the stress tensor and the heat flux tend to those of the compressible Navier-Stokes equations

$$\sigma \to \mu \left(\nabla \mathbf{v} + \nabla \mathbf{v}^T - \frac{2}{3}(\nabla \cdot \mathbf{v})\mathbf{I} \right) \qquad \text{and} \qquad \mathbf{q} \to -\lambda \nabla T, \tag{30}$$

with transport coefficients $\mu = \mu(\tau_1, c_s)$ and $\lambda = \lambda(\tau_2, \alpha)$ related to the relaxation times τ_1 and τ_2 and to the propagation speeds c_s and α, respectively. For a complete derivation, see References [17,18]. In the opposite limit $\tau_1 \to \infty$ the model describes an ideal elastic solid with large deformations. This means that elastic solids as well as viscous fluids can be described with the aid of the same mathematical model. At this point we stress that numerically we always solve the unified *first order hyperbolic* PDE system in Equations (24)–(28), even in the stiff relaxation limit in Equation (30), when the compressible Navier-Stokes-Fourier system is retrieved asymptotically. We emphasize that we never need to discretize any parabolic terms, since the hyperbolic system in Equations (24)–(28) with algebraic relaxation source terms fits perfectly into the framework of Equation (1).

In the Figure 4 we show numerical results obtained in Reference [17] for a viscous heat conducting shock wave and the comparison with the exact solution of the compressible Navier-Stokes equations.

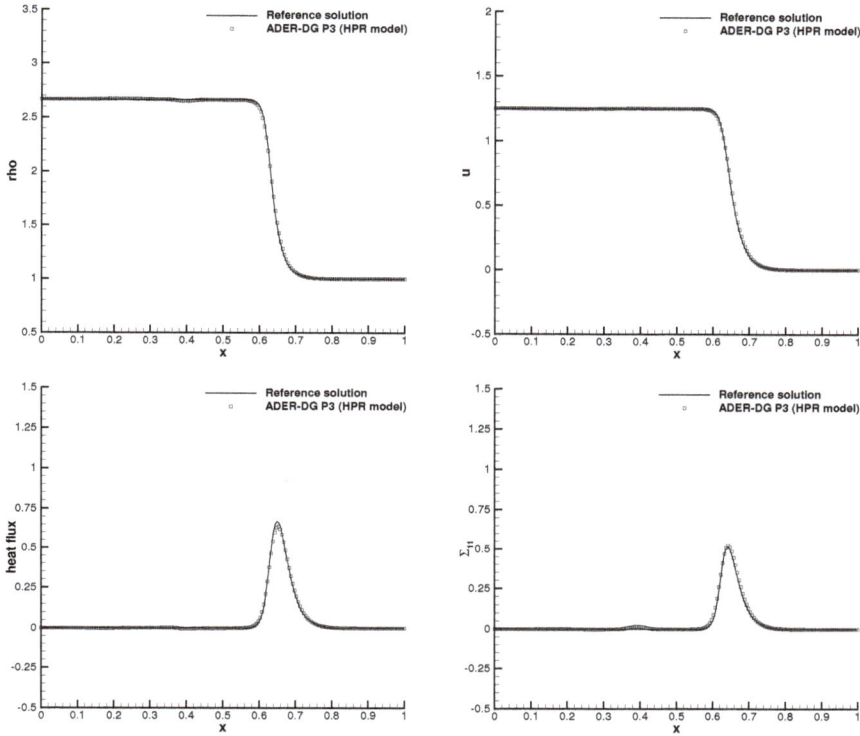

Figure 4. Viscous heat conducting shock. Comparison of the exact solution of the compressible Navier-Stokes equations with the numerical solution of the GPR model based on ADER-DG $P3$ schemes. Density profile (**top left**), velocity profile (**top right**), heat flux (**bottom left**) and stress σ_{11} (**bottom right**).

3.4. The Equations of Ideal General Relativistic Magnetohydrodynamics (GRMHD)

A very challenging PDE system is given by the equations of ideal general relativistic magnetohydrodynamics (GRMHD). The governing PDE are a result of the Einstein field equations and can be written in compact covariant notation as follows:

$$\nabla_\mu T^{\mu\nu} = 0, \quad \text{and} \quad \nabla_\mu {}^*F^{\mu\nu} = 0 \quad \text{and} \quad \nabla_\mu(\rho u^\mu) = 0, \tag{31}$$

where ∇_μ is the usual covariant derivative operator, $T^{\mu\nu}$ is the energy-momentum tensor, ${}^*F^{\mu\nu}$ is the Faraday tensor and u^μ is the four-velocity. The compact equations above can be expanded into a so-called 3+1 formalism, which can be cast into the form of Equation (1), see References [57,121,122] for more details. The final evolution system involves nine field variables plus the 10 quantities of the background space–time, which is supposed to be stationary here. A numerical convergence study for the large amplitude Alfvén wave test problem described in Reference [122] solved in the domain $\Omega = [0, 2\pi]^3$ up to $t = 1$ and carried out with high order ADER-DG schemes in Reference [57] is reported in the Table 2 below, where we also show a direct comparison with high order Runge-Kutta discontinuous Galerkin schemes. We observe that the ADER-DG schemes are competitive with RKDG methods, even for this very complex system of hyperbolic PDE. The results reported in Table 2 refer to the non-vectorized version of the code. Further significant performance improvements are expected from a carefully vectorized implementation of the GRMHD equations, in particular concerning the vectorization of the cumbersome conversion of the vector of conservative variables to the vector of primitive variables, i.e., the function $\mathbf{V} = \mathbf{V}(\mathbf{Q})$. For the GRMHD system $\mathbf{V}$ cannot be computed analytically in terms of $\mathbf{Q}$, but requires the iterative solution of one nonlinear scalar algebraic equation together with the computation of the roots of a third order polynomial, see Reference [122] for details. In our vectorized implementation of the PDE, we have therefore in particular vectorized the primitive variable recovery via a direct implementation in AVX intrinsics. We have furthermore made use of careful auto-vectorization via the compiler for the evaluation of the physical flux function and for the non-conservative product. Thanks to this vectorization effort, on one single CPU core of an Intel i9-7900X Skylake test workstation with 3.3 GHz nominal clock frequency and using AVX 512 the CPU time necessary for a single degree of freedom update (TDU) for a fourth order ADER-DG scheme ($N = 3$) could be reduced to TDU = 2.3 µs for the GRMHD equations in three space dimensions.

Table 2. Accuracy and cost comparison between ADER-DG and RKDG schemes of different orders for the GRMHD equations in three space dimensions. The errors refer to the variable B_y. The table also contains total wall clock times (WCT) measured in seconds using 512 MPI ranks of the SuperMUC phase 1 system at the LRZ in Garching, Germany.

N_x	L_2 Error	L_2 Order	WCT [s]	N_x	L_2 Error	L_2 Order	WCT [s]
ADER-DG ($N = 3$)				RKDG ($N = 3$)			
8	7.6396×10^{-4}		0.093	8	8.0909×10^{-4}		0.107
16	1.7575×10^{-5}	5.44	1.371	16	2.2921×10^{-5}	5.14	1.394
24	6.7968×10^{-6}	2.34	6.854	24	7.3453×10^{-6}	2.81	6.894
32	1.0537×10^{-6}	6.48	21.642	32	1.3793×10^{-6}	5.81	21.116
ADER-DG ($N = 4$)				RKDG ($N = 4$)			
8	6.6955×10^{-5}		0.363	8	6.8104×10^{-5}		0.456
16	2.2712×10^{-6}	4.88	5.696	16	2.3475×10^{-6}	4.86	6.666
24	3.3023×10^{-7}	4.76	28.036	24	3.3731×10^{-7}	4.78	29.186
32	7.4728×10^{-8}	5.17	89.271	32	7.7084×10^{-8}	5.13	87.115
ADER-DG ($N = 5$)				RKDG ($N = 5$)			
8	5.2967×10^{-7}		1.090	8	5.7398×10^{-7}		1.219
16	7.4886×10^{-9}	6.14	16.710	16	8.1461×10^{-9}	6.14	17.310
24	7.1879×10^{-10}	5.78	84.425	24	7.7634×10^{-10}	5.80	83.777
32	1.2738×10^{-10}	6.01	263.021	32	1.3924×10^{-10}	5.97	260.859

As second test problem we present the results obtained for the Orszag-Tang vortex system in flat Minkowski spacetime, where the GRMHD equations reduce to the special relativistic MHD equations. The initial condition is given by

$$\left(\rho, u, v, w, p, B_x, B_y, B_z \right) = \left(1, -\frac{3}{4\sqrt{2}} \sin y, \frac{3}{4\sqrt{2}} \sin x, 0, 1, -\sin y, \sin 2x, 0 \right),$$

and we set the adiabatic index to $\Gamma = 4/3$. The computational domain is $\Omega = [0, 2\pi]^2$ and is discretized with a dynamically adaptive AMR grid. For this test we chose the P_5 version of the ADER-DG scheme with FV subcell limiter and the rest mass density as indicator function for AMR, i.e., $\varphi(\mathbf{Q}) = \rho$. Figure 5 shows 1D cuts through the numerical solution at time $t = 2$ and at $y = 0.01$, while Figure 6 shows the numerical results for the AMR-grid with limiter-status map (blue cells are unlimited, while limited cells are highlighted in red), together with Schlieren images for the rest-mass density at time $t = 2$. The same simulation has been repeated with different refinement estimator functions χ that tell the AMR algorithm where and when to refine and to coarsen the mesh: (i) A simple first order derivative estimator χ_1 based on discrete gradients of the indicator function $\varphi(\mathbf{Q})$, (ii) the classical second order derivative estimator χ_2 based on Reference [123], (iii) a novel estimator χ_3 based on the action of the *a posteriori* subcell finite volume limiter, i.e., the mesh is refined where the limiter is active (iv) a multi-resolution estimator χ_4 based on the difference in L_∞ norm of the discrete solution on two different refinement levels ℓ and $\ell - 1$. The reference solution is obtained on a uniform fine grid corresponding to the finest refinement level, i.e., a uniform composed of 270×270 elements. The results shown in Figure 6 clearly show that the numerical results obtained by means of different refinement estimator functions are comparable with each other and thus the proposed AMR algorithm is robust with respect to the particular choice of the mesh.

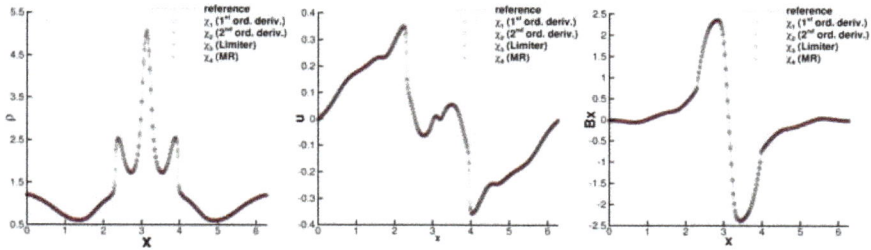

Figure 5. Results for the GRMHD Orszag-Tang vortex problem in flat space–time (SRMHD) at $t = 2$ obtained with ADER-DG-$\mathbb{P}_5$ schemes supplemented with *a posteriori* subcell finite volume limiter and using different refinement estimator functions χ. A set of 1D cuts taken at $y = 10^{-2}$ are shown. From (**left**) to (**right**): the rest-mass density, the velocity u and the magnetic field component B_x. One can note an excellent agreement between the reference solution and the ones obtained on different AMR grids.

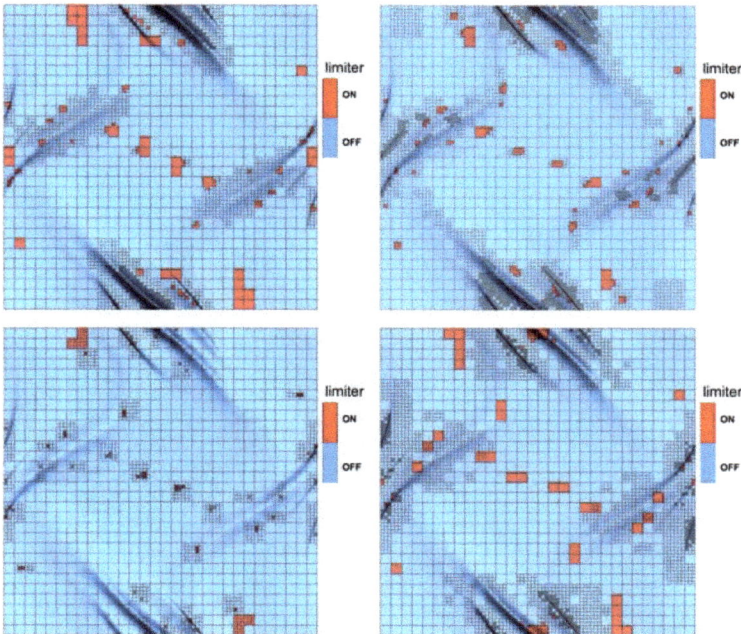

Figure 6. Results for the GRMHD Orszag-Tang vortex problem in flat space–time (SRMHD) at $t = 2$ obtained with ADER-DG-$\mathbb{P}_5$ schemes, supplemented with *a posteriori* subcell finite volume limiter and using different refinement estimator functions χ. (i) first order-derivative estimator χ_1 (**top left**); (ii) second-order derivative estimator χ_2 (**top right**); (iii) a new limiter-based estimator χ_3 (row two, **left**) and (iv) a new multi-resolution estimator χ_4 based on the difference between the discrete solution on two adjacent refinement levels (row two, **right**).

As a last test case we simulate a stationary neutron star in three space dimensions using the Cowling approximation, i.e., assuming a fixed *static* background spacetime. The initial data for the matter and the spacetime are both compatible with the Einstein field equations and are given by the solution of the Tolman–Oppenheimer–Volkoff (TOV) equations, which constitute a nonlinear ODE system in the radial coordinate that can be numerically solved up to any precision at the aid of a fourth order Runge-Kutta scheme using a very fine grid. We setup a stable nonrotating TOV star without magnetic field and with central rest mass density $\rho(0,0) = 1.28 \times 10^{-3}$ and adiabatic exponent $\Gamma = 2$ in a computational domain $\Omega = [-10,+10]^3$ discretized with a fourth order ADER-DG scheme ($N = 3$) using 32^3 elements, which corresponds to 128^3 spatial degrees of freedom. The pressure in the atmosphere outside the compact object is set to $p_{atm} = 10^{-13}$. We run the simulation until a final time of $t = 1000$ and measure the L_∞ error norms of the rest mass density and the pressure against the exact solution, which is given by the initial condition. The error measured at $t = 1000$ for the rest mass density is $L_\infty(\rho) = 1.553778 \times 10^{-5}$ while the error for the pressure is $L_\infty(p) = 1.605334 \times 10^{-7}$. The simulation was carried out with the vectorized version of the code on 512 CPU cores of the SuperMUC phase 2 system (based on AVX2) and required only 3010 s of wallclock time. The same simulation with the established finite difference GR code WhiskyTHC [124] required 8991 s of wall clock time on the same machine with the same spatial mesh resolution and the same number of CPU cores. The time series of the relative error of the central rest mass density in the origin of the domain is plotted in the left panel of Figure 7. At the final time $t = 1000$, the relative error of the central rest mass density is still below 0.1%. In the right panel of Figure 7 we show the contour surfaces of the pressure at the final time $t = 1000$. In Figure 8 we show a 1D cut along the x axis, comparing the numerical solution at time $t = 1000$ with the exact one. We note that the numerical scheme is very accurate, but it is *not well-balanced* for the GRMHD equations, i.e., the method *cannot* preserve the stationary equilibrium solution of the TOV equations *exactly* at the discrete level. Therefore, further work along the lines of research reported recently in Reference [86] for the Euler equations with Newtonian gravity are needed, extending the framework of well-balanced methods [79,80,125] also to general relativity. Finally, in Figure 9 we compare the exact and the numerical solution at time $t = 1000$ in the x–y plane.

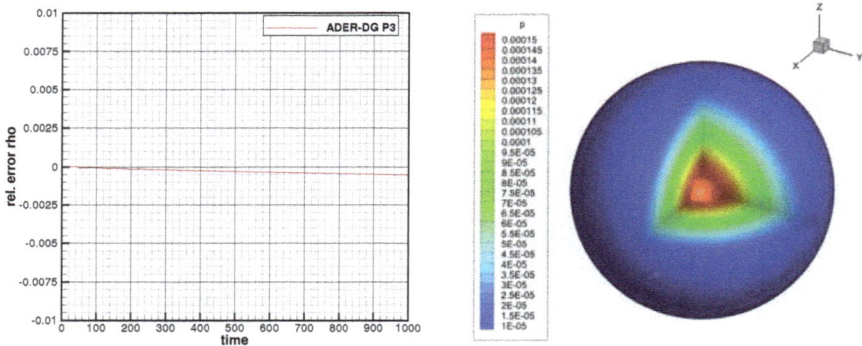

Figure 7. Computational results for a stable 3D neutron star. Time series of the relative error of the central rest mass density $(\rho(0,t) - \rho(0,0))/\rho(0,0)$ (**left**) and 3D view of of the pressure contour surfaces at time $t = 1000$ (**right**).

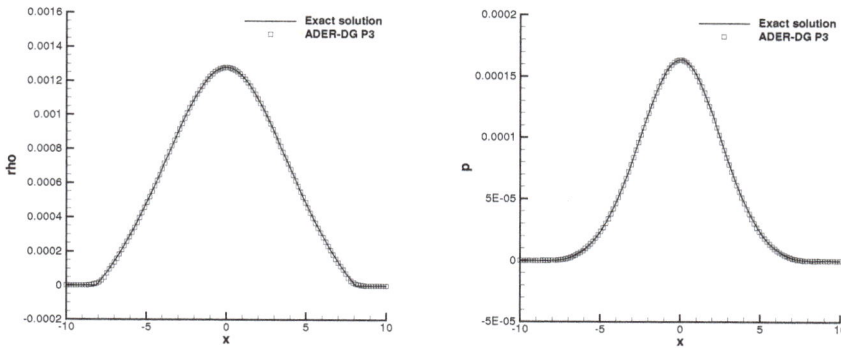

Figure 8. Computational results for a stable 3D neutron star. Comparison of the numerical solution with the exact one at time $t = 1000$ on a 1D cut along the x-axis for the rest mass density (**left**) and the pressure (**right**).

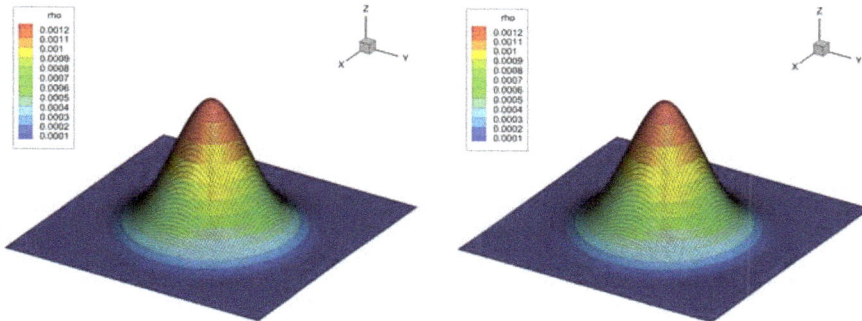

Figure 9. Computational results for a stable 3D neutron star. Cut through the x–y plane with pressure on the z axis and rest mass density contour colors. Exact solution (**left**) and numerical solution at time $t = 1000$ (**right**).

3.5. A Strongly Hyperbolic First Order Reduction of the CCZ4 Formulation of the Einstein Field Equations (FO-CCZ4)

The last PDE system under consideration here are the Einstein field equations that describe the evolution of dynamic spacetimes. Here we consider the so-called CCZ4 formulation [126], which is based on the Z4 formalism that takes into account the involutions (stationary differential constraints) inherent in the Einstein equations via an augmented system similar to the generalized Lagrangian multiplier (GLM) approach of Dedner et al. [127] that takes care of the stationary divergence-free constraint of the magnetic field in the MHD equations. In compact covariant notation the undamped Z4 Einstein equations in vacuum, which can be derived from the Einstein–Hilbert action integral associated with the Z4 Lagrangian $\mathcal{L} = g^{\mu\nu}\left(R_{\mu\nu} + 2\nabla_\mu Z_\nu\right)$, read

$$R_{\mu\nu} + \nabla_{(\mu} Z_{\nu)} = 0, \tag{32}$$

where $g^{\mu\nu}$ is the 4-metric of the spacetime, $R_{\mu\nu}$ is the 4-Ricci tensor and the 4-vector Z_ν accounts for the stationary constraints of the Einstein equations, as already mentioned before. After introducing the usual 3+1 ADM split of the 4-metric as

$$ds^2 = -\alpha^2 dt^2 + \gamma_{ij} \left(dx^i + \beta^i dt \right) \left(dx^j + \beta^j dt \right), \tag{33}$$

the equations can be cast into a time-dependent system of 25 partial differential equations that involve first order derivatives in time and both first and second order derivatives in space, see Reference [126]. Nevertheless, the system is *not* dissipative, but a rather unusual formulation of a wave equation, see Reference [128]. In the expression above, α denotes the so-called lapse, β^i is the spatial shift vector and γ_{ij} is the spatial metric. In the original form presented in Reference [126], the PDE system does *not* fit into the formalism given by Equation (1). After the introduction of 33 auxiliary variables, which are the spatial gradients of some of the 25 primary evolution quantities, it is possible to derive a first order reduction of the system that contains a total of 58 evolution quantities. However, a naive procedure of converting the original second order evolution system into a first order system leads only to a *weakly hyperbolic* formulation, which is not suitable for numerical simulations since the initial value problem is not well posed in this case. Only after adding suitable first and second order ordering constraints, which arise from the definition of the auxiliary variables, it is possible to obtain a provably strongly hyperbolic and thus well-posed evolution system, denoted by FO-CCZ4 in the following. For all details of the derivation, the strong hyperbolicity proof and numerical results achieved with high order ADER-DG schemes, the reader is referred to Reference [129]. In order to give an idea about the complexity of the Einstein field equations, it should be mentioned that one single evaluation of the FO-CCZ4 system requires about 20,000 floating point operations! In order to obtain still a computationally efficient implementation, the entire PDE system has been carefully vectorized using blocks of the size VECTORLENGTH, so that in the end a level of 99.9% of vectorization of the code has been reached. Using a fourth order ADER-DG scheme ($N = 3$) the time per degree of freedom update (TDU) metric per core on a modern workstation with Intel i9-7900X CPU that supports the novel AVX 512 instructions is TDU = 4.7 µs.

4. Strong MPI Scaling Study for the FO-CCZ4 System

A major focus of this paper is the efficient implementation of ADER-DG schemes for high performance computing (HPC) on massively parallel distributed memory supercomputers. For this purpose, we have very recently carried out a systematic study of the strong MPI scaling efficiency of our new high order fully-discrete one-step ADER-DG schemes on the Hazel Hen supercomputer of the HLRS center in Stuttgart, Germany, using from 720 up to 180,000 CPU cores. We have furthermore carried out a systematic comparison with conventional Runge-Kutta DG schemes using the SuperMUC phase 1 system of the LRZ center in Munich, Germany.

As already discussed before, the particular feature of ADER-DG schemes compared to traditional Runge-Kutta DG schemes (RKDG) is that they are intrinsically *communication-avoiding* and *cache-blocking*, which makes them particularly well suited for modern massively parallel distributed memory supercomputers. As governing PDE system for the strong scaling test the novel first-order reduction of the CCZ4 formulation of the 3+1 Einstein field equations has been been adopted [129]. We recall that FO-CCZ4 is a very large nonlinear hyperbolic PDE system that contains 58 evolution quantities.

The first strong scaling study on the SuperMUC phase 1 system uses 64 to 64,000 CPU cores. The test problem was the gauge wave problem [129] setup on the 3D domain $\Omega = [-0.5, 0.5]^3$. For the test we have compared a fourth order ADER-DG scheme ($N = 3$) with a fourth order accurate RKDG scheme on a uniform Cartesian grid composed of 120^3 elements. It has to be stressed, that when using 64,000 CPU cores for this setup each CPU has to update only $3^3 = 27$ elements. The wall clock time as a function of the used number of CPU cores (nCPU) and the obtained parallel efficiency with respect to an ideal linear scaling are reported in the left panel of Figure 10. We find that ADER-DG schemes provide a better parallel efficiency than RKDG schemes, as expected.

The second strong scaling study has been performed on the Hazel-Hen supercomputer, using 720 to 180,000 CPU cores. Again we have used a fourth order accurate ADER-DG scheme ($N = 3$), this time using a uniform grid of $200 \times 180 \times 180$ elements, solving again the 3D gauge wave benchmark problem detailed in Reference [129]. The measured wall-clock-times (WCT) as a function of the employed number of CPU cores, as well as the corresponding parallel scaling-efficiency are shown in Figure 10. The results depicted in Figure 10 clearly show that our new ADER-DG schemes **scale very well** up to 90,000 CPU cores with a parallel efficiency **greater than 95%**, and up to 180,000 cores with a parallel efficiency that is still greater than **93%**. Furthermore, the code was instrumented with manual FLOP counters in order to measure the floating point performance quantitatively. The full machine run on **180,000 CPU cores** of Hazel Hen took place on 7 May 2018. During the run, each core has provided an average performance of 8.2 GFLOPS, leading to a total of **1.476 PFLOPS** of sustained performance. To our knowledge, this was the largest test run ever carried out with high order ADER-DG schemes for nonlinear hyperbolic systems of partial differential equations. For large runs with sustained petascale performance of ADER-DG schemes for linear hyperbolic PDE systems on unstructured tetrahedral meshes, see Reference [105].

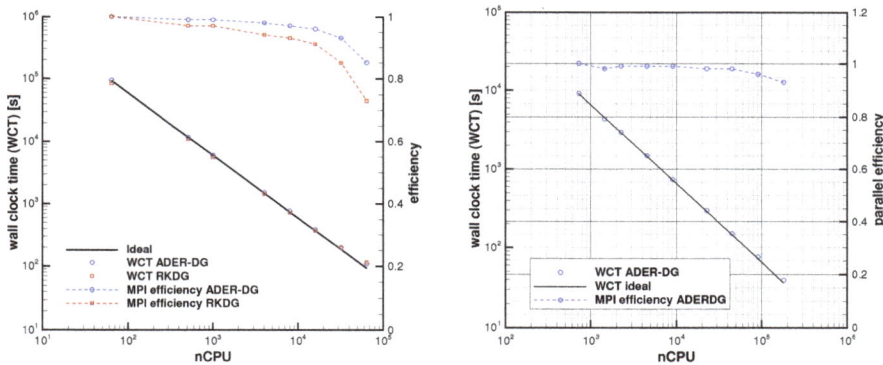

Figure 10. Strong MPI scaling study of ADER-DG schemes for the novel FO-CCZ4 formulation of the Einstein field equations recently proposed in Reference [129]. (**Left**) comparison of ADER-DG schemes with conventional Runge-Kutta DG schemes from 64 to 64,000 CPU cores on the SuperMUC phase 1 system of the LRZ supercomputing center (Garching, Germany). (**Right**) strong scaling study from 720 to 180,000 CPU cores, including a full machine run on the Hazel Hen supercomputer of HLRS (Stuttgart, Germany) with ADER-DG schemes (**right**). Even on the full machine we observe still more than 90% of parallel efficiency.

5. Conclusions

In this paper we have presented an efficient implementation of high order ADER-DG schemes on modern massively parallel supercomputers using the ExaHyPE engine. The key ingredients are the communication-avoiding and cache-blocking properties of ADER-DG, together with an efficient vectorization of the high level user functions that provide the evaluation of the physical fluxes $\mathbf{F}(\mathbf{Q})$, of the non-conservative products $\mathbf{B}(\mathbf{Q}) \cdot \nabla \mathbf{Q}$ and of the algebraic source terms $\mathbf{S}(\mathbf{Q})$. The engine is highly versatile and flexible and allows to solve a very broad spectrum of different hyperbolic PDE systems in a very efficient and highly scalable manner. In order to support this claim, we have provided a rather large set of different numerical examples solved with ADER-DG schemes. To show the excellent parallel scalability of the ADER-DG method, we have provided strong scaling results on 64 to 64,000 CPU cores including a detailed and quantitative comparison with RKDG schemes. We have furthermore shown strong scaling results of the vectorized ADER-DG implementation for the FO-CCZ4 formulation of the Einstein field equations using 720 to 180,000 CPU cores of the Hazel Hen supercomputer at the HLRS in Stuttgart, Germany, where a sustained performance of more than one petaflop has been reached.

Future research in ExaHyPE will concern an extension of the GPR model to full general relativity, able to describe nonlinear elastic and plastic solids as well as viscous and ideal fluids in one single governing PDE system. We furthermore plan an implementation of the FO-CCZ4 system [129] directly based on AVX intrinsics, in order to further improve the performance of the scheme and to reduce computational time. The final aim of our developments are the simulation of ongoing nonlinear dynamic rupture processes during earthquakes, as well as the inspiral and merger of binary neutron star systems and the associated generation of gravitational waves. Although both problems seem to be totally different and unrelated, it is indeed possible to write the mathematical formulation of both applications under the same form of a hyperbolic system given by Equation (1) and thus to solve both problems within the same computer software.

Author Contributions: Conceptualization and Methodology: M.D., M.B. and T.W.; Software: M.D., T.W., F.F. and M.T.; Validation: F.F. and M.T.; Strong MPI Scaling Study: F.F.; Writing original draft and editing: M.D.; Supervision and Project Administration: M.B., T.W. and M.D.; Funding Acquisition: M.B., T.W. and M.D.

Funding: This research was funded by the European Union's Horizon 2020 Research and Innovation Programme under the project *ExaHyPE*, grant no. 671698 (call FETHPC-1-2014). The first three authors also acknowledge funding from the Italian Ministry of Education, University and Research (MIUR) in the frame of the Departments of Excellence Initiative 2018–2022 attributed to DICAM of the University of Trento. M.D. also acknowledges support from the University of Trento in the frame of the Strategic Initiative *Modeling and Simulation*.

Acknowledgments: The authors acknowledge the support of the HLRS supercomputing center in Stuttgart, Germany, for awarding access to the HazelHen supercomputer, as well as of the Leibniz Rechenzentrum (LRZ) in Munich, Germany for awarding access to the SuperMUC supercomputer. In particular, the authors are very grateful for the technical support provided by Björn Dick (HLRS) and Nicolay Hammer (LRZ). M.D., F.F. and M.T. are all members of the Italian INdAM Research group GNCS.

Conflicts of Interest: The authors declare no conflict of interest.

References

1. Riemann, B. *Über die Fortpflanzung ebener Luftwellen von Endlicher Schwingungsweite*; Göttinger Nachrichten: Göttingen, Germany, 1859; Volume 19.
2. Riemann, B. Über die Fortpflanzung ebener Luftwellen von endlicher Schwingungsweite. *Abhandlungen der Königlichen Gesellschaft der Wissenschaften zu Göttingen* **1860**, *8*, 43–65.
3. Noether, E. Invariante Variationsprobleme. In *Nachrichten der königlichen Gesellschaft der Wissenschaften zu Göttingen, Mathematisch-Physikalische Klasse*; Weidmannsche Buchhandlung: Berlin, Germany, 1918; pp. 235–257.
4. Courant, R.; Friedrichs, K.; Lewy, H. Über die partiellen Differenzengleichungen der mathematischen Physik. *Math. Annal.* **1928**, *100*, 32–74. [CrossRef]
5. Courant, R.; Isaacson, E.; Rees, M. On the solution of nonlinear hyperbolic differential equations by finite differences. *Commun. Pure Appl. Math.* **1952**, *5*, 243–255. [CrossRef]
6. Courant, R.; Hilbert, D. *Methods of Mathematical Physics*; John Wiley and Sons, Inc.: New York, NY, USA, 1962.
7. Courant, R.; Friedrichs, K.O. *Supersonic Flows and Shock Waves*; Springer: Berlin, Germany, 1976.
8. Friedrichs, K. Symmetric positive linear differential equations. *Commun. Pure Appl. Math.* **1958**, *11*, 333–418. [CrossRef]
9. Godunov, S. An interesting class of quasilinear systems. *Dokl. Akad. Nauk SSSR* **1961**, *139*, 521–523.
10. Friedrichs, K.; Lax, P. Systems of conservation equations with a convex extension. *Proc. Natl. Acad. Sci. USA* **1971**, *68*, 1686–1688. [CrossRef] [PubMed]
11. Godunov, S.; Romenski, E. Thermodynamics, conservation laws, and symmetric forms of differential equations in mechanics of continuous media. In *Computational Fluid Dynamics Review 95*; John Wiley: New York, NY, USA, 1995; pp. 19–31.

12. Godunov, S.; Romenski, E. *Elements of Continuum Mechanics and Conservation Laws*; Kluwer Academic/Plenum Publishers: New York, NY, USA, 2003.

13. Godunov, S. Symmetric form of the magnetohydrodynamic equation. *Numer. Methods Mech. Contin. Med.* **1972**, *3*, 26–34.

14. Godunov, S.; Romenski, E. Nonstationary equations of the nonlinear theory of elasticity in Euler coordinates. *J. Appl. Mech. Tech. Phys.* **1972**, *13*, 868–885. [CrossRef]

15. Romenski, E. Hyperbolic systems of thermodynamically compatible conservation laws in continuum mechanics. *Math. Comput. Model.* **1998**, *28*, 115–130. [CrossRef]

16. Peshkov, I.; Romenski, E. A hyperbolic model for viscous Newtonian flows. *Contin. Mech. Thermodyn.* **2016**, *28*, 85–104. [CrossRef]

17. Dumbser, M.; Peshkov, I.; Romenski, E.; Zanotti, O. High order ADER schemes for a unified first order hyperbolic formulation of continuum mechanics: Viscous heat-conducting fluids and elastic solids. *J. Comput. Phys.* **2016**, *314*, 824–862. [CrossRef]

18. Dumbser, M.; Peshkov, I.; Romenski, E.; Zanotti, O. High order ADER schemes for a unified first order hyperbolic formulation of Newtonian continuum mechanics coupled with electro-dynamics. *J. Comput. Phys.* **2017**, *348*, 298–342. [CrossRef]

19. Boscheri, W.; Dumbser, M.; Loubère, R. Cell centered direct Arbitrary-Lagrangian-Eulerian ADER-WENO finite volume schemes for nonlinear hyperelasticity. *Comput. Fluids* **2016**, *134-135*, 111–129. [CrossRef]

20. Neumann, J.V.; Richtmyer, R.D. A method for the numerical calculation of hydrodynamic shocks. *J. Appl. Phys.* **1950**, *21*, 232–237. [CrossRef]

21. Godunov, S.K. A finite difference Method for the Computation of discontinuous solutions of the equations of fluid dynamics. *Math. USSR Sbornik* **1959**, *47*, 357–393.

22. Kolgan, V.P. Application of the minimum-derivative principle in the construction of finite-difference schemes for numerical analysis of discontinuous solutions in gas dynamics. *Trans. Central Aerohydrodyn. Inst.* **1972**, *3*, 68–77. (In Russian)

23. Van Leer, B. Towards the Ultimate Conservative Difference Scheme II: Monotonicity and conservation combined in a second order scheme. *J. Comput. Phys.* **1974**, *14*, 361–370. [CrossRef]

24. van Leer, B. Towards the Ultimate Conservative Difference Scheme V: A second Order sequel to Godunov's Method. *J. Comput. Phys.* **1979**, *32*, 101–136. [CrossRef]

25. Harten, A.; Osher, S. Uniformly high–order accurate nonoscillatory schemes I. *SIAM J. Num. Anal.* **1987**, *24*, 279–309. [CrossRef]

26. Jiang, G.; Shu, C. Efficient Implementation of Weighted ENO Schemes. *J. Comput. Phys.* **1996**, *126*, 202–228. [CrossRef]

27. Cockburn, B.; Shu, C.W. TVB Runge-Kutta local projection discontinuous Galerkin finite element method for conservation laws II: General framework. *Math. Comput.* **1989**, *52*, 411–435.

28. Cockburn, B.; Lin, S.Y.; Shu, C. TVB Runge-Kutta local projection discontinuous Galerkin finite element method for conservation laws III: One dimensional systems. *J. Comput. Phys.* **1989**, *84*, 90–113. [CrossRef]

29. Cockburn, B.; Hou, S.; Shu, C.W. The Runge-Kutta local projection discontinuous Galerkin finite element method for conservation laws IV: The multidimensional case. *Math. Comput.* **1990**, *54*, 545–581.

30. Cockburn, B.; Shu, C.W. The Runge-Kutta discontinuous Galerkin method for conservation laws V: Multidimensional systems. *J. Comput. Phys.* **1998**, *141*, 199–224. [CrossRef]

31. Cockburn, B.; Shu, C.W. The Local Discontinuous Galerkin Method for Time-Dependent Convection Diffusion Systems. *SIAM J. Numer. Anal.* **1998**, *35*, 2440–2463. [CrossRef]

32. Cockburn, B.; Shu, C.W. Runge-Kutta Discontinuous Galerkin Methods for Convection-Dominated Problems. *J. Sci. Comput.* **2001**, *16*, 173–261. [CrossRef]

33. Shu, C. High order WENO and DG methods for time-dependent convection-dominated PDEs: A brief survey of several recent developments. *J. Comput. Phys.* **2016**, *316*, 598–613. [CrossRef]

34. Dumbser, M.; Balsara, D.; Toro, E.; Munz, C. A Unified Framework for the Construction of One-Step Finite–Volume and discontinuous Galerkin schemes. *J. Comput. Phys.* **2008**, *227*, 8209–8253. [CrossRef]

35. Dumbser, M.; Castro, M.; Parés, C.; Toro, E. ADER Schemes on Unstructured Meshes for Non–Conservative Hyperbolic Systems: Applications to Geophysical Flows. *Comput. Fluids* **2009**, *38*, 1731–1748. [CrossRef]

36. Dumbser, M.; Zanotti, O.; Loubère, R.; Diot, S. A posteriori subcell limiting of the discontinuous Galerkin finite element method for hyperbolic conservation laws. *J. Comput. Phys.* **2014**, *278*, 47–75. [CrossRef]

37. Zanotti, O.; Fambri, F.; Dumbser, M.; Hidalgo, A. Space-time adaptive ADER discontinuous Galerkin finite element schemes with a posteriori sub-cell finite volume limiting. *Comput. Fluids* **2015**, *118*, 204–224. [CrossRef]

38. Dumbser, M.; Loubère, R. A simple robust and accurate a posteriori sub-cell finite volume limiter for the discontinuous Galerkin method on unstructured meshes. *J. Comput. Phys.* **2016**, *319*, 163–199. [CrossRef]

39. Titarev, V.; Toro, E. ADER: Arbitrary High Order Godunov Approach. *J. Sci. Comput.* **2002**, *17*, 609–618. [CrossRef]

40. Toro, E.; Titarev, V. Solution of the generalized Riemann problem for advection-reaction equations. *Proc. R. Soc. Lond.* **2002**, *458*, 271–281. [CrossRef]

41. Titarev, V.; Toro, E. ADER schemes for three-dimensional nonlinear hyperbolic systems. *J. Comput. Phys.* **2005**, *204*, 715–736. [CrossRef]

42. Toro, E.F.; Titarev, V.A. Derivative Riemann solvers for systems of conservation laws and ADER methods. *J. Comput. Phys.* **2006**, *212*, 150–165. [CrossRef]

43. Bungartz, H.; Mehl, M.; Neckel, T.; Weinzierl, T. The PDE framework Peano applied to fluid dynamics: An efficient implementation of a parallel multiscale fluid dynamics solver on octree-like adaptive Cartesian grids. *Comput. Mech.* **2010**, *46*, 103–114. [CrossRef]

44. Weinzierl, T.; Mehl, M. Peano-A traversal and storage scheme for octree-like adaptive Cartesian multiscale grids. *SIAM J. Sci. Comput.* **2011**, *33*, 2732–2760. [CrossRef]

45. Khokhlov, A. Fully Threaded Tree Algorithms for Adaptive Refinement Fluid Dynamics Simulations. *J. Comput. Phys.* **1998**, *143*, 519–543. [CrossRef]

46. Baeza, A.; Mulet, P. Adaptive mesh refinement techniques for high-order shock capturing schemes for multi-dimensional hydrodynamic simulations. *Int. J. Numer. Methods Fluids* **2006**, *52*, 455–471. [CrossRef]

47. Colella, P.; Dorr, M.; Hittinger, J.; Martin, D.F.; McCorquodale, P. High-order finite-volume adaptive methods on locally rectangular grids. *J. Phys. Conf. Ser.* **2009**, *180*, 012010. [CrossRef]

48. Bürger, R.; Mulet, P.; Villada, L. Spectral WENO schemes with Adaptive Mesh Refinement for models of polydisperse sedimentation. *ZAMM J. Appl. Math. Mech. Z. Math. Mech.* **2013**, *93*, 373–386. [CrossRef]

49. Ivan, L.; Groth, C. High-order solution-adaptive central essentially non-oscillatory (CENO) method for viscous flows. *J. Comput. Phys.* **2014**, *257*, 830–862. [CrossRef]

50. Buchmüller, P.; Dreher, J.; Helzel, C. Finite volume WENO methods for hyperbolic conservation laws on Cartesian grids with adaptive mesh refinement. *Appl. Math. Comput.* **2016**, *272*, 460–478. [CrossRef]

51. Semplice, M.; Coco, A.; Russo, G. Adaptive Mesh Refinement for Hyperbolic Systems based on Third-Order Compact WENO Reconstruction. *J. Sci. Comput.* **2016**, *66*, 692–724. [CrossRef]

52. Shen, C.; Qiu, J.; Christlieb, A. Adaptive mesh refinement based on high order finite difference WENO scheme for multi-scale simulations. *J. Comput. Phys.* **2011**, *230*, 3780–3802. [CrossRef]

53. Dumbser, M.; Zanotti, O.; Hidalgo, A.; Balsara, D. ADER-WENO Finite Volume Schemes with Space-Time Adaptive Mesh Refinement. *J. Comput. Phys.* **2013**, *248*, 257–286. [CrossRef]

54. Dumbser, M.; Hidalgo, A.; Zanotti, O. High Order Space-Time Adaptive ADER-WENO Finite Volume Schemes for Non-Conservative Hyperbolic Systems. *Comput. Methods Appl. Mech. Eng.* **2014**, *268*, 359–387. [CrossRef]

55. Zanotti, O.; Fambri, F.; Dumbser, M. Solving the relativistic magnetohydrodynamics equations with ADER discontinuous Galerkin methods, a posteriori subcell limiting and adaptive mesh refinement. *Mon. Not. R. Astron. Soc.* **2015**, *452*, 3010–3029. [CrossRef]

56. Fambri, F.; Dumbser, M.; Zanotti, O. Space-time adaptive ADER-DG schemes for dissipative flows: Compressible Navier-Stokes and resistive MHD equations. *Comput. Phys. Commun.* **2017**, *220*, 297–318. [CrossRef]

57. Fambri, F.; Dumbser, M.; Köppel, S.; Rezzolla, L.; Zanotti, O. ADER discontinuous Galerkin schemes for general-relativistic ideal magnetohydrodynamics. *Mon. Not. R. Astron. Soc.* **2018**, *477*, 4543–4564. [CrossRef]

58. Berger, M.J.; Oliger, J. Adaptive Mesh Refinement for Hyperbolic Partial Differential Equations. *J. Comput. Phys.* **1984**, *53*, 484–512. [CrossRef]

59. Berger, M.J.; Jameson, A. Automatic adaptive grid refinement for the Euler equations. *AIAA J.* **1985**, *23*, 561–568. [CrossRef]

60. Berger, M.J.; Colella, P. Local adaptive mesh refinement for shock hydrodynamics. *J. Comput. Phys.* **1989**, *82*, 64–84. [CrossRef]

61. Leveque, R. Clawpack Software. Available online: http://depts.washington.edu/clawpack/ (accessed on 23 August 2018).
62. Berger, M.J.; LeVeque, R. Adaptive mesh refinement using wave-propagation algorithms for hyperbolic systems. *SIAM J. Numer. Anal.* **1998**, *35*, 2298–2316. [CrossRef]
63. Bell, J.; Berger, M.; Saltzman, J.; Welcome, M. Three-dimensional adaptive mesh refinement for hyperbolic conservation laws. *SIAM J. Sci. Comput.* **1994**, *15*, 127–138. [CrossRef]
64. Quirk, J. A parallel adaptive grid algorithm for computational shock hydrodynamics. *Appl. Numer. Math.* **1996**, *20*, 427–453. [CrossRef]
65. Coirier, W.; Powell, K. Solution-adaptive Cartesian cell approach for viscous and inviscid flows. *AIAA J.* **1996**, *34*, 938–945. [CrossRef]
66. Deiterding, R. A parallel adaptive method for simulating shock-induced combustion with detailed chemical kinetics in complex domains. *Comput. Struct.* **2009**, *87*, 769–783. [CrossRef]
67. Lopes, M.M.; Deiterding, R.; Gomes, A.F.; Mendes, O.; Domingues, M.O. An ideal compressible magnetohydrodynamic solver with parallel block-structured adaptive mesh refinement. *Comput. Fluids* **2018**, *173*, 293–298. [CrossRef]
68. Dezeeuw, D.; Powell, K.G. An Adaptively Refined Cartesian Mesh Solver for the Euler Equations. *J. Comput. Phys.* **1993**, *104*, 56–68. [CrossRef]
69. Balsara, D. Divergence-free adaptive mesh refinement for magnetohydrodynamics. *J. Comput. Phys.* **2001**, *174*, 614–648. [CrossRef]
70. Teyssier, R. Cosmological hydrodynamics with adaptive mesh refinement. A new high resolution code called RAMSES. *Astron. Astrophys.* **2002**, *385*, 337–364. [CrossRef]
71. Keppens, R.; Nool, M.; Tóth, G.; Goedbloed, J.P. Adaptive Mesh Refinement for conservative systems: Multi-dimensional efficiency evaluation. *Comput. Phys. Commun.* **2003**, *153*, 317–339. [CrossRef]
72. Ziegler, U. The NIRVANA code: Parallel computational MHD with adaptive mesh refinement. *Comput. Phys. Commun.* **2008**, *179*, 227–244. [CrossRef]
73. Mignone, A.; Zanni, C.; Tzeferacos, P.; van Straalen, B.; Colella, P.; Bodo, G. The PLUTO Code for Adaptive Mesh Computations in Astrophysical Fluid Dynamics. *Astrophys. J. Suppl. Ser.* **2012**, *198*, 7. [CrossRef]
74. Cunningham, A.; Frank, A.; Varnière, P.; Mitran, S.; Jones, T.W. Simulating Magnetohydrodynamical Flow with Constrained Transport and Adaptive Mesh Refinement: Algorithms and Tests of the AstroBEAR Code. *Astrophys. J. Suppl. Ser.* **2009**, *182*, 519. [CrossRef]
75. Keppens, R.; Meliani, Z.; van Marle, A.; Delmont, P.; Vlasis, A.; van der Holst, B. Parallel, grid-adaptive approaches for relativistic hydro and magnetohydrodynamics. *J. Comput. Phys.* **2012**, *231*, 718–744. [CrossRef]
76. Porth, O.; Olivares, H.; Mizuno, Y.; Younsi, Z.; Rezzolla, L.; Moscibrodzka, M.; Falcke, H.; Kramer, M. The black hole accretion code. *Comput. Astrophys. Cosmol.* **2017**, *4*, 1–42. [CrossRef]
77. Dubey, A.; Almgren, A.; Bell, J.; Berzins, M.; Brandt, S.; Bryan, G.; Colella, P.; Graves, D.; Lijewski, M.; Löffler, F.; et al. A survey of high level frameworks in block-structured adaptive mesh refinement packages. *J. Parallel Distrib. Comput.* **2014**, *74*, 3217–3227. [CrossRef]
78. Stroud, A. *Approximate Calculation of Multiple Integrals*; Prentice-Hall Inc.: Englewood Cliffs, NJ, USA, 1971.
79. Castro, M.; Gallardo, J.; Parés, C. High-order finite volume schemes based on reconstruction of states for solving hyperbolic systems with nonconservative products. Applications to shallow-water systems. *Math. Comput.* **2006**, *75*, 1103–1134. [CrossRef]
80. Parés, C. Numerical methods for nonconservative hyperbolic systems: A theoretical framework. *SIAM J. Numer. Anal.* **2006**, *44*, 300–321. [CrossRef]
81. Rhebergen, S.; Bokhove, O.; van der Vegt, J. Discontinuous Galerkin finite element methods for hyperbolic nonconservative partial differential equations. *J. Comput. Phys.* **2008**, *227*, 1887–1922. [CrossRef]
82. Dumbser, M.; Hidalgo, A.; Castro, M.; Parés, C.; Toro, E. FORCE Schemes on Unstructured Meshes II: Non–Conservative Hyperbolic Systems. *Comput. Methods Appl. Mech. Eng.* **2010**, *199*, 625–647. [CrossRef]
83. Müller, L.O.; Parés, C.; Toro, E.F. Well-balanced high-order numerical schemes for one-dimensional blood flow in vessels with varying mechanical properties. *J. Comput. Phys.* **2013**, *242*, 53–85. [CrossRef]
84. Müller, L.; Toro, E. Well-balanced high-order solver for blood flow in networks of vessels with variable properties. *Int. J. Numer. Methods Biomed. Eng.* **2013**, *29*, 1388–1411. [CrossRef] [PubMed]

85. Gaburro, E.; Dumbser, M.; Castro, M. Direct Arbitrary-Lagrangian-Eulerian finite volume schemes on moving nonconforming unstructured meshes. *Comput. Fluids* **2017**, *159*, 254–275. [CrossRef]

86. Gaburro, E.; Castro, M.; Dumbser, M. Well balanced Arbitrary-Lagrangian-Eulerian finite volume schemes on moving nonconforming meshes for the Euler equations of gasdynamics with gravity. *Mon. Not. R. Astron. Soc.* **2018**, *477*, 2251–2275. [CrossRef]

87. Toro, E.; Hidalgo, A.; Dumbser, M. FORCE Schemes on Unstructured Meshes I: Conservative Hyperbolic Systems. *J. Comput. Phys.* **2009**, *228*, 3368–3389. [CrossRef]

88. Dumbser, M.; Toro, E.F. A Simple Extension of the Osher Riemann Solver to Non-Conservative Hyperbolic Systems. *J. Sci. Comput.* **2011**, *48*, 70–88. [CrossRef]

89. Castro, M.; Pardo, A.; Parés, C.; Toro, E. On some fast well-balanced first order solvers for nonconservative systems. *Math. Comput.* **2010**, *79*, 1427–1472. [CrossRef]

90. Dumbser, M.; Toro, E.F. On Universal Osher–Type Schemes for General Nonlinear Hyperbolic Conservation Laws. *Commun. Comput. Phys.* **2011**, *10*, 635–671. [CrossRef]

91. Einfeldt, B.; Roe, P.L.; Munz, C.D.; Sjogreen, B. On Godunov-type methods near low densities. *J. Comput. Phys.* **1991**, *92*, 273–295. [CrossRef]

92. Dumbser, M.; Balsara, D. A New, Efficient Formulation of the HLLEM Riemann Solver for General Conservative and Non-Conservative Hyperbolic Systems. *J. Comput. Phys.* **2016**, *304*, 275–319. [CrossRef]

93. Dumbser, M.; Enaux, C.; Toro, E. Finite Volume Schemes of Very High Order of Accuracy for Stiff Hyperbolic Balance Laws. *J. Comput. Phys.* **2008**, *227*, 3971–4001. [CrossRef]

94. Dumbser, M.; Zanotti, O. Very High Order PNPM Schemes on Unstructured Meshes for the Resistive Relativistic MHD Equations. *J. Comput. Phys.* **2009**, *228*, 6991–7006. [CrossRef]

95. Harten, A.; Engquist, B.; Osher, S.; Chakravarthy, S. Uniformly high order essentially non-oscillatory schemes, III. *J. Comput. Phys.* **1987**, *71*, 231–303. [CrossRef]

96. Dumbser, M.; Käser, M.; Titarev, V.; Toro, E. Quadrature-Free Non-Oscillatory Finite Volume Schemes on Unstructured Meshes for Nonlinear Hyperbolic Systems. *J. Comput. Phys.* **2007**, *226*, 204–243. [CrossRef]

97. Taube, A.; Dumbser, M.; Balsara, D.; Munz, C. Arbitrary High Order Discontinuous Galerkin Schemes for the Magnetohydrodynamic Equations. *J. Sci. Comput.* **2007**, *30*, 441–464. [CrossRef]

98. Hidalgo, A.; Dumbser, M. ADER Schemes for Nonlinear Systems of Stiff Advection-Diffusion-Reaction Equations. *J. Sci. Comput.* **2011**, *48*, 173–189. [CrossRef]

99. Jackson, H. On the eigenvalues of the ADER-WENO Galerkin predictor. *J. Comput. Phys.* **2017**, *333*, 409–413. [CrossRef]

100. Zanotti, O.; Dumbser, M. Efficient conservative ADER schemes based on WENO reconstruction and space-time predictor in primitive variables. *Comput. Astrophys. Cosmol.* **2016**, *3*, 1. [CrossRef]

101. Owren, B.; Zennaro, M. Derivation of efficient, continuous, explicit Runge–Kutta methods. *SIAM J. Sci. Stat. Comput.* **1992**, *13*, 1488–1501. [CrossRef]

102. Gassner, G.; Dumbser, M.; Hindenlang, F.; Munz, C. Explicit one-step time discretizations for discontinuous Galerkin and finite volume schemes based on local predictors. *J. Comput. Phys.* **2011**, *230*, 4232–4247. [CrossRef]

103. Heinecke, A.; Pabst, H.; Henry, G. LIBXSMM: A High Performance Library for Small Matrix Multiplications. Technical Report, SC'15: The International Conference for High Performance Computing, Networking, Storage and Analysis, Austin (Texas), 2015. Available online: https://github.com/hfp/libxsmm (accessed on 23 August 2018).

104. Breuer, A.; Heinecke, A.; Bader, M.; Pelties, C. Accelerating SeisSol by generating vectorized code for sparse matrix operators. *Adv. Parallel Comput.* **2014**, *25*, 347–356.

105. Breuer, A.; Heinecke, A.; Rettenberger, S.; Bader, M.; Gabriel, A.; Pelties, C. Sustained petascale performance of seismic simulations with SeisSol on SuperMUC. *Lect. Notes Comput. Sci. (LNCS)* **2014**, *8488*, 1–18.

106. Charrier, D.; Weinzierl, T. Stop talking to me—A communication-avoiding ADER-DG realisation. *arXiv* **2018**, arXiv:1801.08682.

107. Boscheri, W.; Dumbser, M. Arbitrary–Lagrangian–Eulerian Discontinuous Galerkin schemes with a posteriori subcell finite volume limiting on moving unstructured meshes. *J. Comput. Phys.* **2017**, *346*, 449–479. [CrossRef]

108. Clain, S.; Diot, S.; Loubère, R. A high-order finite volume method for systems of conservation laws— Multi-dimensional Optimal Order Detection (MOOD). *J. Comput. Phys.* **2011**, *230*, 4028–4050. [CrossRef]

109. Diot, S.; Clain, S.; Loubère, R. Improved detection criteria for the Multi-dimensional Optimal Order Detection (MOOD) on unstructured meshes with very high-order polynomials. *Comput. Fluids* **2012**, *64*, 43–63. [CrossRef]

110. Diot, S.; Loubère, R.; Clain, S. The MOOD method in the three-dimensional case: Very-High-Order Finite Volume Method for Hyperbolic Systems. *Int. J. Numer. Methods Fluids* **2013**, *73*, 362–392. [CrossRef]

111. Loubère, R.; Dumbser, M.; Diot, S. A New Family of High Order Unstructured MOOD and ADER Finite Volume Schemes for Multidimensional Systems of Hyperbolic Conservation Laws. *Commun. Comput. Phys.* **2014**, *16*, 718–763. [CrossRef]

112. Levy, D.; Puppo, G.; Russo, G. Central WENO schemes for hyperbolic systems of conservation laws. *M2AN Math. Model. Numer. Anal.* **1999**, *33*, 547–571. [CrossRef]

113. Levy, D.; Puppo, G.; Russo, G. Compact central WENO schemes for multidimensional conservation laws. *SIAM J. Sci. Comput.* **2000**, *22*, 656–672. [CrossRef]

114. Dumbser, M.; Boscheri, W.; Semplice, M.; Russo, G. Central weighted ENO schemes for hyperbolic conservation laws on fixed and moving unstructured meshes. *SIAM J. Sci. Comput.* **2017**, *39*, A2564–A2591. [CrossRef]

115. Hu, C.; Shu, C. Weighted essentially non-oscillatory schemes on triangular meshes. *J. Comput. Phys.* **1999**, *150*, 97–127. [CrossRef]

116. Sedov, L. *Similarity and Dimensional Methods in Mechanics*; Academic Press: New York, NY, USA, 1959.

117. Kamm, J.; Timmes, F. *On Efficient Generation of Numerically Robust Sedov Solutions*; Technical Report LA-UR-07-2849; LANL: Los Alamos, NM, USA, 2007.

118. Tavelli, M.; Dumbser, M.; Charrier, D.; Rannabauer, L.; Weinzierl, T.; Bader, M. A simple diffuse interface approach on adaptive Cartesian grids for the linear elastic wave equations with complex topography. *arXiv* **2018**, arXiv:1804.09491.

119. Dumbser, M.; Käser, M. An arbitrary high-order discontinuous Galerkin method for elastic waves on unstructured meshes—II. The three-dimensional isotropic case. *Geophys. J. Int.* **2006**, *167*, 319–336. [CrossRef]

120. Godunov, S.K.; Zabrodin, A.V.; Prokopov, G.P. A Difference Scheme for Two-Dimensional Unsteady Aerodynamics. *J. Comp. Math. Math. Phys. USSR* **1961**, *2*, 1020–1050.

121. Antón, L.; Zanotti, O.; Miralles, J.A.; Martí, J.M.; Ibáñez, J.M.; Font, J.A.; Pons, J.A. Numerical 3+1 general relativistic magnetohydrodynamics: A local characteristic approach. *Astrophys. J.* **2006**, *637*, 296. [CrossRef]

122. Zanna, L.D.; Zanotti, O.; Bucciantini, N.; Londrillo, P. ECHO: An Eulerian Conservative High Order scheme for general relativistic magnetohydrodynamics and magnetodynamics. *Astron. Astrophys.* **2007**, *473*, 11–30. [CrossRef]

123. Löhner, R. An adaptive finite element scheme for transient problems in CFD. *Comput. Methods Appl. Mech. Eng.* **1987**, *61*, 323–338. [CrossRef]

124. Radice, D.; Rezzolla, L.; Galeazzi, F. High-order fully general-relativistic hydrodynamics: New approaches and tests. *Class. Quantum Gravity* **2014**, *31*, 075012. [CrossRef]

125. Bermúdez, A.; Vázquez, M. Upwind methods for hyperbolic conservation laws with source terms. *Comput. Fluids* **1994**, *23*, 1049–1071. [CrossRef]

126. Alic, D.; Bona-Casas, C.; Bona, C.; Rezzolla, L.; Palenzuela, C. Conformal and covariant formulation of the Z4 system with constraint-violation damping. *Phys. Rev. D* **2012**, *85*, 064040. [CrossRef]

127. Dedner, A.; Kemm, F.; Kröner, D.; Munz, C.D.; Schnitzer, T.; Wesenberg, M. Hyperbolic Divergence Cleaning for the MHD Equations. *J. Comput. Phys.* **2002**, *175*, 645–673. [CrossRef]

128. Gundlach, C.; Martin-Garcia, J. Symmetric hyperbolic form of systems of second-order evolution equations subject to constraints. *Phys. Rev. D* **2004**, *70*, 044031. [CrossRef]

129. Dumbser, M.; Guercilena, F.; Köppel, S.; Rezzolla, L.; Zanotti, O. Conformal and covariant Z4 formulation of the Einstein equations: Strongly hyperbolic first–order reduction and solution with discontinuous Galerkin schemes. *Phys. Rev. D* **2018**, *97*, 084053. [CrossRef]

![axioms logo] *axioms*

MDPI

Article

On a Class of Hermite-Obreshkov One-Step Methods with Continuous Spline Extension

Francesca Mazzia [1,*,†] [ID] and **Alessandra Sestini** [2,†] [ID]

1   Dipartimento di Informatica, Università degli Studi di Bari Aldo Moro, 70125 Bari, Italy
2   Dipartimento di Matematica e Informatica U. Dini, Università di Firenze, 50134 Firenze, Italy;
    alessandra.sestini@unifi.it
*   Correspondence: francesca.mazzia@uniba.it; Tel.: +39-080-5443291
†   Member of the INdAM Research group GNCS.

Received: 23 May 2018; Accepted: 15 August 2018; Published: 20 August 2018

Abstract: The class of A-stable symmetric one-step Hermite–Obreshkov (HO) methods introduced by F. Loscalzo in 1968 for dealing with initial value problems is analyzed. Such schemes have the peculiarity of admitting a multiple knot spline extension collocating the differential equation at the mesh points. As a new result, it is shown that these maximal order schemes are conjugate symplectic up to order $p + r$, where $r = 2$ and p is the order of the method, which is a benefit when the methods have to be applied to Hamiltonian problems. Furthermore, a new efficient approach for the computation of the spline extension is introduced, adopting the same strategy developed for the BS linear multistep methods. The performances of the schemes are tested in particular on some Hamiltonian benchmarks and compared with those of the Gauss–Runge–Kutta schemes and Euler–Maclaurin formulas of the same order.

Keywords: initial value problems; one-step methods; Hermite–Obreshkov methods; symplecticity; B-splines; BS methods

1. Introduction

We are interested in the numerical solution of the Cauchy problem, that is the first order Ordinary Differential Equation (ODE),

$$\mathbf{y}'(t) = \mathbf{f}(\mathbf{y}(t)), \ t \in [t_0, t_0 + T], \tag{1}$$

associated with the initial condition:

$$\mathbf{y}(t_0) = \mathbf{y}_0, \tag{2}$$

where $\mathbf{f} : \mathbb{R}^m \to \mathbb{R}^m, m \geq 1$, is a $C^{R-1}, R \geq 1$, function on its domain and $\mathbf{y}_0 \in \mathbb{R}^m$ is assigned. Note that there is no loss of generality in assuming that the equation is autonomous. In this context, here, we focus on one-step Hermite–Obreshkov (HO) methods ([1], p. 277). Unlike Runge–Kutta schemes, a high order of convergence is obtained with HO methods without adding stages. Clearly, there is a price for this because total derivatives of the $\mathbf{f}$ function are involved in the difference equation defining the method, and thus, a suitable smoothness requirement for $\mathbf{f}$ is necessary. Multiderivative methods have been considered often in the past for the numerical treatment of ODEs, for example also in the context of boundary value methods [2], and in the last years, there has been a renewed interest in this topic, also considering its application to the numerical solution of differential algebraic equations; see, e.g., [3–8]. Here, we consider the numerical solution of Hamiltonian problems which in canonical form can be written as follows:

$$\mathbf{y}' = J \nabla H(\mathbf{y}), \qquad \mathbf{y}(t_0) = \mathbf{y}_0 \in \mathbb{R}^{2\ell}, \tag{3}$$

with:

$$
\mathbf{y} = \begin{pmatrix} \mathbf{q} \\ \mathbf{p} \end{pmatrix}, \quad \mathbf{q}, \mathbf{p} \in \mathbb{R}^{\ell}, \quad J = \begin{pmatrix} O & I_{\ell} \\ -I_{\ell} & O \end{pmatrix}, \tag{4}
$$

where $\mathbf{q}$ and $\mathbf{p}$ are the generalized coordinates and momenta, $H : \mathbb{R}^{2\ell} \to \mathbb{R}$ is the Hamiltonian function and I_{ℓ} stands for the identity matrix of dimension ℓ. Note that the flow $\varphi_t : \mathbf{y}_0 \to \mathbf{y}(t)$ associated with the dynamical system (3) is symplectic; this means that its Jacobian satisfies:

$$
\frac{\partial \varphi_t(\mathbf{y})^{\top}}{\partial \mathbf{y}} J \frac{\partial \varphi_t(\mathbf{y})}{\partial \mathbf{y}} = J, \quad \forall \mathbf{y} \in \mathbb{R}^{2\ell}. \tag{5}
$$

A one-step numerical method $\Phi_h : \mathbb{R}^{2\ell} \to \mathbb{R}^{2\ell}$ with stepsize h is symplectic if the discrete flow $\mathbf{y}_{n+1} = \Phi_h(\mathbf{y}_n), n \geq 0$, satisfies:

$$
\frac{\partial \Phi_h(\mathbf{y})^{\top}}{\partial \mathbf{y}} J \frac{\partial \Phi_h(\mathbf{y})}{\partial \mathbf{y}} = J, \quad \forall \mathbf{y} \in \mathbb{R}^{2\ell}. \tag{6}
$$

Two numerical methods Φ_h, Ψ_h are conjugate to each other if there exists a global change of coordinates χ_h, such that:

$$
\Psi_h = \chi_h \circ \Phi_h \circ \chi_h^{-1}
$$

with $\chi_h(\mathbf{y}) = \mathbf{y} + O(h)$ uniformly for $\mathbf{y}$ varying in a compact set and $\circ$ denoting a composition operator [9]. A method which is conjugate to a symplectic method is said to be conjugate symplectic, this is a less strong requirement than symplecticity, which allows the numerical solution to have the same long-time behavior of a symplectic method. Observe that the conjugate symplecticity here refers to a property of the discrete flow of the two numerical methods; this should be not confused with the group of conjugate symplectic matrices, the set of matrices $M \in \mathbb{C}^{2\ell}$ that satisfy $M^H J M = J$, where H means Hermitian conjugate [10].

A more relaxed property, shared by a wider class of numerical schemes, is a generalization of the conjugate-symplecticity property, introduced in [11]. A method $\mathbf{y}_1 = \Psi_h(\mathbf{y}_0)$ of order p is conjugate-symplectic up to order $p + r$, with $r \geq 0$, if a global change of coordinates $\chi_h(\mathbf{y}) = \mathbf{y} + O(h^p)$ exists such that $\Psi_h = \chi_h \circ \Phi_h \circ \chi_h^{-1}$, with the map Ψ_h satisfying

$$
\frac{\partial \Psi_h(\mathbf{y})^{\top}}{\partial \mathbf{y}} J \frac{\partial \Psi_h(\mathbf{y})}{\partial \mathbf{y}} = J + O(h^{p+r+1}). \tag{7}
$$

A consequence of property (7) is that the method $\Psi_h(\mathbf{y})$ nearly conserves all quadratic first integrals and the Hamiltonian function over time intervals of length $O(h^{-r})$ (see [11]).

Recently, the class of Euler–Maclaurin methods for the solution of Hamiltonian problems has been analyzed in [12,13] where the conjugate symplecticity up to order $p + 2$ of the p-th order methods was proven.

In this paper, we consider the symmetric one-step HO methods, which were analyzed in [14,15] in the context of spline applications. We call them BSHO methods, since they are connected to B-Splines, as we will show. BSHO methods have a formulation similar to that of the Euler–Maclaurin formulas, and the order two and four schemes of the two families are the same. As a new result, we prove that BSHO methods are conjugate symplectic schemes up to order $p + 2$, as is the case for the Euler–Maclaurin methods [12,13], and so, both families are suited to the context of geometric integration.

BSHO methods are also strictly related to BS methods [16,17], which are a class of linear multistep methods also based on B-splines suited for addressing boundary value problems formulated as first order differential problems. Note that also BS methods were firstly studied in [14,15], but at that time, they were discarded in favor of BSHO methods since; when used as initial value methods, they are not convergent. In [16,17], the same schemes have been studied as boundary value methods, and they have been recovered in particular in connection with boundary value problems. As for the BSHO methods, the discrete solution generated by a BS method can be easily extended to a continuous spline collocating the differential problem at the mesh points [18]. The idea now is to rely on B-splines with multiple inner knots in order to derive one-step HO schemes. The inner knot multiplicity is strictly connected to the number of derivatives of **f** involved in the difference equations defining the method and consequently with the order of the method. The efficient approach introduced in [18] dealing with BS methods for the computation of the collocating spline extension is here extended to BSHO methods, working with multiple knots. Note that we adopt a reversed point of view with respect to [14,15] because we assume to have already available the numerical solution generated by the BSHO methods and to be interested in an efficient procedure for obtaining the B-spline coefficients of the associated spline.

The paper is organized as follows. In Section 2, one-step symmetric HO methods are introduced, focusing in particular on BSHO methods. Section 3 is devoted to proving that BSHO methods are conjugate symplectic methods up to order $p + 2$. Then, Section 4 first shows how these methods can be revisited in the spline collocation context. Successively, an efficient procedure is introduced to compute the B-spline form of the collocating spline extension associated with the numerical solution produced by the R-th BSHO, and it is shown that its convergence order is equal to that of the numerical solution. Section 6 presents some numerical results related to Hamiltonian problems, comparing them with those generated by Euler–Maclaurin and Gauss–Runge–Kutta schemes of the same order.

2. One-Step Symmetric Hermite–Obreshkov Methods

Let $t_i, i = 0, \ldots, N$, be an assigned partition of the integration interval $[t_0, t_0 + T]$, and let us denote by $\mathbf{u}_i$ an approximation of $\mathbf{y}(t_i)$. Any one-step symmetric Hermite–Obreshkov (HO) method can be written as follows, clearly setting $\mathbf{u}_0 := \mathbf{y}_0$,

$$\mathbf{u}_{n+1} = \mathbf{u}_n + \sum_{j=1}^{R} h_n^j \, \beta_j^{(R)} \left(\mathbf{u}_n^{(j)} - (-1)^j \mathbf{u}_{n+1}^{(j)} \right), \quad n = 0, \ldots, N-1, \tag{8}$$

where $h_n := t_{n+1} - t_n$ and where $\mathbf{u}_r^{(j)}$, for $j \geq 1$, denotes the total $(j-1)$-th derivative of **f** with respect to t computed at $\mathbf{u}_r$,

$$\mathbf{u}_r^{(j)} := \frac{d^{j-1}\mathbf{f}}{dt^{j-1}}(\mathbf{y}(t))|_{\mathbf{u}_r}, \quad j = 1, \ldots, R. \tag{9}$$

Note that $\mathbf{u}_r^{(j)} \approx \mathbf{y}^{(j)}(t_r)$, and on the basis of (1), the analytical computation of the j-th derivative $\mathbf{y}^{(j)}$ involves a tensor of order j. For example, $\mathbf{y}^{(2)}(t) = \frac{d\mathbf{f}}{dt}(\mathbf{y}(t)) = \frac{\partial \mathbf{f}}{\partial \mathbf{y}}(\mathbf{y}(t)) \mathbf{f}(\mathbf{y}(t))$ (where $\frac{\partial \mathbf{f}}{\partial \mathbf{y}}$ becomes the Jacobian $m \times m$ matrix of **f** with respect to **y** when $m > 1$). As a consequence, it is $\mathbf{u}_r^{(2)} = \frac{\partial \mathbf{f}}{\partial \mathbf{y}}(\mathbf{u}_r) \mathbf{f}(\mathbf{u}_r)$. We observe that the definition in (14) implies that only $\mathbf{u}_{n+1}$ is unknown in (8), which in general is a nonlinear vector equation in $\mathbb{R}^m$ with respect to it.

For example, the one-step Euler–Maclaurin [1] formulas of order $2s$ with $s \in \mathbb{N}, s \geq 1$,

$$\mathbf{u}_{n+1} = \mathbf{u}_n + \frac{h_n}{2} \left(\mathbf{u}_n^{(1)} + \mathbf{u}_{n+1}^{(1)} \right) + \sum_{i=1}^{s-1} h_n^{2i} \frac{b_{2i}}{(2i)!} \left(\mathbf{u}_n^{(2i)} - \mathbf{u}_{n+1}^{(2i)} \right), \quad n = 0, \ldots, N-1, \tag{10}$$

(where the b_{2i} denote the Bernoulli numbers, which are reported in Table 2) belong to this class of methods. These methods will be referred to in the following with the label EMHO (Euler–Maclaurin Hermite–Obreshkov).

Here, we consider another class of symmetric HO methods that can be obtained by defining as follows the polynomial P_{2R},

$$P_{2R}(x) := \frac{x^R(x-1)^R}{(2R)!} \tag{11}$$

appearing in ([1], Lemma 13.3), the statement of which is reported in Lemma 1.

Lemma 1. *Let R be any positive integer and P_{2R} be a polynomial of exact degree 2R. Then, the following one-step linear difference equation,*

$$\sum_{j=0}^{2R} h_n^j \, \mathbf{u}_{n+1}^{(j)} P_{2R}^{(2R-j)}(0) = \sum_{j=0}^{2R} h_n^j \, \mathbf{u}_n^{(j)} P_{2R}^{(2R-j)}(1)$$

defines a multiderivative method of order 2R.

Referring to the methods obtainable by Lemma 1, if in particular the polynomial P_{2R} is defined as in (11), then we obtain the class of methods in which we are interested here. They can be written as in (8) with,

$$\beta_j^{(R)} := \frac{1}{j!} \frac{R(R-1)\ldots(R-j+1)}{(2R)(2R-1)\ldots(2R-j+1)} \tag{12}$$

which are reported in Table 1, for $R = 1, \ldots, 5$. In particular, for $R = 1$ and $R = 2$, we obtain the trapezoidal rule and the Euler–Maclaurin method of order four, respectively.

Table 1. Symmetric one-step B-Spline Hermite–Obreshkov (BSHO) coefficients.

R	$\beta_1^{(R)}$	$\beta_2^{(R)}$	$\beta_3^{(R)}$	$\beta_4^{(R)}$	$\beta_5^{(R)}$
1	$\frac{1}{2}$				
2	$\frac{1}{2}$	$\frac{1}{12}$			
3	$\frac{1}{2}$	$\frac{1}{10}$	$\frac{1}{120}$		
4	$\frac{1}{2}$	$\frac{3}{28}$	$\frac{1}{84}$	$\frac{1}{1680}$	
5	$\frac{1}{2}$	$\frac{1}{9}$	$\frac{1}{72}$	$\frac{1}{1008}$	$\frac{1}{30240}$

These methods were originally introduced in the spline collocation context, dealing in particular with splines with multiple knots [14,15], as we will show in Section 4. We call them BSHO methods since we will show that they can be obtained dealing in particular with the standard B-spline basis. The stability function of the R-th one-step symmetric BSHO method is the rational function corresponding to the (R, R)-Padé approximation of the exponential function, as is that of the same order Runge–Kutta–Gauss method ([19], p. 72). It has been proven that methods with this stability function are A-stable ([19], Theorem 4.12). For the proof of the statement of the following corollary, which will be useful in the sequel, we refer to [15],

Corollary 1. *Let us assume that $\mathbf{f} \in C^{2R+1}(\mathcal{D})$, where $\mathcal{D} := \{\mathbf{y} \in \mathbb{R}^m \mid \exists t \in [t_0, t_0 + T]$ such that $\|\mathbf{y} - \mathbf{y}(t)\|_2 \le L_b\}$, with $L_b > 0$. Then, there exists a positive constant h_b such that if $\max_{0 \le n \le N-1} h_n =: h < h_b$ and $\{\mathbf{u}_i\}_{i=0}^N$ denotes the related numerical solution produced by the R-th one-step symmetric BSHO method in (8)–(12), it is:*

$$\|\mathbf{u}_i^{(j)} - \mathbf{y}_i^{(j)}\| = O(h^{2R}), \quad j = 1, \ldots, R, \quad i = 0, \ldots, N.$$

3. Conjugate Symplecticity of the Symmetric One-Step BSHO Methods

Following the lines of the proof given in [13], in this section, we prove that one-step symmetric BSHO methods are conjugate symplectic schemes up to order $2R + 2$. The following lemma, proved in [20], is the starting point of the proof, and it makes use of the B-series integrator concept. On this concern, referring to [9] for the details, here, we just recall that a B-series integrator is a numerical method that can be expressed as a formal B-series, that is it has a power series in the time step in which each term is a sum of elementary differentials of the vector field and where the number of terms is allowed to be infinite.

Lemma 2. *Assume that Problem (1) admits a quadratic first integral* $Q(\mathbf{y}) = \mathbf{y}^T S \mathbf{y}$ *(with S denoting a constant symmetric matrix) and that it is solved by a B-series integrator* $\Phi_h(\mathbf{y})$. *Then, the following properties, where all formulas have to be interpreted in the sense of formal series, are equivalent:*

(a) $\Phi_h(\mathbf{y})$ *has a modified first integral of the form* $\tilde{Q}(\mathbf{y}) = Q(\mathbf{y}) + hQ_1(\mathbf{y}) + h^2 Q_2(\mathbf{y}) + \dots$ *where each* $Q_i(\cdot)$ *is a differential functional;*
(b) $\Phi_h(\mathbf{y})$ *is conjugate to a symplectic B-series integrator.*

We observe that Lemma 2 is used in [21] to prove the conjugate symplecticity of symmetric linear multistep methods. Following the lines of the proof given in [13], we can actually prove that the R-th one-step symmetric BSHO method is conjugate symplectic up to order $2R + 2$. With similar arguments of [13] we prove the following theorem, showing that the map $\mathbf{y}_1 = \Psi_h(\mathbf{y}_0)$ associated with the BSHO method is such that $\Psi_h(\mathbf{y}) = \Phi_h(\mathbf{y}) + O(h^{2R+3})$, where $\mathbf{y}_1 = \Phi_h(\mathbf{y}_0)$ is a suitable conjugate symplectic B-series integrator.

Theorem 1. *The map* $\mathbf{u}_1 = \Psi_h(\mathbf{u}_0)$ *associated with the one-step method (8) admits a B-series expansion and is conjugate to a symplectic B-series integrator up to order $2R + 2$.*

Proof. The existence of a B-series expansion for $\mathbf{y}_1 = \Psi_h(\mathbf{y}_0)$ is directly deduced from [22], where a B-series representation of a generic multi-derivative Runge-Kutta method has been obtained. By defining the two characteristic polynomials of the trapezoidal rule:

$$\rho(z) := z - 1, \qquad \sigma(z) := \frac{1}{2}(z+1),$$

and the shift operator $E(\mathbf{u}_n) := \mathbf{u}_{n+1}$, the R-th method described in (8) reads,

$$\rho(E)\mathbf{u}_n = \sum_{k=1}^{\lceil R/2 \rceil} 2\beta_{2k-1}^{(R)} h^{2k-1} \sigma(E)\mathbf{u}_n^{(2k-1)} - \sum_{k=1}^{\lfloor R/2 \rfloor} \beta_{2k}^{(R)} h^{2k} \rho(E)\mathbf{u}_n^{(2k)}. \tag{13}$$

Observe that $\mathbf{u}_i^{(j)}$, for $j \geq 1$, denotes the $(j-1)$-th Lie derivative of $\mathbf{f}$ computed at $\mathbf{u}_i$,

$$\mathbf{u}_i^{(j)} := D_{j-1}\mathbf{f}(\mathbf{u}_i), \quad j = 1, \dots, R, \tag{14}$$

where $D_0 = I$ is the identity operator and $D_k\mathbf{f}(\mathbf{z})$ is defined as the k-th total derivative of $\mathbf{f}(\mathbf{y}(t))$ computed at $\mathbf{y}(t) = \mathbf{z}$, where for the computation of the total derivative it is assumed that $\mathbf{y}$ satisfies the differential equation in (1). Note that we use the subscript to define the Lie operator to avoid confusion with the same order classical derivative operator in the following denoted as D^k. With this clarification on the definition of $\mathbf{u}_i^{(j)}$, we now consider a function $\mathbf{v}(t)$, a stepsize h and the shift operator $E_h(\mathbf{v}(t)) := \mathbf{v}(t+h)$, and we look for a continuous function $\mathbf{v}(t)$ that satisfies (13) in the sense of formal series (a series where the number of terms is allowed to be infinite), using the relation $E_h = \sum_{j=0}^{\infty} \frac{h^j}{j!} D^j \equiv e^{hD}$ where $D = D^1$ is the classical derivative operator,

$$\rho(e^{hD})\mathbf{v}(t) = \sum_{k=1}^{\lceil R/2 \rceil} 2\beta_{2k-1}^{(R)} h^{2k-1} \sigma(e^{hD}) D_{2k-2}\mathbf{f}(\mathbf{v}(t)) - \sum_{k=1}^{\lfloor R/2 \rfloor} \beta_{2k}^{(R)} h^{2k} \rho(e^{hD}) D_{2k-1}\mathbf{f}(\mathbf{v}(t)).$$

By multiplying both sides of the previous equation by $D\rho(e^{hD})^{-1}$, we obtain:

$$D\mathbf{v}(t) = hD\rho(e^{hD})^{-1}\sigma(e^{hD}) \sum_{k=0}^{\lceil R/2 \rceil - 1} 2\beta_{2k+1}^{(R)} h^{2k} D_{2k}\mathbf{f}(\mathbf{v}(t)) - \sum_{k=1}^{\lfloor R/2 \rfloor} \beta_{2k}^{(R)} h^{2k} DD_{2k-1}\mathbf{f}(\mathbf{v}(t)). \tag{15}$$

Now, since Bernoulli numbers define the Taylor expansion of the function $z/(e^z - 1)$ and $b_0 = 1, b_1 = -1/2$ and $b_j = 0$ for the other odd j, we have:

$$\frac{z\sigma(e^z)}{\rho(e^z)} = \frac{1}{2}\frac{z(e^z + 1)}{e^z - 1} = \frac{z}{e^z - 1} + \frac{z}{2} = 1 + \sum_{j=1}^{\infty} \frac{b_{2j}}{(2j)!} z^{2j}.$$

Thus, we can write (15) as

$$\dot{\mathbf{v}}(t) = \left(\left(I + \sum_{j=1}^{\infty} \frac{b_{2j}}{(2j)!} h^{2j} D^{2j} \right) \left(I + \sum_{k=1}^{\lceil R/2 \rceil - 1} 2\beta_{2k+1}^{(R)} h^{2k} D_{2k} \right) - \sum_{k=1}^{\lfloor R/2 \rfloor} \beta_{2k}^{(R)} h^{2k} DD_{2k-1} \right) \mathbf{f}(\mathbf{v}(t)).$$

Adding and subtracting terms involving the classical derivative operator D^{2k}, D^{2k-1}, we get

$$\dot{\mathbf{v}}(t) = \left(\left(I + \sum_{j=1}^{\infty} \frac{b_{2j}}{(2j)!} h^{2j} D^{2j} \right) \right.$$
$$\left(I + \sum_{k=1}^{\lceil R/2 \rceil - 1} 2\beta_{2k+1}^{(R)} h^{2k} D^{2k} + \sum_{k=1}^{\lceil R/2 \rceil - 1} 2\beta_{2k+1}^{(R)} h^{2k}(D_{2k} - D^{2k}) \right)$$
$$\left. - \sum_{k=1}^{\lfloor R/2 \rfloor} \beta_{2k}^{(R)} h^{2k} DD^{2k-1} - \sum_{k=1}^{\lfloor R/2 \rfloor} \beta_{2k}^{(R)} h^{2k} D(D_{2k-1} - D^{2k-1}) \right) \mathbf{f}(\mathbf{v}(t)).$$

that we recast as

$$\dot{\mathbf{v}}(t) = \left(\left(I + \sum_{j=1}^{\infty} \frac{b_{2j}}{(2j)!} h^{2j} D^{2j} \right) \left(I + \sum_{k=1}^{\lceil R/2 \rceil - 1} 2\beta_{2k+1}^{(R)} h^{2k} D^{2k} \right) \right. \tag{16}$$
$$\left. - \sum_{k=1}^{\lfloor R/2 \rfloor} \beta_{2k}^{(R)} h^{2k} D^{2k} \right) \mathbf{f}(\mathbf{v}(t))$$
$$+ \left(\left(I + \sum_{j=1}^{\infty} \frac{b_{2j}}{(2j)!} h^{2j} D^{2j} \right) \left(\sum_{k=1}^{\lceil R/2 \rceil - 1} 2\beta_{2k+1}^{(R)} h^{2k}(D_{2k} - D^{2k}) \right) \right.$$
$$\left. - \sum_{k=1}^{\lfloor R/2 \rfloor} \beta_{2k}^{(R)} h^{2k} D(D_{2k-1} - D^{2k-1}) \right) \mathbf{f}(\mathbf{v}(t)).$$

Since $\mathbf{v}(t) = \mathbf{y}(t) + O(h^{2R})$, due to the regularity conditions on the function $\mathbf{f}$, we see that $(D^i - D_i)\mathbf{f}(\mathbf{v}(t)) = O(h^{2R}), i = 1, \ldots, R - 1$ and hence the solution $\mathbf{v}(t)$ of (16) is $O(h^{2R+2})$-close to the solution of the following initial value problem

$$\dot{\mathbf{w}}(t) = \mathbf{f}(\mathbf{w}(t)) + \sum_{j=R}^{\infty} \delta_j h^{2j} D^{2j}\mathbf{f}(\mathbf{w}(t)), \tag{17}$$

with:

$$\delta_j := \sum_{k=0}^{\lceil R/2 \rceil - 1} \frac{b_{2(j-k)}}{(2(j-k))!} 2\beta_{2k+1}^{(R)}, \quad j \geq R.$$

that has been derived from (16) by neglecting the sums containing the derivatives D_{2k}, D_{2k-1}. Observe that $\delta_j = 0$ for $j = 1, \ldots, R - 1$, since the method is of order $2R$ (see [9], Theorem 3.1, page 340). We may interpret (17) as the modified equation of a one-step method $\mathbf{y}_1 = \Phi_h(\mathbf{y}_0)$, where Φ_h is evidently the time-h flow associated with (17). Expanding the solution of (17) in Taylor series, we get the modified initial value differential equation associated with the numerical scheme by coupling (17) with the initial condition $\mathbf{w}(t_0) = \mathbf{y}_0$. Thus, Φ_h is a B-series integrators. The proof of the conjugated symplecticity of Φ_h follows exactly the same steps of the analogous proof in Theorem 1 of [13]. Since $\Psi_h(\mathbf{y}) = \Phi_h(\mathbf{y}) + O(h^{2R+3})$ and Φ_h is conjugate-symplectic, the result follows using the same global change of coordinates $\chi_h(\mathbf{y})$ associated to Φ_h. □

In Table 2, we report the coefficients δ_R for $R \leq 5$ and the corresponding Bernoulli numbers. We can observe that the truncation error in the modified initial value problem is smaller than the one of the EMHO methods of the same order, which is equal to $b_i/i!$ (see [13]). The conjugate symplecticity up to order $2R + 2$ property of a numerical scheme makes it suitable for the solution of Hamiltonian problems. A well-known pair of conjugate symplectic methods is composed by the trapezoidal and midpoint rules. Observe that the trapezoidal rule belongs to both the classes BSHO and EMHO of multiderivative methods, and its characteristic polynomial plays an important role in the proof of Theorem 1.

Table 2. Coefficients of the modified differential equations and Bernoulli numbers.

R	1	2	3	4	5
δ_R	$\frac{b_2}{2!}$	$\frac{b_4}{4!}$	$\frac{3}{10}\frac{b_6}{6!}$	$\frac{1}{21}\frac{b_8}{8!}$	$\frac{1}{210}\frac{b_{10}}{10!}$
b_{2R}	$\frac{1}{6}$	$-\frac{1}{30}$	$\frac{1}{42}$	$-\frac{1}{30}$	$\frac{5}{66}$

4. The Spline Extension

A (vector) Hermite polynomial of degree $2R + 1$ interpolating both $\mathbf{u}_n$ and $\mathbf{u}_{n+1}$ respectively at t_n and t_{n+1} together with assigned derivatives $\mathbf{u}_n^{(k)}, \mathbf{u}_{n+1}^{(k)}, k = 1, \ldots, R$, can be computed using the Newton interpolation formulas with multiple nodes. On the other hand, in his Ph.D. thesis [15], Loscalzo proved that a polynomial of degree $2R$ verifying the same conditions exists if and only if (8) is fulfilled with the β coefficients defined as in (12). Note that, since the polynomial of degree $2R + 1$ fulfilling these conditions is always unique and its principal coefficient is given by the generalized divided difference $u[t_n, \ldots, t_n, t_{n+1}, \ldots, t_{n+1}]$ of order $2R + 1$ associated with the given R-order Hermite data, the n-th condition in (8) holds iff this coefficient vanishes. If all the conditions in (8) are fulfilled, it is possible to define a piecewise polynomial, the restriction to $[t_n, t_{n+1}]$ of which coincides with this polynomial, and it is clearly a C^R spline of degree $2R$ with breakpoints at the mesh points. Now, when the definition given in (14) is used together with the assumption $\mathbf{u}_0 = \mathbf{y}_0$, the conditions in (8) become a multiderivative one-step scheme for the numerical solution of (1). Thus, the numerical solution $\mathbf{u}_n, n = 0, \ldots, N$ it produces and the associated derivative values defined as in (14) can be associated with the above-mentioned $2R$ degree spline extension. Such a spline collocates the differential equation at the mesh points with multiplicity R, that is it verifies the given differential equation and also the equations $\mathbf{y}^{(j)}(t) = \frac{d^{(j-1)}(\mathbf{f} \circ \mathbf{y})}{dt^{j-1}}(t), j = 2, \ldots, R$ at the mesh points. This piecewise representation of the spline is that adopted in [15]. Here, we are interested in deriving its more compact B-spline representation. Besides being more compact, this also allows us to clarify the connection between BSHO and BS methods previously introduced in [16–18]. For this aim, let us introduce some necessary notation. Let S_{2R}, be the space of C^R $2R$-degree splines with breakpoints at $t_i, i = 0, \ldots, N$, where $t_0 < \cdots < t_N = t_0 + T$. Since we relate to the B-spline basis, we need to introduce the associated extended knot vector:

$$\mathcal{T} := \{\tau_{-2R}, \ldots, \tau_{-1}, \tau_0, \ldots, \tau_{(N-1)R}, \tau_{(N-1)R+1}, \tau_{(N-1)R+2} \cdots, \tau_{(N+1)R+1}\}, \tag{18}$$

where:

$$\tau_{-2R} = \cdots = \tau_0 = t_0,$$
$$\tau_{(n-1)R+1} = \cdots = \tau_{nR} = t_n, \qquad\qquad n = 1, \ldots, N-1,$$
$$\tau_{(N-1)R+1} = \cdots = \tau_{(N+1)R+1} = t_N,$$

which means that all the inner breakpoints have multiplicity R in $\mathcal{T}$ and both t_0 and t_N have multiplicity $2R + 1$. The associated B-spline basis is denoted as $B_i, i = -2R, \ldots, (N-1)R$ and the dimension of S_{2R} as D, with $D := (N+1)R + 1$.

The mentioned result proven by Loscalzo is equivalent to saying that, if the β coefficients are defined as in (12), any C^R spline of degree $2R$ with breakpoints at the mesh points fulfills the relation in (8), where $u_n^{(j)}$ denotes the j-th spline derivative at t_n. In turn, this is equivalent to saying that such a relation holds for any element of the B-spline basis of S_{2R}. Thus, setting $\boldsymbol{\alpha} := (-1; 1)^T \in \mathbb{R}^2$ and $\boldsymbol{\beta}^{(i)} := (\beta_i^{(R)}; -(-1)^i \beta_i^{(R)}) \in \mathbb{R}^2, i = 1, \ldots, R$, considering the local support of the B-spline basis, we have that $(\boldsymbol{\alpha}; \boldsymbol{\beta}^{(1)}; \ldots; \boldsymbol{\beta}^{(R)})$, where the punctuation mark ";" means vertical catenation (to make a column-vector), can be also characterized as the unique solution of the following linear system,

$$G^{(n)} \left(\boldsymbol{\alpha}; \boldsymbol{\beta}^{(1)}; \ldots; \boldsymbol{\beta}^{(R)} \right) = \mathbf{e}_{2R+2}, \tag{19}$$

where $\mathbf{e}_{2R+2} = (0; \ldots; 0; 1) \in \mathbb{R}^{2R+2}$ and:

$$G^{(n)} := \begin{bmatrix} A_1^{(n)T} & -h_n A_2^{(n)T} & -h_n^2 A_3^{(n)T} & \ldots & -h_n^R A_{R+1}^{(n)T} \\ (0,0) & (1,1) & (0,0) & \ldots & (0,0) \end{bmatrix}, \tag{20}$$

with $A_1^{(n)}, A_2^{(n)}, \ldots A_{R+1}^{(n)}$ defined as,

$$A_{j+1}^{(n)} := \begin{bmatrix} B_{(n-2)R}^{(j)}(t_n), & \ldots, & B_{nR}^{(j)}(t_n) \\ B_{(n-2)R}^{(j)}(t_{n+1}), & \ldots, & B_{nR}^{(j)}(t_{n+1}) \end{bmatrix}_{2 \times (2R+1)} \tag{21}$$

where $B_i^{(j)}$ denotes the j-th derivative of B_i. Note that the last equation in (19), $2\beta_1^{(R)} = 1$, is just a normalization condition.

In order to prove the non-singularity of the matrix $G^{(n)}$, we need to introduce the following definition,

Definition 1. *Given a non-decreasing set of abscissas* $\Theta := \{\theta_i\}_{i=0}^M$, *we say that a function* g_1 *agrees with another function* g_2 *at* Θ *if* $g_1^{(j)}(\theta_i) = g_2^{(j)}(\theta_i)$, $j = 0, \ldots, m_i - 1, i = 0, \ldots, M$, *where* m_i *denotes the multiplicity of* θ_i *in* Θ.

Then, we can formulate the following proposition,

Proposition 1. *The* $(2R+2) \times (2R+2)$ *matrix* $G^{(n)}$ *defined in* (20) *and associated with the B-spline basis of* S_{2R} *is nonsingular.*

Proof. Observe that the restriction to $I_n = [t_n, t_{n+1}]$ of the splines in S_{2R} generates Π_{2R} since there are no inner knots in I_n. Then, restricting to I_n, Π_{2R} can be also generated by the B-splines of S_{2R} not vanishing in I_n, that is from $B_{(n-2)R}, \ldots, B_{nR}$. Since the polynomial in Π_{2R} agreeing with a given function in:

$$\Theta = \{\overbrace{t_n, \ldots, t_n}^{R+1}, \overbrace{t_{n+1}, \ldots, t_{n+1}}^{R}\},$$

is unique, it follows that also the corresponding $(2R + 1) \times (2R + 1)$ matrix collocating the spline basis active in I_n is nonsingular. Such a matrix is the principal submatrix of $G^{(n)T}$ of order $2R + 1$. Thus now, considering that the restriction to I_n of any function in S_{2R} is a polynomial of degree $2R$, we prove by reductio ad absurdum that the last row of $G^{(n)}$ cannot be a linear combination of the other rows. In fact, in the opposite case, there would exist a polynomial P of degree $2R$ such that $P(t_n) = P(t_{n+1}) = 0$, $P'(t_n) = P'(t_{n+1}) = -1$, and $P^{(j)}(t_n) = P^{(j)}(t_{n+1}) = 0$, $j = 2, \ldots, R$. Considering the specific interpolation conditions, this P does not fulfill the n-th condition in (8). This is absurd, since Loscalzo [15] has proven that such a condition is equivalent to requiring degree reduction for the unique polynomial of degree less than or equal to $2R + 1$, fulfilling $R + 1$ Hermite conditions at both t_n and t_{n+1}. $\square$

Note that this different form for defining the coefficient of the R-th BSHO scheme is analogous to that adopted in [17] for defining a BS method on a general partition. However, in this case, the coefficients of the scheme do not depend on the mesh distribution, so there is no need to determine them solving the above linear system. On the other hand, having proven that the matrix $G^{(n)}$ is nonsingular will be useful in the following for determining the B-spline form of the associated spline extension.

Thus, let us now see how the B-spline coefficients of the spline in S_{2R} associated with the numerical solution generated by the R-th BSHO can be efficiently obtained, considering that the following conditions have to be imposed,

$$\begin{cases} \mathbf{s}_{2R}(t_n) & = \mathbf{u}_n, \\[2mm] \mathbf{s}_{2R}^{(j)}(t_n) & = \mathbf{u}_n^{(j)}, \; j = 1, \ldots, R. \end{cases} \qquad n = 0, \ldots, N. \tag{22}$$

Now, we are interested in deriving the B-spline coefficients $\mathbf{c}_i, i = -2R, \ldots, (N-1)R$, of $\mathbf{s}_{2R}$,

$$\mathbf{s}_{2R}(t) = \sum_{i=-2R}^{(N-1)R} \mathbf{c}_i \, B_i(t), \; t \in [t_0, t_0 + T]. \tag{23}$$

Relying on the representation in (23), all the conditions in (22) can be re-written in the following compact matrix form,

$$(A \otimes I_m) \, \mathbf{c} = (\mathbf{u}_0; \ldots; \mathbf{u}_N; \mathbf{u}_0^{(1)}; \ldots; \mathbf{u}_N^{(1)}; \ldots; \mathbf{u}_0^{(R)}; \ldots; \mathbf{u}_N^{(R)}), \tag{24}$$

where $\mathbf{c} = (\mathbf{c}_{-2R}; \ldots; \mathbf{c}_{(N-1)R}) \in \mathbb{R}^{mD}$, with $\mathbf{c}_j \in \mathbb{R}^m$, I_m is the identity matrix of size $m \times m$, D is the dimension of the spline space previously introduced and where:

$$A := (A_1; A_2; \ldots; A_{R+1}),$$

with each A_ℓ being a $(R + 1)$-banded matrix of size $(N + 1) \times D$ (see Figure 1) with entries defined as follows:

$$(A_\ell)_{i,j} := B_j^{(\ell-1)}(t_i). \tag{25}$$

The following theorem related to the rectangular linear system in (24) ensures that the collocating spline s_{2R} is well defined.

Theorem 2. *The rectangular linear system in (24) has always a unique solution, if the entries of the vector on its right-hand side satisfy the conditions in (8) with the β coefficients given in (12).*

Proof. The proof is analogous to that in [18] (Theorem 1), and it is omitted. $\square$

We now move to introduce the strategy adopted for an efficient computation of the B-spline coefficients of s_{2R}.

Figure 1. Sparsity structure of the matrix A with $N = 8$, $R = 1$ (**left**) and with $N = 8$, $R = 2$ (**right**).

4.1. Efficient Spline Computation

Concerning the computation of the spline coefficient vectors:

$$\mathbf{c}_i, \ i = -(2R), \ldots, (N-1)R,$$

the unique solution of (24) can be computed with several different strategies, which can have very different computational costs and can produce results with different accuracy when implemented in finite arithmetic. Here, we follow the local strategy used in [18]. Taking into account the banded structure of $A_i, i = 1, \ldots, R+1$, we can verify that (24) implies the following relations,

$$\begin{bmatrix} A_1^{(i)} \\ -h_i A_2^{(i)} \\ \vdots \\ -h_i^R A_{R+1}^{(i)} \end{bmatrix} \otimes I_m \ \mathbf{c}^{(i)} \ = \ \mathbf{w}^{(i)}(\mathbf{u}) \tag{26}$$

where $\mathbf{u} = (\mathbf{u}_0; \ldots; \mathbf{u}_N)$, $\mathbf{c}^{(i)} := (\mathbf{c}_{(i-3)R}; \ldots; \mathbf{c}_{(i-1)R}) \in \mathbb{R}^{m(2R+1)}$, $i = 1, \ldots, N$ and:

$$\mathbf{w}^{(i)}(\mathbf{u}) \ := \ (\mathbf{u}_{i-1}; \mathbf{u}_i; -h_i \mathbf{u}_{i-1}^{(1)}; -h_i \mathbf{u}_i^{(1)}; \ldots; -h_i^R \mathbf{u}_{i-1}^{(R)}; -h_i^R \mathbf{u}_i^{(R)}).$$

As a consequence, we can also write that,

$$(G^{(i)T} \otimes I_m) \ \hat{\mathbf{c}}^{(i)} \ = \ \mathbf{w}^{(i)}(\mathbf{u}) \tag{27}$$

where $\hat{\mathbf{c}}^{(i)} := (\mathbf{c}^{(i)}; \mathbf{0}) \in \mathbb{R}^{m(2R+2)}$.

Now, for all integers $r < 2R + 2$, we can define other $R + 1$ auxiliary vectors $\hat{\boldsymbol{\alpha}}_{i,r}^{(R)}, \hat{\boldsymbol{\beta}}_{l,i,r}^{(R)}$, $l = 1, \ldots, R \in \mathbb{R}^2$, defined as the solution of the following linear system,

$$G^{(i)} (\hat{\boldsymbol{\alpha}}_{i,r}^{(R)}; \hat{\boldsymbol{\beta}}_{1,i,r}^{(R)}; \ldots; \hat{\boldsymbol{\beta}}_{R,i,r}^{(R)}) \ = \ \mathbf{e}_r, \tag{28}$$

where $\mathbf{e}_r$ is the r-th unit vector in $\mathbb{R}^{2R+2}$ (that is the auxiliary vectors define the r-th column of the inverse of $G^{(i)}$). Then, we can write,

$$((\hat{\boldsymbol{\alpha}}_{i,r}^{(R)}; \hat{\boldsymbol{\beta}}_{1,i,r}^{(R)}; \ldots; \hat{\boldsymbol{\beta}}_{R,i,r}^{(R)})^T \otimes I_m) \ (G^{(i)T} \otimes I_m) \ \hat{\mathbf{c}}^{(i)} \ = \ (\mathbf{e}_r^T \otimes I_m) \ \hat{\mathbf{c}}^{(i)} \ = \ \mathbf{c}_{(i-3)R+r-1}.$$

From this formula, considering (27), we can conclude that:

$$\mathbf{c}_{(i-3)R+r-1} \ = \ ((\hat{\boldsymbol{\alpha}}_{i,r}^{(R)}; \hat{\boldsymbol{\beta}}_{1,i,r}^{(R)}; \ldots; \hat{\boldsymbol{\beta}}_{R,i,r}^{(R)})^T \otimes I_m) \mathbf{w}^{(i)}(\mathbf{u}) \tag{29}$$

Thus, solving all the systems (28) for $i = 1, \ldots, N$, $r = r_1(i), \ldots, r_2(i)$, with:

$$r_1(i) := \begin{cases} 1 & \text{if} \quad i = 1, \\ R + 1 & \text{if} \quad 1 < i \leq N, \end{cases} \qquad r_2(i) := \begin{cases} 2R & \text{if} \quad 1 \leq i < N, \\ 2R + 1 & \text{if} \quad i = N, \end{cases}$$

all the spline coefficients are obtained. Note that, with this approach, we solve D auxiliary systems, the size of which does not depend on N, using only N different coefficient matrices. Furthermore, only the information at t_{i-1} and t_i is necessary to compute $c_{(i-3)R+r-1}$. Thus, the spline can be dynamically computed at the same time the numerical solution is advanced at a new time value. This is clearly of interest for a dynamical adaptation of the stepsize.

In the following subsection, relying on its B-spline representation, we prove that the convergence order of s_{2R} to y is equal to that of the numerical solution. This result was already available in [15] (see Theorem 4.2 in the reference), but proven with different longer arguments.

4.2. Spline Convergence

Let us assume the following quasi-uniformity requirement for the mesh,

$$M_l \leq \frac{h_i}{h_{i+1}} \leq M_u, \quad i = 0, \ldots, N-1, \tag{30}$$

where M_l and M_u are positive constants not depending on h, with $M_l \leq 1$ and $M_u \geq 1$. Note that this requirement is a standard assumption in the refinement strategies of numerical methods for ODEs. We first prove the following result, that will be useful in the sequel.

Proposition 2. *If* $y \in S_{2R}$ *and so in particular if* y *is a polynomial of degree at most* $2R$, *then:*

$$\mathbf{y}_{n+1} - \mathbf{y}_n - \sum_{j=1}^{R} h_n^j \beta_j^{(R)} \left(\mathbf{y}_n^{(j)} - (-1)^j \mathbf{y}_{n+1}^{(j)} \right) = 0, \quad n = 0, \ldots, N-1,$$

where $\mathbf{y}_n := \mathbf{y}(t_n), \mathbf{y}_n^{(j)} := \frac{d^j \mathbf{y}}{d^j t}(t_n), j = 1, \ldots, R, n = 0, \ldots, N$, *and the spline extension* s_{2R} *coincides with* $\mathbf{y}$.

Proof. The result follows by considering that the divided difference vanishes and, as a consequence, the local truncation error of the methods is null. □

Then, we can prove the following theorem (where for notational simplicity, we restrict to $m = 1$), the statement of which is analogous to that on the convergence of the spline extension associated with BS methods [18]. In the proof of the theorem, we relate to the quasi-interpolation approach for function approximation, the peculiarity of which consists of being a local approach. For example, in the spline context considered here, this means that only a local subset of a given discrete dataset is required to compute a B-spline coefficient of the approximant; refer to [23] for the details.

Theorem 3. *Let us assume that the assumptions on* f *done in Corollary* 1 *hold and that* (30) *holds. Then, the spline extension* s_{2R} *approximates the solution* y *of* (1) *with an error of order* $O(h^{2R})$ *where* $h := \max\limits_{i=0,\ldots,N-1} h_i$.

Proof. Let $\bar{s}_{2R}$ denote the spline belonging to S_{2R} obtained by quasi-interpolating y with one of the rules introduced in Formula (5.1) in [23] by point evaluation functionals. From [23] (Theorem 5.2), under the quasi-uniformity assumption on the mesh distribution, we can derive that such a spline approximates y with maximal approximation order also with respect to all the derivatives, that is,

$$\|\bar{s}_{2R}^{(j)} - y^{(j)}\|_\infty \leq K \|y^{(2R+1)}\|_\infty h^{2R+1-j}, \quad j = 0, \ldots, R, \tag{31}$$

where K is a constant depending only on R, M_l and M_u.

On the other hand, by using the triangular inequality, we can state that:

$$\|s_{2R} - y\|_\infty \leq \|s_{2R} - \bar{s}_{2R}\|_\infty + \|\bar{s}_{2R} - y\|_\infty, \tag{32}$$

Thus, we need to consider the first term on the right-hand side of this inequality. On this concern, because of the partition of unity property of the B-splines, we can write:

$$\|s_{2R} - \bar{s}_{2R}\|_\infty = \| \sum_{i=-2R}^{(N+1)R+1} (c_i - \bar{c}_i) B_i(\cdot) \|_\infty \leq \|\mathbf{c} - \bar{\mathbf{c}}\|_\infty,$$

where $\mathbf{c} := (c_{-2R}; \ldots; c_{(N+1)R+1})$ and $\bar{\mathbf{c}} := (\bar{c}_{-2R}; \ldots; \bar{c}_{(N+1)R+1})$.

Now, for any function $g \in C^{2R}[t_0, t_0 + T]$, we can define the following linear functionals,

$$\lambda_{i,r}(g) := \mathbf{w}^{(i)T}(g)(\hat{\boldsymbol{\alpha}}_{i,r}^{(R)}; \hat{\boldsymbol{\beta}}_{1,i,r}^{(R)}; \ldots; \hat{\boldsymbol{\beta}}_{R,i,r}^{(R)}),$$

where:

$$\mathbf{w}^{(i)}(g) := (g(t_{i-1}); g(t_i); -h_i g'(t_{i-1}); -h_i g'(t_i); \ldots; -h_i^R g^{(R)}(t_{i-1}); -h_i^R g^{(R)}(t_i))$$

and the vector $(\hat{\boldsymbol{\alpha}}_{i,r}^{(R)}; \hat{\boldsymbol{\beta}}_{1,i,r}^{(R)}; \ldots; \hat{\boldsymbol{\beta}}_{R,i,r}^{(R)})$ has been defined in the previous section. Considering from Proposition 2 that $\bar{s}_{2R}$, as well as any other spline belonging to S_{2R} can be written as follows,

$$\bar{s}_{2R}(\cdot) = \sum_{i=1}^{N} \sum_{r=r_1(i)}^{r_2(i)} \lambda_{i,r}(\bar{s}_{2R}) B_{-2R-1+i+r-r_1(i)}(\cdot),$$

from (31), we can deduce that:

$$\bar{\mathbf{c}} = \left(\lambda_{1,r_1(1)}(\bar{s}_{2R}); \ldots; \lambda_{N,r_2(N)}(\bar{s}_{2R})\right) = \left(\lambda_{1,r_1(1)}(y); \ldots; \lambda_{N,r_2(N)}(y)\right) + O(h^{2R+1}).$$

Now, the vector $(\hat{\boldsymbol{\alpha}}_{i,r}^{(R)}; \hat{\boldsymbol{\beta}}_{1,i,r}^{(R)}; \ldots; \hat{\boldsymbol{\beta}}_{R,i,r}^{(R)})$ is defined in (28) as the r-th column of the inverse of the matrix $G^{(i)}$. On the other hand, the entries of such nonsingular matrix do not depend on h, but because of the locality of the B-spline basis and of the R-th multiplicity of the inner knots, only on the ratios $h_j/h_{j+1}, j = i - 1, i$, which are uniformly bounded from below and from above because of (30). Thus, there exists a constant C depending on M_l, M_u and R such that $\| (G^{(i)})^{-1} \| \leq C$, which implies that the same is true for any one of the mentioned coefficient vectors. From the latter, we deduce that for all indices, we find:

$$|c_i - \bar{c}_i| \leq K \|\mathbf{w}^{(i)}(u) - \mathbf{w}^{(i)}(y)\| + O(h^{2R+1}).$$

On the other hand, taking into account the result reported in Corollary 1 besides (31), we can easily derive that $\|\mathbf{w}^{(i)}(u) - \mathbf{w}^{(i)}(y)\| = O(h^{2R})$, which then implies that $\|\mathbf{c} - \bar{\mathbf{c}}\|_\infty = O(h^{2R})$. $\square$

5. Approximation of the Derivatives

The computation of the derivative $\mathbf{u}_n^{(j)}, j \geq 2$, from the corresponding $\mathbf{u}_n$ is quite expensive, and thus, usually, methods not requiring derivative values are preferred. Therefore, as well as for any other multiderivative method, it is of interest to associate with BSHO methods an efficient way to compute the derivative values at the mesh points. We are exploiting a number of possibilities, such as:

- using generic symbolic tools, if the function **f** is known in closed form;
- using a tool of automatic differentiation, like ADiGator, a MATLAB Automatic Differentiation Tool [24];
- using the *Infinity Computer Arithmetic*, if the function **f** is known as a black box [6,7,13];
- approximating it with, for example, finite differences.

As shown in the remainder of this section, when approximate derivatives are used, we obtain a different numerical solution, since the numerical scheme for its identification changes. In this case, the final formulation of the scheme is that of a standard linear multistep method, being still derived from (8) with coefficients in (12), but by replacing derivatives of order higher than one with their approximations. In this section, we just show the relation of these methods with a class of Boundary Value Methods (BVMs), the Extended Trapezoidal Rules (ETRs), linear multistep methods used with boundary conditions [25]. Similar relations have been found in [26] with HO and the equivalent class of the super-implicit methods, which require the knowledge of functions not only at past, but also at future time steps. The ETRs can be derived from BSHO when the derivatives are approximated by finite differences. Let us consider the order four method with $R = 2$. In this case, the first derivative of f could be approximated using central differences:

$$\mathbf{f}'_i \approx \frac{\mathbf{f}_{i+1} - \mathbf{f}_{i-1}}{2h_i}$$

the numerical scheme (8), denoting $\mathbf{u}_i^{(1)} =: \mathbf{f}_i$ and $\mathbf{u}_i^{(2)} =: \mathbf{f}'_i$, is:

$$\mathbf{u}_{i+1} = \mathbf{u}_i + \frac{h}{2}\left(\mathbf{f}_{i+1} + \mathbf{f}_i\right) - \frac{h^2}{12}\left(\mathbf{f}'_{i+1} - \mathbf{f}'_i\right),$$

after the approximation becomes:

$$\mathbf{u}_{i+1} = \mathbf{u}_i + \frac{h}{2}\left(\mathbf{f}_{i+1} + \mathbf{f}_i\right) - \frac{h}{24}\left(\mathbf{f}_{i+2} - \mathbf{f}_i - \mathbf{f}_{i+1} + \mathbf{f}_{i-1}\right),$$

rearranging, we recover the ETR of order four:

$$\mathbf{u}_{i+1} = \mathbf{u}_i + \frac{h}{24}\left(-\mathbf{f}_{i+2} + 13\mathbf{f}_{i+1} + 13\mathbf{f}_i - \mathbf{f}_{i-1}\right).$$

With similar arguments for the method of order six, $R = 3$, by approximating the derivatives with the order four finite differences:

$$\mathbf{f}'_i \approx \frac{1}{h}\left(\frac{1}{12}\mathbf{f}_{i+3} + \frac{2}{3}\mathbf{f}_{i+2} - \frac{2}{3}\mathbf{f}_i + \frac{1}{12}\mathbf{f}_{i-1}\right),$$

and:

$$\mathbf{u}_i^{(3)} =: \mathbf{f}''_i \approx \frac{1}{h^2}\left(-\frac{1}{12}\mathbf{f}_{i+2} + \frac{4}{3}\mathbf{f}_{i+1} - \frac{5}{2}\mathbf{f}_i + \frac{4}{3}\mathbf{f}_{i-1} - \frac{1}{12}\mathbf{f}_{i-2}\right),$$

and rearranging, we obtain the sixth order ETR method:

$$\mathbf{u}_{i+1} = \mathbf{u}_i + \frac{h}{1440}\left(11\mathbf{f}_{i+3} - 93\mathbf{f}_{i+2} + 802\mathbf{f}_{i+1} + 802\mathbf{f}_i - 93\mathbf{f}_{i-1} + 11\mathbf{f}_{i-2}\right).$$

This relation allows us to derive a continuous extension of the ETR schemes using the continuous extension of the BSHO method, just substituting the derivatives by the corresponding approximations. Naturally, a change of the stepsize will now change the coefficients of the linear multistep schemes.

Observe that BVMs have been efficiently used for the solution of boundary value problems in [27], and the BS methods are also in this class [16].

It has been proven in [21] that symmetric linear multistep methods are conjugate symplectic schemes. Naturally, in the context of linear multistep methods used with only initial conditions, this property refers only to the trapezoidal method, but when we solve boundary value problems, the correct use of a linear multistep formula is with boundary conditions; this makes the corresponding formulas stable, with a region of stability equal to the left half plane of $\mathbb{C}$ (see [25]). The conjugate symplecticity of the methods is the reason for their good behavior shown in [28,29] when used in block form and with a sufficiently large block for the solution of conservative problems.

Remark 1. *We recall that, even when approximated derivatives are used, the numerical solution admits a C^R 2R-degree spline extension verifying all the conditions in (24), where all the $\mathbf{u}_n^{(j)}$, $j \geq 2$ appearing on the right-hand side have to be replaced with the adopted approximations. The exact solution of the rectangular system in (24) is still possible, since (8) with coefficients in (12) is still verified by the numerical solution $\mathbf{u}_n$, $n = 0, \ldots, N$, by its derivatives $\mathbf{u}_n^{(1)} = \mathbf{f}(\mathbf{u}_n)$, $n = 0, \ldots, N$ and by the approximations of the higher order derivatives. The only difference in this case is that the continuous spline extension collocates at the breakpoints of just the given first order differential equation.*

6. Numerical Examples

The numerical examples reported here have two main purposes: the first is to show the good behavior of BSHO methods for Hamiltonian problems, showing both the linear growth of the error for long time computation and the conservation of the Hamiltonian. To this end, we compare the methods with the symplectic Gauss–Runge–Kutta methods and with the conjugate symplectic up to order $p + 2$ EMHO methods. On the other hand, we are interested in showing the convergence properties of the spline continuous extensions. Observe that the availability of a continuous extension of the same order of the method is an important property. In fact for high order methods, especially for superconvergent methods like the Gauss ones, it is very difficult to find a good continuous extension. The natural continuous extension of these methods does not keep the same order of accuracy, without adding extra stages [30]. Observe also that a good continuous extension is an important tool, for example for the event location.

We report results of our experiments for BSHO methods of order six and eight. We recall that the order two BSHO method corresponds to the well-known trapezoidal rule, the property of conjugate symplecticity of which is well known (see for example [9]) and the continuous extension by the B-spline of which has been already developed in [18]. The order four BSHO belongs also to the EMHO class, and it has been analyzed in detail in [13].

6.1. Kepler Problem

The first example is the classical Kepler problem, which describes the motion of two bodies subject to Newton's law of gravitation. This problem is a completely integrable Hamiltonian nonlinear dynamical system with two degrees of freedom (see, for details, [31]). The Hamiltonian function:

$$H(q_1, q_2, p_1, p_2) = \frac{1}{2}(p_1^2 + p_2^2) - \frac{1}{\sqrt{q_1^2 + q_2^2}},$$

describes the motion of the body that is not located in the origin of the coordinate systems. This motion is an ellipse in the q_1-q_2 plane, the eccentricity e of which is set using as starting values:

$$q_1(0) = 1 - e, \quad q_2(0) = 0, \quad p_1(0) = 0, \quad p_2(0) = \sqrt{\frac{1 + e}{1 - e}},$$

and with period $\mu := 2\pi$. The first integrals of this problem are: the total energy H, the angular momentum:

$$M(q_1, q_2, p_1, p_2) := q_1 p_2 - q_2 p_1.$$

and the Lenz vector $\mathbf{A} := (A_1, A_2, A_3)^\top$, the components of which are:

$$A_1(q, p) := p_2 M(q, p) - \frac{q_1}{||q||_2}, \quad A_2(q, p) := -p_1 M(q, p) - \frac{q_2}{||q||_2}, \quad A_3(q, p) := 0.$$

Only three of the four first integrals are independent, so, for example, A_2 can be neglected.

As in [13], we set $e = 0.6$ and $h = \mu/200$, and we integrate the problem over 10^3 periods. Setting $\mathbf{y} := (q_1, q_2, p_1, p_2)$, the error $||\mathbf{y}_j - \mathbf{y}_0||_1$ in the solution is computed at specific times fixed equal to multiples of the period, that is at $t_j = 2\pi j$, with $j = 1, 2, \ldots$; the errors in the invariants have been computed at the mesh points $t_n = \pi n$, $n = 1, 3, 5 \ldots$. Figure 2 reports the obtained results for the sixth and eighth order BSHO (dotted line, BSHO6, BSHO8), the sixth order EMHO (solid lines, EMHO6) and the sixth and eighth order Gauss–Runge–Kutta (GRK) (dashed lines, GRK6, GRK8) methods. In the top-left picture, the absolute error of the numerical solution is shown; the top-right picture shows the error in the Hamiltonian function; the error in the angular momentum is drawn in the bottom-left picture, while the bottom-right picture concerns the error in the first component of the Lenz vector. As expected from a symplectic or a conjugate symplectic integrator, we can see a linear drift in the error $||\mathbf{y}_j - \mathbf{y}_0||_1$ as the time increases (top left plot) and in the first component of the Lenz vector (bottom right picture). As well as for the other considered methods, we can see that BSHO methods guarantee a near conservation of the Hamiltonian function and of the angular momentum (other pictures). This latter quadratic invariant is precisely conserved (up to machine precision) by GRK methods due to their symplecticity property. We observe also that, as expected, the error for the BSHO6 method is $\frac{3}{10}$ of the error of the EMHO6 method.

To check the convergence behavior of the continuous extensions, we integrated the problem over 10 periods starting with stepsize $h = \mu/N$, $N = 100$. We computed a reference solution using the order eight method with a halved stepsize, and we computed the maximum absolute error on the doubled grid. The results are reported in Table 3 for the solution and the first derivative and clearly show that the continuous extension respects the theoretical order of convergence.

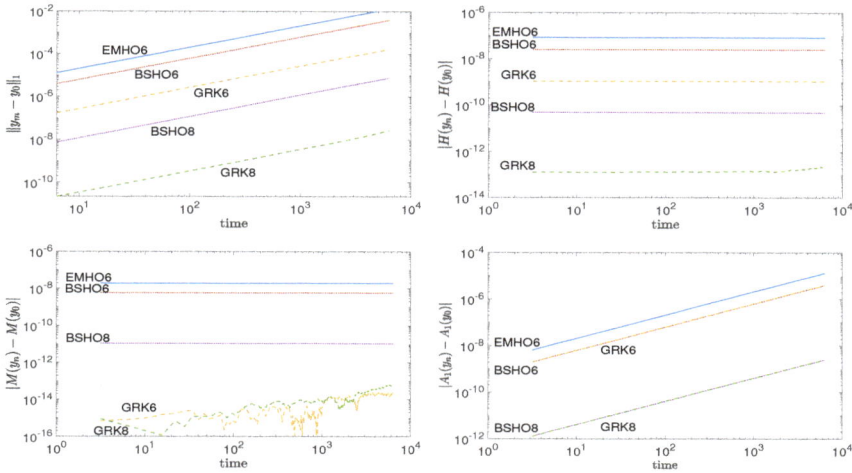

Figure 2. Kepler problem: results for the sixth (BSHO6, red dotted line) and eighth (BSHO8, purple dotted line) order BSHO methods, sixth order Euler–Maclaurin method (EMHO6, blue solid line) and sixth (Gauss–Runge–Kutta (GRK6), yellow dashed line) and eighth (GRK8-green dashed line) order Gauss methods. (**Top-left**) Absolute error of the numerical solution; (**top-right**) error in the Hamiltonian function; (**bottom-left**) error in the angular momentum; (**bottom-right**) error in the second component of the Lenz vector.

Table 3. Kepler problem: maximum absolute error of the numerical solution and its derivative computed for 10 periods.

Order	N	erry	Rate	erry	Rate
4	100	$2.69 \cdot 10^{-1}$		$1.33 \cdot 10^{0}$	
4	200	$1.69 \cdot 10^{-2}$	3.99	$8.50 \cdot 10^{-2}$	3.96
4	400	$1.06 \cdot 10^{-3}$	4.00	$5.30 \cdot 10^{-3}$	4.00
4	800	$6.60 \cdot 10^{-5}$	4.00	$3.31 \cdot 10^{-4}$	4.00
6	100	$1.95 \cdot 10^{-3}$		$9.74 \cdot 10^{-3}$	
6	200	$2.96 \cdot 10^{-5}$	6.03	$1.48 \cdot 10^{-4}$	6.03
6	400	$4.60 \cdot 10^{-7}$	6.00	$2.30 \cdot 10^{-6}$	6.00
6	800	$7.19 \cdot 10^{-9}$	6.00	$3.60 \cdot 10^{-8}$	6.00
8	100	$1.56 \cdot 10^{-5}$		$7.82 \cdot 10^{-5}$	
8	200	$5.75 \cdot 10^{-8}$	8.08	$2.88 \cdot 10^{-7}$	8.08
8	400	$2.17 \cdot 10^{-10}$	8.05	$1.08 \cdot 10^{-9}$	8.05
8	800	$7.62 \cdot 10^{-12}$	4.87	$3.70 \cdot 10^{-11}$	4.44

6.2. Non-Linear Pendulum Problem

As a second example, we consider the dynamics of a pendulum under the influence of gravity. This dynamics is usually described in terms of the angle q that the pendulum forms with its stable rest position:

$$\ddot{q} + \sin q = 0, \tag{33}$$

where $p = \dot{q}$ is the angular velocity. The Hamiltonian function associated with (33) is:

$$H(q, p) = \frac{1}{2}p^2 - \cos q. \tag{34}$$

An initial condition (q_0, p_0) such that $|H(q_0, p_0)| < 1$ gives rise to a periodic solution $y(t) = (q(t), p(t))^\top$ corresponding to oscillations of the pendulum around the straight-down stationary position. In particular, starting at $y_0 = (q_0, 0)^\top$, the period of oscillation may be expressed in terms of the complete elliptical integral of the first kind as:

$$\mu(q_0) = \int_0^1 \frac{dz}{\sqrt{(1 - z^2)(1 - \sin^2(q_0/2)z^2)}}.$$

For the experiments, we choose $q_0 = \pi/2$; thus, the period μ is equal to 7.416298709205487. We use the sixth and eighth order BSHO and GRK methods and the sixth order EMHO method with stepsize $h = \mu/20$ to integrate the problem over $2 \cdot 10^4$ periods. Setting $y = (q, p)$, again, the errors $\|y_j - y_0\|$ in the solution are evaluated at times that are multiples of the period μ, that is for $t_j = \mu j$, with $j = 1, 2, \ldots$; the energy error $H(y_n) - H(y_0)$ has been computed at the mesh points $t_n = 11hn, n = 1, 2, \ldots$. Figure 3 reports the obtained results. In the left plot, we can see that, for all the considered methods, the error in the solution grows linearly as time increases. A near conservation of the energy function is observable in both pictures on the right. The amplitudes of the bounded oscillations are similar for both methods, confirming the good long-time behavior properties of BSHO methods for the problem at hand. To check the convergence behavior of the continuous extensions, we integrated the problem over 10 periods starting with stepsize $h = \mu/N, N = 10$. We computed a reference solution using the order eight method with a halved stepsize, and we compute the maximum absolute error on the doubled grid. The results are reported in Table 4 for the solution and the first derivative and clearly show, also for this example, that the continuous extension respects the theoretical order of convergence.

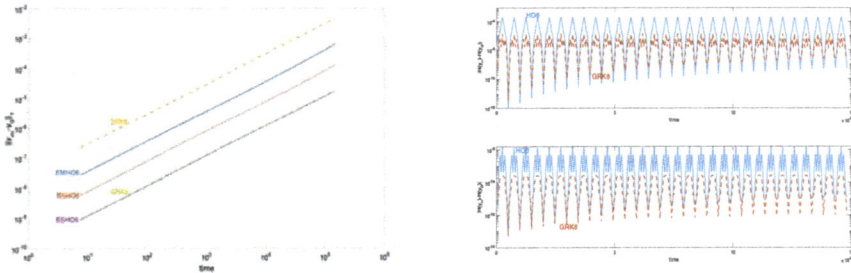

Figure 3. Nonlinear pendulum problem: results for the Hermite–Obreshkov method of order six and eight (BSHO6, red, and BSHO8, purple dotted lines), for the sixth order Euler–Maclaurin (EMHO6, blue solid line) and Gauss methods (GRK6, yellow, and GRK8, green dashed lines) applied to the pendulum problem. (**Left**) plot: absolute error of the numerical solution; (**upper-right**) and (**bottom-right**) plots: error in the Hamiltonian function for the sixth order and eighth order integrators, respectively.

Table 4. Nonlinear pendulum problem: Maximum absolute error of the numerical solution and its derivative computed for 10 periods.

Order	N	erry	Rate	erry	Rate
4	10	$1.26 \cdot 10^{-2}$		$1.28 \cdot 10^{-2}$	
4	20	$9.02 \cdot 10^{-4}$	3.81	$1.10 \cdot 10^{-3}$	3.53
4	40	$5.73 \cdot 10^{-5}$	3.97	$6.60 \cdot 10^{-5}$	4.06
4	80	$3.58 \cdot 10^{-6}$	4.00	$4.52 \cdot 10^{-6}$	3.86
6	10	$2.65 \cdot 10^{-4}$		$2.82 \cdot 10^{-4}$	
6	20	$1.36 \cdot 10^{-6}$	7.59	$5.77 \cdot 10^{-6}$	5.61
6	40	$2.07 \cdot 10^{-8}$	6.04	$1.15 \cdot 10^{-8}$	5.65
6	80	$3.21 \cdot 10^{-10}$	6.01	$1.81 \cdot 10^{-9}$	5.98
8	10	$2.56 \cdot 10^{-5}$		$2.61 \cdot 10^{-5}$	
8	20	$1.53 \cdot 10^{-8}$	10.7	$8.50 \cdot 10^{-8}$	8.26
8	40	$6.14 \cdot 10^{-11}$	7.96	$4.02 \cdot 10^{-10}$	7.72
8	80	$3.01 \cdot 10^{-13}$	7.67	$1.56 \cdot 10^{-12}$	8.01

7. Conclusions

In this paper, we have analyzed the BSHO schemes, a class of symmetric one-step multi-derivative methods firstly introduced in [14,15] for the numerical solution of the Cauchy problem. As a new result, we have proven that these are conjugate symplectic schemes up to order $2R + 2$, thus suited to the context of geometric integration. Moreover, an efficient approach for the computation of the B-spline form of the spline extending the numerical solution produced by any BSHO method has been presented. The spline associated with the R-th BSHO method collocates the differential equation at the mesh points with multiplicity R and approximates the solution of the considered differential problem with the same accuracy $O(h^{2R})$ characterizing the numerical solution. The relation between BSHO schemes and symmetric linear multistep methods when the derivatives are approximated by finite differences has also been pointed out.

Future related work will consist in studying the possibility of associating with the BSHO schemes a dual quasi-interpolation approach, as already done dealing with the *BS* linear multistep methods in [16,18,32].

Author Contributions: Conceptualization, F.M. and A.S.. Formal analysis, F.M. and A.S. Investigation, F.M. and A.S. Methodology, F.M. and A.S. Writing, original draft, F.M. and A.S. Writing, review and editing, F.M. and A.S.

Funding: Supported by INdAM through GNCS 2018 research projects.

Acknowledgments: We thank Felice Iavernaro for helpful discussions related to this research and the anonymous referees for their careful reading and useful remarks.

Conflicts of Interest: The authors declare no conflict of interest.

References

1. Hairer, E.; Nørsett, S.; Wanner, G. *Solving Ordinary Differential Equations I. Nonstiff Problems*, 2nd ed.; Springer: Berlin, Germany, 1993.
2. Ghelardoni, P.; Marzulli, P. Stability of some boundary value methods for IVPs. *Appl. Numer. Math.***1995**, *18*, 141–153. [CrossRef] [CrossRef]
3. Estévez Schwarz, D.; Lamour, R. A new approach for computing consistent initial values and Taylor coefficients for DAEs using projector-based constrained optimization. *Numer. Algorithms* **2018**, *78*, 355–377. [CrossRef]
4. Baeza, A.; Boscarino, S.; Mulet, P.; Russo, G.; Zorío, D. Approximate Taylor methods for ODEs. *Comput. Fluids* **2017**, *159*, 156–166. [CrossRef] [CrossRef]
5. Skvortsov, L. A fifth order implicit method for the numerical solution of differential-algebraic equations. *Comput. Math. Math. Phys.* **2015**, *55*, 962–968. [CrossRef] [CrossRef]

6. Amodio, P.; Iavernaro, F.; Mazzia, F.; Mukhametzhanov, M.; Sergeyev, Y. A generalized Taylor method of order three for the solution of initial value problems in standard and infinity floating-point arithmetic. *Math. Comput. Simul* **2017**, *141*, 24–39. [CrossRef] [CrossRef]

7. Sergeyev, Y.; Mukhametzhanov, M.; Mazzia, F.; Iavernaro, F.; Amodio, P. Numerical methods for solving initial value problems on the Infinity Computer. *Int. J. Unconv. Comput.* **2016**, *12*, 3–23.

8. Butcher, J.; Sehnalová, P. Predictor-corrector Obreshkov pairs. *Computing* **2013**, *95*, 355–371. [CrossRef] [CrossRef]

9. Hairer, E.; Lubich, C.; Wanner, G. *Geometric Numerical Integration. Structure-Preserving Algorithms for Ordinary Differential Equations*, 2nd ed.; Springer: Berlin, Germany, 2006.

10. Mackey, D.S.; Mackey, N.; Tisseur, F. Structured tools for structured matrices. *Electron. J. Linear Algebra* **2003**, *10*, 106–145. [CrossRef] [CrossRef]

11. Hairer, E.; Zbinden, C.J. On conjugate symplecticity of B-series integrators. *IMA J. Numer. Anal.* **2013**, *33*, 57–79. [CrossRef]

12. Iavernaro, F.; Mazzia, F. Symplecticity properties of Euler–Maclaurin methods. In Proceedings of the AIP Conference, Thessaloniki, Greece, 25–30 September 2017; Volume 1978.

13. Iavernaro, F.; Mazzia, F.; Mukhametzhanov, M.; Sergeyev, Y. Conjugate symplecticity of Euler–Maclaurin methods and their implementation on the Infinity Computer. *arXiv* **2018**, arXiv:1807.10952. *Applied Numerical Mathematics.* in press.

14. Loscalzo, F. An introduction to the application of spline functions to initial value problems. In *Theory and Applications of Spline Functions*; Academic Press: New York, NY, USA, 1969; pp. 37–64.

15. Loscalzo, F. On the Use of Spline Functions for the Numerical Solution Of Ordinary Differential Equations, Ph.D. Thesis, University of Wisconsin, Madison, WI, USA, 1968.

16. Mazzia, F.; Sestini, A.; Trigiante, D. B-spline linear multistep methods and their continuous extensions. *SIAM J. Numer. Anal.* **2006**, *44*, 1954–1973. [CrossRef] [CrossRef]

17. Mazzia, F.; Sestini, A.; Trigiante, D. BS linear multistep methods on non-uniform meshes. *J. Numer. Anal. Ind. Appl. Math.* **2006**, *1*, 131–144.

18. Mazzia, F.; Sestini, A.; Trigiante, D. The continuous extension of the B-spline linear multistep methods for BVPs on non-uniform meshes. *Appl. Numer. Math.* **2009**, *59*, 723–738. [CrossRef] [CrossRef]

19. Hairer, E.; Wanner, G. *Solving Ordinary Differential Equations II. Stiff and Differential Algebraic Problems*, 2nd ed.; Springer: Berlin, Germany, 1996.

20. Chartier, P.; Faou, E.; Murua, A. An algebraic approach to invariant preserving integrators: The case of quadratic and hamiltonian invariants. *Numer. Math.* **2006**, *103*, 575–590. [CrossRef] [CrossRef]

21. Hairer, E. Conjugate-symplecticity of linear multistep methods. *J. Comput. Math.* **2008**, *26*, 657–659.

22. Hairer, E.; Murua, A.; Sanz-Serna, J. The non-existence of symplectic multi-derivative Runge–Kutta methods. *BIT* **1994**, *34*, 80–87. [CrossRef] [CrossRef]

23. Lyche, T.; Shumaker, L.L. Local spline approximation methods. *J. Approx. Theory* **1975**, *15*, 294–325. [CrossRef] [CrossRef]

24. Weinstein, M.J.; Rao, A.V. Algorithm 984: ADiGator, a toolbox for the algorithmic differentiation of mathematical functions in MATLAB using source transformation via operator overloading. *ACM Trans. Math. Softw.* **2017**, *44*, 21:1–21:25. [CrossRef] [CrossRef]

25. Brugnano, L.; Trigiante, D. *Solving Differential Problems by Multistep Initial and Boundary Value Methods*; CRC Press: Boca Raton, FL, USA, 2016.

26. Neta, B.; Fukushima, T. Obrechkoff versus super-implicit methods for the solution of first- and second-order initial value problems. *Comput. Math. Appl.* **2003**, *45*, 383–390. [CrossRef] [CrossRef]

27. Mazzia, F.; Trigiante, D. A hybrid mesh selection strategy based on conditioning for boundary value ODE problems. *Numer. Algorithms* **2004**, *36*, 169–187. [CrossRef] [CrossRef]

28. Iavernaro, F.; Mazzia, F.; Trigiante, D. Multistep methods for conservative problems. *Med. J. Math.* **2005**, *2*, 53–69. [CrossRef] [CrossRef]

29. Mazzia, F.; Pavani, R. Symmetric block BVMs for the solution of conservative systems. In Proceedings of the AIP Conference Proceedings, Rhodes, Greece, 21–27 September 2013; Volume 1558, pp. 738–741.

30. Shampine, L.F.; Jay, L.O. Dense Output. In *Encyclopedia of Applied and Computational Mathematics*; Springer: Berlin, Germany, 2015; pp. 339–345.

31. Brugnano, L.; Iavernaro, F. *Line Integral Methods for Conservative Problems*; Monographs and Research Notes in Mathematics; Gordon and Breach Science Publishers: Amsterdam, The Netherlands, 1998.
32. Mazzia, F.; Sestini, A. The BS class of Hermite spline quasi-interpolants on nonuniform knot distributions. *BIT Numer. Math.* **2009**, *49*, 611–628. [CrossRef] [CrossRef]

axioms

MDPI

Review

Line Integral Solution of Differential Problems

Luigi Brugnano [1] and **Felice Iavernaro** [2,*]

1   Dipartimento di Matematica e Informatica "U. Dini", Università di Firenze, Viale Morgagni 67/A, 50134 Firenze, Italy; luigi.brugnano@unifi.it
2   Dipartimento di Matematica, Università di Bari, Via Orabona 4, 70125 Bari, Italy
*   Correspondence: felice.iavernaro@uniba.it; Tel.: +39-080-544-2703

Received: 4 May 2018; Accepted: 28 May 2018; Published: 1 June 2018

Abstract: In recent years, the numerical solution of differential problems, possessing constants of motion, has been attacked by imposing the vanishing of a corresponding line integral. The resulting methods have been, therefore, collectively named (*discrete*) *line integral methods*, where it is taken into account that a suitable numerical quadrature is used. The methods, at first devised for the numerical solution of Hamiltonian problems, have been later generalized along several directions and, actually, the research is still very active. In this paper we collect the main facts about line integral methods, also sketching various research trends, and provide a comprehensive set of references.

Keywords: conservative problems; Hamiltonian problems; energy-conserving methods; Poisson problems; Hamiltonian Boundary Value Methods; HBVMs; line integral methods; constrained Hamiltonian problems; Hamiltonian PDEs; highly oscillatory problems

MSC: 65P10; 65L06; 65L05

1. Introduction

The numerical solution of differential problems in the form

$$\dot{y}(t) = f(y(t)), \quad t \geq 0, \qquad y(0) = y_0 \in \mathcal{D} \subseteq \mathbb{R}^m, \tag{1}$$

is needed in a variety of applications. In many relevant instances, the solution has important *geometric properties* and the name *geometric integrator* has been coined to denote a numerical method able to preserve them (see, e.g., the monographs [1–4]). Often, the geometric properties of the vector field are summarized by the presence of *constants of motion*, namely functions of the state vector which are conserved along the solution trajectory of (1). For this reason, in such a case one speaks about a *conservative problem*. For sake of simplicity, let us assume, for a while, that there exists only one constant of motion, say

$$C(y(t)) \equiv C(y_0), \qquad \forall t \geq 0, \qquad \forall y_0 \in \mathcal{D}, \tag{2}$$

along the solution $y(t)$ of (1). Hereafter, we shall assume that both $f : \mathcal{D} \to \mathbb{R}^m$ and $C : \mathcal{D} \to \mathbb{R}$ are suitably smooth (e.g., analytical). In order for the conservation property (2) to hold, one requires that

$$\frac{d}{dt} C(y(t)) = \nabla C(y(t))^\top \dot{y}(t) = \nabla C(y(t))^\top f(y(t)) = 0.$$

Consequently, one obtains the equivalent condition

$$\nabla C(y)^\top f(y) = 0, \qquad \forall y \in \mathcal{D}. \tag{3}$$

However, the conservation property (2) can be equivalently restated through the vanishing of a line integral:

$$C(y(t)) - C(y_0) = C(y(t)) - C(y(0)) = \int_0^t \nabla C(y(\tau))^\top \dot{y}(\tau) \mathrm{d}\tau = 0, \qquad \forall t \geq 0. \qquad (4)$$

In fact, since y satisfies (1), one obtains that the integrand is given by

$$\nabla C(y(\tau))^\top f(y(\tau)) \equiv 0,$$

because of (3). On the other hand, if one is interested in obtaining an approximation to y, ruled by a discrete-time dynamics with time-step h, one can look for any path σ joining y_0 to $y_1 \approx y(h)$, i.e.,

$$\sigma(0) = y_0, \qquad \sigma(h) = y_1, \qquad (5)$$

and such that

$$C(y_1) - C(y_0) \equiv C(\sigma(h)) - C(\sigma(0)) = \int_0^h \nabla C(\sigma(t))^\top \dot{\sigma}(t) \mathrm{d}t = h \int_0^1 \nabla C(\sigma(ch))^\top \dot{\sigma}(ch) \mathrm{d}c = 0. \quad (6)$$

Definition 1. *The path σ satisfying (5) and (6) defines a* line integral method *providing an approximation y_1 to $y(h)$ such that $C(y_1) = C(y_0)$.*

Obviously, the process is then repeated on the interval $[h, 2h]$, starting from y_1, and so on. We observe that the path σ now provides the vanishing of the line integral in (6) without requiring the integrand be identically zero. This, in turn, allows much more freedom during the derivation of such methods. In addition to this, it is important to observe that one cannot, in general, directly impose the vanishing of the integral in (6) since, in most cases, the integrand function does not admit a closed form antiderivative. Consequently, in order to obtain a *ready to use numerical method*, the use of a suitable quadrature rule is mandatory.

Since we shall deal with *polynomial paths σ*, it is natural to look for an interpolatory quadrature rule defined by the abscissae and weights (c_i, b_i), $i = 1, \ldots, k$. In order to maximize the order of the quadrature, i.e., $2k$, we place the abscissae at the zeros of the kth shifted and scaled Legendre polynomial P_k (i.e., $P_k(c_i) = 0$, $i = 1, \ldots, k$). Such polynomials provide an orthonormal basis for functions in $L^2[0, 1]$:

$$\deg(P_i) = i, \qquad \int_0^1 P_i(x) P_j(x) \mathrm{d}x = \delta_{ij}, \qquad \forall i, j = 0, 1, \ldots, \qquad (7)$$

with δ_{ij} denoting the Kronecker delta. Consequently, (6) becomes

$$C(y_1) - C(y_0) = h \int_0^1 \nabla C(\sigma(ch))^\top \dot{\sigma}(ch) \mathrm{d}c \approx h \sum_{i=1}^k b_i \nabla C(\sigma(c_i h))^\top \dot{\sigma}(c_i h) = 0. \qquad (8)$$

Definition 2. *The path σ satisfying (5) and (8) defines a* discrete line integral method *providing an approximation y_1 to $y(h)$ such that $C(y_1) \approx C(y_0)$, within the accuracy of the quadrature rule.*

As is clear, if C is such that the quadrature in (8) is exact, then the method reduces to the line integral method satisfying (5) and (6), exactly conserving the invariant. In the next sections we shall make the above statements more precise and operative.

Line integral methods were at first studied to derive energy-conserving methods for Hamiltonian problems: a coarse idea of the methods can be found in [5,6]; the first instances of such methods were then studied in [7–9]; later on, the approach has been refined in [10–13] and developed in [14–18]. Further generalizations, along several directions, have been considered in [19–28]: in particular, in [19]

Hamiltonian boundary value problems have been considered, which are not covered in this review. The main reference on line integral methods is given by the monograph [1].

With these premises, the paper is organized as follows: in Section 2 we shall deal with the numerical solution of Hamiltonian problems; Poisson problems are then considered in Section 3; constrained Hamiltonian problems are studied in Section 4; Hamiltonian partial differential equations (PDEs) are considered in Section 5; highly oscillatory problems are briefly discussed in Section 6; at last, Section 7 contains some concluding remarks.

2. Hamiltonian Problems

A canonical Hamiltonian problem is in the form

$$\dot{y} = J\nabla H(y), \qquad y(0) = y_0 \in \mathbb{R}^{2m}, \qquad J = \begin{pmatrix} & I_m \\ -I_m & \end{pmatrix}, \tag{9}$$

with H the *Hamiltonian function* and, in general, I_r hereafter denoting the identity matrix of dimension r. Because of the skew-symmetry of J, one readily verifies that H is a constant of motion for (9):

$$\frac{\mathrm{d}}{\mathrm{d}t} H(y) = \nabla H(y)^\top \dot{y} = \nabla H(y)^\top J \nabla H(y) = 0.$$

For isolated mechanical systems, H has the physical meaning of the total energy, so that it is often referred to as the *energy*. When solving (9) numerically, it is quite clear that this conservation property becomes paramount to get a correct simulation of the underlying phenomenon. The first successful approach in the numerical solution of Hamiltonian problems has been the use of *symplectic integrators*. The characterization of a symplectic Runge-Kutta method

$$\begin{array}{c|c} c & A \\ \hline & b^\top \end{array}$$

is based on the following algebraic property of its Butcher tableau [29,30] (see also [31])

$$\Omega A + A^\top \Omega = bb^\top, \qquad \Omega = \mathrm{diag}(b), \tag{10}$$

which is tantamount to the conservation of any quadratic invariant of the continuous problem.

Under appropriate assumptions, symplectic integrators provide a bounded Hamiltonian error over long time intervals [2], whereas generic numerical methods usually exhibit a drift in the numerical Hamiltonian. Alternatively, one can look for *energy conserving methods* (see, e.g., [32–39]). We here sketch the *line integral solution* to the problem. According to (5) and (6) with $C = H$, let us set

$$\dot{\sigma}(ch) = \sum_{j=0}^{s-1} P_j(c)\gamma_j, \qquad c \in [0,1], \tag{11}$$

where the coefficients $\gamma_j \in \mathbb{R}^{2m}$ are at the moment unknown. Integrating term by term, and imposing the initial condition, yields the following polynomial of degree s:

$$\sigma(ch) = y_0 + h \sum_{j=0}^{s-1} \int_0^c P_j(x)\mathrm{d}x \, \gamma_j, \qquad c \in [0,1]. \tag{12}$$

By defining the approximation to $y(h)$ as

$$y_1 := \sigma(h) = y_0 + h\gamma_0, \tag{13}$$

where we have taken into account the orthonormality conditions (7), so that $\int_0^1 P_j(x)\,dx = \delta_{j0}$, one then obtains that the conditions (5) are fulfilled. In order to fulfil also (6) with $C = H$, one then requires, by taking into account (11)

$$
\begin{aligned}
H(y_1) - H(y_0) &= H(\sigma(h)) - H(\sigma(0)) = h\int_0^1 \nabla H(\sigma(ch))^\top \dot{\sigma}(ch)\,dc \\
&= h\sum_{j=0}^{s-1}\left[\int_0^1 P_j(c)\nabla H(\sigma(ch))\,dc\right]^\top \gamma_j = 0.
\end{aligned}
$$

This latter equation is evidently satisfied, due to the skew-symmetry of J, by setting

$$
\gamma_j = J\int_0^1 P_j(c)\nabla H(\sigma(ch))\,dc, \qquad j = 0,\ldots,s-1. \tag{14}
$$

Consequently, (12) becomes

$$
\sigma(ch) = y_0 + h\sum_{j=0}^{s-1}\int_0^c P_j(x)\,dx \int_0^1 P_j(c)J\nabla H(\sigma(ch))\,dc, \qquad c \in [0,1], \tag{15}
$$

which, according to ([12], Definition 1), is the *master functional equation* defining σ. Consequently, the conservation of the Hamiltonian is assured. Next, we discuss the order of accuracy of the obtained approximation, namely the difference $\sigma(h) - y(h)$: this will be done in the next section, by using the approach defined in [18].

2.1. Local Fourier Expansion

By introducing the notation

$$
f = J\nabla H, \qquad \gamma_j(\sigma) = \int_0^1 P_j(c)f(\sigma(ch))\,dc, \qquad j \geq 0, \tag{16}
$$

one has that (9) can be written, on the interval $[0,h]$, as

$$
\dot{y}(ch) = \sum_{j\geq 0} P_j(c)\gamma_j(y), \qquad c \in [0,1], \tag{17}
$$

with $\gamma_j(y)$ defined according to (16), by formally replacing σ with y. Similarly,

$$
f(\sigma(ch)) = \sum_{j\geq 0} P_j(c)\gamma_j(\sigma), \qquad c \in [0,1], \tag{18}
$$

with the polynomial σ in (15) satisfying, by virtue of (11) and (14), the differential equation:

$$
\dot{\sigma}(ch) = \sum_{j=0}^{s-1} P_j(c)\gamma_j(\sigma), \qquad c \in [0,1]. \tag{19}
$$

The following result can be proved (see [18], Lemma 1).

Lemma 1. *Let* $g : [0,h] \to V$, *with V a vector space, admit a Taylor expansion at 0. Then, for all* $j \geq 0$:

$$
\int_0^1 P_j(c)g(ch)\,dc = O(h^j).
$$

Moreover, let us denote by $y(t, \tilde{t}, \tilde{y})$ the solution of the ODE-IVPs

$$\dot{y}(t) = f(y(t)), \qquad t \geq \tilde{t}, \qquad y(\tilde{t}) = \tilde{y},$$

and by $\Phi(t, \tilde{t})$ the fundamental matrix solution of the variational problem associated to it. We recall that

$$\frac{\partial}{\partial \tilde{y}} y(t, \tilde{t}, \tilde{y}) = \Phi(t, \tilde{t}), \qquad \frac{\partial}{\partial \tilde{t}} y(t, \tilde{t}, \tilde{y}) = -\Phi(t, \tilde{t}) f(\tilde{y}). \tag{20}$$

We are now in the position to prove the result concerning the accuracy of the approximation (13).

Theorem 1. $\sigma(h) - y(h) = O(h^{2s+1})$ *(in other words, the polynomial σ defines an approximation procedure of order 2s).*

Proof. One has, by virtue of (5), (16)–(19), Lemma 1, and (20):

$$\sigma(h) - y(h) = y(h, h, \sigma(h)) - y(h, 0, \sigma(0)) = \int_0^h \frac{\mathrm{d}}{\mathrm{d}t} y(h, t, \sigma(t)) \mathrm{d}t$$

$$= \int_0^h \left[\frac{\partial}{\partial \tilde{t}} y(h, \tilde{t}, \sigma(t)) \Big|_{\tilde{t}=t} + \frac{\partial}{\partial \tilde{y}} y(h, t, \tilde{y}) \Big|_{\tilde{y}=\sigma(t)} \dot{\sigma}(t) \right] \mathrm{d}t = \int_0^h \Phi(h, t) \left[\dot{\sigma}(t) - f(\sigma(t)) \right] \mathrm{d}t$$

$$= h \int_0^1 \Phi(h, ch) \left[\dot{\sigma}(ch) - f(\sigma(ch)) \right] \mathrm{d}c = h \int_0^1 \Phi(h, ch) \left[\sum_{j=0}^{s-1} P_j(c) \gamma_j(\sigma) - \sum_{j \geq 0} P_j(c) \gamma_j(\sigma) \right] \mathrm{d}c$$

$$= -h \sum_{j \geq s} \underbrace{\int_0^1 P_j(c) \Phi(h, ch) \mathrm{d}c}_{=O(h^j)} \underbrace{\gamma_j(\sigma)}_{=O(h^j)} = O(h^{2s+1}). \qquad \square$$

2.2. Hamiltonian Boundary Value Methods

Quoting Dahlquist and Björk [40], p. 521, *as is well known, even many relatively simple integrals cannot be expressed in finite terms of elementary functions, and thus must be evaluated by numerical methods.* In our framework, this obvious statement means that, in order to obtain a numerical method from (15), we need to approximate the integrals appearing in that formula by means of a suitable quadrature procedure. In particular, as anticipated above, we shall use the Gauss-Legendre quadrature of order $2k$, whose abscissae and weights will be denoted by (c_i, b_i) (i.e., $P_k(c_i) = 0$, $i = 1, \ldots, k$). Hereafter, we shall obviously assume $k \geq s$. In so doing, in place of (15), one obtains a (possibly different) polynomial,

$$u(ch) = y_0 + h \sum_{j=0}^{s-1} \int_0^c P_j(x) \mathrm{d}x \sum_{\ell=1}^k b_\ell P_j(c_\ell) f(u(c_\ell h)), \qquad c \in [0, 1], \tag{21}$$

where (see (16)):

$$\hat{\gamma}_j := \sum_{\ell=1}^k b_\ell P_j(c_\ell) f(u(c_\ell h)) = \int_0^1 P_j(c) f(u(ch)) \mathrm{d}c + \Delta_j(h) \equiv \gamma_j(u) + \Delta_j(h), \tag{22}$$

with $\Delta_j(h)$ the quadrature error. Considering that the quadrature is exact for polynomial integrands of degree $2k - 1$, one has:

$$\Delta_j(h) = \begin{cases} 0, & \text{if } H \text{ is a polynomial of degree } \nu \leq (2k + s - 1 - j)/s, \\ O(h^{2k-j}), & \text{otherwise.} \end{cases} \tag{23}$$

As a consequence, $u \equiv \sigma$ if H is a polynomial of degree $\nu \leq 2k/s$. In such a case, $H(u(h)) - H(u(0)) \equiv H(\sigma(h)) - H(\sigma(0)) = 0$, i.e., the energy is exactly conserved. Differently, one has, by virtue of (22) and Lemma 1:

$$
\begin{aligned}
H(u(h)) - H(u(0)) &= h \int_0^1 \nabla H(u(ch))^\top \dot{u}(ch) \mathrm{d}c = h \int_0^1 \nabla H(u(ch))^\top \left[\sum_{j=0}^{s-1} P_j(c) \hat{\gamma}_j \right] \mathrm{d}c \\
&= h \sum_{j=0}^{s-1} \left[\int_0^1 P_j(c) \nabla H(u(ch)) \mathrm{d}c \right]^\top \sum_{\ell=1}^k b_\ell P_j(c_\ell) f(u(c_\ell h)) = h \sum_{j=0}^{s-1} \gamma_j(u)^\top J \left[\gamma_j(u) + \Delta_j(h) \right] \\
&= h \sum_{j=0}^{s-1} \underbrace{\gamma_j(u)^\top}_{=O(h^j)} J \underbrace{\Delta_j(h)}_{=O(h^{2k-j})} = O(h^{2k+1}).
\end{aligned}
$$

The result of Theorem 1 continues to hold for u. In fact, by using arguments similar to those used in the proof of that theorem, one has, by taking into account (22) and that $k \geq s$:

$$
\begin{aligned}
u(h) - y(h) &= y(h, h, u(h)) - y(h, 0, u(0)) = \int_0^h \frac{\mathrm{d}}{\mathrm{d}t} y(h, t, u(t)) \mathrm{d}t \\
&= \int_0^h \left[\frac{\partial}{\partial \tilde{t}} y(h, \tilde{t}, u(t)) \Big|_{\tilde{t}=t} + \frac{\partial}{\partial \tilde{y}} y(h, t, \tilde{y}) \Big|_{\tilde{y}=u(t)} \dot{u}(t) \right] \mathrm{d}t = \int_0^h \Phi(h, t) \left[\dot{u}(t) - f(u(t)) \right] \mathrm{d}t \\
&= h \int_0^1 \Phi(h, ch) \left[\dot{u}(ch) - f(u(ch)) \right] \mathrm{d}c = h \int_0^1 \Phi(h, ch) \left[\sum_{j=0}^{s-1} P_j(c) \hat{\gamma}_j - \sum_{j \geq 0} P_j(c) \gamma_j(u) \right] \mathrm{d}c \\
&= h \int_0^1 \Phi(h, ch) \left[\sum_{j=0}^{s-1} P_j(c) (\gamma_j(u) + \Delta_j(h)) - \sum_{j \geq 0} P_j(c) \gamma_j(u) \right] \mathrm{d}c \\
&= h \sum_{j=0}^{s-1} \underbrace{\int_0^1 P_j(c) \Phi(h, ch) \mathrm{d}c}_{=O(h^j)} \underbrace{\Delta_j(u)}_{=O(h^{2k-j})} - h \sum_{j \geq s} \underbrace{\int_0^1 P_j(c) \Phi(h, ch) \mathrm{d}c}_{=O(h^j)} \underbrace{\gamma_j(u)}_{=O(h^j)} = O(h^{2s+1}).
\end{aligned}
$$

Definition 3. *The polynomial u defined at (21) defines a Hamiltonian Boundary Value Method (HBVM) with parameters k and s, in short HBVM(k, s).*

Actually, by observing that in (21) only the values of u at the abscissae are needed, one obtains, by setting $Y_i := u(c_i h)$, and rearranging the terms:

$$
Y_i = y_0 + h \sum_{j=1}^k \left[b_j \sum_{\ell=0}^{s-1} \int_0^{c_i} P_\ell(x) \mathrm{d}x P_\ell(c_j) \right] f(Y_j), \qquad i = 1, \ldots, k, \tag{24}
$$

with the new approximation given by

$$
y_1 := u(h) = y_0 + h \sum_{i=1}^k b_i f(Y_i). \tag{25}
$$

Consequently, we are speaking about the k-stage Runge-Kutta method with Butcher tableau given by:

$$
\begin{array}{c|c}
c & \mathcal{I}_s \mathcal{P}_s^\top \Omega \\
\hline
& b^\top
\end{array} \tag{26}
$$

with

$$
c = \begin{pmatrix} c_1 \\ \vdots \\ c_k \end{pmatrix}, \qquad b = \begin{pmatrix} b_1 \\ \vdots \\ b_k \end{pmatrix}, \qquad \Omega = \begin{pmatrix} b_1 & & \\ & \ddots & \\ & & b_k \end{pmatrix}, \tag{27}
$$

and

$$
\mathcal{P}_s = \begin{pmatrix} P_0(c_1) & \cdots & P_{s-1}(c_1) \\ \vdots & & \vdots \\ P_0(c_k) & \cdots & P_{s-1}(c_k) \end{pmatrix}, \quad \mathcal{I}_s = \begin{pmatrix} \int_0^{c_1} P_0(x)\mathrm{d}x & \cdots & \int_0^{c_1} P_{s-1}(x)\mathrm{d}x \\ \vdots & & \vdots \\ \int_0^{c_k} P_0(x)\mathrm{d}x & \cdots & \int_0^{c_k} P_{s-1}(x)\mathrm{d}x \end{pmatrix} \in \mathbb{R}^{k \times s}. \tag{28}
$$

The next result summarizes the properties of HBVMs sketched above, where we also take into account that the abscissae are symmetrically distributed in the interval $[0, 1]$ (we refer to [1,18] for full details).

Theorem 2. *For all $k \geq s$, a HBVM(k,s) method:*

- *is symmetric and $y_1 - y(h) = O(h^{2s+1})$;*
- *when $k = s$ it becomes the s-stage Gauss collocation Runge-Kutta method;*
- *it is energy conserving when the Hamiltonian H is a polynomial of degree not larger than $2k/s$;*
- *conversely, one has $H(y_1) - H(y_0) = O(h^{2k+1})$.*

We conclude this section by showing that, for HBVM(k,s), whichever is the value $k \geq s$ considered, the discrete problem to be solved has dimension s, *independently of k*. This fact is of paramount importance, in view of the use of relatively large values of k, which are needed, in order to gain a (at least practical) energy conservation. In fact, even for non polynomial Hamiltonians, one obtains a practical energy conservation, once the $O(h^{2k+1})$ Hamiltonian error falls, by choosing k large enough, within the round-off error level.

By taking into account the stage Equation (24), and considering that $Y_i = u(c_i h)$, one has that the stage vector can be written as

$$
u(ch) = e \otimes y_0 + h\mathcal{I}_s \mathcal{P}_s^\top \Omega \otimes I_{2m} f(u(ch)) =: e \otimes y_0 + h\mathcal{I}_s \otimes I_{2m} \hat{\gamma}, \tag{29}
$$

where

$$
e = \begin{pmatrix} 1 \\ \vdots \\ 1 \end{pmatrix} \in \mathbb{R}^k, \qquad u(ch) = \begin{pmatrix} u(c_1 h) \\ \vdots \\ u(c_k h) \end{pmatrix}, \qquad f(u(ch)) = \begin{pmatrix} f(u(c_1 h)) \\ \vdots \\ f(u(c_k h)) \end{pmatrix}, \tag{30}
$$

and

$$
\hat{\gamma} \equiv \begin{pmatrix} \hat{\gamma}_0 \\ \vdots \\ \hat{\gamma}_{s-1} \end{pmatrix} = \mathcal{P}_s^\top \Omega \otimes I_{2m} f(u(ch)), \tag{31}
$$

is the block vector (of dimension s) with the coefficients (22) of the polynomial u in (21). By combining (29) and (31), one then obtains the discrete problem, equivalent to (24),

$$
\hat{\gamma} = \mathcal{P}_s^\top \Omega \otimes I_{2m} f \left(e \otimes y_0 + h\mathcal{I}_s \otimes I_{2m} \hat{\gamma} \right), \tag{32}
$$

having (block) dimension s, *independently of k*. Once this has been solved, one verifies that the new approximation (25) turns out to be given by (compare also with (13)):

$$
y_1 := u(h) = y_0 + h\hat{\gamma}_0. \tag{33}
$$

Next section will concern the efficient numerical solution of the discrete problem

$$G(\hat{\gamma}) := \hat{\gamma} - \mathcal{P}_s^\top \Omega \otimes I_{2m} f(e \otimes y_0 + h\mathcal{I}_s \otimes I_{2m}\hat{\gamma}) = 0, \tag{34}$$

generated by the application of a HBVM(k, s) method. In fact, a straightforward fixed-point iteration,

$$\hat{\gamma}^{\ell+1} = \mathcal{P}_s^\top \Omega \otimes I_{2m} f\left(e \otimes y_0 + h\mathcal{I}_s \otimes I_{2m}\hat{\gamma}^\ell\right), \qquad \ell = 0, 1, \ldots,$$

may impose severe stepsize limitations. On the other hand, the application of the simplified Newton iteration for solving (34) reads, by considering that (see (27) and (28))

$$\mathcal{P}_s^\top \Omega \mathcal{I}_s = X_s \equiv \begin{pmatrix} \xi_0 & -\xi_1 & & \\ \xi_1 & 0 & \ddots & \\ & \ddots & \ddots & -\xi_{s-1} \\ & & \xi_{s-1} & 0 \end{pmatrix}, \qquad \xi_i = \frac{1}{2\sqrt{|4i^2 - 1|}}, \quad i = 0, \ldots, s-1, \tag{35}$$

and setting $f'(y_0)$ the Jacobian of f evaluated at y_0:

$$\left[I_s \otimes I_{2m} - hX_s \otimes f'(y_0)\right](\hat{\gamma}^{\ell+1} - \hat{\gamma}^\ell) = -G(\hat{\gamma}^\ell), \qquad \ell = 0, 1, \ldots. \tag{36}$$

This latter iteration, in turn, needs to factor a matrix whose size is s times larger than that of the continuous problem. This can represent an issue, when large-size problems are to be solved and/or large values of s are considered.

2.3. Blended Implementation of HBVMs

We here sketch the main facts concerning the so called *blended implementation* of HBVMs, a Newton-like iteration alternative to (36), which only requires to factor a matrix having the same size as that of the continuous problem, thus resulting into a much more efficient implementation of the methods [1,16]. This technique derives from the definition of *blended implicit methods*, which have been at first considered in [41,42], and then developed in [43–45]. Suitable blended implicit methods have been implemented in the Fortran codes BIM [46], solving stiff ODE-IVPs, and BIMD [47], also solving linearly implicit DAEs up to index 3. The latter code is also available on the *Test Set for IVP Solvers* [48] (see also [49]), and turns out to be among the most reliable and efficient codes currently available for solving stiff ODE-IVPs and linearly implicit DAEs. It is worth mentioning that, more recently, the blended implementation of RKN-type methods has been also considered [50].

Let us then consider the iteration (36), which requires the solution of linear systems in the form

$$\left[I_s \otimes I_{2m} - hX_s \otimes f'(y_0)\right] x = \eta. \tag{37}$$

By observing that matrix X_s defined at (35) is nonsingular, we can consider the *equivalent* linear system

$$\rho_s \left[X_s^{-1} \otimes I_{2m} - hI_s \otimes f'(y_0)\right] x = \rho_s X_s^{-1} \otimes I_{2m} \eta =: \eta_1, \tag{38}$$

where ρ_s is a positive parameter to be determined. For this purpose, let

$$f'(y_0) = V\Lambda V^{-1},$$

be the Jordan canonical form of $f'(y_0)$. For simplicity, we shall assume that Λ is diagonal, and let λ be any of its diagonal entries. Consequently, the two linear systems (37) and (38), projected in the invariant subspace corresponding to that entry, respectively become

$$[I_s - qX_s]x = \eta, \qquad \rho_s[X_s^{-1} - qI_s]x = \eta_1, \qquad q = h\lambda, \tag{39}$$

again being equivalent to each other (i.e., having the same solution $x \in \mathbb{R}^s$). We observe that the coefficient matrix of the former system is $I_s + O(q)$, when $q \approx 0$, whereas that of the latter one is $-\rho_s q(I_s + O(q^{-1}))$, when $|q| \gg 1$. Consequently, one would like to solve the former system, when $q \approx 0$, and the latter one, when $|q| \gg 1$. This can be done automatically by considering a *weighting function* $\theta(q)$ such that

$$\theta(0) = I_s, \qquad \text{and} \qquad \theta(q) \to O, \quad \text{as} \quad q \to \infty, \tag{40}$$

then considering the *blending* of the two equivalent systems (39) with weights $\theta(q)$ and $I_s - \theta(q)$, respectively:

$$M(q)x = \eta(q), \tag{41}$$
$$M(q) = \theta(q)[I_s - qX_s] + (I_s - \theta(q))\rho_s[X_s^{-1} - qI_s], \qquad \eta(q) = \theta(q)\eta + (I_s - \theta(q))\eta_1.$$

In particular, (40) can be accomplished by choosing

$$\theta(q) := I_s \cdot (1 - \rho_s q)^{-1} \equiv I_s \cdot (1 - h\rho_s \lambda)^{-1}. \tag{42}$$

Consequently, one obtains that

$$M(q) = I_s + O(q), \quad q \approx 0, \qquad \text{and} \qquad M(q) = -\rho_s q(I_s + O(q^{-1})), \quad |q| \gg 1.$$

As a result, one can consider the following splitting for solving the problem:

$$N(q)x = (N(q) - M(q))x + \eta(q), \qquad N(q) = I_s \cdot (1 - \rho_s q) \equiv \theta(q)^{-1}.$$

The choice of the scalar parameter ρ_s is then made in order to optimize the convergence properties of the corresponding iteration. According to the analysis in [42], we consider

$$\rho_s = \min_{\mu \in \sigma(X_s)} |\mu|, \tag{43}$$

where, as is usual, $\sigma(X_s)$ denotes the spectrum of matrix X_s. A few values of ρ_s are listed in Table 1.

Table 1. A few values of the parameter defined at (43).

s	1	2	3	4	5	6	7	8	9	10
ρ_s	0.5	0.2887	0.1967	0.1475	0.1173	0.09710	0.08265	0.07185	0.06348	0.05682

Coming back to the original iteration (36), one has that the weighting function (42) now becomes

$$\Theta = I_s \otimes [I_{2m} - h\rho_s f'(y_0)]^{-1} =: I_s \otimes \Sigma, \tag{44}$$

which requires to factor only the matrix

$$I_{2m} - h\rho_s f'(y_0) \in \mathbb{R}^{2m \times 2m},$$

having the same size as that of the continuous problem, thus obtaining the following *blended iteration* for HBVMs:

$$\eta^\ell = -G(\gamma^\ell), \quad \eta_1^\ell = \rho_s X_s^{-1} \otimes I_{2m} \eta^\ell, \quad \hat{\gamma}^{\ell+1} = \hat{\gamma}^\ell + I_s \otimes \Sigma \left[\eta_1^\ell + I_s \otimes \Sigma (\eta^\ell - \eta_1^\ell) \right], \quad \ell = 0, 1, \dots. \tag{45}$$

It is worth mentioning that:

- in the special case of separable Hamiltonian problems, the blended implementation of the methods can be made even more efficient, since the discrete problem can be cast in terms of the generalized coordinates only (see [16] or ([1], Chapter 4));
- the coding of the blended iteration becomes very high-performance by considering a matrix formulation of (45) (see, e.g., ([1], Chapter 4.2.2) or [51]). As matter of fact, it has been actually implemented in the Matlab code hbvm, which is freely available on the internet at the url [52].

In order to give evidence of the usefulness of energy conservation, let us consider the solution of the well-known pendulum problem, with Hamiltonian

$$H(q, p) = \frac{1}{2} p^2 - \cos(q). \tag{46}$$

When considering the trajectory starting at [1,53]

$$q(0) = 0, \quad p(0) = 1.99999, \tag{47}$$

one obtains a periodic solution of period $T \approx 28.57109480185544$. In Table 2 we list the obtained results when solving the problem over 10 periods, by using a stepsize $h = T/n$, with HBVM(6,3) and HBVM(3,3) (i.e., the symplectic 3-stage Gauss collocation method), for increasing values of n. As one may see, even though both methods are 6th order accurate, nevertheless, HBVM(6,3) becomes (practically) energy-conserving as soon as $n \geq 40$, whereas HBVM(3,3) does not. One clearly sees that, for the problem at hand, the energy-conserving method is pretty more accurate than the non conserving one.

Table 2. Solution error e_y ($y = (q, p)^\top$) and Hamiltonian error e_H when solving problem (46)–(47) with stepsize $h = T/n$.

	HBVM(6,3)			**HBVM(3,3)**		
n	e_y	Rate	e_H	e_y	Rate	e_H
20	5.12×10^{-3}	—	2.78×10^{-8}	9.13×10^1	—	1.37×10^{-3}
30	2.60×10^{-4}	7.4	1.05×10^{-11}	3.80	7.8	5.18×10^{-4}
40	1.41×10^{-4}	2.1	0.00	2.93	0.9	1.11×10^{-5}
50	3.65×10^{-5}	6.1	2.22×10^{-16}	3.13	-0.3	1.05×10^{-5}
60	1.22×10^{-5}	6.0	0.00	2.88	0.5	2.93×10^{-6}
70	4.88×10^{-6}	5.9	0.00	1.81	3.0	1.00×10^{-6}
80	2.27×10^{-6}	5.7	2.22×10^{-16}	9.06×10^{-1}	5.2	5.24×10^{-7}
90	1.15×10^{-6}	5.8	2.22×10^{-16}	4.53×10^{-1}	5.9	1.06×10^{-7}
100	6.23×10^{-7}	5.8	1.11×10^{-16}	2.40×10^{-1}	6.0	1.74×10^{-8}

2.4. Energy and QUadratic Invariants Preserving (EQUIP) Methods

According to Theorem 2, when $k = s$, HBVM(s,s) reduces to the s-stage Gauss method. For such a method, one has, with reference to (26)–(28) and (35) with $k = s$,

$$\mathcal{P}_s X_s = \mathcal{I}_s, \quad \mathcal{P}_s^{-1} = \mathcal{P}_s^\top \Omega, \tag{48}$$

so that the Butcher matrix in (26) becomes the W-transformation [54] of the s-stage Gauss method, i.e., the Butcher matrix is $A = \mathcal{P}_s X_s \mathcal{P}_s^{-1}$. Moreover, since $A = \mathcal{P}_s X_s \mathcal{P}_s^\top \Omega$, the method is easily verified to be symplectic, because of (10). In fact, by setting in general $e_i \in \mathbb{R}^s$ the ith unit vector, and with reference to the vector e defined in (30) with $k = s$, one has:

$$\Omega A + A^\top \Omega = \Omega \mathcal{P}_s (X_s + X_s^\top) \mathcal{P}_s^\top \Omega = \Omega \mathcal{P}_s (e_1 e_1^\top) \mathcal{P}_s^\top \Omega = \Omega e e^\top \Omega = bb^\top.$$

We would arrive at the very same conclusion if we replace X_s by

$$X_s(\alpha) := X_s - \alpha V, \qquad V^\top = -V, \qquad \alpha \in \mathbb{R}.$$

In particular, if the parameter α is small enough, matrix αV will act as a perturbation of the underlying Gauss formula and the question is whether it is possible to choose α, at each integration step, such that the resulting integrator may be energy conserving. In order for $V \neq O$, we need to assume, hereafter, $s \geq 2$. In particular, by choosing

$$V = e_2 e_1^\top - e_1 e_2^\top,$$

it is possible to show [26] that the scalar parameter α can be chosen, at each integration step, such that, when solving the Hamiltonian problem (9) with a sufficiently small stepsize h:

- $\alpha = O(h^{2s-2})$,
- the method retains the order $2s$ of the original s-stage Gauss method,
- $H(y_1) = H(y_0)$.

This fact is theoretically intriguing, since this means that we have a kind of state-dependent Runge-Kutta method, defined by the Butcher tableau

$$\begin{array}{c|c} c & \mathcal{P}_s[X_s - \alpha(e_2 e_1^\top - e_1 e_2^\top)]\mathcal{P}_s^{-1} \\ \hline & b^\top \end{array}, \qquad b = \begin{pmatrix} b_1 \\ \vdots \\ b_s \end{pmatrix}, \qquad c = \begin{pmatrix} c_1 \\ \vdots \\ c_s \end{pmatrix}, \tag{49}$$

which is energy conserving and is defined, at each integration step, by a symplectic map, given by a small perturbation of that of the underlying s-stage Gauss method. Consequently, EQUIP methods do not infringe the well-known result about the nonexistence of energy conserving symplectic numerical methods [55,56]. Since the symplecticity condition (10) is equivalent to the conservation of all quadratic invariants of the problem, these methods have been named *Energy and QUadratic Invariants Preserving* (*EQUIP*) methods [26,57]. It would be interesting to study the extent to which the solutions generated by an EQUIP method inherit the good long time behavior of the associated Gauss integrator with reference to the nearly conservation property of further non-quadratic first integrals. This investigation would likely involve a backward error analysis approach, similar to that done in [2], and up to now remains an open question.

For a thorough analysis of such methods we refer to [26,53]. In the next section, we sketch their line-integral implementation when solving Poisson problems, a wider class than that of Hamiltonian problems.

3. Poisson Problems

Poisson problems are in the form

$$\dot{y} = B(y)\nabla H(y), \qquad y(0) = y_0 \in \mathbb{R}^n, \qquad B(y)^\top = -B(y). \tag{50}$$

When $B(y) \equiv J$ as defined in (9), then one retrieves canonical Hamiltonian problems. As in that case, since $B(y)$ is skew-symmetric, then H, still referred to as the *Hamiltonian*, is conserved:

$$\frac{d}{dt} H(y) = \nabla H(y)^\top \dot{y} = \nabla H(y)^\top B(y)\nabla H(y) = 0.$$

Moreover, any scalar function $C(y)$ such that $\nabla C(y)^\top B(y) = \mathbf{0}^\top$ is also conserved, since:

$$\frac{\mathrm{d}}{\mathrm{d}t} C(y) = \nabla C(y)^\top \dot{y} = \underbrace{\nabla C(y)^\top B(y)}_{=\mathbf{0}^\top} \nabla H(y) = 0.$$

C is called a *Casimir function* for (50). Consequently, all possible Casimirs and the Hamiltonian H are conserved quantities for (50). In the sequel, we show that EQUIP methods can be conveniently used for solving such problems. As before, the scalar parameter α in (49) is selected in such a way that the Hamiltonian H is conserved. Moreover, since the Butcher matrix in (49) satisfies (10), then all quadratic Casimirs turn out to be conserved as well. The conservation of all quadratic invariants, in turn, is an important property as it has been observed in [58].

Let us then sketch the choice of the parameter α to gain energy conservation (we refer to [53] for full details). By setting

$$\boldsymbol{\phi}_i \equiv \begin{pmatrix} \phi_{i0} \\ \vdots \\ \phi_{i,s-1} \end{pmatrix} := X_s^{-1} e_i, \qquad i = 1, 2,$$

one has that the Butcher matrix in (49) can be written as

$$A = \mathcal{P}_s X_s [I_s - \alpha(\boldsymbol{\phi}_2 e_1^\top - \boldsymbol{\phi}_1 e_2^\top)] \mathcal{P}_s^\top \Omega.$$

Consequently, by denoting $f(y) = B(y)\nabla H(y)$, and setting $Y_i := u(c_i h)$, $i = 1, \dots, s$, the stages of the method, one obtains that the polynomial $u(ch)$ is given by

$$u(ch) = y_0 + h \sum_{j=0}^{s-1} \int_0^c P_j(x)\mathrm{d}x \, [\hat{\gamma}_j - \alpha(\phi_{2j}\hat{\gamma}_0 - \phi_{1j}\hat{\gamma}_1)], \qquad c \in [0, 1], \tag{51}$$

with the (block) vectors $\hat{\gamma}_j$ formally defined as in (22) with $k = s$. In vector form, one has then (compare with (31)):

$$\hat{\gamma} \equiv \begin{pmatrix} \hat{\gamma}_0 \\ \vdots \\ \hat{\gamma}_{s-1} \end{pmatrix} = \mathcal{P}_s^\top \Omega \otimes I_n f(u(ch)). \tag{52}$$

We observe that, from (51), one also obtains:

$$\dot{u}(ch) = \sum_{j=0}^{s-1} P_j(c) \, [\hat{\gamma}_j - \alpha(\phi_{2j}\hat{\gamma}_0 - \phi_{1j}\hat{\gamma}_1)], \qquad c \in [0, 1]. \tag{53}$$

Nevertheless, the new approximation, still given by

$$y_1 = y_0 + h \sum_{i=1}^{s} b_i f(Y_i) \equiv y_0 + h\hat{\gamma}_0, \tag{54}$$

now differs from

$$u(h) = y_0 + h[\hat{\gamma}_0 - \alpha(\phi_{20}\hat{\gamma}_0 - \phi_{10}\hat{\gamma}_1)] \equiv y_1 - \alpha h(\phi_{20}\hat{\gamma}_0 - \phi_{10}\hat{\gamma}_1). \tag{55}$$

Consequently, in order to define a path joining y_0 to y_1, to be used for imposing energy-conservation by zeroing a corresponding line-integral, we can consider the polynomial path made up by u plus

$$w(c) = u(h) + c\alpha h(\phi_{20}\hat{\gamma}_0 - \phi_{10}\hat{\gamma}_1) \qquad \Rightarrow \qquad \dot{w}(c) \equiv \alpha h(\phi_{20}\hat{\gamma}_0 - \phi_{10}\hat{\gamma}_1), \qquad c \in [0, 1]. \tag{56}$$

As a result, by considering that $w(1) = y_1$, $w(0) = u(h)$, $u(0) = y_0$, we shall choose α such that (see (51)–(56)):

$$H(y_1) - H(y_0) \;=\; H(w(1)) - H(w(0)) + H(u(h)) - H(u(0))$$

$$=\; \int_0^1 \nabla H(w(c))^\top \dot{w}(c)\,\mathrm{d}c + h \int_0^1 \nabla H(u(ch))^\top \dot{u}(ch)\,\mathrm{d}c = 0. \tag{57}$$

In more details, by defining the vectors

$$\rho_j(u) = \int_0^1 P_j(c) \nabla H(u(ch))\,\mathrm{d}c, \quad j = 0, \dots, s-1, \qquad \bar{\rho}_0(w) = \int_0^1 \nabla H(w(c))\,\mathrm{d}c, \tag{58}$$

and resorting to the usual line integral argument, it is possible to prove the following result ([53], Theorem 2).

Theorem 3. (57) *holds true, provided that*

$$\alpha = \frac{\sum_{j=0}^{s-1} \rho_j(u)^\top \hat{\gamma}_j}{(\rho_0(u) - \bar{\rho}_0(w))^\top (\phi_{20}\hat{\gamma}_0 - \phi_{10}\hat{\gamma}_1) + \sum_{j=1}^{s-1} \rho_j(u)^\top (\phi_{2j}\hat{\gamma}_0 - \phi_{1j}\hat{\gamma}_1)}. \tag{59}$$

As in the case of HBVMs, however, we shall obtain a practical numerical method only provided that the integrals in (58) are suitably approximated by means of a quadrature which, as usual, we shall choose as the Gauss-Legendre formula of order $2k$. In so doing, one obtains an $EQUIP(k,s)$ *method*. The following result can be proved ([53], Theorem 8).

Theorem 4. *Under suitable regularity assumptions on both $B(y)$ and $H(y)$, one has that for all $k \geq s$, the EQUIP(k,s) method has order $2s$, conserves all quadratic invariants and, moreover,*

$$H(y_1) - H(y_0) = \begin{cases} 0, & \text{if } H \text{ is a polynomial of degree } v \leq 2k/s, \\ O(h^{2k+1}), & \text{otherwise.} \end{cases}$$

We observe that, for EQUIP(k,s), even a not exact conservation of the energy may result in a much better error growth, as the next example shows. We consider the Lotka-Volterra problem [53], which is in the form (50) with

$$y = \begin{pmatrix} y_1 \\ y_2 \end{pmatrix}, \qquad B(y) = \begin{pmatrix} 0 & y_1 y_2 \\ -y_1 y_2 & 0 \end{pmatrix}, \qquad H(y) = a \log y_1 - y_1 + b \log y_2 - y_2. \tag{60}$$

By choosing the following parameters and initial values,

$$a = 1, \qquad b = 2, \qquad y_1(0) = y_2(0) = 0.1, \tag{61}$$

one obtains a periodic solution of period $T \approx 7.720315563434113$. If we solve the problems (60) and (61) with the EQUIP(6,3) and the 3-stage Gauss methods with stepsize $h = T/50$ over 100 periods, we obtain the error growths, in the numerical solution, depicted in Figure 1. As one may see, the EQUIP(6,3) method (which only approximately conserves the Hamiltonian), exhibits a *linear* error growth; on the contrary, the 3-stage Gauss method (which exhibits a drift in the numerical Hamiltonian) has a *quadratic* error growth. Consequently, there is numerical evidence that EQUIP methods can be conveniently used for numerically solving Poisson problems (a further example can be found in [53]).

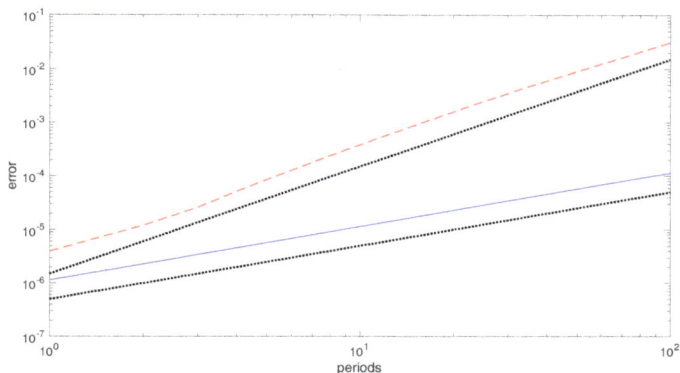

Figure 1. Error growth over 100 periods when solving problems (60) and (61) by using the EQUIP(6,3) method (blue solid line) and the 3-stage Gauss method (red dashed line) with stepsize $h = T/50 \approx 0.15$. The dotted lines show the linear and quadratic error growths.

4. Constrained Hamiltonian Problems

In this section, we report about some recent achievements concerning the line integral solution of constrained Hamiltonian problems with holonomic constraints [23]. This research is at a very early stage and, therefore, it is foreseeable that new results will follow in the future.

To begin with, let us consider the separable problem defined by the Hamiltonian

$$H(q, p) = \frac{1}{2} p^\top M^{-1} p + U(q), \qquad q, p \in \mathbb{R}^m, \tag{62}$$

with M a symmetric and positive definite matrix, subject to the $\nu < m$ holonomic constraints

$$g(q) = 0 \in \mathbb{R}^\nu. \tag{63}$$

We shall assume that the points are regular for the constraints, so that $\nabla g(q)$ has full column rank and, therefore, the $\nu \times \nu$ matrix $\nabla g(q)^\top M^{-1} \nabla g(q)$ is nonsingular. By introducing the vector of the Lagrange multipliers $\lambda \in \mathbb{R}^\nu$, problems (62) and (63) can be equivalently cast in Hamiltonian form by defining the augmented Hamiltonian

$$\hat{H}(q, p, \lambda) = H(q, p) + \lambda^\top g(q), \tag{64}$$

thus obtaining the equivalent constrained problem:

$$\dot{q} = M^{-1} p, \qquad \dot{p} = -\nabla U(q) - \nabla g(q)\lambda, \qquad g(q) = 0, \qquad t \geq 0, \tag{65}$$

subject to the initial conditions

$$q(0) = q_0, \qquad p(0) = p_0, \qquad \text{such that} \qquad g(q_0) = 0, \qquad \nabla g(q_0)^\top M^{-1} p_0 = 0. \tag{66}$$

We observe that the first requirement ($g(q_0) = 0$) obviously derives from the given constraints. The second, in turn, derives from

$$0 = \dot{g}(q) = \nabla g(q)^\top \dot{q} = \nabla g(q)^\top M^{-1} p,$$

which has to be satisfied by the solution of (65). These latter constraints are sometimes referred to as the *hidden constraints*. A formal expression of the vector of the Lagrange multipliers can be obtained by further differentiating the previous expression, thus giving

$$\left[\nabla g(q)^\top M^{-1} \nabla g(q) \right] \lambda = \nabla^2 g(q) \left(M^{-1} p, M^{-1} p \right) - \nabla g(q)^\top M^{-1} \nabla U(q),$$

which is well defined, because of the assumption that $\nabla g(q)^\top M^{-1} \nabla g(q)$ is nonsingular. Consequently,

$$\lambda = \left[\nabla g(q)^\top M^{-1} \nabla g(q) \right]^{-1} \left[\nabla^2 g(q) \left(M^{-1} p, M^{-1} p \right) - \nabla g(q)^\top M^{-1} \nabla U(q) \right] =: \lambda(q, p), \quad (67)$$

where the notation $\lambda(q, p)$ means that the vector λ is a function of q and p. It is easily seen that both the two Hamiltonians (62) and (64) are conserved along the solution of the problems (65) and (66), and assume the same value. For numerically solving the problem, we shall consider a discrete mesh with time-step h, i.e., $t_n = nh$, $n = 0, 1, \ldots$, looking for approximations

$$q_n \approx q(t_n), \qquad p_n \approx p(t_n), \qquad \lambda_n \approx \lambda(q(t_n), p(t_n)), \qquad n = 0, 1, \ldots,$$

such that, starting from (q_n, p_n), one arrives at (q_{n+1}, p_{n+1}) by choosing λ_n in order for:

$$H(q_{n+1}, p_{n+1}) = H(q_n, p_n), \qquad g(q_{n+1}) = 0, \qquad \nabla g(q_{n+1})^\top M^{-1} p_{n+1} = O(h^2). \quad (68)$$

Consequently, the new approximation conserves the Hamiltonan and exactly satisfies the constraints, but only approximately the hidden contraints. In particular, we shall consider a piecewise constant approximation of the vector of the Lagrange multipliers λ, i.e., λ_n is assumed to be constant on the interval $[t_n, t_{n+1}]$. In other words, we consider the sequence of problems (compare with (65)), for $n = 0, 1, \ldots$:

$$\dot{u} = M^{-1} v, \qquad \dot{v} = -\nabla U(u) - \nabla g(u) \lambda_n, \qquad t \in [t_n, t_{n+1}], \qquad u(t_n) = q_n, \qquad v(t_n) = p_n, \quad (69)$$

where the constant vector λ_n is chosen in order to satisfy the constraints at t_{n+1}. That is, such that the new approximations, defined as

$$q_{n+1} := u(t_{n+1}), \qquad p_{n+1} := v(t_{n+1}), \quad (70)$$

satisfy (68). The reason for choosing λ_n as a constant vector stems from the following result, which concerns the augmented Hamiltonian (64).

Theorem 5. *For all $\lambda_n \in \mathbb{R}^\nu$, the solution of (69) and (70) satisfies $\hat{H}(q_{n+1}, p_{n+1}, \lambda_n) = \hat{H}(q_n, p_n, \lambda_n)$.*

Proof. In fact, denoting by $\hat{H}_q(q, p, \lambda)$ the gradient of $\hat{H}$ with respect to the q variables, and similarly for $\hat{H}_p$, the usual line integral argument provides:

$$\hat{H}(q_{n+1}, p_{n+1}, \lambda_n) - \hat{H}(q_n, p_n, \lambda_n) = \hat{H}(u(t_{n+1}), v(t_{n+1}), \lambda_n) - \hat{H}(u(t_n), v(t_n), \lambda_n)$$

$$= \int_{t_n}^{t_{n+1}} \frac{d}{dt} \hat{H}(u(t), v(t), \lambda_n) dt = \int_{t_n}^{t_{n+1}} \left\{ \hat{H}_q(u(t), v(t), \lambda_n)^\top \dot{u}(t) + \hat{H}_p(u(t), v(t), \lambda_n)^\top \dot{v}(t) \right\} dt$$

$$= \int_{t_n}^{t_{n+1}} \left\{ [\nabla U(u(t)) + \nabla g(u(t)) \lambda_n]^\top M^{-1} v(t) - v(t)^\top M^{-1} [\nabla U(u(t)) + \nabla g(u(t)) \lambda_n] \right\} dt = 0.$$

□

Consequently, one obtains that

$$g(q_{n+1}) = g(q_n) \qquad \Leftrightarrow \qquad H(q_{n+1}, p_{n+1}) = H(q_n, p_n),$$

i.e., energy conservation is equivalent to satisfy the constraints.

In order to fulfil the constraints, we shall again resort to a line integral argument. For this purpose, we need to define the Fourier coefficients (along the Legendre basis (7)) of the functions appearing at the right-hand sides in (69):

$$\gamma_j(v, t_n) = M^{-1} \int_0^1 P_j(c) v(t_n + ch) \mathrm{d}c, \qquad \psi_j(u, t_n) = \int_0^1 P_j(c) \nabla U(u(t_n + ch)) \mathrm{d}c,$$

$$\rho_j(u, t_n) = \int_0^1 P_j(c) \nabla g(u(t_n + ch)) \mathrm{d}c, \qquad j = 0, 1, \dots, \tag{71}$$

so that, in particular,

$$u(t_n + ch) = q_n + h \sum_{j \geq 0} \int_0^c P_j(x) \mathrm{d}x \gamma_j(v, t_n), \tag{72}$$

$$v(t_n + ch) = p_n - h \sum_{j \geq 0} \int_0^c P_j(x) \mathrm{d}x \left[\psi_j(u, t_n) + \rho_j(u, t_n) \lambda_n \right], \qquad c \in [0, 1].$$

We also need the following result.

Lemma 2. *With reference to matrix X_s defined in (35), one has, for all $s = 1, 2, \dots$:*

$$\int_0^1 P_j(c) \int_0^c P_i(x) \mathrm{d}x \mathrm{d}c = (X_s)_{j+1, i+1}, \qquad i, j = 0, \dots, s - 1.$$

Proof. See, e.g., ([23], Lemma 2). □

We are now in the position of deriving a formal expression for the constant approximation to the vector of the Lagrange multipliers through the usual line integral approach:

$$\begin{aligned} 0 &= g(q_{n+1}) - g(q_n) = g(u(t_{n+1})) - g(u(t_n)) = \int_{t_n}^{t_{n+1}} \frac{\mathrm{d}}{\mathrm{d}t} g(u(t)) \mathrm{d}t = \int_{t_n}^{t_{n+1}} \nabla g(u(t))^\top \dot{u}(t) \mathrm{d}t \\ &= h \int_0^1 \nabla g(t_n + ch)^\top \dot{u}(t_n + ch) \mathrm{d}c = h \int_0^1 \nabla g(t_n + ch)^\top \sum_{j \geq 0} P_j(c) \gamma_j(v, t_n) \mathrm{d}c \\ &= h \sum_{j \geq 0} \rho_j(u, t_n)^\top \gamma_j(v, t_n) = h \sum_{j \geq 0} \rho_j(u, t_n)^\top M^{-1} \int_0^1 P_j(c) v(t_n + ch) \mathrm{d}c \\ &= h \sum_{j \geq 0} \rho_j(u, t_n)^\top M^{-1} \int_0^1 P_j(c) \left\{ p_n - h \sum_{i \geq 0} \int_0^c P_i(x) \mathrm{d}x \left[\psi_i(u, t_n) + \rho_i(u, t_n) \lambda_n \right] \right\} \mathrm{d}c \\ &= h \rho_0(u, t_n)^\top M^{-1} p_n - h^2 \sum_{i,j \geq 0} \rho_j(u, t_n)^\top M^{-1} \left[\psi_i(u, t_n) + \rho_i(u, t_n) \lambda_n \right] \int_0^1 P_j(c) \int_0^c P_i(x) \mathrm{d}x \mathrm{d}c. \end{aligned}$$

Because of Lemma 2, one then obtains that (see (35)),

$$\left(\xi_0 \rho_0(u, t_n)^\top M^{-1} \rho_0(u, t_n) + \sum_{j \geq 1} \xi_j \left[\rho_j(u, t_n)^\top M^{-1} \rho_{j-1}(u, t_n) - \rho_{j-1}(u, t_n)^\top M^{-1} \rho_j(u, t_n) \right] \right) \lambda_n$$

$$= h^{-1} \rho_0(u, t_n)^\top M^{-1} p_n - \tag{73}$$

$$\left(\xi_0 \rho_0(u, t_n)^\top M^{-1} \psi_0(u, t_n) + \sum_{j \geq 1} \xi_j \left[\rho_j(u, t_n)^\top M^{-1} \psi_{j-1}(u, t_n) - \rho_{j-1}(u, t_n)^\top M^{-1} \psi_j(u, t_n) \right] \right).$$

The following result can be proved [23].

Theorem 6. *The Equation* (73) *is consistent with* (67) *and is well defined for all sufficiently small $h > 0$. The sequence (q_n, p_n, λ_n) generated by* (69)–(73) *satisfies, for all $n = 0, 1, \dots$:*

$$q_n = q(t_n) + O(h^2), \qquad p_n = p(t_n) + O(h^2), \qquad \lambda_n = \lambda(q(t_n), p(t_n)) + O(h),$$

$$\nabla g(q_n)^\top M^{-1} p_n = O(h^2), \qquad g(q_n) = 0, \qquad H(q_n, p_n) = H(q_0, p_0).$$

Moreover, in the case where $\lambda(q(t), p(t)) \equiv \bar{\lambda}$, for all $t \geq 0$, then $q_n \equiv q(t_n)$, $p_n \equiv p(t_n)$, $\lambda_n \equiv \bar{\lambda}$.

At this point, in order to obtain a numerical method, the following two steps need to be done:

1. truncate the infinite series in (72) to finite sums, say up to $j = s - 1$. In so doing, the expression of λ_n changes accordingly, since in (73) the infinite sums will consequently arrive up to $j = s - 1$;
2. approximate the integrals in (71), for $j = 0, \dots, s - 1$. As usual, we shall consider a Gauss-Legendre formula of order $2k$, with $k \geq s$, thus obtaining corresponding approximations which we denote, for $n = 0, 1, \dots$,

$$\gamma_{j,n}, \qquad \psi_{j,n}, \qquad \rho_{j,n}, \qquad j = 0, \dots, s - 1.$$

In so doing, it can be seen that one retrieves the usual HBVM(k, s) method defined in Section 2, applied for solving (69), coupled with the following equation for λ_n, representing the approximation of (73):

$$\left(\xi_{0,n} \rho_{0,n}^\top M^{-1} \rho_{0,n} + \sum_{j=1}^{s-1} \xi_j \left[\rho_{j,n}^\top M^{-1} \rho_{j-1,n} - \rho_{j-1,n}^\top M^{-1} \rho_{j,n} \right] \right) \lambda_n$$

$$= h^{-1} \rho_{0,n}^\top M^{-1} p_n - \left(\xi_0 \rho_{0,n}^\top M^{-1} \psi_{0,n} + \sum_{j=1}^{s-1} \xi_j \left[\rho_{j,n}^\top M^{-1} \psi_{j-1,n} - \rho_{j-1,n}^\top M^{-1} \psi_{j,n} \right] \right). \tag{74}$$

The following result can be proved [23].

Theorem 7. *For all sufficiently small stepsizes h, the HBVM(k, s) method coupled with* (74), *used for solving* (65) *and* (66) *over a finite interval, is well defined and symmetric. It provides a sequence of approximations (q_n, p_n, λ_n) such that (see* (67)):*

$$q_n = q(t_n) + O(h^2), \ p_n = p(t_n) + O(h^2), \ \lambda_n = \lambda(q(t_n), p(t_n)) + O(h), \ \nabla g(q_n)^\top M^{-1} p_n = O(h^2),$$

$$g(q_n) = \begin{cases} 0, & \text{if } g \text{ is a polynomial of degree not larger that } 2k/s, \\ O(h^{2k}), & \text{otherwise,} \end{cases}$$

$$H(q_n, p_n) - H(q_0, p_0) = \begin{cases} 0, & \text{if } g \text{ is a polynomial of degree not larger that } 2k/s, \\ O(h^{2k}), & \text{otherwise.} \end{cases}$$

Moreover, in the case where $\lambda(q(t), p(t)) \equiv \bar{\lambda}$, for all $t \geq 0$, then

$$q_n = q(t_n) + O(h^{2s}), \ p_n = p(t_n) + O(h^{2s}), \ \lambda_n = \lambda(q(t_n), p(t_n)) + O(h^{2s}), \ \nabla g(q_n)^\top M^{-1} p_n = O(h^{2s}).$$

It is worth mentioning that, even in the case where the vector of the Lagrange multipliers is not constant, so that all HBVM(k, s) are second-order accurate for all $s \geq 1$, the choice $s > 1$ generally provides a much smaller solution error.

We refer to [23] for a number of examples of application of HBVMs to constrained Hamiltonian problems. We here only provide the application of HBVM(4,4) (together with (74) with $s = 4$) for solving the so called *conical pendulum* problem. In more details, let us have a pendulum of unit mass connected to a fixed point (the origin) by a massless rod of unit length. The initial conditions are such that the motion is periodic of period T and takes place in the horizontal plane $q_3 = z_0$. Normalizing the acceleration of gravity, the augmented Hamiltonian is given by

$$\hat{H}(q, p, \lambda) = \frac{1}{2} p^\top p + e_3^\top q + \lambda (q^\top q - 1) \equiv H(q, p) + \lambda g(q),$$

where $p = \dot{q}$ and $e_i \in \mathbb{R}^3$ is the ith unit vector. Choosing $q(0) = 2^{-\frac{1}{2}}(e_1 - e_3)$, $p(0) = 2^{-\frac{1}{4}} e_2$, results in

$$T = 2^{\frac{3}{4}} \pi, \qquad z_0 = -2^{-\frac{1}{2}}, \qquad \lambda \equiv \bar{\lambda} = 2^{-\frac{1}{2}}.$$

Since the augmented Hamiltonian $\hat{H}$ is quadratic and λ is constant, any HBVM(s, s) method, coupled with (74), is energy and constraint conserving, and of order $2s$. If we apply the HBVM(4,4) method for solving the problem over 10 periods by using the stepsize $h = T/N$, $N = 10, 20, 40$, it turns out the $\lambda_n \equiv \bar{\lambda}$, $g(q_n) = 0$, $\nabla g(q_n)^\top p_n = 0$, and $H(q_n, p_n) = H(q_0, p_0)$ within the round-off error level, for all $n = 0, 1, \ldots, 10N$. On the other hand, the corresponding solution errors, after 10 periods, turn out to be given by 4.9944×10^{-8}, 1.9676×10^{-10}, 7.3944×10^{-13}, thus confirming the order 8 of convergence of the resulting method.

5. Hamiltonian PDEs

Quoting [4], p. 187, *the numerical solution of time dependent PDEs may often be conceived as consisting of two parts. First the spatial derivatives are discretized by finite differences, finite elements, spectral methods, etc. to obtain a system of ODEs, with t as the independent variable. Then this system of ODEs is integrated numerically. If the PDEs are of Hamiltonian type, one may insist that both stages preserve the Hamiltonian structure. Thus the space discretization should be carried out in such a way that the resulting system of ODEs is Hamiltonian (for a suitable Poisson bracket).* This approach has been systematically used for solving a number of Hamiltonian PDEs by using HBVMs [1,20,22,59,60] and this research is still under development. We here sketch te main facts for the simplest possible example, provided by the semilinear wave equation,

$$u_{tt}(x, t) = u_{xx}(x, t) - f'(u(x, t)), \qquad (x, t) \in (a, b) \times (0, \infty), \tag{75}$$

with f' the derivative of f, coupled with the initial conditions

$$u(x, 0) = \phi_0(x), \qquad u_t(x, 0) = \phi_1(x), \qquad x \in [a, b], \tag{76}$$

and periodic boundary conditions. We shall assume that f, ϕ_0, ϕ_1 are regular enough (the last two functions, as periodic functions). By setting (hereafter, when not necessary, we shall avoid the arguments of the functions)

$$v = u_t, \qquad z = \begin{pmatrix} u \\ v \end{pmatrix}, \qquad J_2 = \begin{pmatrix} 0 & 1 \\ -1 & 0 \end{pmatrix}, \tag{77}$$

so that $v(x, 0) = \phi_1(x)$, $x \in [a, b]$, the problem can be cast into Hamiltonian form as

$$z_t = J_2 \frac{\delta \mathcal{H}}{\delta z}, \qquad \frac{\delta \mathcal{H}}{\delta z} = \left(\frac{\delta \mathcal{H}}{\delta u}, \frac{\delta \mathcal{H}}{\delta v} \right)^\top, \tag{78}$$

i.e.,

$$u_t = v, \qquad v_t = u_{xx} - f'(u), \tag{79}$$

where $\frac{\delta\mathcal{H}}{\delta z}$ is the vector of the functional derivatives [2,3,22] of the *Hamiltonian functional*

$$\mathcal{H}[u,v](t) = \frac{1}{2}\int_a^b \left[v^2(x,t) + u_x^2(x,t) + 2f(u(x,t))\right]dx =: \int_a^b E(x,t)dx. \tag{80}$$

Because of the periodic boundary conditions, this latter functional turns out to be conserved. In fact, by considering that

$$E_t = vv_t + u_xu_{xt} + f'(u)u_t = v(u_{xx} - f'(u)) + u_xv_x + f'(u)v = vu_{xx} + u_xv_x = (u_xv)_x,$$

one obtains:

$$\dot{\mathcal{H}}[u,v](t) = \int_a^b E_t(x,t)dx = [u_x(x,t)v(x,t)]_{x=a}^b = 0,$$

because of the periodicity in space. Consequently,

$$\mathcal{H}[u,v](t) = \mathcal{H}[u,v](0), \qquad \forall t \geq 0.$$

Moreover, since $u(x,t)$ is periodic for $x \in [a,b]$, we can expand it in space along the following slight variant of the Fourier basis,

$$c_j(x) = \sqrt{\frac{2-\delta_{j0}}{b-a}}\cos\left(2\pi j\frac{x-a}{b-a}\right), \quad j \geq 0, \qquad s_j(x) = \sqrt{\frac{2}{b-a}}\sin\left(2\pi j\frac{x-a}{b-a}\right), \quad j \geq 1, \tag{81}$$

so that, for all allowed i,j:

$$\int_a^b c_i(x)\,c_j(x)dx = \delta_{ij} = \int_a^b s_i(x)\,s_j(x)dx, \qquad \int_a^b c_i(x)\,s_j(x)dx = 0. \tag{82}$$

In so doing, for suitable time dependent coefficients $\gamma_j(t), \eta_j(t)$, one obtains the expansion:

$$u(x,t) = c_0(x)\gamma_0(t) + \sum_{j\geq 1}\left[c_j(x)\gamma_j(t) + s_j(x)\eta_j(t)\right] \equiv \boldsymbol{\omega}(x)^\top \boldsymbol{q}(t), \tag{83}$$

having introduced the infinite vectors

$$\begin{aligned}
\boldsymbol{\omega}(x) &= \left(\ c_0(x),\ c_1(x),\ s_1(x),\ c_2(x),\ s_2(x),\ \ldots\ \right)^\top, \\
\boldsymbol{q}(t) &= \left(\ \gamma_0(t),\ \gamma_1(t),\ \eta_1(t),\ \gamma_2(t),\ \eta_2(t),\ \ldots\ \right)^\top.
\end{aligned} \tag{84}$$

By considering that

$$\int_a^b \boldsymbol{\omega}(x)\boldsymbol{\omega}(x)^\top dx = I, \tag{85}$$

the identity operator, and introducing the infinite matrix (see (77))

$$D = \left(\frac{2\pi}{b-a}\right)\begin{pmatrix} 0 & & & \\ & 1\cdot J_2^\top & & \\ & & 2\cdot J_2^\top & \\ & & & 3\cdot J_2^\top \\ & & & & \ddots \end{pmatrix}, \tag{86}$$

so that $\omega'(x) = D\omega(x)$ and $\omega''(x) = D^2\omega(x) = -D^\top D\omega(x)$, one then obtains that (79) can be rewritten as the infinite system of ODEs:

$$\dot{q}(t) = p(t), \qquad \dot{p}(t) = -D^\top Dq(t) - \int_a^b \omega(x)f'(\omega(x)^\top q(t))dx, \qquad t > 0, \tag{87}$$

subject to the initial conditions (see (76))

$$q(0) = \int_a^b \omega(x)\phi_0(x)dx, \qquad p(0) = \int_a^b \omega(x)\phi_1(x)dx. \tag{88}$$

The following result is readily established.

Theorem 8. *Problem* (87) *is Hamiltonian, with Hamiltonian*

$$H(q, p) = \frac{1}{2}\left(p^\top p + q^\top D^\top Dq\right) + \int_a^b f(\omega(x)^\top q)dx. \tag{89}$$

This latter is equivalent to the Hamiltonian functional (80), *via the expansion* (83).

Proof. The first part of the proof is straightforward. Concerning the second part, one has, by virtue of (85):

$$\int_a^b v^2(x,t)dx = \int_a^b \dot{q}(t)^\top \omega(x)\omega(x)^\top \dot{q}(t)dx = p^\top \underbrace{\int_a^b \omega(x)\omega(x)^\top dx}_{=I} p = p^\top p.$$

Similarly, by considering that $u_x(x,t) = \omega(x)^\top Dq(t)$, one obtains:

$$\int_a^b u_x^2(x,t)dx = \int_a^b q(t)^\top D^\top \omega(x)\omega(x)^\top Dq(t)dx = q^\top D^\top \underbrace{\int_a^b \omega(x)\omega(x)^\top dx}_{=I} Dq = q^\top D^\top Dq.$$

□

In order to obtain a computational procedure, one needs to truncate the infinite series in (83) to a finite sum, i.e.,

$$u(x,t) \approx c_0(x)\gamma_0(t) + \sum_{j=1}^N \left[c_j(x)\gamma_j(t) + s_j(x)\eta_j(t)\right] =: u_N(x,t). \tag{90}$$

Clearly, such an approximation no longer satisfies the wave Equation (79). Nevertheless, in the spirit of Fourier–Galerkin methods [61], by requiring that the residual be orthogonal to the functional subspace

$$V_N = \text{span}\{c_0(x), c_1(x), s_1(x), \ldots, c_N(x), s_N(x)\}$$

containing the approximation $u(x,t)$ for all times $t \geq 0$, one obtains a finite dimensional ODE system, formally still given by (87) and (88), upon replacing the involved infinite vectors and matrices, previously defined in (84) and (86), with the following finite dimensional ones (of dimension $2N + 1$):

$$\omega(x) = \begin{pmatrix} c_0(x) \\ c_1(x) \\ s_1(x) \\ \vdots \\ c_N(x) \\ s_N(x) \end{pmatrix}, \quad q(t) = \begin{pmatrix} \gamma_0(x) \\ \gamma_1(x) \\ \eta_1(x) \\ \vdots \\ \gamma_N(x) \\ \eta_N(x) \end{pmatrix}, \quad D = \left(\frac{2\pi}{b-a}\right)\begin{pmatrix} 0 \\ & 1 \cdot J_2^\top \\ & & 2 \cdot J_2^\top \\ & & & \ddots \\ & & & & N \cdot J_2^\top \end{pmatrix}. \tag{91}$$

Moreover, the result of Theorem 8 continues formally to hold for the finite dimensional problem, with the sole exception that now the Hamiltonian (89) only yields an approximation to the continuous functional (80). Nevertheless, it is well known that, under suitable regularity assumptions on f and the initial data, this truncated version converges exponentially to the original functional (80), as $N \to \infty$ (this phenomenon is usually referred to as *spectral accuracy*). Since problem (87) is Hamiltonian, one can use HBVM(k,s) methods for solving it. It is worth mentioning that, in so doing, the blended implementation of the methods can be made extremely efficient, by considering that:

- an accurate approximation in space usually requires the use of large values of N;
- as a consequence, in most cases one has

$$[2\pi N/(b-a)]^2 = \|D^\top D\| \gg \|f'\|,$$

in a suitable neighbourhood of the solution.

Consequently, the Jacobian matrix of (87) can be approximated by the linear part alone, i.e., as

$$\begin{pmatrix} & I \\ -D^\top D & \end{pmatrix} \equiv \begin{pmatrix} & I \\ D^2 & \end{pmatrix} \in \mathbb{R}^{4N+2}. \tag{92}$$

This implies that matrix Σ involved in the definition of Θ in (44) becomes (all the involved matrices are diagonal and have dimension $2N+1$):

$$\Sigma = \begin{pmatrix} D_1 & D_2 \\ D_3 & \bar{D} \end{pmatrix}, \quad \bar{D} = (I - (h\rho_s D)^2)^{-1}, \quad D_1 = I + (h\rho_s D)^2 \bar{D}, \quad D_2 = h\rho_s \bar{D}, \quad D_3 = h\rho_s D^2 \bar{D},$$

where ρ_s is the parameter defined in (43), and h is the used time-step. As a result, Σ:

- is constant for all time steps, so that it has to be computed only once;
- it has a block diagonal structure (and, in particular, $\bar{D}$ is positive definite).

The above features make the resulting blended iteration (45) inexpensive, thus allowing the use of large values of N and large time-steps h. As an example, let us solve the *sine-Gordon* equation [22],

$$u_{tt} = u_{xx} - \sin(u), \quad (x,t) \in [-50, 50] \times [0, 100], \quad u(x,0) = 0, \quad u_t(x,0) = \frac{4}{\gamma}\mathrm{sech}\left(\frac{x}{\gamma}\right), \tag{93}$$

with $\gamma > 0$, whose solution is, by considering the value $\gamma = 1.5$,

$$u(x,t) = 4\arctan\left(\frac{\sin t\sqrt{1-\gamma^{-2}}}{\sqrt{\gamma^2 - 1}}\mathrm{sech}\left(\frac{x}{\gamma}\right)\right).$$

In such a case, the value $N = 300$ in (90) and (91) is sufficient to obtain an error in the spatial semi-discretization comparable with the round-off error level. Then, we solve in time the semi-discrete problem (87), of dimension $2N + 2 = 602$, by using the HBVM(k,s) methods. In Table 3 we list the obtained maximum errors in the computed solution, by using a time-step $h = 1$, along with the corresponding Hamiltonian errors and execution times, for various choices of (k,s) (in particular, for $k = s, s = 1, 2, 3$, we have the s-stage Gauss collocation methods, which are symplectic but not energy conserving) (all numerical tests have been performed on a laptop with a 2.2 GHz dual core i7 processor, 8 GB of memory, and running Matlab R2017b). From the listed results, one sees that HBVM$(2s,s)$ methods become energy-conserving for $s \geq 4$ and the error decreases until the round-off error level, for $s = 10$, with a computational time 10 times larger than that of the implicit mid-point rule (obtained for $k = s = 1$) which, however, has a solution error 10^{13} times larger.

Table 3. Solution error e_u and Hamiltonian error e_H when solving problem (93) with stepsize $h = 1$ and $N = 300$ in (90), by using HBVM(k, s) methods.

(k, s)	e_u	e_H	Time (s)
(1,1)	5.57	7.02×10^{-1}	0.64
(2,2)	9.97×10^{-1}	1.09×10^{-1}	1.73
(3,3)	8.39×10^{-2}	9.36×10^{-3}	2.62
(6,3)	1.05×10^{-2}	8.53×10^{-8}	2.98
(8,4)	3.89×10^{-4}	8.88×10^{-15}	3.42
(10,5)	1.44×10^{-5}	7.11×10^{-15}	4.64
(12,6)	5.31×10^{-7}	5.33×10^{-15}	4.79
(14,7)	1.92×10^{-8}	5.33×10^{-15}	4.97
(16,8)	6.87×10^{-10}	5.33×10^{-15}	4.46
(18,9)	2.47×10^{-11}	5.33×10^{-15}	5.52
(20,10)	9.06×10^{-13}	5.33×10^{-15}	6.46

6. Highly Oscillatory Problems

The Hamiltonian system of ODEs (87) is a particular instance of problems in the form

$$\ddot{q} + A^2 q + \nabla f(q) = 0, \qquad t \geq 0, \qquad q(0) = q_0, \ \dot{q}(0) = \dot{q}_0 \in \mathbb{R}^m, \tag{94}$$

where, without loss of generality, A is a symmetric and positive definite matrix and f is a scalar function such that

$$\omega := \|A\| \gg \|\nabla f\|, \tag{95}$$

in a neighbourhood of the solution. Moreover, hereafter we consider the 2-norm, so that ω equals the largest eigenvalue of A (in general, any convenient upper bound would suffice). The problem is Hamiltonian, with Hamiltonian

$$H(q, \dot{q}) = \frac{1}{2} \left(\|\dot{q}\|^2 + \|Aq\|^2 \right) + f(q). \tag{96}$$

Problems in the form (94) satisfying (95) are named *(multi-frequency) highly oscillatory problems*, since they are ruled by the linear part, possessing large (possibly different) complex conjugate eigenvalues. Such problems have been investigated since many years, starting from the seminal papers [62,63], and we refer to the monograph [64] for more recent findings. A common feature of the methods proposed so far, however, is that of requiring the use of time-steps h such that $h\omega < 1$, either for stability and/or accuracy requirements. We here sketch the approach recently defined in [27], relying on the use of HBVMs, which will allow the use of stepsizes h without such a restriction.

To begin with, and for analysis purposes, let us recast problem (94) in first order form, by setting $p = A^{-1}\dot{q}$, as

$$\dot{y} = J_2 \otimes Ay + J_2 \otimes I_m \tilde{f}(y), \qquad y = \begin{pmatrix} q \\ p \end{pmatrix}, \quad \tilde{f}(y) = \begin{pmatrix} A^{-1}\nabla f(q) \\ 0 \end{pmatrix}, \tag{97}$$

with J_2 the 2×2 skew-symmetric matrix defined in (77). As is usual in this context, one considers at first the linear part of (97),

$$\dot{y} = J_2 \otimes Ay, \qquad y(0) = y_0 := \begin{pmatrix} q_0 \\ p_0 \end{pmatrix}, \qquad p_0 = A^{-1}\dot{q}_0. \tag{98}$$

The solution of (98), $y(t) = e^{J_2 \otimes A\,t} y_0$, on the interval $[0, h]$ is readily seen to be given, according to the local Fourier expansion described in Section 2.1, by:

$$y(ch) = y_0 + h \sum_{j \geq 0} \int_0^c P_j(x) \mathrm{d}x \hat{\gamma}_j(y), \quad c \in [0,1], \qquad \hat{\gamma}_j(y) = J_2 \otimes A \int_0^1 P_j(\tau) y(\tau h) \mathrm{d}\tau. \tag{99}$$

However, when using a finite precision arithmetic with machine epsilon u, the best we can do is to approximate

$$y(ch) \doteq \sigma_0(ch) = y_0 + h \sum_{j=0}^{s_0-1} \int_0^c P_j(x) \mathrm{d}x \hat{\gamma}_j(\sigma_0), \qquad c \in [0,1], \tag{100}$$

where hereafter $\doteq$ means "equal within round-off error level", provided that

$$\|\hat{\gamma}_j(y)\| < u \cdot \max_{i < s_0} \|\hat{\gamma}_i(y)\|, \qquad \forall j \geq s_0. \tag{101}$$

In fact, further terms in the infinite series in (99) would provide a negligible contribution, in the used finite precision arithmetic. By defining the function

$$g(j, \omega h) := \sqrt{\frac{(2j+1)\pi}{\omega h}} \left| J_{j+\frac{1}{2}} \left(\frac{\omega h}{2} \right) \right|, \qquad j = 0, 1, \ldots, \tag{102}$$

with $J_{j+\frac{1}{2}}(\cdot)$ the Bessel functions of the first kind, it can be shown ([27], Criterion 1) that the requirement (101) is accomplished by requiring

$$g(s_0, \omega h) < u \cdot \max_{j < s_0} g(j, \omega h). \tag{103}$$

In so doing, one implicitly defines a function φ_u, depending on the machine epsilon u, such that

$$s_0 = \varphi_u(\omega h). \tag{104}$$

The plot of such a function is depicted in Figure 2 for the double precision IEEE: as one may see, the function is well approximated by the line [27]

$$24 + 0.7 \cdot \omega h. \tag{105}$$

Next, one consider the whole problem (97), whose solution, on the interval $[0, h]$, can be written as

$$y(ch) = y_0 + h \sum_{j \geq 0} \int_0^c P_j(x) \mathrm{d}x \gamma_j(y), \quad c \in [0,1], \tag{106}$$

$$\gamma_j(y) = \int_0^1 P_j(\tau) \left[J_2 \otimes Ay(\tau h) + J_2 \otimes I_m \tilde{f}(y(\tau h)) \right] \mathrm{d}\tau. \tag{107}$$

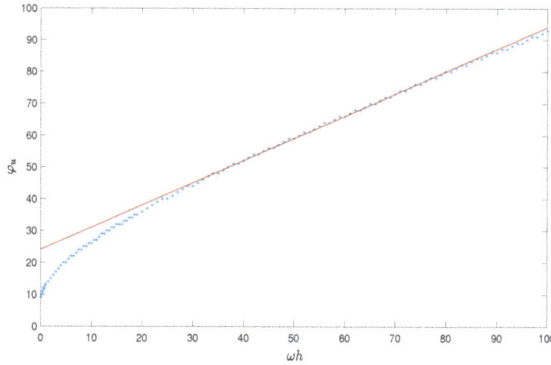

Figure 2. Plot of the function φ_u defined in (104) versus ωh (blue plus line), for the double precision IEEE, together with its linear approximation (105) (red solid line).

As observed before, when using a finite precision arithmetic with machine epsilon u, the best we can do is to approximate (106) with a polynomial:

$$y(ch) \doteq \sigma(ch) = y_0 + h \sum_{j=0}^{s-1} \int_0^c P_j(x)\mathrm{d}x \gamma_j(\sigma), \qquad c \in [0,1], \tag{108}$$

provided that

$$\|\gamma_j(y)\| < u \cdot \max_{i<s} \|\gamma_i(y)\|, \qquad \forall j \geq s. \tag{109}$$

Assuming the *ansatz* $\|\tilde{f}(y(ch))\| \sim \|e^{\nu I_2 \otimes A\,ch}\|$, for a suitable $\nu \geq 1$ (essentially, one requires that ∇f is well approximated by a polynomial of degree ν), the requirement (109) is accomplished by choosing ([27], Criterion 2)

$$s = \varphi_u(\nu\omega h), \tag{110}$$

where φ_u is the same function considered in (104). By taking into account that $\nu \geq 1$ and that φ_u is an increasing funcion (see Figure 2), one then obtains that $s \geq s_0$.

Next, one has to approximate the Fourier coefficients (see (107)) $\gamma_j(\sigma)$, $j = 0, \ldots, s-1$, needed in (108): for this purpose, we consider the Gauss-Legendre quadrature formula of order $2k$. In order to gain full machine accuracy, when using the double precision IEEE, according to [27], we choose

$$k = \max\{20, s+2\}. \tag{111}$$

In so doing, we arrive to a HBVM(k,s) method. For such a method, the blended iteration (45) can be made extremely efficient by approximating the Jacobian of (97) with its linear part (this has been already done when solving Hamiltonian PDEs, see (92)), so that the matrix Σ in (44) has to be computed only once. Moreover, since we are going to use relatively large stepsizes h, the initial guess for the vector $\hat{\gamma}^0$ in (45) is chosen, by considering the polynomial σ_0 defined in (100) derived from the linear problem (98), as:

$$\hat{\gamma}^0 = \begin{pmatrix} \bar{\gamma}_0(\sigma_0) \\ \vdots \\ \bar{\gamma}_{s_0-1}(\sigma_0) \\ \mathbf{0} \end{pmatrix}, \qquad \mathbf{0} \in \mathbb{R}^{(s-s_0)2m}.$$

The name *spectral HBVM with parameters* (k, s, s_0) has been coined in [27] to denote the resulting method, in short $SHBVM(k, s, s_0)$. In order to show its effectiveness, we report here some numerical results obtained by solving the Duffing equation

$$\ddot{q} = -(\kappa^2 + \beta^2)q + 2\kappa^2 q^3, \qquad t \in [0, 20], \qquad q(0) = 0, \quad \dot{q}(0) = \beta, \tag{112}$$

with Hamiltonian

$$H(q, \dot{q}) = \frac{1}{2}\left[\dot{q}^2 + (\kappa^2 + \beta^2)q^2 - \kappa^2 q^4\right], \tag{113}$$

and exact solution

$$q(t) = \text{sn}\left(\beta t, (\kappa/\beta)^2\right). \tag{114}$$

Here, sn is the elliptic Jacobi function, with elliptic modulus specified by the second argument. In particular, we choose the values

$$\kappa = 1, \qquad \beta = 10^3, \tag{115}$$

providing a problem in the form (94) to (95), with a corresponding Hamiltonian value $H_0 = 5 \times 10^5$. For solving it, we shall use the SHBVM method with parameters:

$$\omega = \sqrt{\kappa^2 + \beta^2} \approx 10^3, \qquad \nu = 3, \tag{116}$$

and a time-step $h = 20/N$, for various values of N, as specified in Table 4. In that table we also list the corresponding:

- maximum absolute error in the computed solution, e_q;
- the maximum relative error in the numerical Hamiltonian, e_H;
- the value of ωh (which is always much greater than 1);
- the parameters (s_0, s, k), computed according to (104), (110), and (111), respectively;
- the execution time (in sec).

As before, the numerical tests have been performed on a laptop with a 2.2 GHz dual core i7 processor, 8 GB of memory, and running Matlab R2017b.

Table 4. Duffing problem (112)–(115) solved by the SHBVM(k, s, s_0) method with parameters (116) and time-step $h = 20/N$: e_q is the maximum absolute error in the computed solution; e_H is the maximum relative error in the numerical Hamiltonian.

N	ωh	e_q	e_H	(s_0, s, k)	Time (s)
1200	16.7	4.12×10^{-8}	4.44×10^{-16}	(33,59,61)	4.09
1300	15.4	1.41×10^{-8}	4.44×10^{-16}	(32,56,58)	3.32
1400	14.3	3.55×10^{-9}	4.44×10^{-16}	(31,54,56)	3.34
1500	13.3	2.02×10^{-9}	3.33×10^{-16}	(30,52,54)	3.33
1600	12.5	6.66×10^{-10}	2.22×10^{-16}	(29,50,52)	3.39
1700	11.8	8.61×10^{-10}	4.44×10^{-16}	(28,48,50)	3.24
1800	11.1	2.25×10^{-10}	4.44×10^{-16}	(28,47,49)	3.49
1900	10.5	2.20×10^{-10}	4.44×10^{-16}	(27,45,47)	3.59
2000	10.0	2.89×10^{-10}	3.33×10^{-16}	(26,44,46)	3.46

As one may see, the method is always energy conserving and the error, as expected, is uniformly small. Also the execution times are all very small (of the order of 3.5 s). It must be emphasized that classical methods, such as the Gautschi or the Deuflhard method, would require the use of much smaller time-steps, and much larger execution times (we refer to [27] for some comparisons).

7. Conclusions

In this paper, we have reviewed the main facts concerning the numerical solution of conservative problems within the framework of (*discrete*) *line integral methods*. Relevant instances of line integral methods are provided by the energy-conserving Runge-Kutta methods named *Hamiltonian Boundary Value Methods* (*HBVMs*) and the *Energy and QUadratic Invariants Preserving* (*EQUIP*) *methods*, providing efficient geometric integrators for Hamiltonian problems. It is worth mentioning that HBVMs can be easily adapted to efficiently handle Hamiltonian BVPs [19], whereas EQUIP methods can be also used for numerically solving Poisson problems. In this paper, we have also reviewed some active research trends, concerning the application of HBVMs for numerically solving constrained Hamiltonian problems, Hamiltonian PDEs, and highly-oscillatory problems. The effectiveness of HBVMs is emphasized by the availability of an efficient nonlinear iteration for solving the generated discrete problems, relying on the so called *blended implementation* of the methods. Matlab software implementing HBVMs for Hamiltonian problems is available at the url [52].

Acknowledgments: The authors wish to thank Gianmarco Gurioli, Gianluca Frasca Caccia, Cecilia Magherini, and the three anonymous reviewers, for carefully reading the manuscript.

Conflicts of Interest: The authors declare no conflict of interest.

References

1. Brugnano, L.; Iavernaro, F. *Line Integral Methods for Conservative Problems*; Chapman et Hall/CRC: Boca Raton, FL, USA, 2016.
2. Hairer, E.; Lubich, C.; Wanner, G. *Geometric Numerical Integration*, 2nd ed.; Springer: Berlin, Germany, 2006.
3. Leimkuhler, B.; Reich, S. *Simulating Hamiltonian Dynamics*; Cambridge University Press: Cambridge, UK, 2004.
4. Sanz-Serna, J.M.; Calvo, M.P. *Numerical Hamiltonian Problems*; Chapman & Hall: London, UK, 1994.
5. Iavernaro, F.; Trigiante, D. On some conservation properties of the trapezoidal method applied to Hamiltonian systems. In Proceedings of the International Conference on Numerical Analysis and Applied Mathematics (ICNAAM), Rhodes, Greece, 16–20 September 2005; Simos, T.E., Psihoyios, G., Tsitouras, Ch., Eds.; Wiley-Vch Verlag GmbH & Co.: Weinheim, Germany, 2005; pp. 254–257.
6. Iavernaro, F.; Trigiante, D. Discrete Conservative Vector Fields Induced by the Trapezoidal Method. *JNAIAM J. Numer. Anal. Ind. Appl. Math.* **2006**, *1*, 113–130.
7. Iavernaro, F.; Pace, B. *s*-stage trapezoidal methods for the conservation of Hamiltonian functions of polynomial type. *AIP Conf. Proc.* **2007**, *936*, 603–606.
8. Iavernaro, F.; Pace, B. Conservative block-Boundary Value Methods for the solution of polynomial Hamiltonian systems. *AIP Conf. Proc.* **2008**, *1048*, 888–891.
9. Iavernaro, F.; Trigiante, D. High-order Symmetric Schemes for the Energy Conservation of Polynomial Hamiltonian Problems. *JNAIAM J. Numer. Anal. Ind. Appl. Math.* **2009**, *4*, 87–101.
10. Brugnano, L.; Iavernaro, F.; Susca, T. Numerical comparisons between Gauss-Legendre methods and Hamiltonian BVMs defined over Gauss points. *Monogr. Real Acad. Cienc. Zaragoza* **2010**, *33*, 95–112.
11. Brugnano, L.; Iavernaro, F.; Trigiante, D. Hamiltonian BVMs (HBVMs): A family of "drift-free" methods for integrating polynomial Hamiltonian systems. *AIP Conf. Proc.* **2009**, *1168*, 715–718.
12. Brugnano, L.; Iavernaro, F.; Trigiante, D. Hamiltonian Boundary Value Methods (Energy Preserving Discrete Line Integral Methods). *JNAIAM J. Numer. Anal. Ind. Appl. Math.* **2010**, *5*, 17–37.
13. Brugnano, L.; Iavernaro, F.; Trigiante, D. Analisys of Hamiltonian Boundary Value Methods (HBVMs): A class of energy-preserving Runge-Kutta methods for the numerical solution of polynomial Hamiltonian systems. *Commun. Nonlinear Sci. Numer. Simul.* **2015**, *20*, 650–667. [CrossRef]
14. Brugnano, L.; Frasca Caccia, G.; Iavernaro, F. Efficient implementation of Gauss collocation and Hamiltonian Boundary Value Methods. *Numer. Algorithms* **2014**, *65*, 633–650. [CrossRef]
15. Brugnano, L.; Frasca Caccia, G.; Iavernaro, F. Hamiltonian Boundary Value Methods (HBVMs) and their efficient implementation. *Math. Eng. Sci. Aerosp.* **2014**, *5*, 343–411.
16. Brugnano, L.; Iavernaro, F.; Trigiante, D. A note on the efficient implementation of Hamiltonian BVMs. *J. Comput. Appl. Math.* **2011**, *236*, 375–383. [CrossRef]

17. Brugnano, L.; Iavernaro, F.; Trigiante, D. The lack of continuity and the role of Infinite and infinitesimal in numerical methods for ODEs: The case of symplecticity. *Appl. Math. Comput.* **2012**, *218*, 8053–8063. [CrossRef]

18. Brugnano, L.; Iavernaro, F.; Trigiante, D. A simple framework for the derivation and analysis of effective one-step methods for ODEs. *Appl. Math. Comput.* **2012**, *218*, 8475–8485. [CrossRef]

19. Amodio, P.; Brugnano, L.; Iavernaro, F. Energy-conserving methods for Hamiltonian Boundary Value Problems and applications in astrodynamics. *Adv. Comput. Math.* **2015**, *41*, 881–905. [CrossRef]

20. Barletti, L.; Brugnano, L.; Frasca Caccia, G.; Iavernaro, F. Energy-conserving methods for the nonlinear Schrödinger equation. *Appl. Math. Comput.* **2018**, *318*, 3–18. [CrossRef]

21. Brugnano, L.; Calvo, M.; Montijano, J.I.; Ràndez, L. Energy preserving methods for Poisson systems. *J. Comput. Appl. Math.* **2012**, *236*, 3890–3904. [CrossRef]

22. Brugnano, L.; Frasca Caccia, G.; Iavernaro, F. Energy conservation issues in the numerical solution of the semilinear wave equation. *Appl. Math. Comput.* **2015**, *270*, 842–870. [CrossRef]

23. Brugnano, L.; Gurioli, G.; Iavernaro, F.; Weinmüller, E.B. Line integral solution of Hamiltonian systems with holonomic constraints. *Appl. Numer. Math.* **2018**, *127*, 56–77. [CrossRef]

24. Brugnano, L.; Iavernaro, F. Line Integral Methods which preserve all invariants of conservative problems. *J. Comput. Appl. Math.* **2012**, *236*, 3905–3919. [CrossRef]

25. Brugnano, L.; Iavernaro, F.; Trigiante, D. A two-step, fourth-order method with energy preserving properties. *Comput. Phys. Commun.* **2012**, *183*, 1860–1868. [CrossRef]

26. Brugnano, L.; Iavernaro, F.; Trigiante, D. Energy and QUadratic Invariants Preserving integrators based upon Gauss collocation formulae. *SIAM J. Numer. Anal.* **2012**, *50*, 2897–2916. [CrossRef]

27. Brugnano, L.; Montijano, J.I.; Rández, L. On the effectiveness of spectral methods for the numerical solution of multi-frequency highly-oscillatory Hamiltonian problems. *Numer. Algorithms* **2018**. [CrossRef]

28. Brugnano, L.; Sun, Y. Multiple invariants conserving Runge-Kutta type methods for Hamiltonian problems. *Numer. Algorithms* **2014**, *65*, 611–632. [CrossRef]

29. Lasagni, F.M. Canonical Runge-Kutta methods. *Z. Angew. Math. Phys.* **1988**, *39*, 952–953. [CrossRef]

30. Sanz-Serna, J.M. Runge-Kutta schemes for Hamiltonian systems. *BIT Numer. Math.* **1988**, *28*, 877–883. [CrossRef]

31. Burrage, K.; Butcher, J.C. Stability criteria for implicit Runge-Kutta methods. *SIAM J. Numer. Anal.* **1979**, *16*, 46–57. [CrossRef]

32. Betsch, P.; Steinmann, P. Inherently Energy Conserving Time Finite Elements for Classical Mechanics. *J. Comput. Phys.* **2000**, *160*, 88–116. [CrossRef]

33. Betsch, P.; Steinmann, P. Conservation properties of a time FE method. I. Time-stepping schemes for *N*-body problems. *Int. J. Numer. Methods Eng.* **2000**, *49*, 599–638. [CrossRef]

34. Bottasso, C.L. A new look at finite elements in time: A variational interpretation of Runge-Kutta methods. *Appl. Numer. Math.* **1997**, *25*, 355–368. [CrossRef]

35. Celledoni, E.; McLachlan, R.I.; McLaren, D.I.; Owren, B.; Quispel, G.R.W.; Wright, W.M. Energy-preserving Runge-Kutta methods. *M2AN Math. Model. Numer. Anal.* **2009**, *43*, 645–649. [CrossRef]

36. Hairer, E. Energy preserving variant of collocation methods. *JNAIAM J. Numer. Anal. Ind. Appl. Math.* **2010**, *5*, 73–84.

37. McLachlan, R.I.; Quispel, G.R.W.; Robidoux, N. Geometric integration using discrete gradient. *Philos. Trans. R. Soc. Lond. A* **1999**, *357*, 1021–1045. [CrossRef]

38. Quispel, G.R.W.; McLaren, D.I. A new class of energy-preserving numerical integration methods. *J. Phys. A Math. Theor.* **2008**, *41*, 045206. [CrossRef]

39. Tang, Q.; Chen, C.M. Continuous finite element methods for Hamiltonian systems. *Appl. Math. Mech.* **2007**, *28*, 1071–1080. [CrossRef]

40. Dahlquist, G.; Björk, Å. *Numerical Methods in Scientific Computing*; SIAM: Philadelphia, PA, USA, 2008; Volume I.

41. Brugnano, L. Blended Block BVMs (B$_3$VMs): A Family of Economical Implicit Methods for ODEs. *J. Comput. Appl. Math.* **2000**, *116*, 41–62. [CrossRef]

42. Brugnano, L.; Magherini, C. Blended Implementation of Block Implicit Methods for ODEs. *Appl. Numer. Math.* **2002**, *42*, 29–45. [CrossRef]

43. Brugnano, L.; Magherini, C. Blended Implicit Methods for solving ODE and DAE problems, and their extension for second order problems. *J. Comput. Appl. Math.* **2007**, *205*, 777–790. [CrossRef]
44. Brugnano, L.; Magherini, C. Recent advances in linear analysis of convergence for splittings for solving ODE problems. *Appl. Numer. Math.* **2009**, *59*, 542–557. [CrossRef]
45. Brugnano, L.; Magherini, C. Blended General Linear Methods based on Boundary Value Methods in the Generalized BDF family. *JNAIAM J. Numer. Anal. Ind. Appl. Math.* **2009**, *4*, 23–40.
46. Brugnano, L.; Magherini, C. The BiM code for the numerical solution of ODEs. *J. Comput. Appl. Math.* **2004**, *164–165*, 145–158. [CrossRef]
47. Brugnano, L.; Magherini, C.; Mugnai, F. Blended Implicit Methods for the Numerical Solution of DAE Problems. *J. Comput. Appl. Math.* **2006**, *189*, 34–50. [CrossRef]
48. Test Set for IVP Solvers. Available online: https://archimede.dm.uniba.it/~testset/testsetivpsolvers/ (accessed on 4 May 2018).
49. The Codes BiM and BiMD Home Page. Available online: http://web.math.unifi.it/users/brugnano/BiM/index.html (accessed on 4 May 2018).
50. Wang, B.; Meng, F.; Fang, Y. Efficient implementation of RKN-type Fourier collocation methods for second-order differential equations. *Appl. Numer. Math.* **2017**, *119*, 164–178. [CrossRef]
51. Simoncini, V. Computational methods for linear matrix equations. *SIAM Rev.* **2016**, *58*, 377–441. [CrossRef]
52. Line Integral Methods for Conservative Problems. Available online: http://web.math.unifi.it/users/brugnano/LIMbook/ (accessed on 4 May 2018).
53. Brugnano, L.; Gurioli, G.; Iavernaro, F. Analysis of Energy and QUadratic Invariant Preserving (EQUIP) methods. *J. Comput. Appl. Math.* **2018**, *335*, 51–73. [CrossRef]
54. Hairer, E.; Wanner, G. *Solving Ordinary Differential Equations II. Stiff and Differential-Algebraic Problems*, 2nd ed.; Springer: Berlin, Germany, 1996.
55. Chartier, P.; Faou, E.; Murua, A. An algebraic approach to invariant preserving integrators: The case of quadratic and Hamiltonian invariants. *Numer. Math.* **2006**, *103*, 575–590. [CrossRef]
56. Ge, Z.; Marsden, J.E. Lie-Poisson Hamilton-Jacobi theory and Lie-Poisson integrators. *Phys. Lett. A* **1988**, *133*, 134–139.
57. Brugnano, L.; Iavernaro, F.; Trigiante, D. Energy and Quadratic Invariants Preserving Integrators of Gaussian Type. *AIP Conf. Proc.* **2010**, *1281*, 227–230.
58. Sanz-Serna, J.M. Symplectic Runge-Kutta schemes for adjoint equations, automatic differentiation, optimal control, and more. *SIAM Rev.* **2016**, *58*, 3–33. [CrossRef]
59. Brugnano, L.; Gurioli, G.; Sun, Y. Energy-conserving Hamiltonian Boundary Value Methods for the numerical solution of the Korteweg-de Vries equation. *J. Comput. Appl. Math.* **2018**, submitted.
60. Brugnano, L.; Zhang, C.; Li, D. A class of energy-conserving Hamiltonian boundary value methods for nonlinear Schrödinger equation with wave operator. *Commun. Nonlinear Sci. Numer. Simul.* **2018**, *60*, 33–49. [CrossRef]
61. Boyd, J.P. *Chebyshev and Fourier Spectral Methods*, 2nd ed.; Dover Publications Inc.: Mineola, NY, USA, 2001.
62. Deuflhard, P. A study of extrapolation methods based on multistep schemes without parasitic solutions. *Z. Angew. Math. Phys.* **1979**, *30*, 177–189. [CrossRef]
63. Gautschi, W. Numerical integration of ordinary differential equations based on trigonometric polynomials. *Numer. Math.* **1961**, *3*, 381–397. [CrossRef]
64. Wu, X.; Wang, B. *Recent Developments in Structure-Preserving Algorithms for Oscillatory Differential Equations*; Springer: Singapore, 2018.

Article

Refinement Algorithms for Adaptive Isogeometric Methods with Hierarchical Splines

Cesare Bracco [1], Carlotta Giannelli [1,*] and Rafael Vázquez [2,3]

1 Dipartimento di Matematica e Informatica "U. Dini", Università degli Studi di Firenze,
  Viale Morgagni 67/A, 50134 Florence, Italy; cesare.bracco@unifi.it
2 Institute of Mathematics, Ecole Polytechnique Fédérale de Lausanne, Station 8, 1015 Lausanne, Switzerland;
  rafael.vazquez@epfl.ch
3 Istituto di Matematica Applicata e Tecnologie Informatiche "E. Magenes" del CNR, via Ferrata 5,
  27100 Pavia, Italy
* Correspondence: carlotta.giannelli@unifi.it; Tel.: +39-055-275-1407

Received: 11 May 2018; Accepted: 18 June 2018; Published: 21 June 2018

Abstract: The construction of suitable mesh configurations for spline models that provide local refinement capabilities is one of the fundamental components for the analysis and development of adaptive isogeometric methods. We investigate the design and implementation of refinement algorithms for hierarchical B-spline spaces that enable the construction of locally graded meshes. The refinement rules properly control the interaction of basis functions at different refinement levels. This guarantees a bounded number of nonvanishing (truncated) hierarchical B-splines on any mesh element. The performances of the algorithms are validated with standard benchmark problems.

Keywords: isogeometric analysis; adaptive methods; hierarchical splines; THB-splines; local refinement

---

1. Introduction

The design of isogeometric methods for the numerical solution of partial differential equations extends classical finite element techniques by taking into account computer aided design (CAD) methods and standards [1,2]. For this reason, the isogeometric approach naturally encourages a tighter connection between computer aided engineering and design software libraries. While the potential of exploiting a common representation model, as well as the enhanced smoothness of higher order spline schemes, opened the possibility of developing highly accurate methods, the backward CAD compatibility also poses some limits and challenges. One of the most important restrictions is the tensor-product structure of standard multivariate B-spline models, which necessarily prevents the possibility of local mesh refinement. This motivates several authors to advance the study of adaptive spline constructions.

Hierarchical B-splines constitute one of the most promising solutions to easily define adaptive spline basis which preserve the nonnegativity of standard B-splines and facilitate the design of fully automatic schemes [3–7]. When the truncated basis of hierarchical B-spline spaces is considered the overlap of truncated basis functions at different hierarchical levels is reduced and the partition of unity property recovered [8]. Regarding the implementation of hierarchical B-splines, different data structures and algorithms have been already proposed in the literature (see e.g., [9–11] and [12,13]). Particular attention was also devoted to the efficient Bernstein–Bézier evaluation in the hierarchical setting [14,15]. In this work, we employ the structures introduced in [11], which have an implementation in GeoPDEs [16].

A key ingredient for the analysis of adaptive isogeometric methods is the possibility to consider certain class of *admissible* meshes which automatically guarantee a bounded number of non-zero basis functions on each mesh element. In the hierarchical spline setting, admissible meshes also guarantee that the level of the mesh element and the level of any basis function that does not vanish on the element differ at most a fixed value. This kind of restricted hierarchies naturally reduces the overlapping of basis functions introduced at different levels and influences the sparsity patterns of the discretization matrices. Being connected to the number of elements influenced by the support of a function in the hierarchical basis, the development of a refine module for the adaptive isogeometric method based on these observations can properly profit of (truncated) basis functions with reduced support. This paper is devoted to the study of design and implementation aspects for the development of hierarchical refinement strategies that guarantee the control of different classes of admissibility. Our analysis allows the definition and construction of suitable admissible meshes for standard and truncated hierarchical B-splines by considering a unified framework. The theoretical properties of the refinement algorithms and the resulting meshes are thoroughly analyzed and presented together with extensive numerical testing.

The structure of the paper is as follows. Section 2 introduces the hierarchical B-spline model by focusing on the basis construction and the main data structures needed for the implementation. Section 3 presents the algorithms for the construction and refinement of admissible hierarchical meshes. In Section 4, we briefly recall the adaptive isogeometric setting, while Section 5 shows the results obtained by integrating the refinement procedures in an adaptive isogeometric scheme. Finally, Section 6 concludes the paper.

2. The Hierarchical B-Spline Model

We briefly review in this section the construction of (truncated) hierarchical B-splines and introduce the data structures and basic functionalities considered for the implementation of the spline hierarchy. The key components for the design of a software architecture devoted to hierarchical spline structures rely on the storage of the hierarchical mesh as well as on the information related to the adaptive spline space.

2.1. Basis Construction

Let $V^0 \subset V^1 \subset \ldots \subset V^{N-1}$ be a nested sequence of N tensor-product d-variate spline spaces defined on a closed hyper-rectangle D in $\mathbb{R}^d$. For each level ℓ, $\ell = 0, \ldots, N-1$, we consider the tensor-product B-spline basis $\mathcal{B}^\ell$ of degree $\mathbf{p} = (p_1, \ldots, p_d)$ defined on the rectilinear grid G^ℓ. B-splines have local support and satisfy the following properties: local linear independence, non-negativity, partition of unity (see, e.g., [17,18]).

We also consider a nested sequence of closed subsets of D to define the domain hierarchy:

$$\Omega = \Omega^0 \supseteq \Omega^1 \supseteq \ldots \supseteq \Omega^{N-1} \supseteq \Omega^N = \emptyset \quad \text{with} \quad \Omega^\ell = \bigcup_{Q \in \mathcal{R}^{\ell-1}} \overline{Q},$$

where $\mathcal{R}^{\ell-1} \subset G^{\ell-1}$ are the refined elements of level $\ell-1$. The hierarchical mesh is defined as the collection of the open active elements at different levels, namely

$$\mathcal{Q} := \left\{ Q \in \mathcal{G}^\ell, \ell = 0, \ldots, N-1 \right\} \quad \text{with} \quad \mathcal{G}^\ell := \left\{ Q \in G^\ell : Q \subset \Omega^\ell \wedge Q \not\subset \Omega^{\ell+1} \right\}. \tag{1}$$

By following [3,4], a subset of B-splines at different hierarchical levels can be properly selected to construct the hierarchical basis according to the following definition.

Definition 1. *The hierarchical B-spline (HB-spline) basis $\mathcal{H}$ with respect to the mesh $\mathcal{Q}$ is defined as*

$$\mathcal{H}(\mathcal{Q}) := \left\{ \beta^\ell \in \mathcal{B}^\ell : \text{supp}\, \beta^\ell \subseteq \Omega^\ell \wedge \text{supp}\, \beta^\ell \not\subseteq \Omega^{\ell+1}, \ell = 0, \ldots, N-1 \right\}.$$

For any level ℓ, the selection mechanism identifies the set of B-splines of this level whose support is contained in the domain Ω^ℓ but not fully contained in the next domain of the hierarchy, $\Omega^{\ell+1}$.

This part of the domain will be covered by selecting B-splines of levels greater than ℓ. The construction preserves the non-negativity and linear independence of one-level B-splines [3,4]. Obviously, coarser B-splines will interact with elements and refined B-splines of subsequent hierarchical levels. In view of this overlap between basis functions at different levels of detail, the partition of unity property is lost in the hierarchical construction. The truncated basis for hierarchical splines [8] recovers this property by removing the contribution of finer B-splines in the hierarchical basis from coarser ones. This is possible thanks to the two-scale relation between any B-spline of level ℓ and refined B-splines of level $\ell + 1$. More precisely, any $s \in V^\ell \subset V^{\ell+1}$ can be expressed as

$$s = \sum_{\beta \in \mathcal{B}^{\ell+1}} c_{\beta^{\ell+1}}(s)\beta^{\ell+1}, \tag{2}$$

in terms of the coefficients $c_{\beta^{\ell+1}}$. The truncation of s with respect to level $\ell + 1$ simply removes, in the above sum, the B-splines of level $\ell + 1$ having supports fully contained in $\Omega^{\ell+1}$, and that will be included in the basis $\mathcal{H}(\mathcal{Q})$. It is then defined as follows:

$$\text{trunc}^{\ell+1} s := \sum_{\beta^{\ell+1} \in \mathcal{B}^{\ell+1}, \, \text{supp} \, \beta^{\ell+1} \not\subseteq \Omega^{\ell+1}} c_{\beta^{\ell+1}}(s)\beta^{\ell+1}. \tag{3}$$

Definition 2. *The truncated hierarchical B-spline (THB-spline) basis $\mathcal{T}$ with respect to the mesh $\mathcal{Q}$ is defined as*

$$\mathcal{T}(\mathcal{Q}) := \left\{ \text{Trunc}^{\ell+1} \beta^\ell : \beta^\ell \in \mathcal{B}^\ell \cap \mathcal{H}(\mathcal{Q}), \ell = 0, \ldots, N-2 \right\} \bigcup \left\{ \beta^{N-1} : \beta^{N-1} \in \mathcal{B}^{N-1} \cap \mathcal{H}(\mathcal{Q}) \right\},$$

where $\text{Trunc}^{\ell+1} \beta^\ell := \text{trunc}^{N-1}(\text{trunc}^{N-2}(\ldots(\text{trunc}^{\ell+1}(\beta^\ell))\ldots))$, *for any* $\beta^\ell \in \mathcal{B}^\ell \cap \mathcal{H}(\mathcal{Q})$.

THB-splines are non-negative, linearly independent, form a partition of unity, and span the same space of HB-splines [8]. The properties of non-negativity and partition of unity imply the convex hull property, a fundamental concept for geometric modelling applications.

2.2. Data Structures for the Implementation

The implementation of numerical methods based on hierarchical B-splines requires the definition of suitable data structures, which contain all the necessary information for the computation of the basis $\mathcal{H}(Q)$ (or $\mathcal{T}(Q)$). In this work, we employ the data structures introduced in [11], and, for the sake of completeness, we recall their main fields and functionalities, before using them to develop the refinement algorithms. We will need four different data structures: two for tensor-product B-splines, and two for hierarchical B-splines.

The first two-structures regard the computation of tensor-product B-splines. We assume that, for each level ℓ, we have a *mesh structure* with all the information of the rectilinear grid G^ℓ, and a *space structure* with all the required information to define the basis functions of the tensor-product space V^ℓ. In particular, these data structures contain the following methods:

- **get_basis_functions**: given an element $Q \in G^\ell$, compute the indices of the basis functions in $\mathcal{B}^\ell$ that do not vanish in Q;
- **get_cells**: given a function $\beta^\ell \in \mathcal{B}^\ell$, compute the elements Q in the support of β;
- **get_support_extension**: for a given element $Q \in G^\ell$, compute its support extension $\tilde{Q}$, defined as

$$\tilde{Q} := \left\{ Q' \in G^\ell : \exists \beta^\ell \in \mathcal{B}^\ell, \text{supp} \, \beta^\ell \cap Q' \neq \varnothing \wedge \text{supp} \, \beta^\ell \cap Q \neq \varnothing \right\}. \tag{4}$$

This last function can be implemented as a sequential call of the previous two methods.

The next two structures contain the information about the hierarchical B-splines. The first one is the *hierarchical mesh structure*, denoted by MESH in the algorithms of Section 3, which contains all the information about the hierarchical mesh $\mathcal{Q}$. In particular, it includes the following fields:

- the number of levels, N;
- for each level ℓ, a structure for the rectilinear grid G^ℓ;
- for each level ℓ, the list of active elements $\mathcal{G}^\ell$, denoted by E_ℓ^{A} in the algorithms of Section 3;
- the kind of refinement (dyadic, triadic...) between levels.

It also contains two methods, necessary for the development of refinement. They can be briefly described as follows:

- **get_parent_of_cell**: given a cell $Q \in G^\ell$ (or a list of cells), compute the index of its parent, that is, the unique cell $Q' \in G^{\ell-1}$ such that $Q \subset Q'$;
- **get_ancestor_of_cell**: given a cell $Q \in G^\ell$ (or a list of cells) of level ℓ, and given $0 \le k < \ell$, return the unique index of the ancestor of Q of level k, that is, the unique cell $Q' \in G^k$ such that $Q \subset Q'$.

The first method was already presented in [11], while the second one is detailed in the recursive Algorithm 1.

Algorithm 1: get_ancestor_of_cell.

Description: get ancestor of level k for an element Q (or a list of elements) of level $\ell > k$.

   Input: MESH, Q, ℓ, k
1: ancestors ← **get_parent_of_cell** (MESH, ℓ, Q)
2: **if** ($k < \ell - 1$) **then**
3:     ancestors ← **get_ancestor_of_cell** (MESH, ancestors, $\ell - 1$, k)
4: **end if**
   Output: ancestors

The last data structure is the *hierarchical space structure*, which will be denoted by SPACE in the algorithms of Section 3, and which contains the necessary information regarding the basis functions $\mathcal{H}(\mathcal{Q})$, or $\mathcal{T}(\mathcal{Q})$. In particular, it contains the following fields:

- the number of levels, N;
- for each level ℓ, a space structure for the tensor-product space V^ℓ;
- for each level ℓ, the set of active basis functions in $\mathcal{B}^\ell \cap \mathcal{H}(\mathcal{Q})$;
- the coefficients of the two-scale relation (2) between levels ℓ and $\ell + 1$.

This is all the functionality required to implement the refinement algorithms of the following section. We refer the interested reader to [11] for a more detailed description of the data structures, and their use in isogeometric analysis.

3. Admissible Refinement Algorithms

In order to develop the theory for adaptive isogeometric methods, exploiting the reduced support of THB-splines with respect to standard HB-splines, Buffa and Giannelli introduced in [5] the concept of admissible meshes, for which the basis functions acting in one element come from a limited number of levels. The same concept was introduced for HB-splines in [7], limiting the number of levels to two. The two types of admissibility are enclosed in the following definition.

Definition 3. *A mesh $\mathcal{Q}$ is $\mathcal{H}$-admissible (respectively, $\mathcal{T}$-admissible) of class m, with $m \ge 2$, if the basis functions in $\mathcal{H}(\mathcal{Q})$ (resp. $\mathcal{T}(\mathcal{Q})$) which take non-zero values over any element $Q \in \mathcal{Q}$ belong to at most m successive levels.*

Let Q be an active element of level ℓ where at least one hierarchical basis function of the same level is non-zero. When an admissible mesh of class 2 is considered, the active basis functions which are non-zero on Q are only the ones of level $\ell - 1$ and ℓ. They are of levels $\ell - 2, \ell - 1, \ell$ when $m = 3$, and in general the basis functions non-vanishing on $Q \in \mathcal{G}^\ell$ belong to levels $\ell - m + 1, \ell - m + 2, \ldots, \ell$.

References [5] and [7] provide refinement algorithms that generate $\mathcal{T}$-admissible and $\mathcal{H}$-admissible meshes, respectively, limited to class 2 in the second case. Both algorithms follow the same idea: given a set of marked elements, coarse elements in their neighborhood are also refined to enforce the admissibility of the mesh. Before properly defining the neighborhood, we follow [5] and extend the definition of *support extension* of a mesh element, introduced in Equation (4), to the hierarchical setting. We note that the definition is independent on whether we work with HB-splines or THB-splines.

Definition 4. *The multilevel support extension $S(Q, k)$ of an element $Q \in G^\ell$ with respect to level k, with $0 \leq k \leq \ell$, is defined as*

$$S(Q, k) := \left\{ Q' \in G^k : \exists \beta^k \in \mathcal{B}^k, \text{ supp } \beta^k \cap Q' \neq \varnothing \wedge \text{supp } \beta^k \cap Q \neq \varnothing \right\}.$$

To keep the notation as simple as possible, we will also denote by $S(Q, k)$ the region occupied by the closure of elements in $S(Q, k)$. Algorithm 2 details the computation of the multilevel support extension using the data structures and the methods introduced in Section 2.2. We first compute, in case $k < \ell$, the ancestor of level k of the given element, and then compute its corresponding support extension in the tensor-product setting. Notice that the support extension depends on the degree and the regularity of the basis functions, so the hierarchical space structure, which also includes data structures for the tensor-product spaces V^ℓ, must be given as an input.

Algorithm 2: get_multilevel_support_extension.
Description: Multilevel support extension of an element (or list of elements) Q of level ℓ, with respect to level $k < \ell$, for a hierarchical space.

Input: MESH, SPACE, Q, ℓ, k
1: **if** $(k = \ell)$ **then**
2:     extension $\leftarrow$ **get_support_extension** (Q, G^ℓ, V^ℓ)
3: **else**
4:     ancestors $\leftarrow$ **get_ancestor_of_cell** (MESH, Q, ℓ, k)
5:     extension $\leftarrow$ **get_support_extension** (ancestors, G^k, V^k)
6: **end if**
Output: extension

We can now rigorously define the neighborhood, which will depend on the chosen basis. The definition for the THB-splines case follows [5], while the definition for the standard hierarchical B-splines generalizes the definition in [7] to a general m.

Definition 5. *Given an element $Q \in \mathcal{Q} \cap \mathcal{G}^\ell$, its $\mathcal{H}$-neighborhood and its $\mathcal{T}$-neighborhood with respect to m are respectively defined as*

$$\mathcal{N}_{\mathcal{H}}(\mathcal{Q}, Q, m) := \left\{ Q' \in \mathcal{G}^{\ell-m+1} : Q' \in S(Q, \ell - m + 1) \right\},$$
$$\mathcal{N}_{\mathcal{T}}(\mathcal{Q}, Q, m) := \left\{ Q' \in \mathcal{G}^{\ell-m+1} : \exists Q'' \in S(Q, \ell - m + 2), Q'' \subseteq Q' \right\},$$

when $\ell - m + 1 \geq 0$, and $\mathcal{N}_{\mathcal{H}}(\mathcal{Q}, Q, m) = \mathcal{N}_{\mathcal{T}}(\mathcal{Q}, Q, m) = \varnothing$ for $\ell - m + 1 < 0$.

Notice that, given the level ℓ and the admissibility class m, the elements of the neighborhood (either $\mathcal{T}$- or $\mathcal{H}$-) have level $\ell - m + 1$. However, since the multilevel support extensions satisfy $S(Q, \ell - m + 2) \subset S(Q, \ell - m + 1)$, the $\mathcal{T}$-neighborhood is always contained in the $\mathcal{H}$-neighborhood. Moreover, we note that all the elements in the neighborhood are contained in $\mathcal{Q}$, that is, they are all active. We take advantage of this fact in Algorithms 3 and 4, which detail the computation of the $\mathcal{H}$-neighborhood and the $\mathcal{T}$-neighborhood, respectively.

Algorithm 3: get_H-neighborhood.
Description: $\mathcal{H}$-neighborhood of an element Q, of level ℓ, with respect to the admissibility class m.

Input: MESH, SPACE, Q, ℓ, m
1: $k \leftarrow \ell - m + 1$
2: **if** $(k < 0)$ **then**
3:     neighborhood $\leftarrow \emptyset$
4: **else**
5:     extension $\leftarrow$ **get_multilevel_support_extension** (MESH, SPACE, Q, ℓ, k)
6:     neighborhood $\leftarrow$ extension $\cap \mathbb{E}_k^A$
7: **end if**
Output: neighborhood

Algorithm 4: get_T-neighborhood.
Description: $\mathcal{T}$-neighborhood of an element Q, of level ℓ, with respect to the admissibility class m.

Input: MESH, SPACE, Q, ℓ, m
1: $k \leftarrow \ell - m + 2$
2: **if** $(k - 1 < 0)$ **then**
3:     neighborhood $\leftarrow \emptyset$
4: **else**
5:     extension $\leftarrow$ **get_multilevel_support_extension** (MESH, SPACE, Q, ℓ, k)
6:     parents $\leftarrow$ **get_parent_of_cell** (MESH, extension, $k, k - 1$)
7:     neighborhood $\leftarrow$ parents $\cap \mathbb{E}_{k-1}^A$
8: **end if**
Output: neighborhood

To develop the algorithm for refinement with guaranteed admissible meshes, we also need to define, for $\ell = 0, \ldots, N - 1$, the auxiliary subdomains

$$\omega_{\mathcal{H}}^{\ell} := \bigcup \left\{ \overline{Q} \ : \ Q \in G^{\ell} \wedge S(Q, \ell - 1) \subseteq \Omega^{\ell} \right\},$$
$$\omega_{\mathcal{T}}^{\ell} := \bigcup \left\{ \overline{Q} \ : \ Q \in G^{\ell} \wedge S(Q, \ell) \subseteq \Omega^{\ell} \right\},$$

and it clearly holds that $\omega_{\mathcal{H}}^{\ell} \subseteq \omega_{\mathcal{T}}^{\ell}$. Then, we also need to introduce a different concept of admissibility, which is based on these auxiliary subdomains.

Definition 6. *A mesh Q is strictly $\mathcal{H}$-admissible (respectively, strictly $\mathcal{T}$-admissible) of class m if it holds that*

$$\Omega^{\ell} \subseteq \omega_{\mathcal{H}}^{\ell - m + 1}, \qquad (resp. \ \Omega^{\ell} \subseteq \omega_{\mathcal{T}}^{\ell - m + 1}),$$

for $\ell = m, m + 1, \ldots, N - 1$.

The definition of the auxiliary subdomains, and their role in strict admissibility, is better understood with the help of Figure 1. We represent $\omega_{\mathcal{H}}^{1}$ and $\omega_{\mathcal{T}}^{1}$ for different mesh configurations and degrees, and note that, in all cases $\omega_{\mathcal{H}}^{1} \subseteq \omega_{\mathcal{T}}^{1}$, and the difference becomes bigger with higher degree. Moreover, in the examples of the figure, the subdomains Ω^{0} and Ω^{1} do not change between the top and bottom row, so also the auxiliary subdomains $\omega_{\mathcal{H}}^{1}$ and $\omega_{\mathcal{T}}^{1}$ do not change. Finally, for the degree and the class in the caption of each subfigure, we can refine the highlighted region $\omega_{\mathcal{H}}^{1}$ (respectively, $\omega_{\mathcal{T}}^{1}$), adding one more level and maintaining the strict $\mathcal{H}$-admissibility (resp. strict $\mathcal{T}$-admissibility) of the mesh.

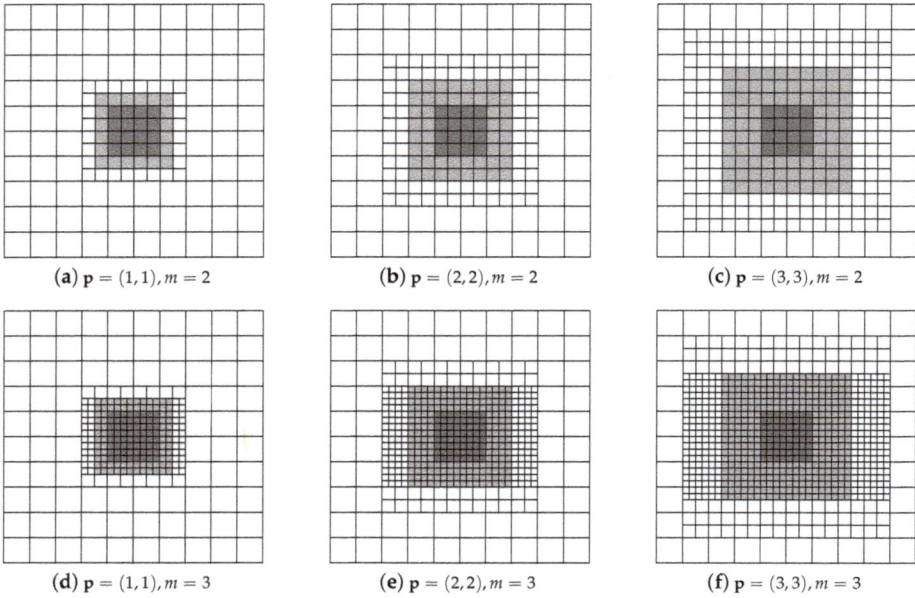

Figure 1. The domains $\omega_{\mathcal{H}}^1$ and $\omega_{\mathcal{T}}^1$ are highlighted in dark gray and light gray, respectively, for different mesh configurations.

The relation between the different admissibility classes is also stated in Proposition 1. One of the results of this proposition requires the following assumption:

$$\text{if } Q \in \mathcal{G}^\ell, \text{ then } \exists \beta \in \mathcal{B}^\ell \cap \mathcal{H}(\mathcal{Q}) : \text{supp } \beta \cap Q \neq \emptyset. \tag{5}$$

This means that, for any mesh element Q, at least one B-spline of the same level of Q, whose support overlaps this element, is active in the hierarchical B-spline basis. The assumption may not be satisfied when refining few adjacent elements, as in the example of Figure 2 below.

Proposition 1. *Let $\mathcal{Q}$ be a hierarchical mesh. We have the following results:*

1. *if $\mathcal{Q}$ is strictly $\mathcal{T}$-admissible of class m, then it is $\mathcal{T}$-admissible of the class m.*
2. *if $\mathcal{Q}$ is strictly $\mathcal{H}$-admissible of class m, then it is $\mathcal{H}$-admissible of the class m.*
3. *if $\mathcal{Q}$ is (strictly) $\mathcal{H}$-admissible of class m, then it is (strictly) $\mathcal{T}$-admissible of the class m.*
4. *if $\mathcal{Q}$ is $\mathcal{H}$-admissible of class m and satisfies assumption (5), then it is strictly $\mathcal{H}$-admissible of the class m.*

Proof. Point 1 was already proved in [5] (Prop. 9). Point 3 is a trivial consequence of Definition 3 and the reduced support of THB-splines for admissible meshes, and a consequence of Definition 6 and the fact that $\omega_{\mathcal{H}}^\ell \subseteq \omega_{\mathcal{T}}^\ell$ for strictly admissible meshes. It remains to prove Points 2 and 4.

Given an active element $Q \in \mathcal{G}^\ell$, by definition of hierarchical B-splines, we know that any finer function in $\mathcal{H}(\mathcal{Q}) \cap \mathcal{B}^k$, with $k > \ell$ vanishing on Q. We need to prove that the same is true for any $k \leq \ell - m$. Since $Q \subseteq \Omega^\ell \subseteq \omega_{\mathcal{H}}^{\ell-m+1}$, by definition of $\omega_{\mathcal{H}}^{\ell-m+1}$, all the functions of level $\ell - m$ that do not vanish on Q have support contained in $\Omega^{\ell-m+1}$, and are not active. Noting that $\Omega^\ell \subseteq \Omega^{\ell-j}$, for $0 \leq j \leq \ell$, we can use the same argument for coarser levels. This proves Point 2.

To prove Point 4, let the mesh element $Q \in \mathcal{G}^\ell$ for $\ell = m, m+1, \ldots, N-1$. We observe that, under assumption (5), since the mesh is $\mathcal{H}$-admissible, the B-splines in $\mathcal{H}(\mathcal{Q})$ whose support overlaps

Q belong to levels $\ell, \ell - 1, \ldots, \ell - m + 1$. Consequently, it does not exist an active B-spline in $\mathcal{H}(\mathcal{Q})$ of level $\ell - m$ whose support overlaps Q. The support extension of Q with respect to level $\ell - m$ is then completely contained in $\Omega^{\ell-m+1}$, and since this holds for any active element of level ℓ, we have $\Omega^\ell \subseteq \omega_\mathcal{H}^{\ell-m+1}$, namely the mesh is strictly $\mathcal{H}$-admissible. $\quad\square$

Points 2 and 4 of Proposition 1 prove the equivalence of $\mathcal{H}$-admissibility and strictly $\mathcal{H}$-admissibility under assumption (5). In general, however, if the hierarchical mesh $\mathcal{Q}$ is $\mathcal{H}$-admissible ($\mathcal{T}$-admissible) of class m, then it is not necessarily strictly $\mathcal{H}$-admissible (respectively, $\mathcal{T}$-admissible). Counterexamples for both cases are shown in Figure 2, where we highlight the elements in $\mathcal{G}^\ell$, and thus, in Ω^ℓ, such that are not contained in $\omega_\mathcal{H}^{\ell-m+1}$. Notice that, in the $\mathcal{H}$-admissible mesh, the highlighted elements are precisely those that do not satisfy (5). In the $\mathcal{T}$-admissible case instead, the highlighted elements of level $\ell = 3$ do not belong to $\omega_\mathcal{T}^{\ell-m+1} = \omega_\mathcal{T}^1$ for $m = 3$.

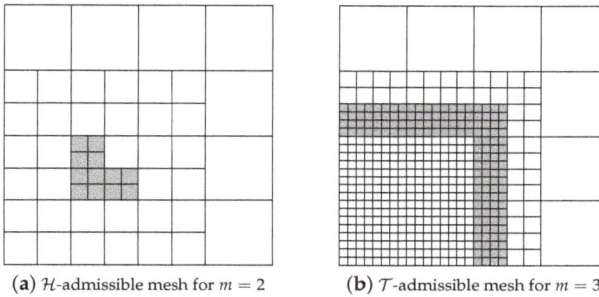

(a) $\mathcal{H}$-admissible mesh for $m = 2$ (b) $\mathcal{T}$-admissible mesh for $m = 3$

Figure 2. Examples of degree $\mathbf{p} = (2, 2)$ of meshes that are admissible, but not strictly admissible.

The last two algorithms of this section adapt to our setting the recursive refinement algorithm in [5], that, given an admissible hierarchical mesh and a set of marked elements, generates a refined $\mathcal{T}$-admissible mesh. Similar to the usage of cell-arrays in Octave/Matlab, we write {marked} when referring to the marked elements of all levels, and marked$_\ell$ when referring to a single level.

The refinement algorithm is presented in Algorithm 5. Given a set of marked (active) elements, in lines 1 to 3, we first enrich this set by a recursive algorithm, explained in detail below, where all active elements in the neighborhood of the marked ones are also marked. Then, in line 4, we refine the hierarchical mesh by refining the updated set of marked elements, and replacing them with their children. This second step can be done as in [11] (Algorithm 2), and is not limited to dyadic refinement. After refining the hierarchical mesh, it is necessary to refine the hierarchical space, that is, to update the list of active basis functions. This last step, which we omit, has been already explained in [11].

Algorithm 5: admissible_refinement.
Description: given a hierarchical mesh and a set of marked elements, generate a refined admissible mesh of class m.

    Input: MESH, SPACE, {marked}, m
1: **for** $\ell = 0, \ldots, N - 1$ **do**
2:     {marked} $\leftarrow$ **mark_recursive** (MESH, SPACE, {marked}, ℓ, m)
3: **end for**
4: MESH $\leftarrow$ **refine_hierarchical_mesh** (MESH, {marked}) $\triangleright$ see Algorithm 2 in [11]
    Output: MESH

Finally, Algorithm 6 is a recursive algorithm that, given a list of marked elements, adds the elements in their neighborhood (either $\mathcal{T}$ or $\mathcal{H}$) to the list. The difference between the two kinds of admissibility only affects the algorithm in the computation of the neighborhood (see Definition 5), in lines 1 to 5. Recursivity is necessary because, to guarantee admissibility, we must also mark the elements in the neighborhood of the newly marked ones.

As we mentioned above, our algorithm is a simple adaption of the one in [5] for $\mathcal{T}$-admissible meshes, while for $\mathcal{H}$-admissible meshes it generalizes to arbitrary $m \geq 2$ the refinement algorithm in [7] (Algorithm 3.1), which only considers the case $m = 2$. Note that $\mathcal{H}$-admissible refinements for $m \geq 2$ were also considered in [19].

Algorithm 6: mark_recursive.
Description: recursive algorithm to mark the elements in the neighborhood of marked ones.

    Input: MESH, SPACE, {marked}, ℓ, m
1: **if** $\mathcal{T}$-admissibility **then**
2:     neighbors $\leftarrow$ **get_T-neighborhood** (MESH, SPACE, marked$_\ell$, ℓ, m)
3: **else if** $\mathcal{H}$-admissibility **then**
4:     neighbors $\leftarrow$ **get_H-neighborhood** (MESH, SPACE, marked$_\ell$, ℓ, m)
5: **end if**
6: **if** (neighbors $\neq \varnothing$) **then**
7:     $k \leftarrow \ell - m + 1$
8:     marked$_k \leftarrow$ marked$_k \cup$ neighbors
9:     {marked} $\leftarrow$ **mark_recursive** (MESH, SPACE, {marked}, k, m)
10: **end if**
    Output: {marked}

The properties of these algorithms were analyzed in [5] and [7,19], for (strictly) $\mathcal{T}$- and $\mathcal{H}$-admissibility, respectively. We report in Lemma 1 and Proposition 2 the proofs of the key results for Algorithm 5, which include both cases in a unified setting.

Lemma 1. *Let $\mathcal{Q}$ be a strictly $\mathcal{H}$-admissible (respectively, strictly $\mathcal{T}$-admissible) hierarchical mesh of class m, associated with a hierarchical space V, let $\mathcal{M} \subseteq \mathcal{Q}$ a set of marked active elements, and let the integer $m \geq 2$. The call to Algorithm 5 in the form*

$$\mathcal{Q}_* = \textbf{admissible\_refinement}(\mathcal{Q}, V, \mathcal{M}, m)$$

terminates and returns a refined mesh $\mathcal{Q}_*$, such that all elements in $\mathcal{Q}_*$ were already active, or are obtained by a single refinement of an element of $\mathcal{Q}$.

Proof. The algorithm terminates because the recursive algorithm ends when the neighborhood is empty, and this condition is reached in a finite number of steps (see lines 1 to 3 in Algorithms 3 and 4). The elements in the input set $\mathcal{M}$ are all active in $\mathcal{Q}$, and the neighborhood only consists of active elements in $\mathcal{Q}$, hence the output set of **mark_recursive** (Algorithm 6) is a list of active elements in $\mathcal{Q}$. Since no further marking is performed, all elements in $\mathcal{Q}_*$ were already active, or are obtained by a single refinement of an element in $\mathcal{Q}$. $\square$

Proposition 2. *Let $\mathcal{Q}$, V, $\mathcal{M}$ and m the input arguments of Algorithm 5, as in Lemma 1, where $\mathcal{Q}$ is strictly $\mathcal{H}$-admissible (respectively, strictly $\mathcal{T}$-admissible) of class m. Then, the algorithm returns a refined hierarchical mesh $\mathcal{Q}_*$, which is strictly $\mathcal{H}$-admissible (resp. strictly $\mathcal{T}$-admissible) of class m.*

Proof. The proof follows the same steps for strictly $\mathcal{H}$-admissible and strictly $\mathcal{T}$-admissible meshes, so we will prove both at once, introducing the notation

$$\omega^\ell \equiv \left\{ \begin{array}{ll} \omega^\ell_{\mathcal{H}} & \text{for strictly } \mathcal{H}\text{-admissible meshes,} \\ \omega^\ell_{\mathcal{T}} & \text{for strictly } \mathcal{T}\text{-admissible meshes,} \end{array} \right.$$

and remarking any other important difference whenever needed.

Let us denote by $\Omega^\ell_* \supseteq \Omega^\ell$, for $\ell = 0, \ldots, N$, the subdomains after refinement. We use the same subindex ($*$) to identify whatever depends on the subdomains, such as the active elements at each level $\mathcal{G}^\ell_*$ as in Equation (1), and the auxiliary subdomains ω^ℓ_*.

Let $Q_* \in \mathcal{G}^\ell_*$, and we have by definition $Q_* \subseteq \Omega^\ell_* \setminus \Omega^{\ell+1}_*$. We need to prove that $Q_* \subseteq \omega^{\ell-m+1}_*$ (with the notation introduced above). There exist two possibilities, depending on whether Q_* was already active or not.

- If $Q_* \in \mathcal{G}^\ell$, then $Q_* \subseteq \Omega^\ell \subseteq \omega^{\ell-m+1}$. Obviously, since $\Omega^k_* \subseteq \Omega^k$ for every k, it holds $\omega^{\ell-m+1} \subseteq \omega^{\ell-m+1}_*$ (both for $\mathcal{H}$- and $\mathcal{T}$-admissible), and we obtain the desired result.
- If $Q_* \notin \mathcal{G}^\ell$, by Lemma 1, it is obtained by a single refinement of an element $Q_r \in \mathcal{G}^{\ell-1}$, thus $Q_* \subseteq Q_r$. The definition of multilevel support extension immediately gives, for any $0 \leq j \leq \ell - 1$,

$$Q_* \subseteq S(Q_*, \ell - j) \subseteq S(Q_r, \ell - j) \subseteq S(Q_r, \ell - j - 1).$$

Moreover, since $\mathcal{Q}$ is strictly admissible we know that $Q_r \subseteq \omega^{\ell-m}$, which combined with the previous equation and the definition of $\omega^{\ell-m}$ yields

$$\begin{array}{ll} S(Q_*, \ell - m) \subseteq S(Q_r, \ell - m - 1) \subseteq \Omega^{\ell-m} & \text{if } \mathcal{Q} \text{ is strictly } \mathcal{H}\text{-admissible,} \\ S(Q_*, \ell - m + 1) \subseteq S(Q_r, \ell - m) \subseteq \Omega^{\ell-m} & \text{if } \mathcal{Q} \text{ is strictly } \mathcal{T}\text{-admissible.} \end{array}$$

According to this, let us introduce $k = m$ for strictly $\mathcal{H}$-admissible meshes, and $k = m - 1$ for strictly $\mathcal{T}$-admissible meshes. Since Q_r was a marked element (either $Q_r \in \mathcal{M}$, or it was marked during the recursive marking), it has been used as an input for **mark_recursive** (Algorithm 6), and as a consequence all the active elements in $\mathcal{G}^{\ell-k-1} \cap S(Q_r, \ell - k - 1)$ have been marked. Hence, $Q_* \subseteq S(Q_*, \ell - k) \subseteq \Omega^{\ell-m+1}_*$, and by definition of $\omega^{\ell-m+1}_*$ the result is proved. $\square$

Proposition 2 guarantees that the strictly admissible nature of the meshes is preserved by Algorithm 5 during the iterative marking and refinement processes of the adaptive loop. Several examples of strictly $\mathcal{H}$- and $\mathcal{T}$-admissible meshes obtained with this algorithm will be shown in Section 5.

Remark 1. *It is important to note that, since the two bases span the same hierarchical space, one could apply the strictly $\mathcal{T}$-admissible refinement with the standard HB-splines (or the strictly $\mathcal{H}$-admissible refinement with THB-splines). In general, the admissibility property is not valid for HB-splines on $\mathcal{T}$-admissible meshes, while it is always valid for THB-splines on $\mathcal{H}$-admissible meshes, as already stated in Proposition 1. The lack of the admissibility property has a negative impact on the sparsity of the matrix, as we will see in the numerical tests of the following section.*

Remark 2. *The linear complexity of Algorithm 6 was proved in [20] for strictly $\mathcal{T}$-admissible meshes, and, subsequently, in [7] and [19] for strictly $\mathcal{H}$-admissible meshes of class $m = 2$ and $m \geq 2$, respectively. The complexity estimates provide an upper bound for the number of elements generated by the adaptive strategy with respect to the number of elements marked for refinement. These results are in line with the estimates obtained in the context of adaptive finite element methods [21,22]. Note that the complexity analysis of the refinement algorithms is a fundamental ingredient to prove the optimality of hierarchical isogeometric methods [6,7].*

4. Adaptive Isogeometric Methods

The hierarchical spaces are a natural choice for the definition of adaptive isogeometric methods [5–7], which are usually based on the adaptive loop

$$\text{SOLVE} \longrightarrow \text{ESTIMATE} \longrightarrow \text{MARK} \longrightarrow \text{REFINE}.$$

In order to illustrate the framework of such kind of methods, let us consider the Poisson model problem with Dirichlet boundary conditions

$$\begin{cases} -\Delta u = f & \text{in } \widehat{\Omega}, \\ u = g & \text{on } \partial\widehat{\Omega}, \end{cases} \tag{6}$$

where $\widehat{\Omega} = \mathbf{F}(\Omega)$ is the physical domain, parametrized by the mapping $\mathbf{F} : \Omega \to \widehat{\Omega}$.

To *solve* the model problem, we consider the hierarchical spline space $V := \text{span}\{\mathcal{H}(\mathcal{Q})\} = \text{span}\{\mathcal{T}(\mathcal{Q})\}$, which gives the discretization space $\widehat{V} := \text{span}\{\beta \circ \mathbf{F}^{-1} : \beta \in V\}$. Then, we determine the solution u_h of the discretized problem (in its variational form)

$$\int_{\widehat{\Omega}} \nabla u_h \cdot \nabla \widehat{v}_h = \int_{\widehat{\Omega}} f \, \widehat{v}_h, \qquad \forall \widehat{v}_h \in \widehat{V}_0 := \{\widehat{w}_h \in \widehat{V} : \widehat{w}_h|_{\partial\widehat{\Omega}} = 0\}, \tag{7}$$

where $u_h = u_0 + u_g$, with $u_0 \in \widehat{V}_0$ and $u_g \in \widehat{V}$ a lifting of an approximation of the boundary function, such that $u_g|_{\partial\widehat{\Omega}} \approx g$, that we obtain by an L^2-projection (see [16]). Note that (7) leads to a linear system whose structure depends on the choice of $\mathcal{H}(\mathcal{Q})$ or $\mathcal{T}(\mathcal{Q})$ as basis of V.

To *estimate* the error, and assuming that the basis functions are at least C^1 continuous, we employ the following element-based residual error estimator [5]. For any $Q \in \mathcal{G}$, the estimator is defined as

$$\varepsilon_Q(u_h) = h_Q \left(\int_{\mathbf{F}(Q)} |f + \Delta u_h|^2 \right)^{1/2}, \tag{8}$$

where $h_Q := \text{diam}(\mathbf{F}(Q))$. We note that other estimators, such as the function based residual estimator in [23], or recovery-based error estimators [24,25] could also be used. At each iteration of the adaptive loop, we compute the initial set of *marked* elements by applying Dörfler's marking strategy [26] with the indicator (8). Finally, in order to get admissible meshes, we *refine* applying the Algorithms of Section 3.

Note that this framework can be easily extended to multipatch configurations with C^0 continuity. Let us assume that the physical domain is composed of several patches $\widehat{\Omega} = \cup_i \widehat{\Omega}_i$, each with its own parametrization $\mathbf{F}_i : \Omega \to \widehat{\Omega}_i$, which overall give a C^0 parametrization. Defining a tensor-product space on each patch with the restrictions of having coinciding knot vectors and control points at the interfaces is enough to get a C^0 spline space [1].

Then, multipatch C^0 hierarchical spline spaces are obtained with the same construction presented in Section 2.1, simply by considering multipatch C^0 spline spaces as elements of the sequence $V^0 \subset V^1 \subset \ldots \subset V^{N-1}$, that is, we have a C^0 multipatch space for each level. The refinement algorithms described in Section 3 do not need any significant modification: only the support extensions, which determine the neighborhoods of the cells, are modified according to the different supports of the multipatch C^0 basis functions (see [11] (Section 3.4) and [27] for more details).

Since the basis functions are only C^0 across the interfaces, the element-based estimator must take into account the jump of the normal derivative, that is, for any $Q \in \mathcal{G}$, the estimator is

$$\varepsilon_Q(u_h) = \left(h_Q^2 \int_{\mathbf{F}(Q)} |f + \Delta u_h|^2 + h_Q \sum_{i \neq j} \int_{\Gamma_{ij} \cap \mathbf{F}(Q)} |\nabla u_h \cdot \mathbf{n}|^2 \right)^{1/2}, \tag{9}$$

where $\Gamma_{ij} := \partial \widehat{\Omega}_i \cap \partial \widehat{\Omega}_j$ denotes the interface between two patches, and $\mathbf{n}$ is the unit normal vector to Γ_{ij}.

Remark 3. *The admissible refinement algorithm can be easily extended to a posteriori estimators based on functions. In this case, one would first mark the elements in the support of the marked functions (see Algorithms 1 and 10 in [11]), and then apply the recursive marking of Algorithm 5 before refining the mesh.*

5. Numerical Results

In this section, we present some numerical examples to show the effectivity of the proposed algorithms, and to compare the different admissibility classes presented in the previous sections. The first numerical test consists of an ad hoc refinement of the unit square, while, in the second numerical test, we perform automatic adaptivity for a Poisson problem with singular solution.

5.1. Diagonal Refinement of the Unit Square

In the first numerical test, we study how the choice of the basis and the admissibility class may affect the matrix of the linear system arising in the isogeometric method. In particular, we are interested in the sparsity pattern and the condition numbers of the matrices.

For our numerical tests, we have chosen a diagonal refinement of the unit square, similar to the one used in [13], as it gives a good compromise between locality of the refinement and an increasing number of basis functions at each step. More precisely, we start from a 4×4 mesh, and at each step refine a strip of $2 \left\lceil \frac{p+1}{2} \right\rceil - 1$ cells centered at the diagonal (compared to $2p + 1$ cells in [13]), which ensures that at each step we add functions of the finest level. The meshes obtained after six levels of refinement are shown in Figures 3–5 for degrees two, three and four, respectively. The degrees of freedom (DOFs) associated to the different meshes are also indicated.

In Tables 1–3, we show the number of nonzeros of the stiffness matrix, and its percentage with respect to the global size of the matrix, after ten refinement steps, for HB-splines and THB-splines considering degrees $\mathbf{p} \equiv (p, p) = (2, 2), (3, 3), (4, 4)$, and for strictly $\mathcal{H}$-admissible and strictly $\mathcal{T}$-admissible hierarchical meshes of classes $m = 2, 3, 4, \infty$, where $m = \infty$ corresponds to refining only the marked elements of the finest level. The reduced support of THB-splines always reduces the number of nonzero entries compared to HB-splines, and this reduction is more evident for $\mathcal{T}$-admissible meshes and for high values of m. For $\mathcal{H}$-admissible meshes, instead, the reduction is not so significant, as the number of functions acting on one element is bounded for HB-splines. We also point out that higher values of m increase the number of nonzero entries with respect to the global size of the matrix, especially for high degree.

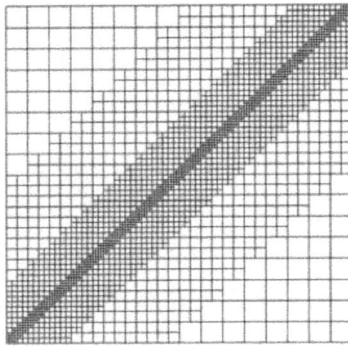

(**a**) $m = 2$, $\mathcal{H}$-admissible (2030 DOFs)

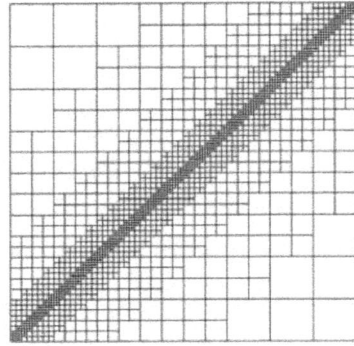

(**b**) $m = 2$, $\mathcal{T}$-admissible (1420 DOFs)

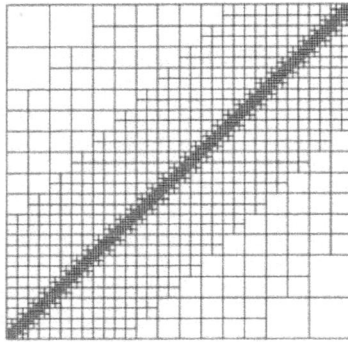

(**c**) $m = 3$, $\mathcal{H}$-admissible (1120 DOFs)

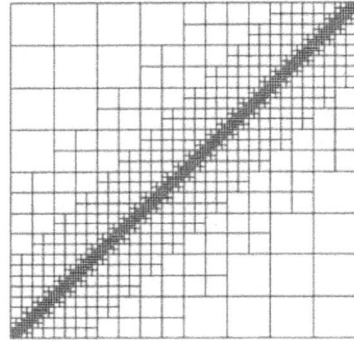

(**d**) $m = 3$, $\mathcal{T}$-admissible (908 DOFs)

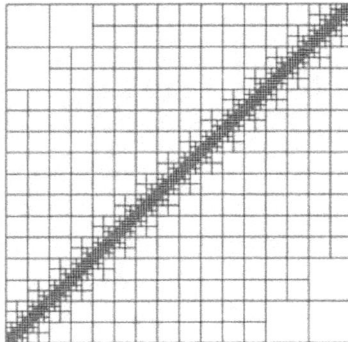

(**e**) $m = 4$, $\mathcal{H}$-admissible (774 DOFs)

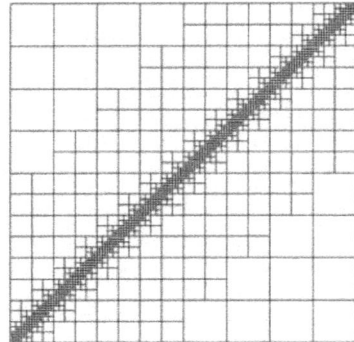

(**f**) $m = 4$, $\mathcal{T}$-admissible (716 DOFs)

Figure 3. Hierarchical meshes obtained with $\mathbf{p} = (2,2)$ and $m = 2, 3, 4$ (from **top** to **bottom**), by using $\mathcal{H}$-admissible (**left**) and $\mathcal{T}$-admissible (**right**) meshes.

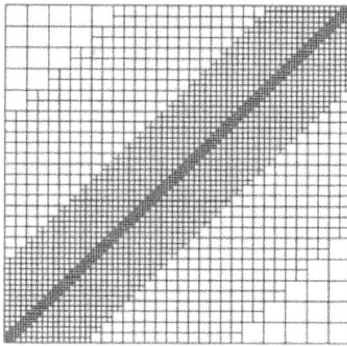

(**a**) $m = 2$, $\mathcal{H}$-admissible (2336 DOFs)

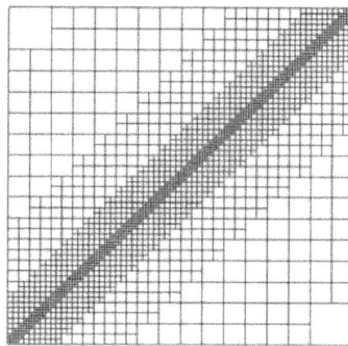

(**b**) $m = 2$, $\mathcal{T}$-admissible (1514 DOFs)

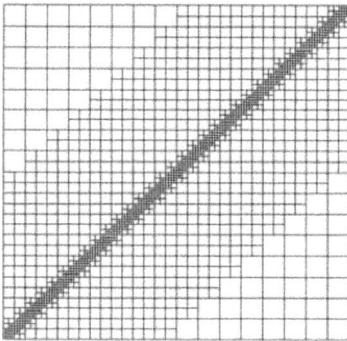

(**c**) $m = 3$, $\mathcal{H}$-admissible (1051 DOFs)

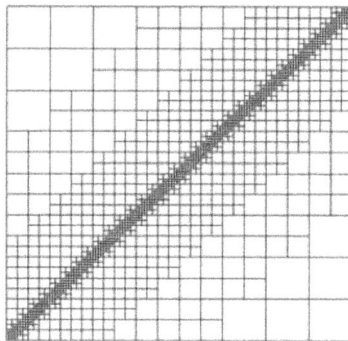

(**d**) $m = 3$, $\mathcal{T}$-admissible (763 DOFs)

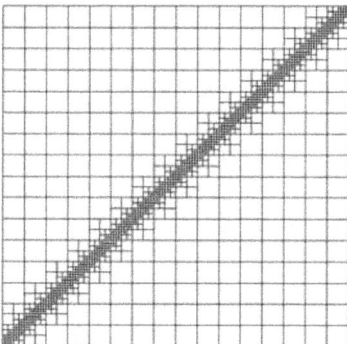

(**e**) $m = 4$, $\mathcal{H}$-admissible (536 DOFs)

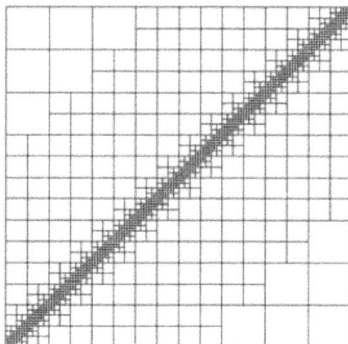

(**f**) $m = 4$, $\mathcal{T}$-admissible (464 DOFs)

Figure 4. Hierarchical meshes obtained with $\mathbf{p} = (3, 3)$ and $m = 2, 3, 4$ (from **top** to **bottom**), by using $\mathcal{H}$-admissible (**left**) and $\mathcal{T}$-admissible (**right**) meshes.

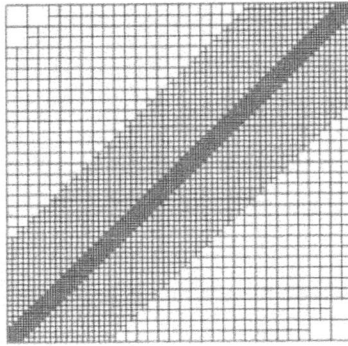

(**a**) $m = 2$, $\mathcal{H}$-admissible (2998 DOFs)

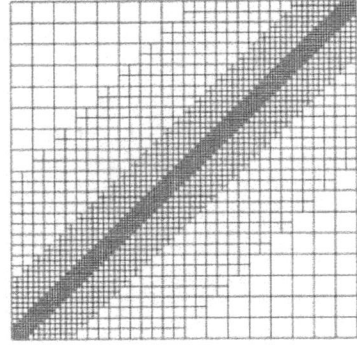

(**b**) $m = 2$, $\mathcal{T}$-admissible (2052 DOFs)

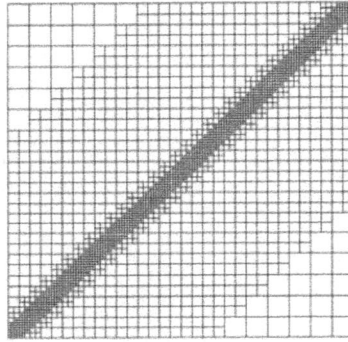

(**c**) $m = 3$, $\mathcal{H}$-admissible (1534 DOFs)

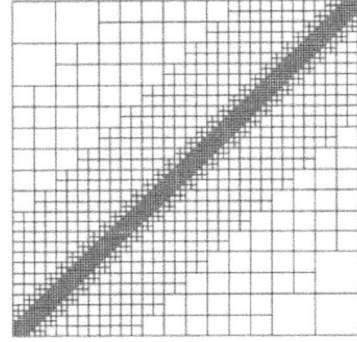

(**d**) $m = 3$, $\mathcal{T}$-admissible (1216 DOFs)

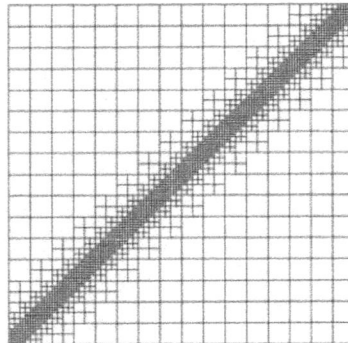

(**e**) $m = 4$, $\mathcal{H}$-admissible (962 DOFs)

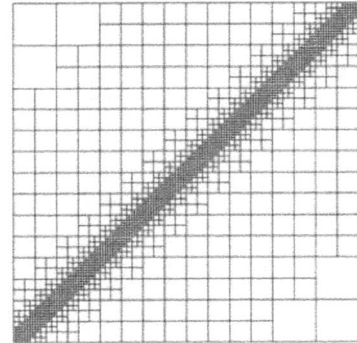

(**f**) $m = 4$, $\mathcal{T}$-admissible (912 DOFs)

Figure 5. Hierarchical meshes obtained with $\mathbf{p} = (4, 4)$ and $m = 2, 3, 4$ (from **top** to **bottom**), by using $\mathcal{H}$-admissible (**left**) and $\mathcal{T}$-admissible (**right**) meshes.

Table 1. Number of nonzeros of the stiffness matrix for $\mathbf{p} = (2, 2)$.

		DOFs	HB	(%)	THB	(%)	THB/HB
	$m = \infty$	8228	808,628	(1.19)	542,548	(0.80)	0.67
$\mathcal{H}$-admissible	$m = 2$	40,058	1,248,786	(0.08)	1,099,583	(0.07)	0.88
$\mathcal{H}$-admissible	$m = 3$	21,028	749,616	(0.17)	627,864	(0.14)	0.84
$\mathcal{H}$-admissible	$m = 4$	14,106	589,834	(0.30)	476,115	(0.24)	0.81
$\mathcal{T}$-admissible	$m = 2$	24,200	990,728	(0.17)	706,113	(0.12)	0.71
$\mathcal{T}$-admissible	$m = 3$	14,664	898,652	(0.42)	478,963	(0.22)	0.53
$\mathcal{T}$-admissible	$m = 4$	11,360	714,736	(0.55)	412,453	(0.32)	0.58

Table 2. Number of nonzeros of the stiffness matrix for $\mathbf{p} = (3, 3)$.

		DOFs	HB	(%)	THB	(%)	THB/HB
	$m = \infty$	2186	156,764	(3.28)	122,728	(2.57)	0.78
$\mathcal{H}$-admissible	$m = 2$	49,940	2,941,926	(0.12)	2,620,770	(0.11)	0.89
$\mathcal{H}$-admissible	$m = 3$	21,227	1,318,125	(0.29)	1,118,981	(0.25)	0.85
$\mathcal{H}$-admissible	$m = 4$	11,064	674,020	(0.55)	571,544	(0.47)	0.85
$\mathcal{T}$-admissible	$m = 2$	26,554	2,087,894	(0.30)	1,486,588	(0.21)	0.71
$\mathcal{T}$-admissible	$m = 3$	12,107	1,466,741	(1.00)	709,261	(0.48)	0.48
$\mathcal{T}$-admissible	$m = 4$	7020	746,362	(1.51)	392,128	(0.80)	0.53

Table 3. Number of nonzeros of the stiffness matrix for $\mathbf{p} = (4, 4)$.

		DOFs	HB	(%)	THB	(%)	THB/HB
	$m = \infty$	8446	1,819,856	(2.55)	1,410,796	(1.98)	0.78
$\mathcal{H}$-admissible	$m = 2$	66,390	6,548,354	(0.15)	5,885,286	(0.13)	0.90
$\mathcal{H}$-admissible	$m = 3$	31,112	3,299,540	(0.34)	2,861,116	(0.30)	0.87
$\mathcal{H}$-admissible	$m = 4$	18,778	2,075,442	(0.59)	1,805,250	(0.51)	0.87
$\mathcal{T}$-admissible	$m = 2$	36,516	4,819,354	(0.36)	3,499,152	(0.26)	0.73
$\mathcal{T}$-admissible	$m = 3$	19,412	3,613,896	(0.96)	2,002,780	(0.53)	0.55
$\mathcal{T}$-admissible	$m = 4$	13,456	2,773,686	(1.53)	1,437,096	(0.79)	0.52

In Figures 6 and 7, we show the computations of the condition number for the mass and the stiffness matrix, respectively, the latter computed after applying Dirichlet homogeneous boundary conditions. The results in Figure 6 show that, for the mass matrix, THB-splines get a lower condition number than HB-splines. Moreover, $\mathcal{H}$-admissible meshes give lower condition numbers than the corresponding $\mathcal{T}$-admissible ones, and lower values of m also produce lower condition numbers, which suggests that limiting the interaction between levels reduces the condition number of the mass matrix. These conclusions cannot be applied to the stiffness matrix. Indeed, the results of Figure 7 do not show any clear advantage of any option, neither for the chosen basis, nor for the admissibility class. We remark that for this particular refinement HB-splines in non-admissible meshes surprisingly provide the lowest condition numbers, which is completely opposite to the behavior for the mass matrix. It should be mentioned that a similar study for THB-splines defined on general (non admissible) hierarchical meshes and strictly admissible meshes of class 2 was already presented in [28]. In that case, there were some advantages on admissible meshes also for the condition number of the stiffness matrix. A better understanding of the problem would require a deeper investigation, which is behind the scope of the present paper.

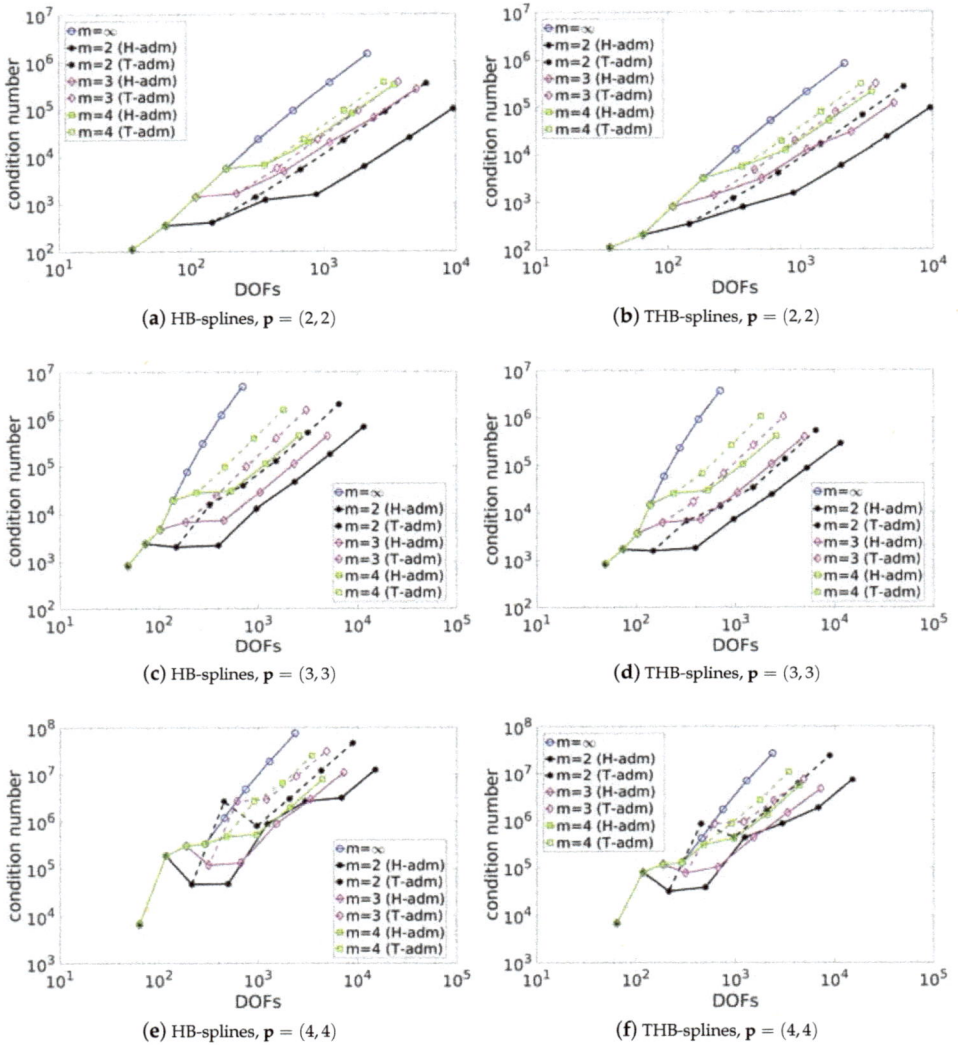

Figure 6. Condition numbers of the mass matrix for HB-splines and THB-splines, for different admissibility classes and different values of the degree.

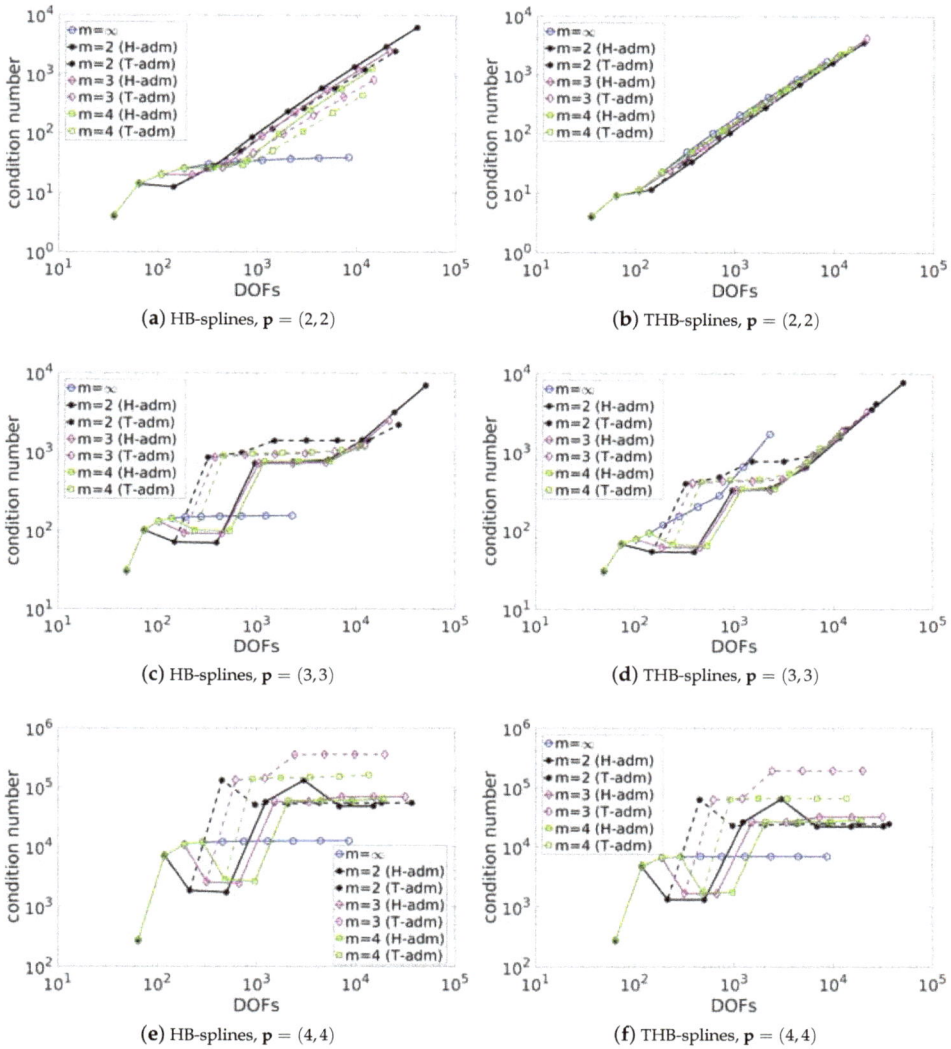

(**a**) HB-splines, **p** = (2, 2)

(**b**) THB-splines, **p** = (2, 2)

(**c**) HB-splines, **p** = (3, 3)

(**d**) THB-splines, **p** = (3, 3)

(**e**) HB-splines, **p** = (4, 4)

(**f**) THB-splines, **p** = (4, 4)

Figure 7. Condition numbers of the stiffness matrix for HB-splines and THB-splines, for different admissibility classes and different values of the degree.

5.2. Adaptive Method

For our second numerical test, we present the results obtained by applying the adaptive isogeometric methods described in Section 4 to the model problem (6) where $f = 0$ and g is the restriction to $\partial\widehat{\Omega}$ of the exact solution

$$u(\rho, \phi) = \rho^{2/3}\sin(2\phi/3),$$

defined in polar coordinates on the curved L-shaped domain shown in Figure 8, which is formed by three patches.

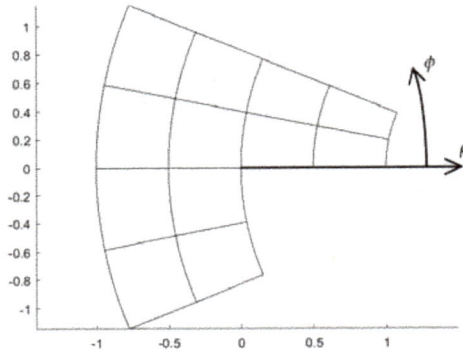

Figure 8. Curved L-shaped domain with the initial mesh mapped on it.

We apply the method with degrees $\mathbf{p} \equiv (p,p) = (2,2), (3,3), (4,4)$ and continuity C^{p-1} inside each patch, and C^0 continuity across the interfaces, using classes of $\mathcal{H}$-admissibility and $\mathcal{T}$-admissibility $m = 2,3,4,\infty$, where $m = \infty$ corresponds to pure Dörfler's marking, without later applying the recursive marking of Section 3. In all of the cases, the inital mesh is a 3-patch mesh with a 2×2 mesh on each patch, and we set a limit of maximum $n = 8$ hierarchical levels. For the Dörfler's marking strategy, we set the value of the parameter $\theta = 0.90$. Figures 9–11 show, for each degree and for each class of admissibility, the differences between $\mathcal{H}$-admissibility and $\mathcal{T}$-admissibility. These figures clearly show how the choice of the parameter m influences the grading of the mesh. As expected, $\mathcal{H}$-admissibility produces more refined meshes than $\mathcal{T}$-admissibility because the $\mathcal{H}$-neighborhood always contains the $\mathcal{T}$-neighborhood.

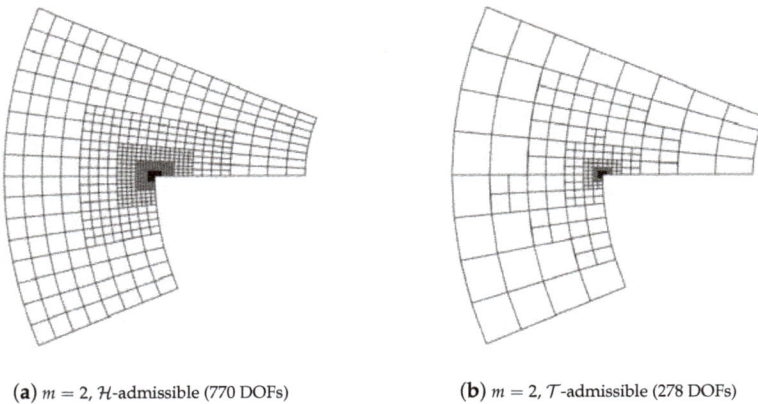

(**a**) $m = 2$, $\mathcal{H}$-admissible (770 DOFs)

(**b**) $m = 2$, $\mathcal{T}$-admissible (278 DOFs)

Figure 9. *Cont.*

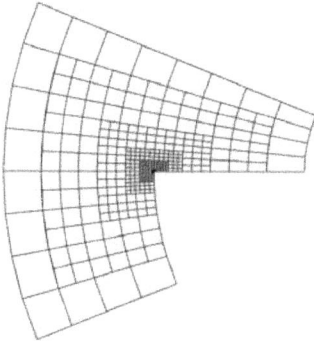

(**c**) $m = 3$, $\mathcal{H}$-admissible (410 DOFs)

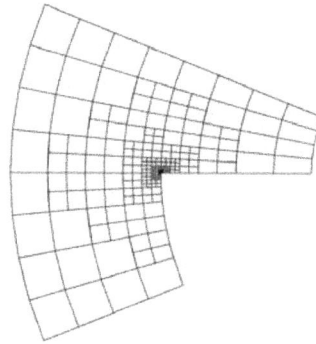

(**d**) $m = 3$, $\mathcal{T}$-admissible (258 DOFs)

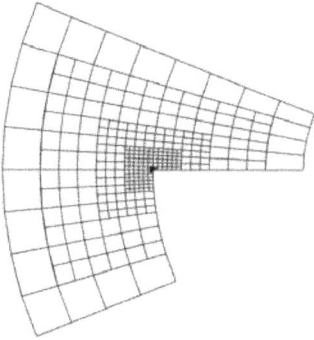

(**e**) $m = 4$, $\mathcal{H}$-admissible (344 DOFs)

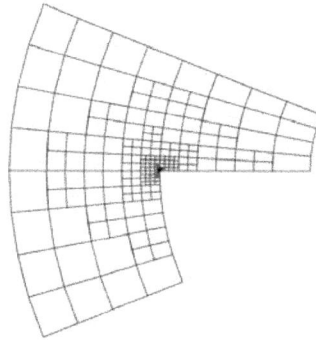

(**f**) $m = 4$, $\mathcal{T}$-admissible (241 DOFs)

Figure 9. Hierarchical meshes obtained with $\mathbf{p} = (2,2)$ and $m = 2, 3, 4$ (from **top** to **bottom**), by using $\mathcal{H}$-admissible (**left**) and $\mathcal{T}$-admissible (**right**) meshes.

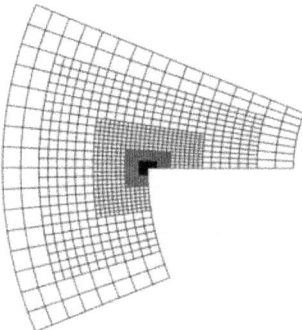

(**a**) $m = 2$, $\mathcal{H}$-admissible (1323 DOFs)

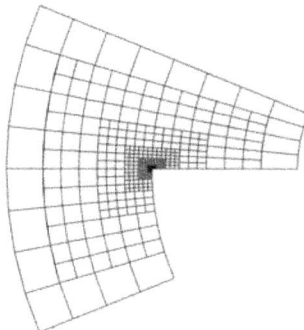

(**b**) $m = 2$, $\mathcal{T}$-admissible (472 DOFs)

Figure 10. *Cont.*

(**c**) $m = 3$, $\mathcal{H}$-admissible (763 DOFs)

(**d**) $m = 3$, $\mathcal{T}$-admissible (275 DOFs)

(**e**) $m = 4$, $\mathcal{H}$-admissible (634 DOFs)

(**f**) $m = 4$, $\mathcal{T}$-admissible (250 DOFs)

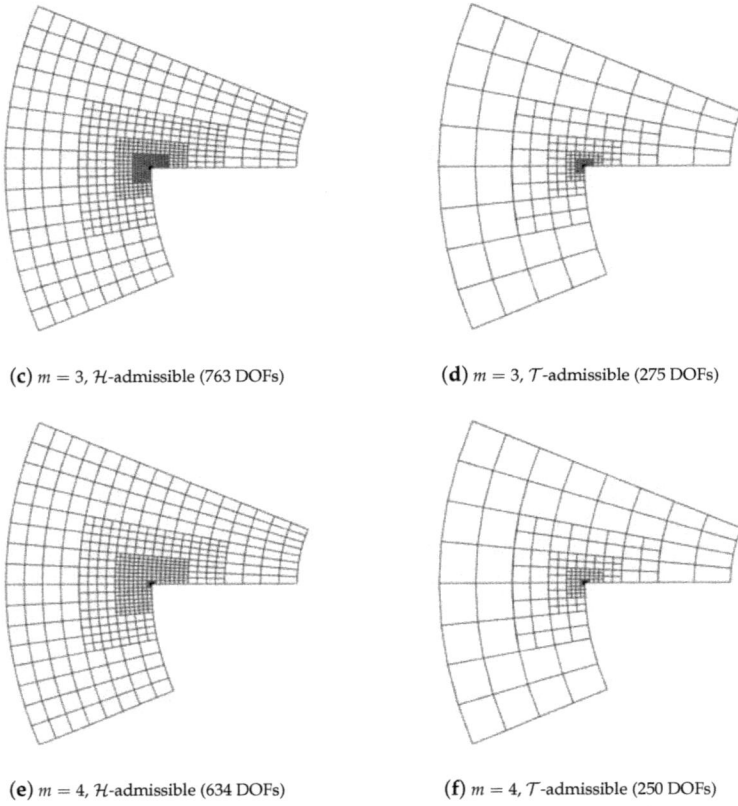

Figure 10. Hierarchical meshes obtained with $\mathbf{p} = (3, 3)$ and $m = 2, 3, 4$ (from **top** to **bottom**), by using $\mathcal{H}$-admissible (**left**) and $\mathcal{T}$-admissible (**right**) meshes.

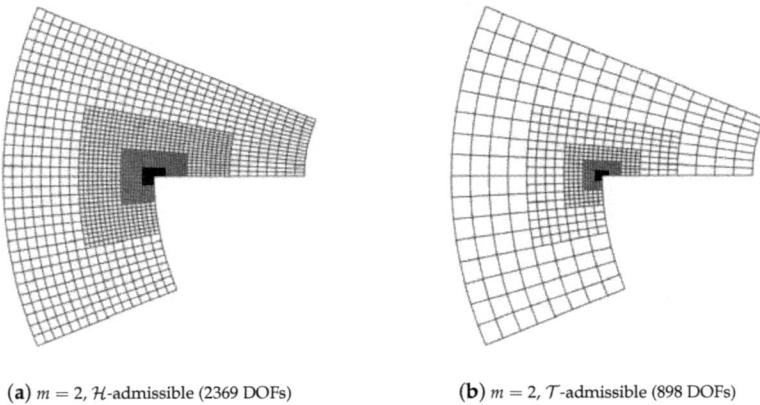

(**a**) $m = 2$, $\mathcal{H}$-admissible (2369 DOFs)

(**b**) $m = 2$, $\mathcal{T}$-admissible (898 DOFs)

Figure 11. *Cont.*

(**c**) $m = 3$, $\mathcal{H}$-admissible (1164 DOFs)

(**d**) $m = 3$, $\mathcal{T}$-admissible (490 DOFs)

(**e**) $m = 4$, $\mathcal{H}$-admissible (859 DOFs)

(**f**) $m = 4$, $\mathcal{T}$-admissible (422 DOFs)

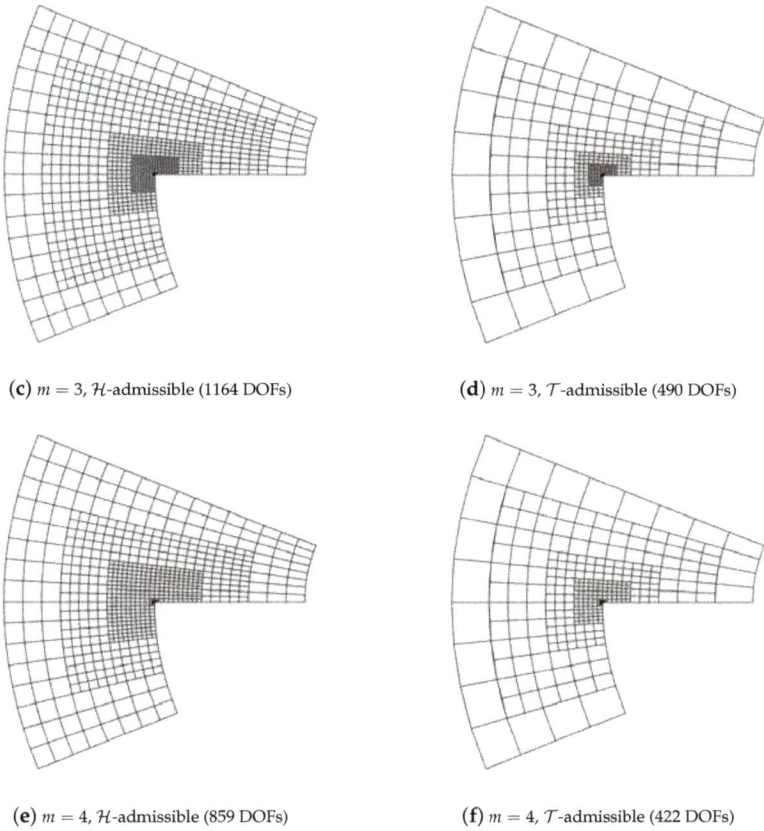

Figure 11. Hierarchical meshes obtained with $\mathbf{p} = (4, 4)$ and $m = 2, 3, 4$ (from **top** to **bottom**), by using $\mathcal{H}$-admissible (**left**) and $\mathcal{T}$-admissible (**right**) meshes.

The importance of the admissibility class is more evident in Figure 12, where we show the convergence of the error in H^1-norm with respect to the number of degrees of freedom. While both $\mathcal{H}$-admissible and $\mathcal{T}$-admissible meshes provide optimal convergence rates, the $\mathcal{T}$-admissible ones require a lower number of degrees of freedom to obtain the same error. This difference between the two classes is particularly evident for higher degree of the basis functions. Obviously, when no additional refinement is considered ($m = \infty$) the refinement is even more localized, but the assumptions of the current theory of adaptive isogeometric methods with hierarchical splines (see [5–7]) are not satisfied.

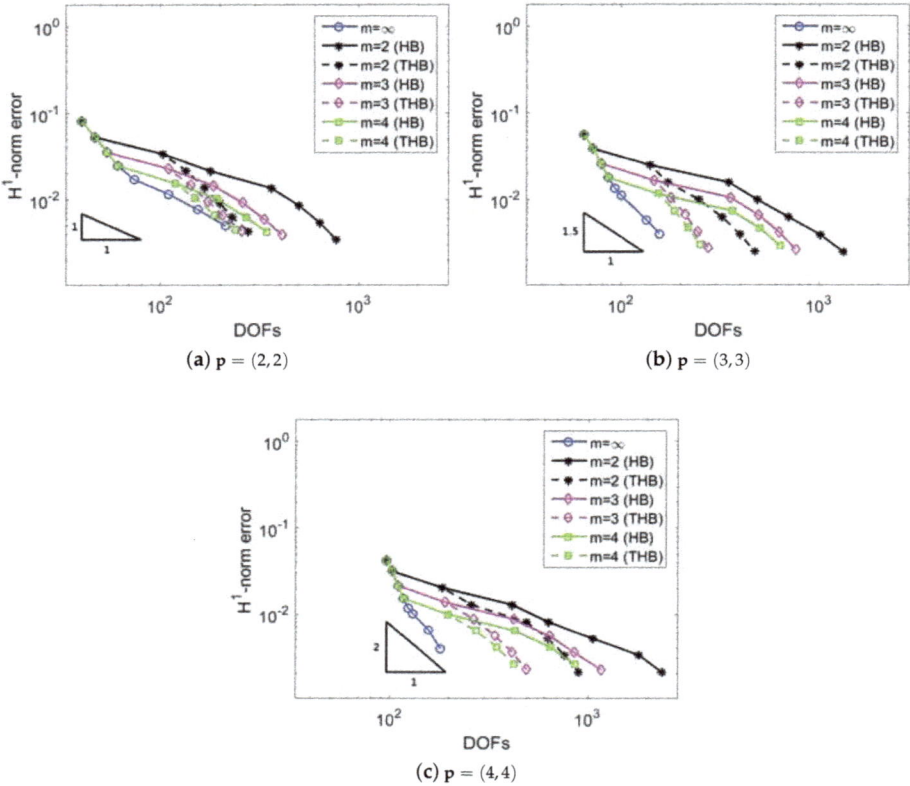

Figure 12. Comparison of the convergence of the H^1-norm error versus degrees of freedom (DOFs) for $\mathcal{H}$-admissible (HB) and $\mathcal{T}$-admissible (THB) meshes, with $\mathbf{p} = (2,2), (3,3), (4,4)$ (from **top** to **bottom**).

6. Conclusions

We presented a general framework for the design and implementation of refinement algorithms with (truncated) hierarchical B-splines. The properties of the admissible mesh configurations obtained with the iterative application of these algorithms were thoroughly analyzed. Note that the structure of hierarchical meshes with a certain class of admissibility can be naturally connected with a corresponding mesh grading, as confirmed by the numerical examples. The truncation mechanism behind the construction of THB-splines influences the strictly $\mathcal{T}$-admissible property leading to more localized refinement possibilities (and in turn to a reduced number of degrees of freedom) than the $\mathcal{H}$-admissible counterpart, that is, for the same admissibility class m. On the other hand, $\mathcal{H}$-admissible meshes guarantee a bounded number of hierarchical B-splines without the need of considering the truncated basis. The numerical examples also confirm the advantages of THB-splines with respect to the sparsity of the discretizations matrices and the condition number of the mass matrix. Concerning the condition number of the stiffness matrix, the situation is more unclear and a deeper study would be required. The comparison between $\mathcal{H}$- and $\mathcal{T}$-admissible refinements was never presented before and opens the path to additional studies on the optimal configuration for the development of hierarchical isogeometric methods.

The refinement algorithms here presented can be properly combined with coarsening algorithms that preserve the admissible nature of the mesh. This is an important aspect for controlling the effect of successive refinement and coarsening of hierarchical meshes in adaptive isogeometric methods (see e.g., [29,30]) and will be the subject matter of a future study.

Author Contributions: Conceptualization, C.B., C.G. and R.V.; Investigation, C.B., C.G. and R.V.; Methodology, C.B., C.G. and R.V.; Software, C.B., C.G. and R.V.; Writing—original draft, C.B., C.G. and R.V.; Writing—Review & Editing, C.B., C.G. and R.V.

Funding: This work was partially supported by the MIUR "Futuro in Ricerca" programme through the project "DREAMS", grant number RBFR13FBI3. Rafael Vázquez has been partially supported by the ERC Advanced Grant "CHANGE", grant number 694515, 2016-2020.

Acknowledgments: The authors are members of the INdAM Research group GNCS. The INdAM support through GNCS and Finanziamenti Premiali SUNRISE is gratefully acknowledged.

Conflicts of Interest: The authors declare no conflict of interest.

References

1. Cottrell, J.A.; Hughes, T.J.R.; Bazilevs, Y. *Isogeometric Analysis: Toward Integration of CAD and FEA*; John Wiley & Sons: Hoboken, NJ, USA, 2009.
2. Hughes, T.J.R.; Cottrell, J.A.; Bazilevs, Y. Isogeometric analysis: CAD, finite elements, NURBS, exact geometry and mesh refinement. *Comput. Methods Appl. Mech. Eng.* **2005**, *194*, 4135–4195. [CrossRef]
3. Kraft, R. Adaptive and Linearly Independent Multilevel B-Splines. In *Surface Fitting and Multiresolution Methods*; Le Méhauté, A., Rabut, C., Schumaker, L.L., Eds.; Vanderbilt University Press: Nashville, TN, USA, 1997; pp. 209–218.
4. Vuong, A.V.; Giannelli, C.; Jüttler, B.; Simeon, B. A hierarchical approach to adaptive local refinement in isogeometric analysis. *Comput. Methods Appl. Mech. Eng.* **2011**, *200*, 3554–3567. [CrossRef]
5. Buffa, A.; Giannelli, C. Adaptive isogeometric methods with hierarchical splines: Error estimator and convergence. *Math. Models Methods Appl. Sci.* **2016**, *26*, 1–25. [CrossRef]
6. Buffa, A.; Giannelli, C. Adaptive isogeometric methods with hierarchical splines: Optimality and convergence rates. *Math. Models Methods Appl. Sci.* **2017**, *27*, 2781–2802. [CrossRef]
7. Gantner, G.; Haberlik, D.; Praetorius, D. Adaptive IGAFEM with optimal convergence rates: Hierarchical B-splines. *Math. Models Methods Appl. Sci.* **2017**, *27*, 2631–2674. [CrossRef]
8. Giannelli, C.; Jüttler, B.; Speleers, H. THB-Splines: The truncated basis for hierarchical splines. *Comput. Aided Geom. Des.* **2012**, *29*, 485–498. [CrossRef]
9. Bornemann, P.; Cirak, F. A subdivision-based implementation of the hierarchical B-spline finite element method. *Comput. Methods Appl. Mech. Eng.* **2013**, *253*, 584–598. [CrossRef]
10. Scott, M.A.; Thomas, D.C.; Evans, E.J. Isogeometric spline forests. *Comput. Methods Appl. Mech. Eng.* **2014**, *269*, 222–264. [CrossRef]
11. Garau, E.; Vázquez, R. Algorithms for the implementation of adaptive isogeometric methods using hierarchical B-splines. *Appl. Numer. Math.* **2018**, *123*, 58–87. [CrossRef]
12. Kiss, G.; Giannelli, C.; Jüttler, B. Algorithms and data structures for truncated hierarchical B-splines. In *Mathematical Methods for Curves and Surfaces*; Floater, M., Lyche, T., Mazure, M.-L., Moerken, K., Schumaker, L.L., Eds.; Lecture Notes in Computer Science; Springer: Berlin/Heidelberg, Germany, 2014; Volume 8177, pp. 304–323.
13. Giannelli, C.; Jüttler, B.; Kleiss, S.K.; Mantzaflaris, A.; Simeon, B.; Špeh, J. THB-splines: An effective mathematical technology for adaptive refinement in geometric design and isogeometric analysis. *Comput. Methods Appl. Mech. Eng.* **2016**, *299*, 337–365. [CrossRef]
14. Hennig, P.; Müller, S.; Kästner, M. Bézier extraction and adaptive refinement of truncated hierarchical NURBS. *Comput. Methods Appl. Mech. Eng.* **2016**, *305*, 316–339. [CrossRef]
15. D'Angella, D.; Kollmannsberger, S.; Rank, E.; Reali, A. Multi-level Bézier extraction for hierarchical local refinement of Isogeometric Analysis. *Comput. Methods Appl. Mech. Eng.* **2018**, *328*, 147–174. [CrossRef]
16. Vázquez, R. A new design for the implementation of isogeometric analysis in Octave and Matlab: GeoPDEs 3.0. *Comput. Math. Appl.* **2016**, *72*, 523–554. [CrossRef]
17. de Boor, C. *A Practical Guide to Splines*, revised ed.; Springer: Berlin/Heidelberg, Germany, 2001.

18. Schumaker, L.L. *Spline Functions: Basic Theory*, 3rd ed.; Cambridge University Press: Cambridge, UK, 2007.
19. Morgenstern, P. Mesh Refinement Strategies for the Adaptive Isogeometric Method. Ph.D. Thesis, Institut für Numerische Simulation, Rheinische Friedrich-Wilhelms-Universität Bonn, Bonn, Germany, 2017.
20. Buffa, A.; Giannelli, C.; Morgenstern, P.; Peterseim, D. Complexity of hierarchical refinement for a class of admissible mesh configurations. *Comput. Aided Geom. Des.* **2016**, *47*, 83–92. [CrossRef]
21. Binev, P.; Dahmen, W.; DeVore, R. Adaptive finite element methods with convergence rates. *Numer. Math.* **2004**, *97*, 219–268. [CrossRef]
22. Stevenson, R. Optimality of a standard adaptive finite element method. *Found. Comput. Math.* **2007**, *7*, 245–269. [CrossRef]
23. Buffa, A.; Garau, E.M. A posteriori error estimators for hierarchical B-spline discretizations. *Math. Models Methods Appl. Sci.* **2016**. [CrossRef]
24. Kumar, M.; Kvamsdal, T.; Johannessen, K.A. Superconvergent patch recovery and a posteriori error estimation technique in adaptive isogeometric analysis. *Comput. Methods Appl. Mech. Eng.* **2017**, *316*, 1086–1156. [CrossRef]
25. Anitescu, C.; Hossain, M.N.; Rabczuk, T. Recovery-based error estimation and adaptivity using high-order splines over hierarchical T-meshes. *Comput. Methods Appl. Mech. Eng.* **2018**, *328*, 638–662. [CrossRef]
26. Dörfler, W. A convergent algorithm for Poisson's equation. *SIAM J. Numer. Anal.* **1996**, *33*, 1106–1124. [CrossRef]
27. Buchegger, F.; Jüttler, B.; Mantzaflaris, A. Adaptively refined multi-patch B-splines with enhanced smoothness. *Appl. Math. Comput.* **2016**, *272*, 159–172.
28. Hennig, P.; Kästner, M.; Morgenstern, P.; Peterseim, D. Adaptive mesh refinement strategies in isogeometric analysis—A computational comparison. *Comput. Methods Appl. Mech. Eng.* **2017**, *316*, 424–448. [CrossRef]
29. Lorenzo, G.; Scott, M.A.; Tew, K.; Hughes, T.J.R.; Gomez, H. Hierarchically refined and coarsened splines for moving interface problems, with particular application to phase-field models of prostate tumor growth. *Comput. Methods Appl. Mech. Eng.* **2017**, *319*, 515–548. [CrossRef]
30. Hennig, P.; Ambati, M.; De Lorenzis, L.; Kästner, M. Projection and transfer operators in adaptive isogeometric analysis with hierarchical B-splines. *Comput. Methods Appl. Mech. Eng.* **2018**, *334*, 313–336. [CrossRef]

axioms

MDPI

Article

A Gradient System for Low Rank Matrix Completion

Carmela Scalone [1,*] and Nicola Guglielmi [2]

1   Dipartimento di Ingegneria e Scienze dell'Informazione e Matematica (DISIM), Università dell'Aquila, via Vetoio 1, 67100 L'Aquila, Italy

2   Section of Mathematics, Gran Sasso Science Institute, via Crispi 7, 67100 L'Aquila, Italy; nicola.guglielmi@gssi.it

*   Correspondence: carmela.scalone@graduate.univaq.it; Tel.: +39-086-243-3136

Received: 7 May 2018; Accepted: 18 July 2018; Published: 24 July 2018

Abstract: In this article we present and discuss a two step methodology to find the closest low rank completion of a sparse large matrix. Given a large sparse matrix M, the method consists of fixing the rank to r and then looking for the closest rank-r matrix X to M, where the distance is measured in the Frobenius norm. A key element in the solution of this matrix nearness problem consists of the use of a constrained gradient system of matrix differential equations. The obtained results, compared to those obtained by different approaches show that the method has a correct behaviour and is competitive with the ones available in the literature.

Keywords: low rank completion; matrix ODEs; gradient system

---

1. Introduction

A large class of datasets are naturally stored in matrix form. In many important applications, the challenge of filling a matrix from a sampling of its entries can arise; this is known as the matrix completion problem. Clearly, such a problem needs some additional constraints to be well-posed. One of its most interesting variants is to find the lower rank matrices that best fit the given data. This constrained optimization problem is known as low-rank matrix completion.

Let $M \in \mathbb{R}^{m \times n}$ be a matrix that is only known on a subset, Ω, of its entries. In [1], the authors provided conditions on the sampling of observed entries, such that the problem which arises has a high probability of not being undetermined. The classical mathematical formulation for the low rank matrix completion problem is :

$$\min \ \text{rank}(X)$$
$$s.t. \ P_\Omega(X) = P_\Omega(M)$$

where P_Ω is the projection onto Ω defined as a function

$$P_\Omega : \mathbb{R}^{m \times n} \longrightarrow \mathbb{R}^{m \times n}$$

such that

$$X_{i,j} \longmapsto \begin{cases} X_{i,j} & if \ (i,j) \in \Omega \\ 0 & if \ (i,j) \notin \Omega \end{cases}$$

This approach may seem like the most natural to describe the problem, but it is not very useful in practice, since it is well known to be NP-hard [2]. In [3], the authors stated the problem as

$$\min \ ||X||_*$$
$$s.t. \ P_\Omega(X) = P_\Omega(M)$$

where $|| \ ||_*$ is the nuclear norm of the matrix, which is the sum of its singular values. This is a convex optimization problem and the authors proved that when Ω is sampled uniformly at random and is sufficiently large, the previous relaxation can recover any matrix of rank r with high probability. We will consider the following formulation as in [4,5],

$$\min \frac{1}{2} ||P_\Omega(X) - P_\Omega(M)||_F^2$$

$$s.t. \ \text{rank}(X) = r$$

Notice that, the projection $P_\Omega(X)$ can be written as a Hadamard product. If we identify the subset Ω of the fixed entries with the matrix Ω such that

$$\Omega_{i,j} = \begin{cases} 1 & if \ (i,j) \in \Omega \\ 0 & if \ (i,j) \notin \Omega \end{cases}$$

it is clear that $P_\Omega(X) = \Omega \circ X$. By considering the manifold

$$\mathcal{M}_r = \{X \in \mathbb{R}^{n \times m} : \text{rank}(X) = r\}$$

we can write the problem as

$$\min_{X \in \mathcal{M}_r} \frac{1}{2} ||\Omega \circ (X - M)||_F^2 \tag{1}$$

This approach is based on the assumption of knowing in advance the rank r of the target matrix. A key feature of the problem is that $r \ll \min\{m, n\}$, that translates, from a practical point of view in a small increase of the cost to eventually update r. In [6] is well explained the possibility of estimating the rank (unknown a priori) based on the gap between singular values of the "trimmed" partial target matrix M. Furthermore, the authors highlight that in collaborative filtering applications, r ranged between 10 and 30. In [4], the author employed optimization techniques widely exploiting the structure of smooth Riemaniann manifold of $\mathcal{M}_r$. The same tools are used in [5], where the authors considered the matrix completion in the presence of outliers. In the recent work [7], the authors provide a non convex relaxation approach for matrix completion in presence of non Gaussian noise and or outliers, by employing the correntropy induced losses. In [8], the authors survey on the literature on matrix completions and deal with target matrices, whose entries are affected by a small amount of noise. Recently, the problem became popular thanks to collaborative filtering applications [9,10] and the Netflix problem [11]. It can also be employed in other fields of practical applications such as sensor network localization [12], signal processing [13] and reconstruction of damaged images [14]. A very suggestive use of modeling as low rank matrix completion problem has been done in biomathmatics area, as shown in [15] for gene-disease associations. Applications to minimal representation of discrete systems can be considered as of more mathematical feature [16]. What makes the problem interesting are not just the multiple applications, but also its variants, such as, for example, structured [17] and Euclidean distance matrix cases [18]. In this paper a numerical technique to solve the low rank matrix completion problem is provided, which makes use of a gradient system of matrix ODEs.

2. General Idea : Two-Level Method

Let us write the unknown matrix X of the problem (1) as $X = \varepsilon E$ with $\varepsilon > 0$ and $||E||_F = 1$. For a fixed norm $\varepsilon > 0$, we aim to minimize the functional

$$F_\varepsilon(E) := ||\Omega \circ (\varepsilon E - M)||_F^2 \tag{2}$$

constrained by $E \in \mathcal{M}_r$ and $||E||_F = 1$ (see [19]). By computing the stationary point of a suitable differential equation, we will find a local minimum E_ε of the functional. Setting $f(\varepsilon) = F_\varepsilon(E_\varepsilon)$, we will look for the minimum value of ε, say ε_*, such that $f(\varepsilon^*) = 0$, by using a Newton-like method. The behaviour of $f(\varepsilon)$ in a left neighbourhood of ε^* is well understood. For $\varepsilon \geq \varepsilon^*$ it is more challenging.

We discuss two possible scenarios; ε^* can be a strict local minimum point or $f(\varepsilon)$ can become identically zero when ε exceeds ε^*. The two situations depend on the rank constraint and on the sparsity pattern. To motivate our assumption, we now present two simple illustrative examples. Suppose that we aim to recover the matrix

$$M = \begin{pmatrix} 1 & 1 & * \\ 1 & * & 1 \\ * & 1 & * \end{pmatrix}.$$

If we constrain the problem by imposing that the rank of the solution has to be equal to 1, we have a strict point of minimum for $\varepsilon^* = 3$ and the optimal rank-1 matrix that fits perfectly the given entries of M is

$$S = \begin{pmatrix} 1 & 1 & 1 \\ 1 & 1 & 1 \\ 1 & 1 & 1 \end{pmatrix}.$$

If we consider the problem of recovering the matrix

$$Y = \begin{pmatrix} 1 & 1 & 1 \\ 1 & * & 1 \\ * & * & 1 \end{pmatrix}$$

requiring that the solution has to be of rank 2, we have that the solutions of minimal norm $\varepsilon^* = 2.6458$ are

$$X_1 = \begin{pmatrix} 1 & 1 & 1 \\ 1 & 1 & 1 \\ 0 & 0 & 1 \end{pmatrix}, \quad X_2 = \begin{pmatrix} 1 & 1 & 1 \\ 1 & 0 & 1 \\ 1 & 0 & 1 \end{pmatrix}$$

However, for all $\varepsilon > \varepsilon^*$, we have a point of minimum of $f(\varepsilon)$. To understand this behaviour, we can intuitively think that there are a lot of "possibilities" to realize a rank-2 matrix "filling" the unknown entries of Y. For example,

$$X_1(\alpha, \beta) = \left\{ \begin{pmatrix} 1 & 1 & 1 \\ 1 & 1 & 1 \\ \alpha & \beta & 1 \end{pmatrix} \right\}_{\alpha, \beta \in \mathbb{R}, \alpha \neq 1 \, \vee \, \beta \neq 1}, X_1(\alpha, \beta) = \left\{ \begin{pmatrix} 1 & 1 & 1 \\ 1 & \alpha & 1 \\ 1 & \beta & 1 \end{pmatrix} \right\}_{\alpha, \beta \in \mathbb{R}, \alpha \neq 1 \, \vee \, \beta \neq 1}$$

are families of solutions of the problem. In the Figure 1, we show the graphics of $f(\varepsilon)$ for the two problems considered.

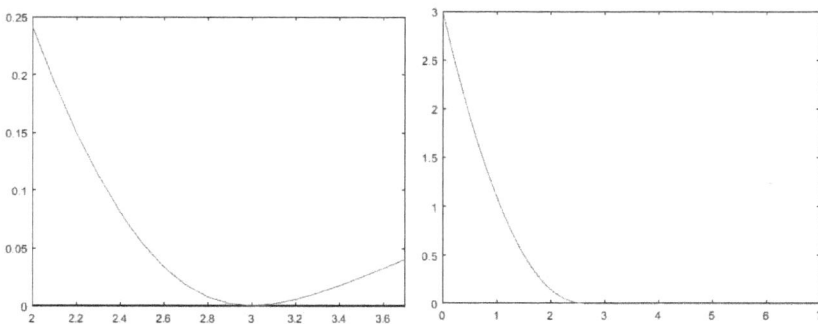

Figure 1. The figure on the left represents the graphic of $f(\varepsilon)$ when we consider the problem of recovering M by a rank-1 matrix. The figure in the right shows that $f(\varepsilon)$ is identically equals to zero, when $\varepsilon \geq \varepsilon^*$, if we require the rank of the solution to be equal to 2, when we compete Y.

The paper is structured as follows. In Sections 3 and 5 we discuss the two level method designed to solve the problem (1). A characterization of local extremizers for the functional (2) is given in Section 4. In Section 6 we present a suitable splitting method for rank-r matricial ODEs employed in the context of the inner iteration. Finally, numerical experiments are showed in Section 7.

3. Differential Equation for E

3.1. Minimizing $F_\varepsilon(E)$ for Fixed ε

Suppose that $E(t)$ is a smooth matrix valued function of t. (We omit the argument t of $E(t)$). Our goal is to find an optimal direction $\dot{E} = Z$ (see [19,20]) such that the functional (2) is characterized by the maximal local decrease, in a way that the matrix E remains in the manifold $\mathcal{M}_r$.

To deal with this goal, we differentiate (2) with respect to t

$$\frac{d}{dt}F_\varepsilon(E) = \frac{d}{dt}\frac{1}{2}||\Omega \circ (\varepsilon E - M)||_F^2 = \frac{1}{2}\frac{d}{dt}\left\langle \Omega \circ (\varepsilon E - M), \Omega \circ (\varepsilon E - M) \right\rangle = \varepsilon\langle \Omega \circ \dot{E}, \Omega \circ (\varepsilon E - M)\rangle$$

Setting

$$G := \Omega \circ (\varepsilon E - M) \tag{3}$$

and since by definition of Ω, $\Omega \circ \Omega = \Omega$, it is clear that $\Omega \circ G = G$.

Thus, we have

$$\langle \Omega \circ \dot{E}, \Omega \circ G \rangle = \langle \Omega \circ \dot{E}, G \rangle = \sum_{i,j}\Omega_{ij}\dot{E}_{ij}G_{ij} = \sum_{i,j}\dot{E}_{ij}\Omega_{ij}G_{ij} = \langle \dot{E}, \Omega \circ G \rangle = \langle \dot{E}, G \rangle$$

Hence, we have

$$\frac{d}{dt}F_\varepsilon(E(t)) = \varepsilon\langle \dot{E}, G \rangle \tag{4}$$

which identifies G as the free gradient of the functional. We have now to include the constraint $||E||_F^2 = 1$. By differentiation

$$\frac{d}{dt}||E||_F^2 = 0 \Rightarrow \langle E, \dot{E} \rangle = 0$$

we gain a linear constraint for $\dot{E}$. By virtue of the rank constraint in (5), we must guarantee that the motion of E remains in the manifold $\mathcal{M}_r$ for all t. In order to get it, we require the derivative $\dot{E}$ to lie in the tangent space in E to $\mathcal{M}_r$, for all t. These considerations led us to the following optimization problem.

$$Z_* = \underset{||Z||_F=1, \langle E,Z\rangle=0, Z \in \mathcal{T}_E\mathcal{M}_r}{\arg\min} \langle Z, G \rangle \tag{5}$$

where $\mathcal{T}_E\mathcal{M}_r$ denotes the tangent space to $\mathcal{M}_r$ at E. The constraint $||Z||_F = 1$ is simply introduced to get a unique direction Z. In the following, we will denote by $P_E(\cdot)$ the orthogonal projection on $\mathcal{T}_E\mathcal{M}_r$.

3.2. Rank-r Matrices and Their Tangent Matrices

See [21]. Every real rank-r matrix E of dimension $n \times m$ can be written in the form

$$E = USV^T \tag{6}$$

where $U \in \mathbb{R}^{n\times r}$ and $V \in \mathbb{R}^{m\times r}$ have orthonormal columns, i.e.,

$$U^T U = I_r, \quad V^T V = I_r$$

and nonsingular $S \in \mathbb{R}^{r\times r}$. In particular, when S is diagonal, we find the SVD. The decomposition (6) is not unique; simply replacing U by $\overline{U} = UP$ and V by $\hat{V} = VQ$ with orthogonal matrices $P, Q \in \mathbb{R}^{r\times r}$

and S by $\hat{S} = P^T S Q$, we get the same matrix $E = USV^T = \hat{U}\hat{S}\hat{V}^T$. However, we can make the decomposition unique in the tangent space. For all E in the manifold $\mathcal{M}_r$, let us consider the tangent space $\mathcal{T}_E\mathcal{M}_r$. It is a linear space and every tangent matrix is of the form

$$\dot{E} = \dot{U}SV^T + U\dot{S}V^T + US\dot{V}^T$$

where $\dot{S} \in \mathbb{R}^{r\times r}$, $U^T\dot{U}$ and $V^T\dot{V}$ are skew-symmetric $r \times r$ matrices. $\dot{S}, \dot{U}, \dot{V}$ are uniquely determined by $\dot{E}$ and U, V, S by imposing the gauge conditions

$$U^T\dot{U} = 0, \quad V^T\dot{V} = 0$$

We consider the following important result from [21], thanks to which it is possible to obtain a formula for the projection of a matrix onto the tangent space to a rank-r matrices.

Lemma 1. *The orthogonal projection onto the tangent space $\mathcal{T}_E\mathcal{M}_r$ at $E = USV^T \in \mathcal{M}_r$ is given by*

$$P_E(Z) = Z - (I - UU^T)Z(I - VV^T)$$

for $Z \in \mathbb{R}^{n\times m}$.

3.3. Steepest Descent Dynamics

Lemma 2. *Let $E \in \mathbb{R}^{n\times m}$ be a real matrix of unit Frobenius norm, such that it is not proportional to $P_E(G)$. Then, the solution of (5) is given by*

$$\mu Z_* = -P_E(G) + \langle E, P_E(G)\rangle E \tag{7}$$

where μ is the reciprocal of the Frobenius norm on the right-hand side.

Proof. Let be $E_\perp = \{Z \in \mathbb{R}^{n\times m} : \langle E, Z\rangle = 0\}$. The function $\langle Z, G\rangle$ is an inner product and the feasible region $\mathcal{R} = E_\perp \cap \mathcal{T}_E\mathcal{M}_r$ is a linear subspace, since it is intersection of subspaces. By observing that the inner product with a given vector is minimized over a subspace by orthogonally projecting the vector onto the subspace, we can say that the solution of (5) is a matrix proportional to the normalized orthogonal projection of the free gradient G onto $\mathcal{R}$. Therefore,

$$P_\mathcal{R}(G) = P_{E_\perp}(P_E(G)) = P_E(G) - \frac{\langle E, P_E(G)\rangle}{\langle E, E\rangle}P_E(G)$$

Note that $P_\mathcal{R}(G) = P_{E_\perp}(P_E(G))$, since $P_{E_\perp}$ and P_E commute. Since $||E||_F = 1$, we have that the solution is given by (7). □

The Expression (4), jointly with the Lemma 2 suggest to consider the following gradient system for $F_\varepsilon(E)$

$$\dot{E} = -P_E(G) + \langle E, P_E(G)\rangle E \tag{8}$$

To get the differential equation in a form involving the factors in $E = USV^T$, we use the following result.

Lemma 3 (See [21]). *For $E = USV^T \in \mathcal{M}_r$, with nonsingular $S \in \mathbb{R}^{r\times r}$ and with $U \in \mathbb{R}^{n\times r}$ and $\mathbb{R}^{n\times r}$ having orthonormal columns, the equation $\dot{E} = P_E(Z)$ is equivalent to $\dot{E} = \dot{U}SV^T + U\dot{S}V^T + US\dot{V}^T$, where*

$$\begin{cases} \dot{S} = U^TZV, \\ \dot{U} = (I - UU^T)ZVS^{-1}, \\ \dot{V} = (I - VV^T)Z^TVS^{-T} \end{cases}$$

In our case $Z = -P_E(G) + \langle E, P_E(G) \rangle E$, this yields that the differential Equation (8) for $E = USV^T$ is equivalent to the following system of differential equations for S, V, U,

$$\begin{cases} \dot{S} = U^T(-P_E(G) + \langle E, P_E(G) \rangle E)V, \\ \dot{U} = (I - UU^T)(-P_E(G) + \langle E, P_E(G) \rangle E)VS^{-1}, \\ \dot{V} = (I - VV^T)(-P_E(G) + \langle E, P_E(G) \rangle E)^T VS^{-T} \end{cases} \tag{9}$$

The following monotonicity result is an immediate consequence of the fact that the differential equation is the gradient flow for F_ϵ on the manifold of matrices, of fixed rank r, and unit norm.

Theorem 1. *Let be $E(t) \in \mathcal{M}_r$ a solution of unit Frobenius norm of the matrix differential Equation (8). Then*

$$\frac{d}{dt}F_\epsilon(E(t)) \leq 0$$

Proof. By Cauchy-Schwarz inequality,

$$|\langle E, P_E(G) \rangle| \leq ||E||_F ||P_E(G)||_F = ||P_E(G)||_F$$

Therefore, using (8),

$$\frac{d}{dt}F_\epsilon(E(t)) = \epsilon \langle \dot{E}, G \rangle = \epsilon \langle -P_E(G) + \langle E, \langle P_E(G) \rangle E, G \rangle = \epsilon \left(-\langle P_E(G), G \rangle + \langle E, P_E(G) \rangle \langle E, G \rangle \right) =$$

$$\epsilon \left(-\langle P_E(G), P_E(G) \rangle + \langle E, P_E(G) \rangle^2 \right) = \epsilon \left(-||P_E(G)||_F^2 + \langle E, P_E(G) \rangle \right) \leq 0$$

$\square$

4. Stationary points

Since we are interested to minimize $F_\epsilon(E)$, we focus on the equilibria of (8), which represents local minima of (2).

Lemma 4. *The following statements are equivalent along the solutions of (8):*

(a) $\dfrac{d}{dt}F_\epsilon(E) = 0.$

(b) $\dot{E} = 0.$

(c) E *is a real multiple of* $P_E(G)$.

Proof. From the expression (4), clearly (b) implies (a).

Supposing (c), we can write $P_E(G) = \alpha E$ with $\alpha \in \mathbb{R}$ and by substitution in (8) we get

$$\dot{E} = -\alpha E + \langle E, \alpha E \rangle E = \alpha \left(-E + ||E||_F E \right) = \alpha \left(-E + E \right) = 0$$

that is (b). So, it remains to show that (a) implies (c).

Note that:

$$\frac{d}{dt}F_\epsilon(E(t)) = \epsilon \langle \dot{E}, G \rangle = \epsilon \langle -P_E(G) + \langle E, P_E(G) \rangle E, G \rangle$$

$$= \epsilon \langle -P_E(G), G \rangle + \langle E, P_E(G) \rangle \langle E, G \rangle = \epsilon \left(-||P_E(G)||_F^2 + \langle E, P_E(G) \rangle^2 \right)$$

So, since $\epsilon > 0$, we have

$$\frac{d}{dt}F_\epsilon(E(t)) = 0 \iff -||P_E(G)||_F^2 + \langle E, P_E(G) \rangle^2 = 0$$

the last equality holds only if $E = \alpha P_E(G)$ for $\alpha \in \mathbb{R}$, that is (c). □

The following result characterizes the local extremizers.

Theorem 2. *Let $E_* \in M_r$ be a real matrix of unit Frobenius norm. Then, the following two statements are equivalent:*

(a) *Every differentiable path $E(t) \in M_r$ (for small $t \geq 0$) with $||E||_F = 1$ and $E(0) = E_*$ satisfies*

$$\frac{d}{dt} F_\varepsilon(E(t)) \geq 0$$

(b) *There exists a $\gamma > 0$ such that*

$$E_* = -\gamma P_E(G)$$

Proof. The strategy of the proof is similar to [22]. Assume that (a) does not hold. Then, there exists a path $E(t) \in M_r$ through E_* such that $\frac{d}{dt} F_\varepsilon(E(t))\mid_{t=0} < 0$. Thus, Lemma 2 shows that also the solution path of (8) passing through E_* is such a path. So E_* is not a stationary point of (8), and according to the Lemma 4, it is not a real multiple of $P_E(G)$.

If E_* is not a multiple of $P_E(G)$, then E_* is not a stationary point of (8) and Theorem 1 and Lemma 4 ensure that $\frac{d}{dt} F_\varepsilon(E(t)) \leq 0$ along the solution path of (8). If $E_* = \gamma P_E(G)$ with $\gamma \geq 0$, we can consider the path $E(t) = (1-t)E_*$ for small $t \geq 0$. This path is such that

$$||E||_F = ||(1-t)E_*||_F = |1-t|\,||E_*||_F \leq 1$$

and,

$$\frac{d}{dt} E(t) = -E_*$$

So, we have

$$\frac{d}{dt} F_\varepsilon(E(t)) = \varepsilon\langle \dot{E}, G\rangle = -\varepsilon\langle E_*, G\rangle = -\varepsilon\gamma\langle P_E(G), G\rangle =$$
$$-\varepsilon\gamma\langle P_E(G), P_E(G)\rangle = -\varepsilon\gamma||P_E(G)||_F^2 < 0$$

in contradiction with (a). □

5. Numerical Solution of Rank-r Matrix Differential Equation

We have seen that the matrix ODE (8) is equivalent to the system (9), involving the factors of the decomposition (6). In (9), the inverse of S appear. Therefore, when S is nearly singular, problems of stability can arise, working with a standard numerical methods for ODEs . To avoid this difficulties, we employ the first order projector-splitting integrator of [23]. The algorithm directly approximate the solution of the Equation (8). It starts from the normalized rank r matrix $E_0 = U_0 S_0 V_0^*$ at the time t_0, obtained by the SVD of the matrix to recover. At the time $t_1 = t_0 + h$, one step of the method works as follows

---

Projector-splitting integrator

Data: $E_0 = U_0 S_0 V_0^*$, $G_0 = G(E_0)$ % G is the free gradient (3)

Result: E_1

begin ;

Set $K_1 = U_0 S_0 - G_0 V_0$;

Compute $U_1 \widehat{S_1} = K_1$;   % QR factorization
   % U_1 orthonormal columns
   % $\widehat{S_1}$ $r \times r$ matrix

Set $\overline{S_0} = \widehat{S_1} + h\, U_1^T G_0 V_0$;

Set $L_1 = V_0 \overline{S_0}^T - h\, G_0^T U_1$;

Compute $V_1 S_1^T = L_1$;   % QR factorization
   % V_1 orthonormal columns
   % S_1 $r \times r$ matrix

Set $\widehat{E_1} = U_1 S_1 V_1^T$;

Normalize $E_1 = \dfrac{\widehat{E_1}}{||E_1||_F}$;

E_1 is taken as approximation to $E(t_1)$. All the nice features of the integrator are presented in [23], but it is already clear that, there is no matrix inversion in the steps of the algorithm.

6. Iteration on ε

In this section we show the outer iteration to manage ε (see [22]).

6.1. Qualitative Tools

For every fixed $\varepsilon > 0$, the gradient system (8) returns a stationary point $E(\varepsilon)$ of unit Frobenius norm that is a local minimum of F_ε.

Setting $f(\varepsilon) = F_\varepsilon(E(\varepsilon))$, our purpose is to solve the problem

$$\min\{\varepsilon > 0 : f(\varepsilon) = 0\}$$

employing a Newton-like method. We assume that $E(\varepsilon)$ is a smooth function of ε, so that, also the function $f(\varepsilon) = F_\varepsilon(E(\varepsilon))$ is differentiable with respect to ε. Let us focus on its derivative,

$$f'(\varepsilon) = \frac{d}{d\varepsilon} F_\varepsilon(E(\varepsilon)) = \frac{d}{d\varepsilon} ||\Omega \circ (\varepsilon E - M)||_F^2 = \frac{d}{d\varepsilon} \Big\langle \Omega \circ (\varepsilon E - M), \Omega \circ (\varepsilon E - M) \Big\rangle =$$

$$\Big\langle \frac{d}{d\varepsilon}(\Omega \circ (\varepsilon E - M)), \Omega \circ (\varepsilon E - M) \Big\rangle = \Big\langle \Omega \circ (\frac{d}{d\varepsilon}(\varepsilon E(\varepsilon))), \Omega \circ (\varepsilon E - M) \Big\rangle =$$

$$\Big\langle \frac{d}{d\varepsilon}(\varepsilon E(\varepsilon)), \Omega \circ (\varepsilon E - M) \Big\rangle = \langle E(\varepsilon) + \varepsilon E'(\varepsilon), G \rangle$$

If we denote $\varepsilon^* = \min\{\varepsilon > 0 : f(\varepsilon) = 0\}$.

By the expression of the free gradient (3), it is clear that

$$0 = f(\varepsilon^*) = \frac{1}{2}||\Omega \circ (\varepsilon^* E^* - M)||^2 = \frac{1}{2}||G(\varepsilon^*)||^2 \Leftrightarrow G(\varepsilon^*) = 0$$

Therefore,

$$f'(\varepsilon^*) = \langle E(\varepsilon^*) + \varepsilon^* E'(\varepsilon*), G(\varepsilon^*) \rangle = \langle E(\varepsilon^*) + \varepsilon^* E'(\varepsilon^*), 0) \rangle = 0$$

This means that ε^* is a double root for $f(\varepsilon)$.

6.2. Numerical Approximation of ε^*

The presence of the double root ensures that $f(\varepsilon)$ is convex for $\varepsilon \leq \varepsilon^*$, therefore, the classical Newton method will approach ε^* from the left. In this case, we are not able to find an analytical formulation for the derivative $f'(\varepsilon)$, so we approximate it with backward finite differences.

Algorithm for computing ε^*

Data: the matrix M to recover is given, tol, $\varepsilon_0, \varepsilon_1$ such that $f(\varepsilon_0) > 0, f(\varepsilon_1) > 0$
Result: $\varepsilon_*, E(\varepsilon_*)$
begin ;
$k = 1$;
while $|\varepsilon_k - \varepsilon_{k-1}| > tol$ **do**

 Compute $f'_k \approx \dfrac{f_k - f_{k-1}}{\varepsilon_k - \varepsilon_{k-1}}$;

 Update $\varepsilon_{k+1} = \varepsilon_k - \dfrac{f_k}{f'_k}$;

 Compute $f_{k+1} = f(\varepsilon_{k+1})$;
 $k = k + 1$; ;

end
$\varepsilon_k = \varepsilon^*$;

---

7. Numerical Experiments

In the following experiments we randomly generate some matrices of low rank. As in [3,4], r is the fixed rank, we generate two random matrices $A_L, A_R \in \mathbb{R}^{n \times r}$ with i.i.d. standard Gaussian entries. We build the matrix $A = A_L A_R^T$ and generate a uniformly distributed sparsity pattern Ω. We work on the matrix M, that is the matrix resulting from the projection of A onto the pattern Ω. In this way we are able to compare the accuracy of the matrix solution of our code with the true solution A. As stopping criteria for the integrator of the ODE in the inner level, we use

$$||\Omega \circ (X - M)||_F / ||M||_F < tol$$

where *tol* is an input tolerance parameter together with a maximum number of iterations and a minimum value for the integrator stepsize. We provide a stepsize control that reduces the step h with a factor γ (the default value is 1.25), when the functional is not decreasing, but increases the step as $h\gamma$ when the value of the objective decreases with respect to the previous iteration. Some computational results are shown in the Tables 1 and 2. In particular, they show the values of the cost function evaluated in ε^*, computed by the outer method, thanks to which, the accuracy of the method when we recover matrices of different rank and different dimension is highlight.

Table 1. Computational results from recovering three matrices of different dimensions and 30% of known entries. The rank is fixed to be 10.

Dim	$f(\varepsilon_*)$	Err	Iter
2000×300	1.0944×10^{-24}	1.8960×10^{-12}	6
2000×650	2.0948×10^{-24}	2.5971×10^{-12}	5
2000×1000	1.3071×10^{-23}	6.4837×10^{-12}	7

Table 2. Computational results from recovering three matrices of different ranks and 30% of known entries. The dimension is always 1000×1000.

Rank	$f(\varepsilon_*)$	Err	Iter
10	3.6533×10^{-24}	1.0079×10^{-12}	10
20	7.6438×10^{-24}	4.9793×10^{-12}	9
30	6.4411×10^{-24}	6.6683×10^{-12}	9

7.1. Computational Variants of the Outer Level

Observe that the presence of the double root would allow us to use a modified Newton iteration (from the left)

$$\varepsilon_{k+1} = \varepsilon_k - 2\frac{f(\varepsilon_k)}{f'(\varepsilon_k)}$$

getting quadratic convergence. Since our purpose is to find an upper bound for ε_*, if it should happen that $\varepsilon_k > \varepsilon_*$, we need a bisection iteration to preserve the approximation from the left. Furthermore, we can observe that, if we indicate by $g(\varepsilon) = ||G||_F$, where G is defined in (3), it is clear that $f(\varepsilon) = \frac{1}{2}g^2(\varepsilon)$, therefore they have common zeros. This allows us to employ the function $g(\varepsilon)$ instead of $f(\varepsilon)$ in the outer level, joining classical Newton and bisection. In practice, this results to be the most efficient approach. Tables 3 and 4 show the behaviours of the two alternative approaches on a test matrix M of dimension 150×150, $\approx$50% of known entries and rank 15.

Table 3. Computational results obtained by coupling modified Newton method and bisection.

Iter	ε	$f(\varepsilon)$
1	$4.666972133737625 \times 10^2$	145.0744
2	$4.684932414610240 \times 10^2$	122.9347
3	$4.884388614255482 \times 10^2$	0.2481
4	$4.885195311133194 \times 10^2$	0.2074
5	$4.893418324967433 \times 10^2$	4.1842×10^{-4}
$\vdots$	$\vdots$	$\vdots$
18	$4.893805094395407 \times 10^2$	9.4499×10^{-21}
19	$4.893805094397171 \times 10^2$	5.8554×10^{-24}
20	$4.893805094397174 \times 10^2$	5.3080×10^{-24}

Table 4. Computational results got by employing the function $g(\varepsilon)$.

Iter	ε	$g(\varepsilon)$
1	$4.664656983250553 \times 10^2$	17.2083
2	$4.880638859063273 \times 10^2$	0.9850
3	$4.893752126994476 \times 10^2$	0.0040
4	$4.893805081705695 \times 10^2$	9.4925×10^{-7}
5	$4.893805094397189 \times 10^2$	2.0713×10^{-12}
6	$4.893805094397218 \times 10^2$	2.2306×10^{-13}
7	$4.893805094397220 \times 10^2$	3.2750×10^{-13}

We tested also the classic Newton method on M. The comparison is summarized in Tables 5 and 6.

Table 5. Comparison between different approach to the outer iteration. The table show the number of iterations done by each method and the optimal values of the cost function.

Method	Iter	$f(\varepsilon_*)$
N2	20	4.9734×10^{-24}
g	7	5.3630×10^{-26}
N	70	9.3812×10^{-26}

Table 6. Comparison between different approach to the outer iteration. The table show the real error and the time.

Method	Err	Time
N2	5.2117×10^{-13}	27.684 s
g	4.4562×10^{-13}	3.6980 s
N	3.7471×10^{-13}	57.648 s

The accuracy is the same for all the choices, but in the case of selecting g instead of f, the computational cost is sharply reduced, both in terms of number of iterations and in terms of timing.

7.2. Experiments with Quasi Low Rank Matrices

The following simulations are devoted to check the "robustness" of the method with respect to small perturbations of the singular values. More precisely, we consider a rank r matrix A, built as introduced in this section, and we perturbe it in order to get a matrix A_P of almost rank r. In other words, we aim to get a matrix A_P that has r significant singular values, whereas the remaining ones become very small. Let $A = U\Sigma V^T$ be the SVD decomposition of A, therefore Σ is diagonal with only the first r diagonal values different form zero. If $\hat{\Sigma}$ is the diagonal matrix such that the first r diagonal entries are zero and the remaining ones (all or a part of them) are put equal to random small values, we build A_P as

$$A_P = U(\Sigma + \hat{\Sigma})V^T$$

where,

$$\Sigma = \begin{pmatrix} \sigma_1 & & & & \\ & \cdot & & \mathbf{0} & \\ & & \sigma_r & & \\ & \mathbf{0} & & 0 & \\ & & & & 0 \end{pmatrix}, \quad \hat{\Sigma} = \begin{pmatrix} 0 & & & & \\ & \cdot & & \mathbf{0} & \\ & & 0 & & \\ & \mathbf{0} & & \hat{\sigma}_r+1 & \\ & & & & \hat{\sigma}_n \end{pmatrix}$$

Table 7 shows the numerical results, obtained by considering a rank 9 matrix A, of size 200 and perturbations of different amplitude.

Table 7. Computational results from recovering three matrices of different ranks and 30% of known entries. The dimension is always 1000×1000.

$\hat{\sigma}_{r+1,r+1}$	ε_*	$f(\varepsilon_*)$	Err
$\approx 1 \times 10^{-4}$	$6.167926656792961 \times 10^2$	2.4002×10^{-5}	0.0037
$\approx 1 \times 10^{-6}$	$6.167937831641854 \times 10^2$	1.9209×10^{-9}	7.2224×10^{-5}
$\approx 1 \times 10^{-8}$	$6.167937831639130 \times 10^2$	1.9211×10^{-13}	2.8729×10^{-7}
0	$6.167937832319908 \times 10^2$	2.1332×10^{-23}	7.7963×10^{-12}

The columns $\hat{\sigma}_{r+1,r+1}$ and *err* contain the orders of magnitude of the greater perturbed singular value, and the values of the real error, computed as the Frobenius distance between A and the optimal matrix, that comes out the code, respectively. As it is natural to expect, the optimal matrix remains close to A, but the error is affected by the perturbations as they grow.

Another interesting example in terms of robustness, when we work with quasi low rank matrices, is given by considering matrices with exponentially decaying singular values. In particular we build a $n \times n$ matrix A, which singular values are given by the sequence $\{exp(-x_i)\}_{i=1,...,n}$ where $x_1 \leq x_2 \leq ... \leq x_n$ are random increasing numbers in an interval $[a, b]$. We build the matrix M to recover, by projecting A onto a sparsity pattern Ω. In the following experiment we fix $n = 100$, $a = 20$ and $b = 1$. The singular values of A range in the interval $[0.3511, 2.0870 \times 10^{-9}]$, and the mask M has about 30% of known elements. We choose the values of rank in the experiments for the completion by considering the changes of order of magnitude of singular values. Table 8 shows the results, in particular, the value of the cost function $f(\varepsilon_*)$ which we compare to f, the one given by the code of [4].

Table 8. The table show the behaviours of the codes when we recover the matrix M fixing different values of the rank, accordingly with the order of magnitude of the singular value of the exact full rank solution A.

r	$\hat{\sigma}_{r+1,r+1}$	$f(\varepsilon_*)$	f
4	$\approx 1 \times 10^{-2}$	0.0061	0033
20	$\approx 1 \times 10^{-3}$	4.98621×10^{-11}	4.2904×10^{-8}
24	$\approx 1 \times 10^{-4}$	8.98261×10^{-32}	6.0490×10^{-16}

7.3. Experiment with Theoretical Limit Number of Samples

In the seminal paper [1], the authors focus on determine a lower bound for the cardinality Ω of the known set of entries, such that it is possible recovering the matrix with high probability. In particular they proved that, most of $n \times n$ matrix of rank r (assumed not too large) can be perfectly recovered solving a convex optimization problem, if $|\Omega| \geq Cn^{1.2}rlogn$, for some positive constant C. Table 9 shows the results when we compare our code with the method in [4]. In particular, we present the best value of the objective functions, the real errors and the computational times. We consider $n = 50$, $r = 3$, therefore, according with the previous bound, we have to set $|\Omega| \geq C\ 1.2832 \times 10^3$. This means that, for $C = 1$, the corresponding target matrix M will have $\approx 51.33\%$ of given entries.

Table 9. The table show the behaviours of the codes when we recover the 50×50 matrix M with $\approx 49.88\%$ of given elements. The rank is 3. Our results are marked by an asterisk.

f	Err	Time
1.951391×10^{-26}	3.37141×10^{-13}	0.066 s
f^*	Err*	Time*
4.08441×10^{-27}	1.0071×10^{-13}	2.28 s

7.4. Behaviour with Respect to Different Ranks

Given a test matrix M, our purpose, in this section, is to understand the behaviour of the cost function, when we set the rank different from the exact one. In particular, we consider a matrix M, with $\approx 44.71\%$ of known elements, of dimension 70×70, and such that the exact rank of the solution is 8. We compute the values of the cost function evaluated in the best fixed rank k approximation (say b^k) and in the solution given by our code (say f^k). For every fixed rank k, the error is given by

$$|b^k - f^k|/||M||_F^2$$

The results are shown in the Table 10(a).

Table 10. The table shows the values of the cost function for different value of the rank and the relative errors. The true rank is 8. The values of the objective for the different ranks are represented on the figure (b).

(a)			(b)
Rank	f	Err	
1	6.73461×10^3	0.0008	
2	5.07831×10^3	0.0086	
3	3.57211×10^3	0.0136	
4	2.35091×10^3	0.0177	
5	1.53641×10^3	0.0029	
6	0.9613	0.0070	
7	0.4461	0.0019	
8	9.1260×10^{-31}	0.22641×10^{-7}	
9	1.6394×10^{-30}	2.76671×10^{-7}	
10	3.9834×10^{-15}	2.83831×10^{-7}	

8. Conclusions

The matrix completion problem consists of recovering a matrix from a few samples of its entries. We formulate the problem as a minimization of the Frobenius distance on the set of the fixed entries, over the manifold of the matrices of fixed rank. In this paper we introduce a numerical technique to deal with this problem. The method works on two levels; the inner iteration computes the fixed norm matrix that best fits the data entries, solving low rank matrix differential equations, the outer iteration optimizes the norm by employing a Newton-like method. A key feature of the method is to avoid the problem of the lack of vector space structure for $\mathcal{M}_r$, moving the dynamics in the tangent space. Numerical experiments show the high accuracy of the method and its robustness with respect to small perturbations of the singular values. However, in presence of very challenging problems it could be suitable to relax tolerance parameters. The method is particularly suited for problems for which guessing the rank is simple. In the field of research on low rank matrix completion, it would be useful to study real databases types of matrices in order to try to establish gaps for the values of the rank. Moreover, since this is a typical context, in which we work with very large matrices, future work could be devoted to develop methods working in parallel. Structured variants, such as nonnegative low rank completions, are suggested from applications. These may be subject of a future work.

Author Contributions: Both authors developed the theoretical part in order to provide the treatment of the problem and obtain a solution. In particular, N.G. suggested the problem and the adopted formulation, and supervised the organisation and the coherence of the work. C.S. focused on the numerical technical part of the work.

Funding: The authors thank INdAM GNCS (Gruppo Nazionale di Calcolo Scientifico) for financial support.

Acknowledgments: The authors thank four anonymous referees for their valuable remarks.

Conflicts of Interest: The authors declare no conflict of interest.

References

1. Candés, E.J.; Recht, B. Exact Matrix Completion via Convex Optimization. *Found. Comput. Math.* **2009**, *9*, 717–772. [CrossRef]
2. Gillis, N.; Glineur, F. Low-rank matrix approximation with weights or missing data is NP-hard. *SIAM J. Matrix Anal. Appl.* **2011**, *4* 1149–1165. [CrossRef]
3. Cai, J.F.; Candés, E.J.; Shen, Z. A singular value thresholding algorithm for matrix completion. *SIAM J. Optim.* **2010**, *20*, 1956–1982. [CrossRef]

4. Vandereycken, B. Low-Rank matrix completion by Riemaniann optimization. *SIAM J. Optim.* **2013**, *23*, 367–384. [CrossRef]

5. Cambier, L.; Absil, P.-A. Robust Low-Rank Matrix Completion by Riemannian Optimization. *SIAM J. Sci. Comput.* **2016**, *38*, 440–460. [CrossRef]

6. Keshavan, R.H.; Montanari, A.; Oh, S. Matrix Completion from a Few Entries. *IEEE Trans. Inf. Theory* **2010**, *38*, 440–460. [CrossRef]

7. Yang, Y.; Feng, Y.; Suykens, J.A.K. Correntropy Based Matrix Completion. *Entropy* **2018**, *20*, 171. [CrossRef]

8. Candés, E.J.; Plan, Y. Matrix completion with noise. *Proc. IEEE* **2010**, *98*, 925–936. [CrossRef]

9. Goldberg, D.; Nichols, D.; Oki, B.M.; Terry, D. Using collaborative filtering to weave an information tapestry. *Commun. ACM* **1992**, *35*, 61–70. [CrossRef]

10. Rodger, A. Toward reducing failure risk in an integrated vehicle health maintenance system: A fuzzy multi-sensor data fusion Kalman filter approach for IVHMS. *Expert Syst. Appl.* **2012**, *39*, 9821–9836. [CrossRef]

11. Bennet, J.; Lanning, S. The netflix prize. In Proceedings of the KKD Cup and Workshop, San Jose, CA, USA, 12 August 2007; p. 35.

12. Nguyen, L.T.; Kim, S.; Shim, B. Localization in the Internet of Things Network: A Low-Rank Matrix Completion Approach. *Sensors* **2016**, *16*, 722. [CrossRef] [PubMed]

13. Wang, X.; Weng, Z. Low-rank matrix completion for array signal processing. In Proceedings of the IEEE International Conference on Acoustics, Speech and Signal Processing (ICASSP), Kyoto, Japan, 25–30 March 2012.

14. Cai, M.; Cao, F.; Tan, Y. Image Interpolation via Low-Rank Matrix Completion and Recovery. *IEEE Trans. Circuits Syst. Video Technol.* **2015**, *25*, 1261–1270.

15. Natarajan, N.; Dhillon, I.S. Inductive Matrix Completion for Predicting Gene-Disease Associations. *Bioinformatics* **2014**, *30*, i60–i68. [CrossRef] [PubMed]

16. Bakonyi, M.; Woederman, H.J. *Matrix Completion, Moments, and Sums of Hermitian Squares*; Princeton University Press: Princeton, NJ, USA, 2011.

17. Markovsky, I.; Usevich, K. Structured Low-Rank Approximation with Missing Data. *SIAM J. Matrix Anal. Appl.* **2013**, *34*, 814–830. [CrossRef]

18. Mishra, B.; Meyer, G.; Sepulchre, R. Low-rank optimization for distance matrix completion. In Proceedings of the 50th IEEE Conference on Decision and Control and European Control Conference, Orlando, FL, USA, 12–15 December 2011.

19. Guglielmi, N.; Lubich, C. Low-rank dynamics for computing extremal points of real pseudospectra. *SIAM J. Matrix Anal. Appl.* **2013**, *34*, 40–66. [CrossRef]

20. Guglielmi, N.; Lubich, C. Differential equations for roaming pseudospectra: paths to extremal points and boundary tracking. *SIAM J. Numer. Anal.* **2011**, *49*, 1194–1209. [CrossRef]

21. Koch, O.; Lubich, C. Dynamical low-rank approximation. *SIAM J. Matrix Anal. Appl.* **2007**, *29*, 434–454. [CrossRef]

22. Guglielmi, N.; Lubich, C.; Mehrmann, V. On the nearest singular matrix pencil. *SIAM J. Matrix Anal. Appl.* **2017**, *38*, 776–806. [CrossRef]

23. Lubich, C.; Oseledets, I.V. A projector-splitting integrator for dynamical low-rank approximation. *Numer. Math.* **2014**, *54*, 171–188. [CrossRef]

axioms

MDPI

Article

A Convex Model for Edge-Histogram Specification with Applications to Edge-Preserving Smoothing

Kelvin C. K. Chan [1],*, Raymond H. Chan [1] and Mila Nikolova [2]

[1] Department of Mathematics, The Chinese University of Hong Kong, Hong Kong; rchan@math.cuhk.edu.hk

[2] Centre de Mathématiques et de Leurs Applications, ENS de Cachan, 94230 Paris, France; nikolova@cmla.ens-cachan.fr

* Correspondence: kelvinckchan@outlook.com; Tel.: +852-6273-7871

Received: 18 June 2018; Accepted: 1 August 2018; Published: 2 August 2018

Abstract: The goal of edge-histogram specification is to find an image whose edge image has a histogram that matches a given edge-histogram as much as possible. Mignotte has proposed a non-convex model for the problem in 2012. In his work, edge magnitudes of an input image are first modified by histogram specification to match the given edge-histogram. Then, a non-convex model is minimized to find an output image whose edge-histogram matches the modified edge-histogram. The non-convexity of the model hinders the computations and the inclusion of useful constraints such as the dynamic range constraint. In this paper, instead of considering edge magnitudes, we directly consider the image gradients and propose a convex model based on them. Furthermore, we include additional constraints in our model based on different applications. The convexity of our model allows us to compute the output image efficiently using either Alternating Direction Method of Multipliers or Fast Iterative Shrinkage-Thresholding Algorithm. We consider several applications in edge-preserving smoothing including image abstraction, edge extraction, details exaggeration, and documents scan-through removal. Numerical results are given to illustrate that our method successfully produces decent results efficiently.

Keywords: edge-histogram; edge-preserving smoothing; histogram specification

---

1. Introduction

Histogram specification is a process where the image histogram is altered such that the histogram of the output image follows a prescribed distribution. It is one of the many important tools in image processing with numerous applications such as image enhancement [1–3], segmentation [4–6] among many others. To match the histogram of an image I to a target histogram $\mathbf{h}$, one can use the following procedure:

1. Compute the probability density function of I, $p(r_j) = \frac{n_j}{n}$, $j = 1, \cdots, L$, where r_j denotes the intensity values, n_j denotes the number of pixels whose value equals to r_j, and n is the total number of pixels in I.
2. With the probability density function $p(r_j)$, compute the cumulative density function $S(r_j)$ of I.
3. With $\mathbf{h}$ serving as a probability density function, compute the cumulative density function $G(s_j)$.
4. For $p = 1, 2, \cdots, L$, replace r_p by s_q, where $S(r_p) = G(s_q)$.

   In the typical 8-bit image setting, $L = 256$ and $r_p = p - 1$, for $p = 1, 2, \cdots, 256$.

   However, the above procedure gives an inexact histogram matching in discrete images, since the number of pixels is far greater than the number of intensity values in discrete images. Therefore, to ensure exact matching of histograms, different algorithms are introduced, see for instance [7–9]. The main idea of these algorithms is to obtain a total order of the pixel values before setting pixel values according to the target histogram.

The goal of edge-histogram specification is to find an image whose edge image has a histogram that matches a given edge-histogram as much as possible. An edge-histogram of an image counts the frequencies of different gradient values in an image and plots the distribution of gradient values. Edge-histograms of natural images have been studied in previous literature [10,11]. It is known that the edge-histograms of natural images contain a sharp peak at 0. This phenomenon is due to the abundance of flat regions in natural images. In natural images, flat regions are seen in the sky, water, walls, and many others. The flat regions in the image produce zero gradients and hence the resulting edge-histogram contains a sharp peak at 0. In addition, as stated in [10], the edge-histograms of natural images have a higher kurtosis and heavier tails compared to a Gaussian distribution with the same mean and variance.

Histogram of gradients has also been used in other fields of image processing. Histogram of oriented gradients (HOG) is used as a feature descriptor for object detection [12–14]. This technique first counts the frequencies of gradient orientation in terms of a histogram in a local region, and combines the histograms of gradients in different regions to a single vector as output.

Unlike HOG, our focus is on matching edge-histograms, and we do not consider local edge-histograms. To match the edge-histogram, the author in [11] proposed the following non-convex model for the problem. Given a discrete input image I of size m-by-n, let $\mathbf{r} \in \mathbb{R}^k$ be a vector storing the pairwise differences

$$r_{s,t} := |I_s - I_t|, \quad s = 1, 2, \cdots mn, \, t \in \mathcal{N}_s, \tag{1}$$

where I_i denotes the value of the i-th pixel of I, and $\mathcal{N}_s$ denotes the set of indices of a fixed shape neighborhood of the s-th pixel, for example, the north, east, south, and west neighboring points of the s-th pixel. Here $k = mn|\mathcal{N}_s|$ with $|\mathcal{N}_s|$ the number of elements in $\mathcal{N}_s$. That is, each entry of the vector $\mathbf{r}$ contains the absolute difference between the values of two neighboring pixels in I. Note that the order of storing the pairwise differences does not matter, as it does not affect the output of histogram specification.

Given a target edge-histogram $\mathbf{h}$, one can apply histogram specification on $\mathbf{r}$ to match $\mathbf{h}$. Denote the output by $\mathbf{d}$:

$$\mathbf{r} = (r_{s_1,t_1}, \cdots, r_{s_k,t_k}) \xrightarrow[\text{specification}]{\text{histogram}} (d_{s_1,t_1}, \cdots, d_{s_k,t_k}) = \mathbf{d}. \tag{2}$$

To ensure exact histogram matching, the author of [11] used the algorithm in [7]. After applying histogram specification as in (2), one gets the output image X by solving the minimization problem

$$\min_X \sum_{s=1}^{mn} \sum_{t \in \mathcal{N}_s} \left((X_s - X_t)^2 - d_{s,t}^2 \right)^2. \tag{3}$$

Model (3) is solved by a conjugate gradient procedure followed by a stochastic local search, please refer to [7,11] for the full details of the algorithm. The non-convex nature of the model hinders the computations, and it is difficult to include additional constraints. In this paper, we propose a convex model that can include additional constraints based on different applications in edge-preserving smoothing.

Edge-preserving smoothing is a popular topic in image processing and computer graphics. Its aim is to suppress insignificant details and keep important edges intact. As an example, the input image in Figure 3a contains textures on the slate and the goal of edge-preserving smoothing is to remove such textures and keep only the object boundaries as in Figure 3c. Numerous methods have been introduced to perform the task. Anisotropic diffusion [15,16] performs smoothing by means of solving a non-linear partial differential equation. Bilateral filtering [17–20] is a method combining domain filters and range filters. They are widely used because of their simplicity. Optimization frameworks such as the weighted least squares (WLS) [21] and TV regularization [22,23] are also introduced. In WLS, a regularization term is added to minimize the horizontal and vertical gradients with corresponding smoothness weights. Recently, models based on l_0-gradient minimization [24–29] have become popular. These models focus on the l_0-norm of the image gradients.

One application of edge-preserving smoothing is scan-through removal. Written or printed documents are usually subjected to various degradations. In particular, two-sided documents can be suffered from the effect of back-to-front interference, known as "see-through", see Figure 10. The problem is especially severe in old documents, which is caused by the bad quality of the paper or ink-bleeding. These effects greatly reduce the readability and hinder optical character recognition. Therefore it is of great importance to remove such interference. However, physical restoration is difficult as it may damage the original contents of the documents, which is clearly undesirable as the contents may be important. Consequently, different approaches in the field of image processing are considered to restore the images digitally.

These approaches can be mainly classified into two classes: Blind and Non-blinds methods. Non-blind methods [30–38] require the information of both sides. These methods usually consist of two steps. First, the two sides of the images are registered. Then, the output image is computed based on the registered images. It is obvious that these methods strongly depend on the quality of registration; therefore, highly accurate registration is needed. However, perfect registration is hard to achieve in practice due to numerous sources of errors including local distortions and scanning errors. Furthermore, information from the back page is not available on some occasions. Therefore, blind methods which do not assume the availability of the back page are also developed in solving the problem, see [39–44].

In this paper, we propose a convex model for applications in edge-preserving smoothing. In our work, we modify the objective function in the non-convex model in [11] so that we only need to solve a convex minimization problem to obtain the output. The simplicity of our model allows us to incorporate different useful constraints such as the dynamic range constraint; the convexity of our model allows us to compute the output efficiently by Fast Iterative Shrinkage-Thresholding Algorithm (FISTA) [45] or Alternating Direction Method of Multipliers (ADMM) [46,47]. We introduce different edge-histograms and suitable constraints in our model, and apply them to different imaging tasks in edge-preserving smoothing, including image abstraction, edge extraction, details exaggeration, and scan-through removal.

The contribution of this paper is twofold. First, we propose a convex variant of the model in [11] which can be solved efficiently. In addition, our model can easily include additional constraints based on different applications. Second, we demonstrate applications of edge-histogram specification in edge-preserving smoothing.

The outline of the paper is as follows: Section 2 describes the proposed convex model, Section 3 presents the applications of our model with numerical results, and conclusions are then presented in Section 4.

2. Our Model

In our model, we do not consider the edge magnitudes as in (1). Instead, we directly consider the image gradients and define **r** with entries

$$r_{s,t} := I_s - I_t.$$

Similar to [11], our model consists of two parts. First, given **r**, we apply histogram specification on **r** to obtain **d** as in (2). In the second part, we solve a minimization problem to obtain the output X. Instead of solving the non-convex model (3), we propose a convex model. In Section 2.1, we present our convex model and its solvers. In Sections 2.2 to 2.4, we apply our model to specific applications in edge-preserving smoothing.

In the following discussions, we consider only grayscale images. For colored images, we apply our method to R, G, B channels separately.

2.1. Proposed Convex Model

Instead of (3), we propose the convex model

$$\min_{X} \sum_{s=1}^{mn} \sum_{t \in \mathcal{N}_s} |(X_s - X_t) - d_{s,t}|^p + \iota_C(X), \tag{4}$$

where $p = 1$ or 2, $\mathbf{C}$ is a convex set to be discussed in Section 2.4, and ι_C denotes the indicator function of $\mathbf{C}$. The choices of p and $\mathbf{C}$ depend on the applications. Let $\mathbf{x}$ be a vector such that its s-th entry is X_s. Then we can rewrite (4) as

$$\min_{\mathbf{x}} ||G\mathbf{x} - \mathbf{d}||_p^p + \iota_C(\mathbf{x}). \tag{5}$$

In our tests, we use $\mathcal{N}_s = \{s_v, s_h\}$, where s_v and s_h denote the indices of the north of north and west neighboring points of the s-th pixel respectively. Hence we can write $G = (G_h, G_v)^T$, where G_h, G_v are the horizontal and vertical backward difference operators. We use periodic boundary condition for pixels outside the boundaries, see ([48], p. 258).

For $p = 2$, model (5) can be solved by FISTA [45]. For $p = 1$, we rewrite (5) as

$$\min_{\mathbf{x},\mathbf{y}} ||\mathbf{y} - \mathbf{d}||_1 + \iota_C(\mathbf{x})$$

$$\text{s.t. } G\mathbf{x} = \mathbf{y},$$

which can be solved by ADMM [46,47].

2.2. Construction of Target Edge-Histogram

For the applications we considered in this paper, one objective is to remove textures in the images where their edges have a small magnitude. As an example, the textures on the slate in Figure 1a produce smaller edge magnitudes compared to the boundaries of the slate and the letters on the slate, see Figure 1b,c. To eliminate those textures, we could set the values of edges with a small magnitude to zero. Hence, in this paper, we propose to use edge-histograms similar to that shown in Figure 2b as target edge-histogram which is obtained by thresholding the input edge-histogram in Figure 2a. In particular, the target edge-histogram is dependent on the input image. We remark here that it is not uncommon to construct the target histogram based on the input. For example, such construction is used in image segmentation [6].

Since we are just thresholding the edges with small values to zero, the edge-histogram specification (2) can be done easily as follows. Given any input image Y, we first compute its gradients $y_{s,t} = Y_s - Y_t$. Then we set

$$z_{s,t} = \begin{cases} y_{s,t} & \text{if } |y_{s,t}| \geq \lambda, \\ 0 & \text{otherwise.} \end{cases} \tag{6}$$

The thresholded $z_{s,t}$, where its histogram is shown in Figure 2b, will be used as the vector $\mathbf{d}$ in (5) to obtain the output $\mathbf{x}$. It is obvious that different λ gives different outputs, see Figure 3. We see that the smoothness of the output increases with λ.

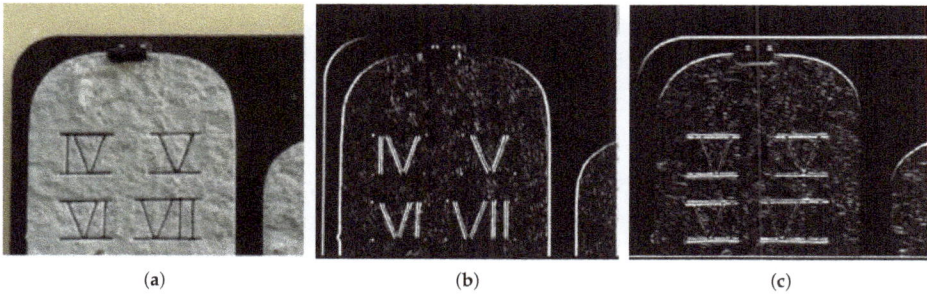

Figure 1. Input image and its horizontal and vertical gradients in absolute values. (**a**) Input; (**b**) absolute value of the horizontal gradient; (**c**) absolute value of the vertical gradient.

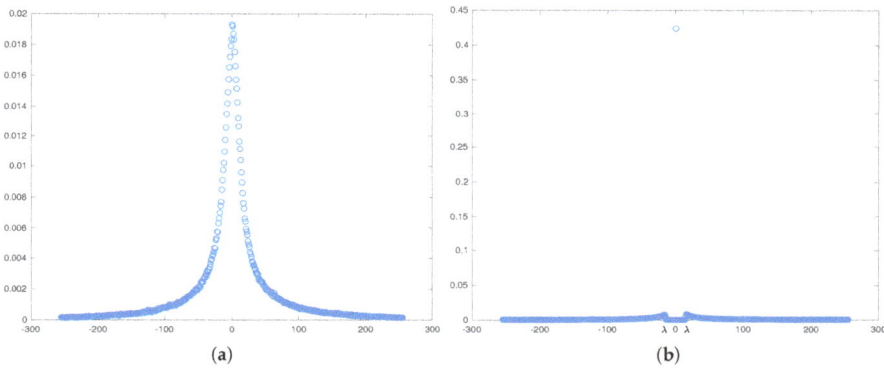

Figure 2. Construction of the target edge-histogram. (**a**) Histogram of gradient $y_{s,t}$; (**b**) histogram of thresholded gradient $z_{s,t}$.

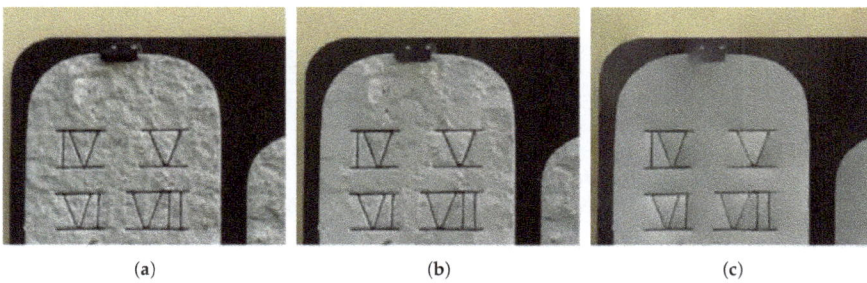

Figure 3. Output of (5) with different λ. (**a**) Input; (**b**) output using $\lambda = 5$; (**c**) output using $\lambda = 15$.

2.3. Gaussian Smoothing and Iterations

Strong textures produce edges with large magnitude, which cannot be eliminated using a thresholded edge-histogram as in Figure 2b. To suppress them, the input image I will first pass through a Gaussian filter with standard deviation σ to get the initial guess $X^{(0)}$. Larger σ will have a greater effect in suppressing strong textures, but at the same time blur the image. Hence, σ should be chosen small enough so that the Gaussian-filtered image is visually equal to I. In our tests, σ is chosen to be less than 1. However, there is not an automatic way to compute the optimal σ. Hence in this model, we leave σ as a parameter to be chosen manually by users. Let $X^{(0)}$ be the Gaussian-filtered image. Whenever such suppression is unnecessary, we set $\sigma = 0$ and hence $X^{(0)} = I$.

As mentioned in Section 2.2, one of our objectives is to map weak edges to zero. This can be done by changing the λ in the thresholded edge-histogram or by solving (5) repeatedly. More specifically, given $X^{(0)}$, we construct $\mathbf{d}$ using (6) and solve (5) to obtain $X^{(1)}$. Then we repeat the process to obtain $X^{(2)}$ and so on. Figure 4 shows a comparison; while we see in Figure 4b that the result after one iteration still contains textures in the grasses, almost all of them are removed after three iterations, see Figure 4c.

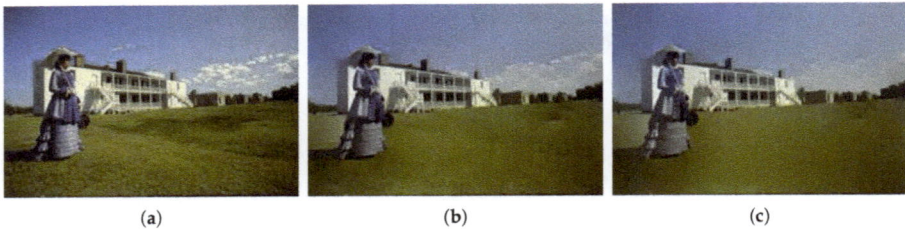

(a)   **(b)**   **(c)**

Figure 4. Output of solving (5) repeatedly with $\lambda = 15$, $\sigma = 0.6$. (**a**) Input image I; (**b**) image after one iteration; (**c**) image after three iterations.

2.4. Convex Set **C**

The model (3) does not consider the dynamic range constraint

$$I_s \in [0, 255], \ \forall s = 1, 2, \cdots, mn. \tag{7}$$

For example, consider the case when one defines $\mathbf{h}$ such that every pixel of $\mathbf{r}$ is doubled. In the absence of (7), it is easy to get an exact solution X of (3) if any one of the pixel values is given. However, there is no guarantee that the pixel values of X lie within $[0, 255]$. Therefore, when X is converted back to the desired dynamic range, either by stretching or clipping, the edge-histogram is no longer preserved. To avoid this, it is better to include the dynamic range constraint in the objective function. Therefore, we use the following constraint in all our applications:

$$\mathbf{C} = \{\mathbf{x} : x_i \in [0, 255], \ \forall i\}. \tag{8}$$

In scan-through removal, we assume the background in books and articles have a lighter intensity than the ink in all color channels. Therefore, in addition to the dynamic range constraint, we also keep the value of the background pixels unchanged. Hence, we set

$$\mathbf{C} = \{\mathbf{x} : x_i \in [0, 255], \ \forall i \text{ and } x_i = X_i^{(0)} \text{ if } x_i \geq \alpha\}, \tag{9}$$

where α is the approximate intensity of the background to be defined in Section 3.4.

3. Applications and Comparisons

Edge-preserving smoothing includes many different applications. In this section, we consider four applications, namely image abstraction, edge extraction, details exaggeration, and scan-through removal. For the first three applications, we use $p = 2$ in (5) and solve it by FISTA with the input image as an initial guess. We compare with four existing methods: bilateral filtering [18], weighted-least square [21], l_0-smoothing [24], and l_0-projection [29]. For the scan-through removal, we use $p = 1$ in (5) and solve it by ADMM with the input image and **d** as the initial guesses. We compare with one blind method [43] and three non-blind methods [32,44,49]. In all applications, the number of iterations is fixed at 3. The values λ and σ vary for different images and will be stated separately.

For the tests below, we select the parameters which give the output image with the best visual quality. Some of the comparison results are obtained directly from the authors' work and some are done by ourselves. For the results done by ourselves, we list out the parameters we have used.

3.1. Image Abstraction

The goal of image abstraction is to remove textures and fine details so that the output looks un-photorealistic. This can be done by solving (5) with constraint (8). As shown in Figure 5, the textures of the objects in the photorealistic input image in Figure 5a is removed and our output in Figure 5f becomes un-photorealistic. We see that our model successfully eliminates almost all object textures and keeps the object boundaries intact. As we see in Figure 5f, the details in the basketball net in our output are kept intact, while it disappears, or almost disappears, in the outputs of other models.

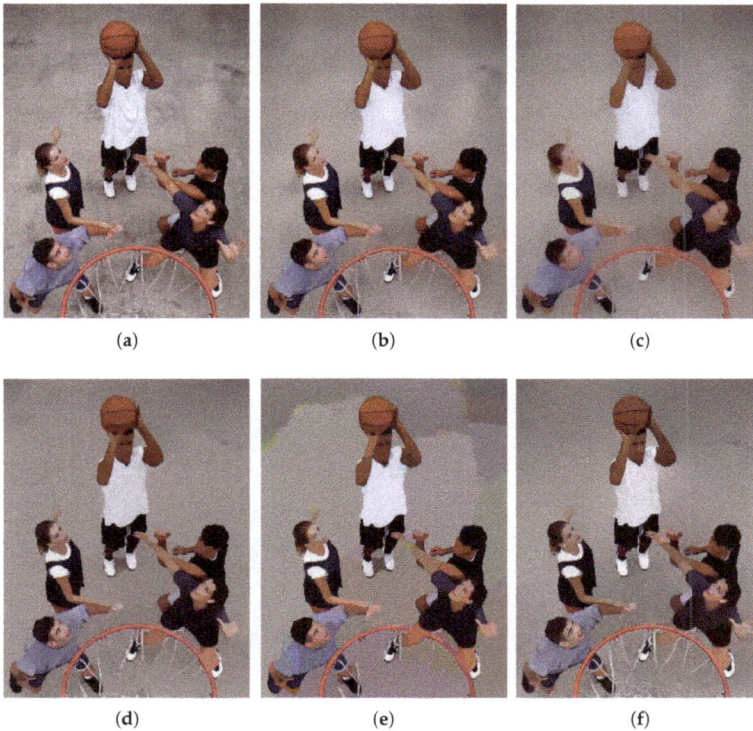

Figure 5. Comparison of our method with other methods in image abstraction. (**a**) Input; (**b**) Bilateral [18] with $\sigma_s = 8$, $\sigma_r = 26$; (**c**) WLS [21] with $\alpha = 1.5$, $\lambda = 0.5$; (**d**) l_0-smoothing (http://www.cse.cuhk.edu.hk/leojia/projects/L0smoothing/ImageSmoothing.htm) [24]; (**e**) l_0-projection [29] with $\alpha = 21749$; (**f**) ours with $\lambda = 15$, $\sigma = 0$.

3.2. Edge Extraction

Object textures are sometimes misclassified as edges during the edge detection process. In order to reduce misclassifications, image abstraction as discussed in the last section can be used to suppress object textures. Given an input image as shown in Figure 6a, objects of less importance such as clouds and grasses can be eliminated by image abstraction. Using our method, a smooth image as in Figure 6f is obtained. Edge detection or segmentation can then be applied to the output image to obtain a result with much fewer distortions. Figure 7 shows the results of applying the Canny edge detector to the grayscale version of Figure 6. We see that while the outputs of other models keep unnecessary details, our model produces a result containing only salient edges, and removes unimportant details.

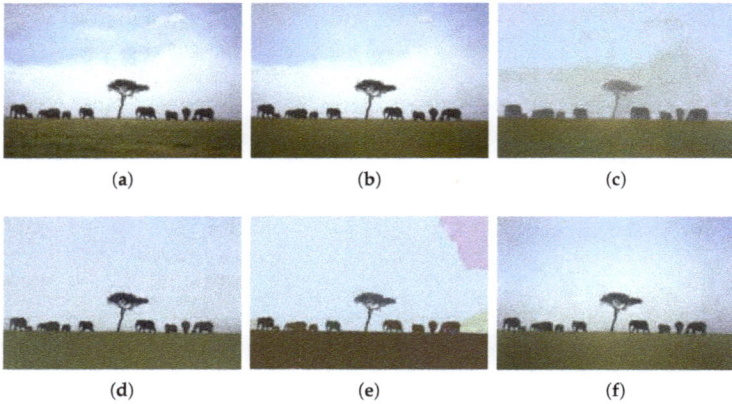

Figure 6. Comparison of our method with other methods in edge extraction. (**a**) Input; (**b**) Bilateral [18] with $\sigma_s = 5$, $\sigma_r = 46$; (**c**) WLS [21] with $\alpha = 2$, $\lambda = 2$; (**d**) l_0-smoothing (http://www.cse.cuhk.edu.hk/leojia/projects/L0smoothing/EdgeEnhancement.htm) [24]; (**e**) l_0-projection [29] with $\alpha = 9264$; (**f**) ours with $\lambda = 10$, $\sigma = 0.7$.

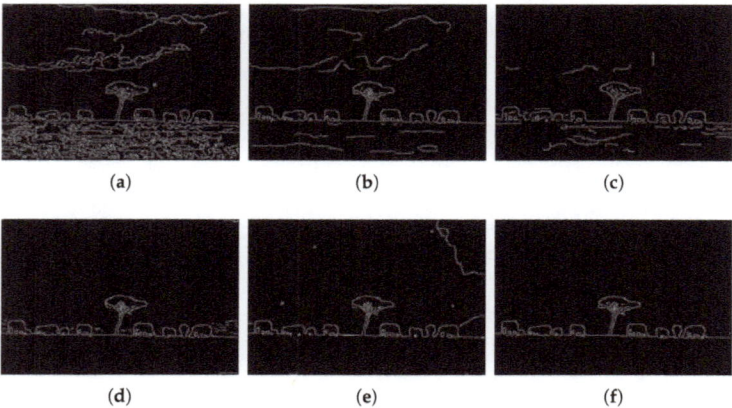

Figure 7. Applying Canny edge detector to the grayscale version of Figure 6. (**a**) Input; (**b**) Bilateral [18]; (**c**) WLS [21]; (**d**) l_0-smoothing [24]; (**e**) l_0-projection [29]; (**f**) ours.

3.3. Details Exaggeration

Details exaggeration is to enhance the fine details in an image as much as possible. Given an input image I, we obtain a smooth image X by our method where the textures in I are removed, see in Figure 8b. As seen in Figure 8c, the image $|I - X|$ has small values in regions with insignificant textures and large values in the parts containing strong textures. By enhancing $(I - X)$ and adding it back to X, a details-exaggerated image J can be obtained, see Figure 8d. Mathematically, we have $J = X + s(I - X)$, where $s > 1$ is a parameter controlling the extent of exaggeration. Figure 9 shows a comparison with the results by other methods. In Figure 9f we see that our model successfully produces a better result with more exaggerated details.

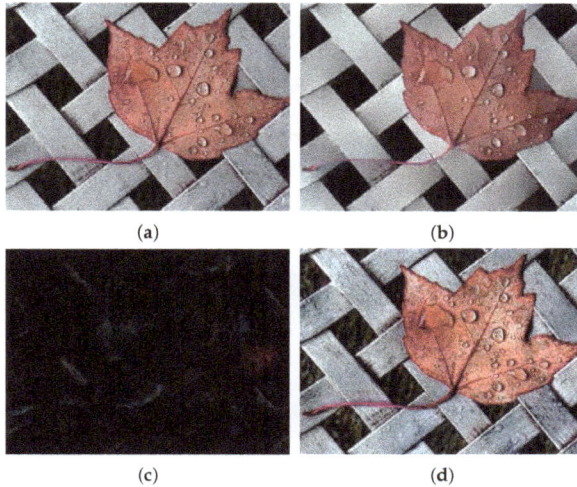

Figure 8. Steps to obtain a details-exaggerated image J. (**a**) Input I; (**b**) output X from (5) with $\lambda = 25$, $\sigma = 0.4$; (**c**) $|I - X|$; (**d**) details-exaggerated image $J = X + 2(I - X)$.

Figure 9. Comparison of our method with other methods in details exaggeration. (**a**) Input; (**b**) Bilateral [18] with $\sigma_s = 17$, $\sigma_r = 20$, $s = 4$; (**c**) WLS (http://www.cs.huji.ac.il/~danix/epd/ MSTM/flower/index.html) [21]; (**d**) l_0-smoothing (http://www.cse.cuhk.edu.hk/leojia/projects/ L0smoothing/ToneMapping.htm) [24]; (**e**) l_0-projection [29] with $\alpha = 127,920$, $s = 2$; (**f**) ours with $\lambda = 13$, $\sigma = 0.7$, $s = 2.5$.

3.4. Scan-Through Removal

Two-sided documents can be suffered from the effect of back-to-front interference, known as "see-through". Usually, "see-through" produces relatively small gradient fluctuations than the main content we want to preserve, see Figure 10. By considering the edges, one can identify interferences and eliminate them.

Recall in (9), we also impose a constraint that background pixels will not be modified. Here background pixels refer to the pixels with values not less than α. To find a suitable α, we first need to locate background regions—regions which contain only insignificant intensity change, i.e., the standard deviation of the intensity within the region should be small. Motivated by this, we design a multi-scale sliding window method to compute a suitable α. A sliding window with size w is used to scan through an input image Y with stride $\lceil w/5 \rceil$. At each location p, the mean intensity m_p of the sliding window is computed and if its standard deviation σ_p is smaller than a parameter $\hat{\sigma}$, m_p will be stored for future selection. After scanning through the whole image, we set α to the largest stored value to avoid choosing regions with purely foreground or interference. If $\sigma_p \geq \hat{\sigma}$ for all p, we replace w by $w/2$ and scan through the image again. At the worst case when $w = 1$, it is equivalent to setting α to the maximum intensity of the image. In our tests, we use $\hat{\sigma} = 3$.

The reason for using a varying window size is that a small window will have a chance of capturing extreme values and a large window will have a chance of failing in capturing pure background. Therefore, we start from a large window and stop once we find at least one region with a small standard deviation. The initial window we use is the largest square window of length $w = 2^\ell$ that can fit inside the given image. Figure 10 shows the windows (red-colored squares) obtained by the procedure above. We see that it successfully locates a pure background. The background level α is the mean intensity of the corresponding square.

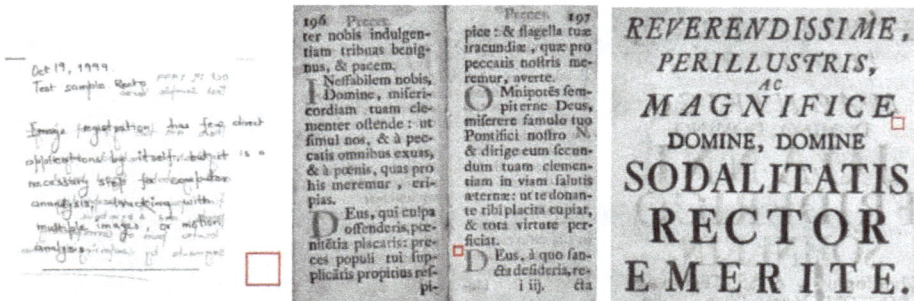

Figure 10. Background region detection for three different inputs. Red-colored squares are the regions located by our procedure.

With α found, we solve our model (5) with constraint (9) to obtain the output. We test our method using the first image in Figure 10. Our output is shown in Figure 11b, where we see that the contents are kept and the back-page interferences are removed. Figure 11a shows a comparison with the blind method from [43]. We also compare our result with three non-blind methods [32,44,49]. For copyright reasons, we can only refer readers to the papers [32,44] to see the resulting images from the three methods. Our method outperforms the blind method and is comparable to the non-blind methods, while these non-blind methods require information from both sides.

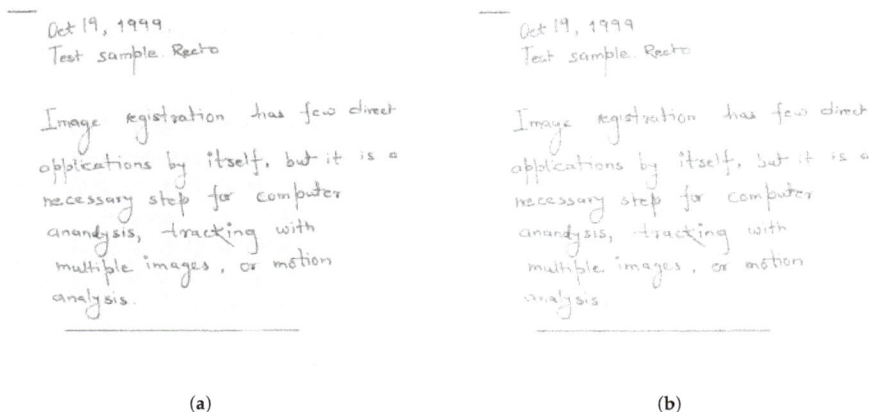

(a)　　　　　　　　　　　　　　　　　　　(b)

Figure 11. Comparison of our method to a blind method in scan-through removal. (**a**) Nishida and Suzuki [43] with $S = 2^7$, $\lambda = 130$; (**b**) ours with $\lambda = 70$, $\alpha = 255$, $\sigma = 0$.

4. Conclusions

We have proposed a convex model with suitable constraints for edge-preserving smoothing tasks including image abstraction, edge extraction, details exaggeration, and documents scan-through removal. Our convex model allows us to solve it efficiently by existing algorithms.

In this paper, because of the special applications we considered, we use only the thresholded histograms as target edge-histograms. In the future, we would investigate more general shapes of edge-histograms and apply them to a wider class of problems.

Author Contributions: Conceptualization, K.C.K.C., R.H.C. and M.N.; Data curation, K.C.K.C. and M.N.; Formal analysis, K.C.K.C., R.H.C. and M.N.; Funding acquisition, R.H.C. and M.N.; Investigation, K.C.K.C., R.H.C. and M.N.; Methodology, K.C.K.C., R.H.C. and M.N.; Resources, K.C.K.C., R.H.C. and M.N.; Software, K.C.K.C. and M.N.; Supervision, R.H.C. and M.N.; Validation, K.C.K.C., R.H.C. and M.N.; Visualization, K.C.K.C. and R.H.C.; Writing—original draft, K.C.K.C.; Writing—review & editing, K.C.K.C. and R.H.C.

Funding: This research is supported by HKRGC Grants No. CUHK14306316, HKRGC CRF Grant C1007-15G, HKRGC AoE Grant AoE/M-05/12, CUHK DAG No. 4053211, and CUHK FIS Grant No. 1907303, and French Research Agency (ANR) under grant No ANR-14-CE27-001 (MIRIAM) and by the Isaac Newton Institute for Mathematical Sciences for support and hospitality during the programme Variational Methods and Effective Algorithms for Imaging and Vision, EPSRC grant no EP/K032208/1.

Conflicts of Interest: The authors declare no conflict of interest.

References

1. Lee, S.; Tseng, C. Color image enhancement using histogram equalization method without changing hue and saturation. In Proceedings of the 2017 IEEE International Conference on Consumer Electronics-Taiwan, Taipei, Taiwan, 12–14 June 2017; pp. 305–306.
2. Lim, S.; Isa, N.; Ooi, C.; Toh, K. A new histogram equalization method for digital image enhancement and brightness preservation. *Signal Image Video Process.* **2015**, *9*, 675–689. [CrossRef]
3. Wang, Y.; Chen, Q.; Zhang, B. Image enhancement based on equal area dualistic sub-image histogram equalization method. *IEEE Trans. Consum. Electron.* **1999**, *45*, 68–75. [CrossRef]
4. Chen, Y.; Chen, D.; Li, Y.; Chen, L. Otsu's thresholding method based on gray level-gradient two-dimensional histogram. In Proceedings of the 2010 2nd International Asia Conference on Informatics in Control, Automation and Robotics, Wuhan, China, 6–7 March 2010; Volume 3, pp. 282–285.

5. Tobias, O.; Seara, R. Image segmentation by histogram thresholding using fuzzy sets. *IEEE Trans. Image Process.* **2002**, *11*, 1457–1465. [CrossRef] [PubMed]

6. Thomas, G. Image segmentation using histogram specification. In Proceedings of the 2008 15th IEEE International Conference on Image Processing, San Diego, CA, USA, 12–15 October 2008; pp. 589–592.

7. Coltuc, D.; Bolon, P.; Chassery, J. Exact histogram specification. *IEEE Trans. Image Process.* **2006**, *15*, 1143–1152. [CrossRef] [PubMed]

8. Sen, D.; Pal, S. Automatic exact histogram specification for contrast enhancement and visual system based quantitative evaluation. *IEEE Trans. Image Process.* **2011**, *20*, 1211–1220. [CrossRef] [PubMed]

9. Nikolova, M.; Wen, Y.; Chan, R. Exact histogram specification for digital images using a variational approach. *J. Math. Imaging Vis.* **2013**, *46*, 309–325. [CrossRef]

10. Zhu, S.; Mumford, D. Prior learning and Gibbs reaction-diffusion. *IEEE Trans. Pattern Anal. Mach. Intell.* **1997**, *19*, 1236–1250.

11. Mignotte, M. An energy-based model for the image edge-histogram specification problem. *IEEE Trans. Image Process.* **2012**, *21*, 379–386. [CrossRef] [PubMed]

12. Dalal, N.; Triggs, B. Histograms of oriented gradients for human detection. In Proceedings of the IEEE Computer Society Conference on Computer Vision and Pattern Recognition, San Diego, CA, USA, 20–25 June 2005; Volume 1, pp. 886–893.

13. Zhu, Q.; Yeh, M.; Cheng, K.; Avidan, S. Fast human detection using a cascade of histograms of oriented gradients. In Proceedings of the IEEE Computer Society Conference on Computer Vision and Pattern Recognition, New York, NY, USA, 17–22 June 2006; Volume 2, pp. 1491–1498.

14. Déniz, O.; Bueno, G.; Salido, J.; De la Torre, F. Face recognition using histograms of oriented gradients. *Pattern Recognit. Lett.* **2011**, *32*, 1598–1603. [CrossRef]

15. Perona, P.; Malik, J. Scale-space and edge detection using anisotropic diffusion. *IEEE Trans. Pattern Anal. Mach. Intell.* **1990**, *12*, 629–639. [CrossRef]

16. Black, M.; Sapiro, G.; Marimont, D.; Heeger, D. Robust anisotropic diffusion. *IEEE Trans. Image Process.* **1998**, *7*, 421–432. [CrossRef] [PubMed]

17. Tomasi, C.; Manduchi, R. Bilateral filtering for gray and color images. In Proceedings of the Sixth International Conference on Computer Vision, Bombay, India, 7 January 1998; pp. 839–846.

18. Paris, S.; Durand, F. A fast approximation of the bilateral filter using a signal processing approach. In Proceedings of the European Conference on Computer Vision, Graz, Austria, 7–13 May 2006; Springer: Berlin/Heidelberg, Germany, 2006; pp. 568–580.

19. Weiss, B. Fast median and bilateral filtering. *ACM Trans. Graph.* **2006**, *25*, 519–526. [CrossRef]

20. Chen, J.; Paris, S.; Durand, F. Real-time edge-aware image processing with the bilateral grid. *ACM Trans. Graph.* **2007**, *26*, 1–9. [CrossRef]

21. Farbman, Z.; Fattal, R.; Lischinski, D.; Szeliski, R. Edge-preserving decompositions for multi-scale tone and detail manipulation. *ACM Trans. Graph.* **2008**, *27*, 1–10. [CrossRef]

22. Rudin, L.; Osher, S.; Fatemi, E. Nonlinear total variation based noise removal algorithms. *Physica D* **1992**, *60*, 259–268. [CrossRef]

23. Chambolle, A. An algorithm for total variation minimization and applications. *J. Math. Imaging Vis.* **2004**, *20*, 89–97.

24. Xu, L.; Lu, C.; Xu, Y.; Jia, J. Image smoothing via L0 gradient minimization. *ACM Trans. Graph.* **2011**, *30*, 1–12.

25. Cheng, X.; Zeng, M.; Liu, X. Feature-preserving filtering with L0 gradient minimization. *Comput. Graph.* **2014**, *38*, 150–157. [CrossRef]

26. Storath, M.; Weinmann, A.; Demaret, L. Jump-sparse and sparse recovery using Potts functionals. *IEEE Trans. Signal Process.* **2014**, *62*, 3654–3666. [CrossRef]

27. Nguyen, R.; Brown, M. Fast and effective L0 gradient minimization by region fusion. In Proceedings of the IEEE International Conference on Computer Vision, Santiago, Chile, 7–13 December 2015; pp. 208–216.

28. Pang, X.; Zhang, S.; Gu, J.; Li, L.; Liu, B.; Wang, H. Improved L0 gradient minimization with L1 fidelity for image smoothing. *PLoS ONE* **2015**, *10*, e0138682. [CrossRef] [PubMed]

29. Ono, S. L0 Gradient Projection. *IEEE Trans. Image Process.* **2017**, *26*, 1554–1564. [CrossRef] [PubMed]

30. Tonazzini, A.; Salerno, E.; Bedini, L. Fast correction of bleed-through distortion in grayscale documents by a blind source separation technique. *Int. J. Doc. Anal. Recognit.* **2007**, *10*, 17–25. [CrossRef]

31. Merrikh-Bayat, F.; Babaie-Zadeh, M.; Jutten, C. Using non-negative matrix factorization for removing show-through. In Proceedings of the International Conference on Latent Variable Analysis and Signal Separation, St. Malo, France, 27–30 September 2010; Springer: Berlin/Heidelberg, Germany, 2010; pp. 482–489.

32. Martinelli, F.; Salerno, E.; Gerace, I.; Tonazzini, A. Nonlinear model and constrained ML for removing back-to-front interferences from recto–verso documents. *Pattern Recognit.* **2012**, *45*, 596–605. [CrossRef]

33. Gerace, I.; Palomba, C.; Tonazzini, A. An inpainting technique based on regularization to remove bleed-through from ancient documents. In Proceedings of the 2016 International Workshop on Computational Intelligence for Multimedia Understanding, Reggio Calabria, Italy, 27–28 October 2016; pp. 1–5.

34. Salerno, E.; Martinelli, F.; Tonazzini, A. Nonlinear model identification and see-through cancelation from recto–verso data. *Int. J. Doc. Anal. Recognit.* **2013**, *16*, 177–187. [CrossRef]

35. Savino, P.; Bedini, L.; Tonazzini, A. Joint non-rigid registration and restoration of recto-verso ancient manuscripts. In Proceedings of the 2016 International Workshop on Computational Intelligence for Multimedia Understanding, Reggio Calabria, Italy, 27–28 October 2016; pp. 1–5.

36. Savino, P.; Tonazzini, A. Digital restoration of ancient color manuscripts from geometrically misaligned recto-verso pairs. *J. Cult. Herit.* **2016**, *19*, 511–521. [CrossRef]

37. Sharma, G. Show-through cancellation in scans of duplex printed documents. *IEEE Trans. Image Process.* **2001**, *10*, 736–754. [CrossRef] [PubMed]

38. Tonazzini, A.; Savino, P.; Salerno, E. A non-stationary density model to separate overlapped texts in degraded documents. *Signal Image Video Process.* **2015**, *9*, 155–164. [CrossRef]

39. Estrada, R.; Tomasi, C. Manuscript bleed-through removal via hysteresis thresholding. In Proceedings of the 2009 10th International Conference on Document Analysis and Recognition, Barcelona, Spain, 26–29 July 2009; pp. 753–757.

40. Tonazzini, A.; Bedini, L.; Salerno, E. Independent component analysis for document restoration. *Doc. Anal. Recognit.* **2004**, *7*, 17–27. [CrossRef]

41. Wolf, C. Document ink bleed-through removal with two hidden markov random fields and a single observation field. *IEEE Trans. Pattern Anal. Mach. Intell.* **2010**, *32*, 431–447. [CrossRef] [PubMed]

42. Sun, B.; Li, S.; Zhang, X.; Sun, J. Blind bleed-through removal for scanned historical document image with conditional random fields. *IEEE Trans. Image Process.* **2016**, *25*, 5702–5712. [CrossRef] [PubMed]

43. Nishida, H.; Suzuki, T. Correcting show-through effects on scanned color document images by multiscale analysis. *Pattern Recognit.* **2003**, *36*, 2835–2847. [CrossRef]

44. Tonazzini, A.; Gerace, I.; Martinelli, F. Multichannel blind separation and deconvolution of images for document analysis. *IEEE Trans. Image Process.* **2010**, *19*, 912–925. [CrossRef] [PubMed]

45. Beck, A.; Teboulle, M. A fast iterative shrinkage-thresholding algorithm for linear inverse problems. *SIAM J. Imaging Sci.* **2009**, *2*, 183–202. [CrossRef]

46. Gabay, D.; Mercier, B. A dual algorithm for the solution of nonlinear variational problems via finite element approximation. *Comput. Math. Appl.* **1976**, *2*, 17–40. [CrossRef]

47. Glowinski, R. *Lectures on Numerical Methods for Non-Linear Variational Problems*; Springer Science & Business Media: Berlin/Heidelberg, Germany, 2008.

48. Gonzales, R.; Woods, R. *Digital Image Processing*; Addison-Welsley: Reading, MA, USA, 1992.

49. Hyvarinen, A. Fast and robust fixed-point algorithms for independent component analysis. *IEEE Trans. Neural Netw.* **1999**, *10*, 626–634. [CrossRef] [PubMed]

MDPI

St. Alban-Anlage 66

4052 Basel

Switzerland

Tel. +41 61 683 77 34

Fax +41 61 302 89 18

www.mdpi.com

Axioms Editorial Office

E-mail: axioms@mdpi.com

www.mdpi.com/journal/axioms

www.ingramcontent.com/pod-product-compliance
Lightning Source LLC
Chambersburg PA
CBHW051717210326
41597CB00032B/5509